Unlocking Statistics
For the
Social Sciences

Unlocking Statistics
For the
Social Sciences

Global Applications Using IBM SPSS

Norma Y. Sinclair

Copyright © 2024 by Norma Y. Sinclair

All rights reserved.

Except as permitted by U.S. copyright law, no part of this work may be reproduced or distributed in any form or by any means, or stored in a database or retrieval system without prior written permission from the copyright owner and the publisher, except in the case of brief quotations in book reviews and certain other noncommercial uses permitted by copyright law.

Ordering Information: Quantity sales. Discounts are available on quantity purchases. For details, contact the author at the email address below.

Printed in the United States of America

Published by CLA Educational Research Clearinghouse, Fort Lauderdale, Florida

Visit the author's website: https://www.keys2statistics.com

To contact the author, email: info@keys2statistics.com

Copy Editor: Carol Rose
Cover and Interior Designer: Ghislain Viau
Indexer: Judy Gordon

Publisher's Cataloging-in-Publication data
Sinclair, Norma Y.
Title: Unlocking Statistics for the Social Sciences: Global Applications Using IBM SPSS/ Norma Y. Sinclair
Library of Congress Control Number: 2024924205
ISBN 979-8-9910305-0-2 (paperback)

First Edition

10 9 8 7 6 5 4 3 2 1

Contents

PREFACE ix

1 INTRODUCTION 1
    1.1 What is Statistics? 1
    1.2 Consumers of Applied Statistics 2
    1.3 The Social Science Context 3
    1.4 Univariate Statistics 3
    1.5 Bivariate Statistics 4
    1.6 Organization of the Book 5
    1.7 Statistical Techniques Included in this Text 7
    1.8 Key Terms 9

2 BASIC CONCEPTS IN RESEARCH DESIGN 10
    2.1 Introduction 10
    2.2 The Scientific Method 11
    2.3 The Research Administration Process 16
    2.4 The Research Plan 29
    2.5 Research Reporting 53
    2.6 Key Terms 53

3 DATA PREPARATION: CLEANSING AND SCREENING 54
    3.1 Introduction 55
    3.2 Coding and Data Entry 56
    3.3 Before Cleaning and Screening Data 61
    3.4 Data Cleaning and Screening 61
    3.5 Key Terms 118

4 BIVARIATE CORRELATION 119
    4.1 Introduction 119
    4.2 The Case of Writing and Mathematics Skills 121
    4.3 Bivariate Correlation Fundamentals 121
    4.4 Planning a Bivariate Correlational Study 124
    4.5 The Pearson Correlation Coefficient 127

    4.6 Calculations for the Pearson Product Moment Correlation 134
    4.7 Correlation and Causation. 139
    4.8 Assumptions of Bivariate Correlation. 140
    4.9 SPSS Correlation Specifications. 142
    4.10 Factors that Influence Correlation 152
    4.11 Intercorrelation Matrices 157
    4.12 Other Correlation Coefficients 158
    4.13 Key Terms 160

5 SIMPLE LINEAR REGRESSION 161
    5.1 Introduction 161
    5.2 The Case of the Height and Weight of College Students 162
    5.3 Simple Linear Regression Fundamentals 163
    5.4 Planning a Simple Linear Regression Study 172
    5.5 Calculations for Simple Linear Regression 176
    5.6 Assumptions of Linear Regression 179
    5.7 SPSS Specifications for Bivariate Regression 184
    5.8 Key Terms. 201

6 MULTIPLE REGRESSION 202
    6.1 Introduction 203
    6.2 Cases for Multiple Regression SPSS Demonstrations. 203
    6.3 Fundamentals of Multiple Regression. 204
    6.4 Planning a Multiple Regression Inquiry 217
    6.5 Data Assumptions of Multiple Regression 219
    6.6 SPSS Procedures for Standard Multiple Regression 221
    6.7 Solving the Regression Equation 230
    6.8 Sample Report (Standard): The Case of the Weight of U.S. Adolescents 230
    6.9 SPSS Specifications for Statistical (Stepwise) Multiple Regression. 231
    6.10 Statistical (Stepwise) Results: The Case of the Weight of U.S. Adolescents 233
    6.11 Sample Report (Stepwise): The Case of the Weight of U.S. Adolescents 238
    6.12 SPSS Specifications for Hierarchical Regression 240
    6.13 Results: The Case of Employee Salaries. 242
    6.14 Sample Report (Hierarchical): The Case of Predicting Employee Salaries 245
    6.19 Key Terms 247

7 ONE-WAY ANALYSIS OF VARIANCE. 248
    7.1 Introduction 248
    7.2 The Case of Differences in Achievement 249
    7.3 One-Way, Between Subjects ANOVA Fundamentals 249
    7.4 Calculations for One-Way Analysis of Variance 256

    7.5 Assumptions of One-Way ANOVA 261
    7.6 When Group Sizes are Unequal 263
    7.7 Calculating Effect Size 264
    7.8 When Group Means Are Significantly Different 265
    7.9 SPSS Procedures for One-Way Analysis of Variance 272
    7.10 Analysis of Variance and Linear Regression 280
    7.11 Other Types of ANOVAs 282
    7.12 Key Terms 282

8 FACTORIAL ANALYSIS OF VARIANCE 283

    8.1 Introduction 284
    8.2 The Case of Differences in Achievement: The Factorial Effect 284
    8.3 Factorial ANOVA Fundamentals 285
    8.4 The Research Plan: The Case of Differences in Achievement 290
    8.5 Assumptions of Factorial ANOVA 293
    8.6 Unequal Sample Sizes 293
    8.7 Calculations for Factorial ANOVA 294
    8.8 SPSS Specifications for Factorial ANOVA 300
    8.9 Factorial ANOVA SPSS Results 305
    8.10 Differences in Achievement Using Actual Test Scores 309
    8.11 Key Terms 313

9 ANALYSIS OF COVARIANCE 314

    9.1 Introduction 314
    9.2 The Case of Appetite After Accounting for Mangoes 315
    9.3 ANCOVA Fundamentals: 316
    9.4 Benefits of ANCOVA 318
    9.5 Research Plan: The Case of Appetite After Accounting for Mangoes 319
    9.6 ANCOVA Assumptions 322
    9.7 SPSS Procedures for One-Way ANCOVA 325
    9.8 Sample Report: The Case of Appetite After Accounting for Mangoes 337
    9.9 Key Terms 337

10 REPEATED MEASURES ANOVA 338

    10.1 Introduction 339
    10.2 The Case of the Sprinting Speeds of Seven Soccer Players 339
    10.3 Repeated Measures ANOVA Fundamentals 340
    10.4 Characteristics of Repeated Measures Designs 342
    10.5 Types of Repeated Measures Designs 343
    10.6 Research Plan: The Case of the Sprinting Speeds of Seven Soccer Players 346
    10.7 Calculations for Repeated Measures ANOVA 348

10.8 Underlying Assumptions of Repeated Measures ANOVA 353
10.9 Benefits and Disadvantages of Repeated Measures ANOVA 355
10.10 Strategies to Counteract Repeated Measures Disadvantages 355
10.11 SPSS Procedures for One-Way Repeated Measures ANOVA 358
10.12 Sample Report: The Case of the Sprinting Speed of Soccer Players 365
10.13 Key Terms 365

11 MIXED DESIGN ANOVA 366

11.1 Introduction 367
11.2 The Case of Olympic Track Athletes' Speed 367
11.3 Simple Mixed Design ANOVA Fundamentals 367
11.4 Characteristics of Simple Mixed Design ANOVA 369
11.5 Types of Mixed Design ANOVAs 370
11.6 Underlying Assumptions of Mixed Design ANOVA 373
11.7 Benefits and Disadvantages of Mixed Design ANOVA 375
11.8 Research Plan: The Case of Olympic Track Athletes' Speed 376
11.9 SPSS Procedures for Mixed Design ANOVA 377
11.10 Sample Report: The Case of Olympic Athletes' Speed 387
11.11 Key Terms 388

12 NONPARAMETRIC TESTS 389

12.1 Introduction 389
12.2 Nonparametric Statistical Analysis Fundamentals 390
12.3 Kruskal-Wallis One-Way ANOVA (H) Test Fundamentals 399
12.4 Nonparametric Correlation: Spearman's Rho and Kendall's Tau 414
12.5 Spearman's Rho 417
12.6 Kendall's Tau 426
12.7 Friedman's Rank 431
12.8 Key Terms 445

APPENDIX A: STATISTICAL TABLES 446

APPENDIX B: DATA FILES 456

REFERENCES 457

INDEX 465

Preface

Statistics Journey

I entered my first statistics course in graduate school with some trepidation. I had been an arts and humanities student since high school in the Caribbean and had never ever considered taking a statistics course of any kind. Nevertheless, the statistics course was a pre-requisite for the terminal degree program I had entered. As I sat down in the back of a cavernous lecture hall among the 80 or so other graduate students and watched as the professor prepared to lecture from the front of the room, I pondered what possessed me to switch from a "safe" terminal degree in the Department of Curriculum and Instruction to one in Evaluation and Measurement in Educational Psychology. Ah yes, I had snagged a delicious, new career opportunity in the Standardized Assessment Bureau at a State Department of Education (think Ministry of Education) in a northeastern state of the United States after a lengthy career in educational publishing, and I was smitten with the new world in which I was engaged.

The statistics course was the beginning of a trove of stimulating new discoveries. One was the essential role of statistics in much of the field of education and other social sciences. In fact, whether or not we realize or accept it, as knowledge continues to increase in the world, the use of statistics increases concomitantly. Statistical methods are crucial in facilitating the use of data to make important decisions, create new discoveries, and form predictions about the world in which we live. Statistics surround us before we exit our mothers' wombs, with data that compare our mothers (age, weight, medical history) and us (gestational age, weight, heart rate, health predictions) with other mothers and babies. In today's data-driven world, statistics remain central to our lives, helping us to better understand the world around us and make informed decisions.

Goals and Distinctive Elements

This text is designed for graduate and advanced undergraduate students in education and the social sciences. Therefore, it is assumed that students completed a basic introductory course in applied statistics. Under that assumption, an important goal of this book is to concentrate on methodology and statistics just beyond the basics found in most introductory statistics courses. In addition, the book is centered within the context of the research process that is typically reported in the methodology and results chapters of a research report, thesis, or dissertation. Therefore, the text includes information on:
- research methods and design, the engine that drives all research studies
- guidelines for choosing appropriate statistics and statistical techniques

- how to evaluate whether the data meet statistical assumptions
- how to conduct data analysis, using SPSS
- how to interpret the results of data analysis
- how to write a report on the results of data analysis

Another goal of the text is to make students more skilled consumers and practitioners of quantitative research by promoting a conceptual understanding of applied statistics. To accomplish that, the text incorporates some unique features:

Establishing connections between previously-learned statistical concepts (e.g., t tests) and upcoming topics covered in this text (e.g., one-way and factorial analysis of variance) is one way of instilling a deep understanding of statistics. This text encourages that process by providing a list of important previously introduced concepts at the beginning of each chapter, along with learning objectives. Each chapter ends with a list of key terms related to the concepts introduced in the chapter.

Although this text assumes that students have completed a basic, introductory statistics course, students needing a review of basic concepts are not abandoned. Rather, the text references and includes introductory statistical concepts (e.g., descriptive statistics and t tests) as part of the discourse around the characteristics of the intermediate statistical procedures covered in this text. This is designed to help students make the necessary linkages to the intermediate statistical concepts introduced and applied in the text, beginning with correlation and linear regression through to the analysis of variance. More about organization and coverage in Chapter 1.

Connecting statistical procedures with research methods and the research process begins in Chapter 2, which provides an overview of research methods and the research process. In addition, each statistical technique is set within the context of a case study that includes the outline of the methodology section of a relevant research plan or dissertation proposal and the associated dataset. For example, the Case of Differences in Achievement examines differences in achievement across different types of schools on Jamaica's high school admissions test. The case study includes a brief description of the case and data for demonstrating statistical analysis as well as the outline of a research plan for the study.

In keeping with the desire to boost students' skills in conducting research, this text leads students through the process of data cleansing and data assumptions testing (Chapter 3). Demonstrations of data analyses also include the testing and interpretation of data assumptions along with corrective actions for assumptions violations to make data ready for central analyses.

Most demonstrations of statistical procedures in this text use data from the Caribbean, Latin America and other global sources. Data used in data analysis demonstrations come from a wide variety of global sources, including track and field data from the 2016 Olympics held in Brazil, soccer sprinting test data from the United Kingdom, achievement test data from Jamaica, World Values Survey data from Haiti and Trinidad and Tobago, as well as adolescent body weight data from the United States. Sprinkled throughout the text are some exercises that are based on fictional studies especially designed for the techniques involved (e.g., The Case of Appetite After Accounting for Mangoes) and to inject some fun into the statistics journey. Data files for all demonstration exercises in this text are available on the following website: https://www.keys2statistics.com. A list of the data files used in this book also appears in Appendix B.

Hands-on demonstration exercises include step-by-step, graphics-supported programming instructions for data analysis for every statistical procedure. Although discussions about data analysis reference several statistical packages (e.g., SAS, R, Minitab and STATA), data analysis instructions are provided using SPSS, one of the most widely-used packages. Nevertheless, students are encouraged to take advantage of the availability of the other statistical packages.

Although this text emphasizes a conceptual understanding of applied statistics rather than memorization of procedures, some students do benefit from a behind-the-scenes view of the computational formulas that produce statistics. Accordingly, the text includes step-by-step instructions of how to computationally derive some statistics (e.g., the F statistic in analysis of variance) alongside the instructions for generating the statistics using SPSS.

To reinforce a conceptual understanding of applied statistics, each demonstration exercise concludes with interpretations of analysis results and a brief apa-style report of data analysis results. Each report includes tables and figures as necessary to support the report.

Acknowledgments

A number of people contributed enormously to the completion of this book: copy editor, Carol Rose; book designer, Ghislain Viau; index compiler, Judy Gordon.

I am truly grateful for the helpful suggestions provided by the reviewers of this text: Mohamed A. Dirir (Connecticut State Department of Education), Richard Mooney (Connecticut State Department of Education), Vivienne A. Quarrie (Northern Caribbean University), Nanibala Paul (Northern Caribbean University), Victoria Anyikwa (Saint Leo University). I owe a debt of gratitude to all of my students over the years who have shared their views of the text with me and have contributed immensely to the improvement of the book.

Finally, I would like to thank the following for permission to reproduce selected statistics tables in Appendix A: The American Statistical Association, The Canadian Journal of Statistics (copyright 1993 by the Statistical Society of Canada).

SPSS screenshots and output are reprinted courtesy of International Business Machines Corporation, © International Business Machines Corporation. IBM, the IBM logo, ibm.com. and SPSS are registered trademarks of International Business Machines Corporation, registered in many jurisdictions worldwide. A current list of IBM trademarks is available at www.ibm.com/legal/copytrade.shtml.

About the Author

Norma Y. Sinclair is an educational psychologist who specializes in measurement, research and evaluation. As a psychometrician, she managed the development, administration, psychometrics research and evaluation of assessments for teachers and students for more than 27 years. Beyond psychometrics, Dr. Sinclair has served on the faculties of the University of New Haven (Connecticut) and Northern Caribbean University (Manchester, Jamaica), teaching intermediate and multivariate statistics as well as research methods.

1

Introduction

"Research is the process of going up alleys to see if they are blind."
— Bartow Bates, President, New Products Inc.

THIS CHAPTER INTRODUCES:

- an overview of statistics in the social sciences
- the types of applied statistics you will learn about
- differences between univariate and bivariate statistics
- differences between inferential and descriptive statistics
- organization and goals of the book

1.1 What is Statistics?

Statistics has been in use far longer than many of us imagine. It could be said to have had its origin in census counts taken thousands of years ago. In the Bible there are more than 30 accounts of censuses taken at different times. The first of two famous census counts occurred when God told Moses to count the Children of Israel as they traveled from Egypt to the Promised Land, just before entering Canaan (New King James Bible, 1982, Numbers 26:2). The other occurred when the Roman Emperor, Caesar Augustus, issued a decree that a census should be taken of the entire Roman world. In obeying that edict, Joseph returned to his birthplace, Bethlehem in Judea, to register with Mary (his betrothed) so they might be counted (New King James Bible, 1982, Luke 2:1-5). While there, Mary gave birth to Jesus.

Of course, the field of statistics is not limited to census counts. Rather, it is a discipline that specializes in the processing and analysis of quantitative data from a target population (Vogt, 2011). As the definition suggests, statistical data is information, values that are measured using numbers or counts; in effect, the data are numerical rather than language-based descriptive data (interviews, focus groups, direct observations) typically used in qualitative research. For example, imagine that a researcher is interested in learning about the anxiety level of the elderly during a disease epidemic. The data representing the anxiety characteristic might be quantitative scores from an anxiety scale or questionnaires that the study participants completed. In this example, the population from which the information is taken may be males or females, people living at home alone or with family, in institutions, or all of the above. In other research enterprises, the population might be animals, organizations, objects, or events. Regardless of the nature of the investigation, core activities of statisticians include using established techniques and procedures to analyze collected quantitative data.

Statistics are also the numeric values used to report the results of the analyses by describing, summarizing and interpreting the data in a way that helps us to make sense of the information as well as make decisions and solve problems. For example, commonly-used, descriptive statistics, such as the average, sum, median, mode, range, or frequency may be used to describe characteristics of the anxiety experienced by the elderly or other vulnerable groups of individuals during a disease pandemic. The term "statistics" may be used to provide summary information about samples of populations. Summary information about whole populations of individuals or things is described as parameters.

To some students, the prospect of taking a statistics course means an expanded opportunity to study mathematics. To others the idea of having a statistics requirement as part of their program of studies induces anxiety. If you are a member of the latter group and your head is swimming right now with visions of mountains of numbers and mathematical algorithms, your anxiety is understandable, but unwarranted. Statistics is used in different ways in various fields of interest. A sportscaster may use statistics to demonstrate the record of an elite soccer player. A mathematician may focus on building statistical theory in mathematical statistics. The setting for this textbook is the social sciences in which applied statistics is the focus. Rather than concentrating on pure mathematical theories, typically, applied statistics is employed by non-mathematicians to help them make sense of, and sometimes intervene in, the world we live in, especially as they relate to people.

1.2 Consumers of Applied Statistics

Everyone uses statistics in some form in their lives. Some knowledge of statistics helps in interpreting weather forecasts (e.g., likelihood that a country will be hit by a hurricane), political and other polls (e.g., likelihood that certain groups of individuals will vote for a political candidate), sports statistics (e.g., characteristics of elite soccer players), and statistics related to maintaining good health (e.g., effectiveness of vaccines).

Most educators are consumers of statistics. A school administrator may use quantitative statistics to plan and make decisions about implementing a new science, technology, engineering and mathematics (STEM) program. Alternatively, a school board may consult relevant statistics that subsequently inform decisions to offer options of online learning to standard education students while recommending face-to-face, in-person learning for special education students. The administrator may not have conducted the research from which the statistics were generated, but knowledge of research methods used and how to interpret the associated statistics are important in making curricular and other education administration decisions.

In institutions of higher education, statistics and statistical techniques play central roles in the lives of faculty and students. Students and faculty members are often expected to conduct literature reviews that include peer-reviewed research journal reports. A fundamental understanding of statistics is indispensable in understanding the research described in articles and evaluating study findings.

Similarly, a knowledge of statistical methods and procedures is crucial for anyone who intends to conduct a quantitative research project. A grounding in statistics supports prospective research investigators in understanding and discriminating amongst related research in the existing literature with the express purpose of informing their methodological decisions. Knowledge of statistics also provides guidance in identifying the appropriate methods and procedures when researchers conduct their own investigations. Alternatively, it may inform researchers of the limits of their understanding and alert them to the nature and type of additional expertise they need to consult.

In summary, reasons to study statistics include being able to:
- act as an informed consumer of research reports
- read and evaluate journal articles and other research reports
- conduct research effectively
- know when outside statistical help is needed

1.3 The Social Science Context

One field of inquiry in which applied statistics is widely used is the social sciences. This field of study primarily specializes in the study of people and their surroundings. Social science researchers employ applied statistics to conduct investigations in these areas. The foundation of research studies in any area of study, and certainly in the social science, is theories. For example, the theory of human capital developed by Becker (1962) and Rosen (1976), in part postulates that employees have a set of skills or abilities which they can improve or accumulate through training and education. The investment in growing and refining skills and abilities contributes to the development of people, organizations, and nations.

With theories serving as the infrastructure for a piece of research, researchers use applied statistics to provide empirical evidence (using the scientific process) to support or refute social science theories such as human capital theory. Accordingly, research may be used to investigate the impact of education on personal achievement and even successful nation-building. For example, in one research study in the Caribbean, Williams (2020) explored whether students participating in STEM programs in Jamaica perform better in mathematics than students who do not. The implications of the outcomes of studies like this one go well beyond individual performance. They can inform stakeholders of constructive directions for curriculum planning and development, long-range planning for effective schools, teacher education programs, and a nation's global competitiveness.

Sounds like a lot of statistics? Not to worry. Many of the studies conducted by social scientists use a limited set of methods of analysis. This text will focus on two of the most frequently used set of techniques in social science—univariate and bivariate statistics or what is traditionally referred to as intermediate statistics.

1.4 Univariate Statistics

To uncover the meaning of univariate statistics, it is helpful to understand that the prefix "uni" means one, as in a single attribute or characteristic, known as a variable. Therefore, univariate statistics may refer to summary descriptive information about a single item. As we discussed in Section 1.1, descriptive univariate statistics is summary information about individuals or things. Typically, these statistics include measures of central tendency (e.g., mean, median, mode); or measures of dispersion (e.g., range or standard deviation); and similar descriptive summaries of a single item, for example, the mean height of 10-year-olds, or the median salary of employees. Notice that in these cases, the attention is solely on a single variable (height or salary) and a single group of individuals (10-year-olds or employees). In addition to focusing on a single group in univariate descriptive statistics, only a single attribute is included in these types of univariate analyses. All other attributes of the group (e.g., gender, ethnicity or socio-economic status) are excluded from analysis.

In this textbook, we will zero in on univariate statistics from another perspective. Tabachnick and Fidell (2007) define univariate statistical tests as the results of analyses involving a single dependent variable. The dependent variable, also called the outcome variable or the variable being measured, may be the same as that which appears in our previous examples—height or salary. Nevertheless, there is more than a single group of study participants. For example, consider the study of the effectiveness of the integrated STEM program on student achievement scores in Jamaica (Williams, 2020). In this study, the primary investigator sought to examine differences in math achievement of male and female students in a STEM and a non-STEM school. This study design included a single dependent variable, Caribbean Secondary Examination Certificate (CSEC) scores, but there were also multiple groups of study participants.

Williams (2020) assigned students to two groups of two categorical (or independent) variables based on their gender (male or female) and the type of school they attended (STEM or non-STEM). In analyzing the data, Williams compared achievement by gender, by school type, as well as combinations of those groupings. Like Williams' study, all of the univariate statistics techniques explored in this volume are designed to analyze a single dependent variable for more than one group of study participants with the objective of comparing differences in a single dependent variable. Analysis of variance is a family of univariate statistics techniques and was the technique used in the Williams investigation. It will also be one of the techniques explored in this text.

1.5 Bivariate Statistics

Contrary to univariate statistics, bivariate statistics describes the analysis of two variables (e.g., height and weight, achievement in mathematics and achievement in reading, or fear of retaliation and level of cooperation with the police). The objective of bivariate analyses is to quantify the relationship between two variables to show the direction and strength of the association. The relationship may be associative or predictive. The two variables may be quantitative (e.g., test scores), categorical (e.g., academic ranking) or a mixture of both.

An associative relationship between two variables describes the extent to which changes in one variable (e.g., population size) are accompanied by corresponding changes in another variable (e.g., crime rate). The changes tend to occur in the same pattern, but they are not necessarily predictive of each other. Various types of correlation coefficients, like Pearson's correlation coefficient, exemplify bivariate statistics that this text will explore.

In addition to quantifying associative relationships, bivariate statistics may be used to quantify predictive relationships in that one variable predicts another. For example, bivariate statistics may be used to quantify the predictive relationship between height and weight. Similar to univariate statistics, in predictive bivariate statistics, there is a single dependent variable (e.g., weight) and a single independent variable (e.g., height). Bivariate regression is a statistical technique used to gauge bivariate predictive relationships.

1.6 Organization of the Book

Selecting the topics covered in a book of this kind is always a difficult process: so many topics so little space and time in which to address them. Nevertheless, with the goals of the text outlined in the preface, topics covered in this volume are limited primarily to the most commonly-used statistics today. They are also among the most enduring statistical methods of the modern era of statistics. The procedures belong to a class of tests described as inferential statistics that help researchers make inferences about a population of cases based on the analysis of a representative sample of those cases.

Descriptive or summary statistics is evident throughout the book. However, it has a secondary role. Every chapter will inevitably include descriptive statistics in the discussion of the statistical procedures as preliminary data analyses necessary for screening and cleansing data, for testing data assumptions and for summarizing data, before moving to the main event—inferential statistical procedures.

1.6.1 Book Sections and Topics

The chapters of this book fall into three categories. The first may be described as the pre-requisites, which include vital information about research design and preparing data for analysis. The second includes basic procedures that are employed when researchers' designs call for studies that examine relationships. The third segment explores statistical tests often employed in implementing causal comparative, experimental, or quasi-experimental research designs.

1.6.1.1 Chapters 1-3: The Fundamentals of Design and Data Preparation

The chapters in this section of the book introduce readers to essential prerequisites of conducting research in the social sciences. Chapter 1 provides a contextual overview of statistics, including the statistics subtypes and specific content covered in this book.

Chapter 2 presents a review of essential understandings of designing research in the social sciences in order to cement the indispensable link between research design and data analysis and the statistical procedures that assist us in that process. This chapter also includes key statistical terms associated with research design and application.

Chapter 3 is devoted to procedures for preparing data for analysis regardless of the statistical procedure that will be conducted. It includes guidelines for data cleansing and data screening as well as how to evaluate statistical assumptions associated with most statistical procedures. All chapters, beginning with Chapter 3, include hands-on practice of selected procedures, using SPSS.

1.6.1.2 Chapters 4-6: Tests Measuring Associative and Predictive Relationships

This segment introduces readers to two foundational statistical data analysis tools that researchers often rely on when conducting correlational studies. Correlational research design embraces research projects in which the researchers attempt to discover or clarify relationships among and between variables (e.g., between height and shoe size, or school absences and grades). Perhaps, the two most commonly-used methods of analysis are bivariate correlation and linear regression analysis.

Chapter 4 introduces the section by exploring the world of bivariate correlation statistics. The exploration in this chapter deals with the concepts and practical application of Pearson product moment correlation which measures the linear relationship between the values of two variables. However, readers are also introduced to other correlational methods, including Spearman's rho (or rank) correlation, Kendall's tau and others.

Chapter 5 eases the reader into the world of making statistical predictions, by introducing simple linear regression. Rather than examining the associative relationship between the values of two variables, this statistical test is used to measure the predictive relationship between two variables, where one variable is the predictor of the other.

Chapter 6 ends this section with an introduction to the basics of multiple linear regression, also described as ordinary least squares regression. Multiple regression is an extension of simple linear regression, but instead of having a single predictor, this statistical test is designed to measure the linear relationship between more than one predictor and a single outcome variable.

1.6.1.3 Chapters 7-12: Tests of Comparison of Mean Differences

The final segment of this book introduces readers to statistical methods often used when quantitative researchers carry out experimental and quasi-experimental design studies. The former establishes cause and effect occurrences if study participants were randomly assigned to the experimental and control groups. In the latter case, random assignment is unavailable and therefore causal relationships may only be explored, but cannot be firmly established when group means are compared.

Chapter 7 in this segment of the book begins with a brief review of the *t*-test and then transitions into more deeply exploring one-way analysis of variance (ANOVA), the first in the analysis of variance family of statistical tests in which there is a single dependent variable. In this case there is also a single independent variable, with more than two groups forming the independent variable. This and all the analysis of variance chapters examine how score or outcome variance helps to identify differences between the groups. Furthermore, this chapter investigates how to statistically identify the source of significant differences that are found.

Chapter 8 continues with the family of analysis of variance tests by adding an introduction to factorial ANOVA which distinguishes itself by having two or more independent variables, each of which comprises at least two groups. Chapter 8 applies all that was learned in Chapter 7 to the factorial ANOVA context as well as investigates the concept of variable interaction.

In Chapter 9, analysis of covariance (ANCOVA) adds the complication of a covariate to the mix discussed in Chapters 7 and 8. Covariates help researchers to add controls to a research project and are especially useful in studies that do not benefit from random assignment. They statistically control for influences that researchers believe may affect the outcome of a study, but cannot control for them in the design of the study itself. All the ANOVAs in Chapters 7 – 9 are also known as between-groups ANOVAs because they are designed to test for differences between groups.

Chapter 10 introduces readers to a slightly different ANOVA. In this statistical test, readers learn to compare the means of the scores study participants receive over a period of time (e.g., pretest and posttest scores) or under different conditions (e.g., essay performance scores by three different raters, or the sprinting speed of soccer players on three different types of fields). This test is called repeated measures ANOVA or within-groups design ANOVA because it is designed to assess the mean differences of repeated performance of participants in a group.

Chapter 11 introduces mixed design ANOVA, which is a test that is used when researchers want an efficient way of measuring differences both between groups and within groups. Accordingly, this test is a combination of repeated measures ANOVA and one-way ANOVA or repeated measures and any of the other ANOVAs described so far.

All of the tests discussed so far are called parametric tests, which are used to analyze data that meet certain assumptions about the associated population. To correctly use these tests certain data assumptions must be met (e.g., the normal distribution of population data). If the assumptions are untenable, then another option is to use nonparametric tests. Chapter 12 introduces readers to four nonparametric tests: the Kruskal-Wallis test, which is equivalent to the one-way ANOVA test; the Friedman's Rank test, which is the nonparametric equivalent of repeated measures ANOVA; and Spearman's rho and Kendall's tau, which are alternatives for Pearson's product moment correlation.

1.6.2 Organization Within Chapters

Although most of the statistical tests described in Section 1.6.1 belong to the same family of statistics—analysis of variance and the general linear model— and share a lot in common, the topics cover a broad range of concepts. Therefore, to enhance comprehension and retention, this book attempts to pull together all the pieces needed in planning and implementing a research project, beginning with project planning to completing and reporting on data analysis. To support this wholistic approach to learning, each chapter, beginning with Chapter 4, is organized to help readers make connections to realistic situations:
- introduction to the statistical technique
- description of fundamentals (characteristics) of the procedure
- introduction of a case study along with data used for practice exercises
- related research plan
- guidelines for testing associated statistical assumptions
- hand-calculations of statistics (selected tests)
- application of the statistics test using SPSS
- sample data analysis report

1.7 Statistical Techniques Included in this Text

The decision tree below, together with Section 1.6.1 above, may be used as an advance organizer or as a device to facilitate review. The final entries for each branch in the diagram are topics or statistical techniques covered in each chapter.

The statistics tests that are highlighted in the diagram are those described in Section 1.6 and will be introduced and explored in the associated chapters described. The remaining statistical techniques are methods probably encountered previously and will be referenced throughout the text in passing.

8 Chapter 1: Introduction

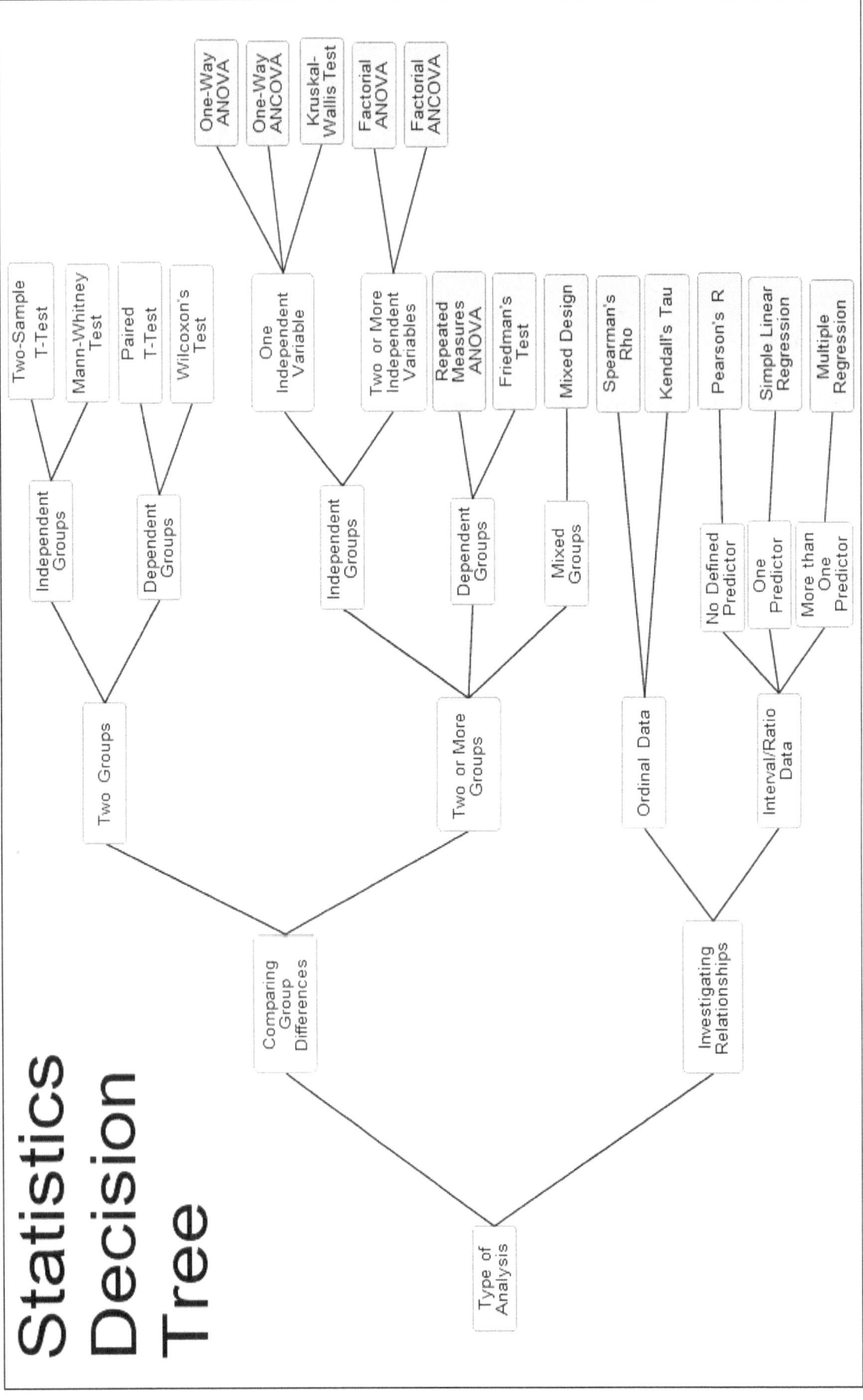

1.8 Key Terms

Bivariate statistics
Categorical variable
Central tendency
Descriptive statistics
Dependent variable
Dispersion
Experimental designs
Frequency

Independent variable
Mean
Median
Mode
Qualitative research
Quantitative data
Quantitative research

Quasi-experimental designs
Range
Standard deviation
Statistics
Theory
Univariate statistics
Variable

2

Basic Concepts in Research Design

> **THIS CHAPTER INTRODUCES**
> - the scientific method
> - the research process
> - quantitative research methods
> - basic research design concepts

2.1 Introduction

Around age two to three, toddlers typically discover the "why" question. Why do I have to eat? Why do I have to brush my teeth? Why is the sky blue? Why do people get sick? Why do people die? Developmental psychologist, Jean Piaget (1936, 1952) refers to this seemingly never-ending phase of children's cognitive development as the preoperational phase when children seek to satisfy their curiosity about their world. And the questions continue throughout most of the preoperational stage of development until children become more aware of methods of answering the questions themselves.

As adults we retain a natural curiosity about the world. After all, according to legend, it was Isaac Newton's curiosity about what caused an apple to fall from a tree in his mother's garden that ultimately led to his creation of the theory of gravity. And as Newton's curiosity led to invention, so our curiosity often leads to new discoveries.

Some of our curiosity-driven creations might be commonplace, like a new way to prepare a meal, or a more effective method of teaching reading to underperforming students. Regardless of the circumstances, we often respond in a similar way—we apply the adage "If at first you don't succeed, try, try and try again." In effect, we use the method of trial and error until we uncover results that satisfy our original intent. This is the essence of the scientific method.

The scientific method is an iterative, systematic method of acquiring knowledge that has been in use since the 17th century or earlier (e.g., Gauch, 2003; Nola & Sankey, 2014). It involves careful observations of your environment, formulating specific questions about what is observed and answering the questions in a way that ensures an empirical process that is objective and consistent.

The scientific method is not restricted to the natural sciences but applies broadly across many different disciplines, including the social sciences. Each academic discipline may use different techniques for collecting and analyzing data. Each component of the scientific method includes practices and procedures that vary across academic disciplines. Nevertheless, in spite of disagreements about the current relevance of the scientific method (Nola & Sankey, 2014), research investigators consider it a vital framework for their studies. Why is it vital? It provides a framework for adding unique contributions to the body of knowledge, and it holds scientists accountable.

Figure 2.1 The Scientific Method

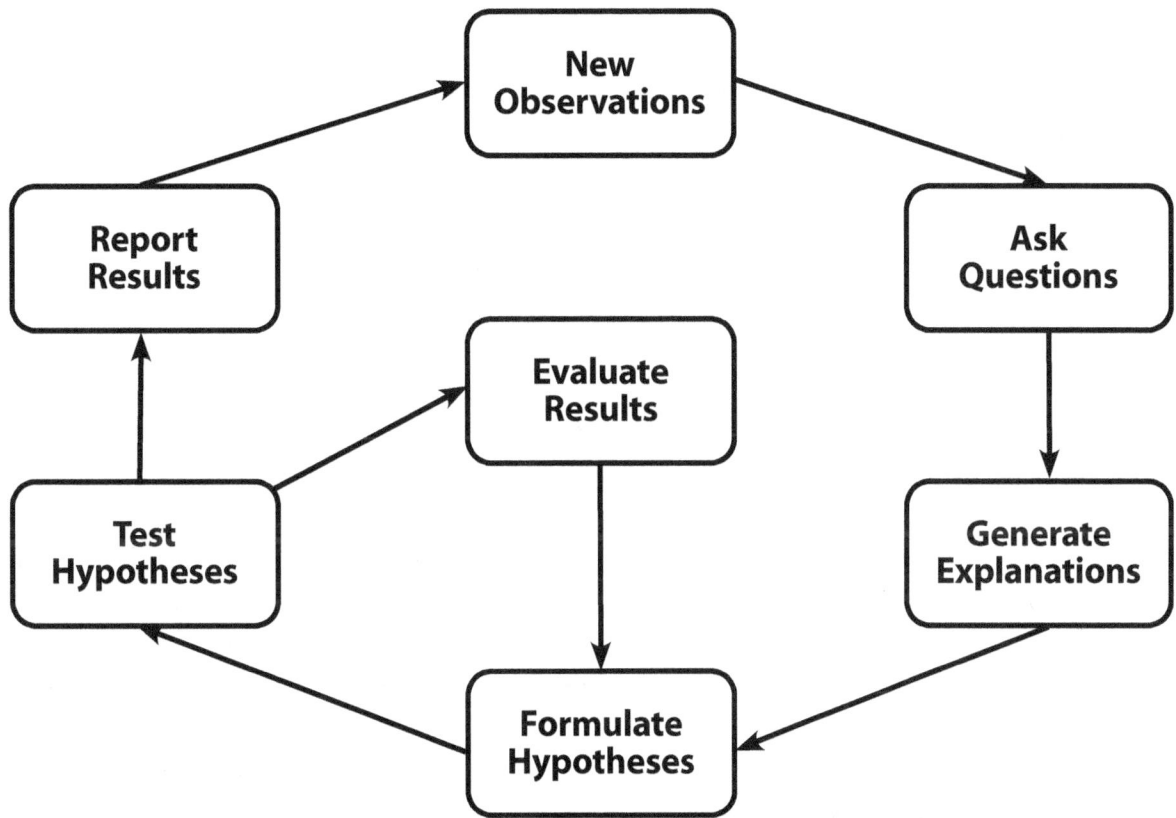

2.2 The Scientific Method

Although the sequence of steps is not the same in every field, the scientific method includes the broad set of principles described in Figure 2.1. Embedded within these general principles are specific procedures and practices characteristic of its application in research in the social sciences.

2.2.1 Observation

The process often begins with observations within your immediate surroundings, observations that boost your curiosity. Your observation might be a result of your own personal experience or a result of watching the conduct of the people around you. For example, perhaps you notice that typically, given the option, people maintain some personal space between themselves and others at events. Your observations lead to pondering why people move away when someone moves toward them. Consider another example in which you observe at your local food store that given a choice between domestically-produced goods and merchandise from other countries, customers tend to select the latter. Your observations lead you to question why customers select foreign goods over local goods.

Sometimes someone else's observations attract your attention; perhaps you watched a televised interview of an expert expounding causes of sports injuries in children, or you read the research findings of an expert in effective school administration. Even overhearing conversations about current events or what people have seen or noticed in their neighborhoods might spark your interest. All of these are differing forms of the first step in the scientific method—observations that raise questions you want to answer.

2.2.2 Asking Questions and Generating Explanations

Imagine that friends offer you something to eat that you have never seen before—several brown, hard-shelled pods. At first glance the pods look somewhat like the pods of the tangy-flavored tamarind fruit, known throughout Asia, Africa, the Caribbean and Central America. You come to the initial conclusion that they are tamarind pods. Filled with curiosity, you select a pod, and upon closer inspection, you discover that the pod is harder than a tamarind pod that you can easily pop open with your fingers. Moreover, the food item is not segmented like a tamarind pod. Rather, the appearance of the pod bears a marked resemblance to your toe. Perhaps, you reason, the item is not a tamarind pod, but a fruit related to the tamarind. Instructed to smash open the pod with a small rock and eat the contents, you comply out of curiosity and immediately regret your hasty action when an overpoweringly unpleasant, dirty socks odor escapes from the pod. Contrary to your friends' delight in eating the fruit, the odor is all it takes for you to lose your appetite and adjust your original conclusion about the pod's resemblance to the tangy, sweet odor of a tamarind.

As you watch your friends eagerly consume the odiferous fruit, you begin to wonder if others share your reaction to the fruit. You begin to take note of other individuals' reaction to the fruit known throughout the Caribbean as well as Central and South America by various monikers, including tinkin' toe (Jamaica, Grenada), locust fruit (Nevis), jumbie tolo (St. Vincent), jatobazeiro (Brazil), and coubaril (Martinique, Dominica). You notice that people around you respond to the fruit in varying ways. In addition, you discover that friends who like novel experiences are likelier to be drawn to tinkin' toe and other unusual foods. That discovery, or more precisely observation, leads you to question; do personality traits determine food preferences?

An internet search of odiferous foods and personality traits leads you to discover the existence of extensive information, studies, and theories related to those topics. Although you do not find studies specifically about tinkin' toe, you find investigations related to personality characteristics and food preferences related to those characteristics. Soon you reach the general conclusion that people who are extroverts like unusual foods. Thus, personality affects food preferences.

The scenario described reflects initial steps in the scientific method of observation, questioning and initial explanation via personal experience. Questioning and potential explanations or generalizations are typical responses to initial exposure to a situation that spurs your curiosity. Your interest in the topic may end with your conclusions or your original explanation. However, if initial explorations and answers are not completely satisfying, the remaining steps in the scientific method provide a framework to dig deeper.

Following your preliminary exploration, the next step in the search for answers typically involves uncovering other factors associated with initial observations and explanation: for example, what additional variables may be associated with tinkin' toe and an extroverted personality? Further inspection of the literature and discussions with members of your social or professional network may reveal that variables linked with tinkin' toe and extroversion include variables such as food odors, food preferences, and personality traits. According to Spence (2022), people with novelty sensation-seeking personalities (extroverts open to new experiences) are linked to a preference for a wide range of foods, including those that are spicy, crunchy, sour, and bitter foods. On the other hand, Spence observed that those who tend to be anxious do not like to try new things, and therefore, have a narrower range of food preferences. Spence concluded that food odor tolerance may be related to length of exposure and experience with the food. Thus, even though a food may smell like dirty socks, previous experience with the food may override concerns about the odor.

With the additional investigations, there are now two possible explanations for your observations:

Personality Traits: People who are sensation-seekers are more likely than high-anxiety individuals to try unusual, novel foods like tinkin' toe.

Environment: People who have been exposed to novel foods in their environment are more likely to consume tinkin' toe than those who have not been exposed to such foods.

These potential explanations might also be called hypotheses —a researcher's proposed explanations of phenomena, which are then tested to see if the explanations are accurate.

2.2.3 Formulating a Measurable Hypothesis

At this point in the scientific method, researchers typically choose one of the explanations they have derived as the focus of their study, leaving the other for investigations in the future. In addition, the selected explanation forms the basis for formulating a measurable hypothesis that can be rejected or retained based on the outcome of a research study. Consider the following characteristics of a measurable hypothesis.

- A hypothesis is a conjecture, an educated guess, based on an investigator's explanation of a phenomenon. It is the framework for deriving a prediction (Section 2.2.4) about the results of a study. The hypothesis is then tested (Section 2.2.5) to determine whether the results support the prediction.
- A hypothesis can be general (for example, odor affects people's food preferences) or specific (People with a sensation-seeking personality are willing to try novel foods like tinkin' toe).
- The selected hypothesis should apply to a situation that is observable and measurable. The most effective hypotheses are those associated with studies in which participants can witness and experience phenomena via the senses in the natural world. For example, through the sense of smell, a person can tell that a food item has an odor that the individual finds unpleasant.
- To be measurable, a hypothesis must be framed so that an investigator can show it to be true or false. That means that it should be possible to reject the hypothesis. For example, a hypothesis that states that Earth is frequently visited by aliens from another planet is one that cannot be tested at this time. Accordingly, it is an untestable hypothesis.

Given the characteristics of a measurable hypothesis, imagine that you have selected the first explanation as the basis for a research study:

> **Explanation 1:** People who are sensation-seekers are likely to try novel, even unusual, foods like tinkin' toe.

A measurable hypothesis for this explanation might be:

> **Hypothesis 1:** There is a difference in the willingness of people with a sensation-seeking personality trait and those with high-anxiety personalities to try novel foods (e.g., locust fruit, goat cheese, and kimchi).

In this study, a researcher might employ published data collection measures designed to (1) identify personality traits and (2) to identify the willingness of study participants to try novel foods.

Consider another hypothesis:

> **Hypothesis 2:** There is a significant relationship between people's level of anxiety and their willingness to try new foods offered to them.

In one study (Otis, 1984), participants were not merely asked to rate their willingness to try different foods, they were offered actual samples of different types of foods. Their levels of anxiety related to trying the foods were subsequently recorded as well as whether they actually tried the novel foods offered to them.

Conducting a research study is an empirical test of the hypothesis that leads to evidence rejecting or failing to reject the hypothesis. In a correctly stated hypothesis, both outcomes must be possible—either to reject the hypothesis or to retain it.

2.2.4 Making a Prediction

Related to formulating a measurable hypothesis is the task of making a prediction about the outcome of the study. The hypothesis is a general statement about the relationship among variables. But it is also critical to the formulation of a prediction, which states a specific, authentic, logical outcome of a hypothesis that is true. A prediction is easily identifiable because it is often written in the form of "if and then" statements, as in, "If my hypothesis is true, and I were to do this test, then this is what I will observe."

Consider a study where teachers introduce two different mathematics instructional methods to two randomly selected groups of students. The central hypothesis is that in-person instruction is more effective than online instruction. The treatment is the type of instruction: students are exposed to, either online or in-person instruction. The prediction is that if the hypothesis is true, then students receiving in-person instruction will receive better grades on a common test than those receiving solely online instruction. Notice the relationship between the general statement of the hypothesis and the specific, logical application in the prediction.

Using the tinkin' toe example, the hypothesis is that individuals with sensation-seeking personality characteristics are more willing to try novel foods than individuals with high-anxiety characteristics. The treatment is to offer people with sensation-seeking traits and those with high-anxiety traits the opportunity to sample novel foods. The prediction is: If the hypothesis is true, then people with sensation-seeking dispositions will try the novel foods, while those with high-anxiety characteristics will not.

2.2.5 Testing a Hypothesis

In the research community, predictions require empirical confirmation. Therefore, after making a specific testable prediction, the researcher proceeds to test the hypothesis to determine whether the associated prediction holds. The process of testing the hypothesis incorporates a series of events. Indeed, it is the phase of a study during which systematic methods designed to test the hypothesis are implemented. It is the basis of much of the methodology component of a research study and much of the curriculum for research methods and quantitative statistics coursework. Activities may involve:

- securing a research design
- acquiring data collection measures (e.g., observation protocols, surveys, academic tests)
- selecting a study sample
- collecting the research data (e.g., achievement test scores, observational ratings),
- analyzing data,
- interpreting data analysis, and
- evaluating the outcome of hypothesis testing.

These are components of the research process, which will be discussed in Section 2.3.

2.2.6 Evaluating Results

Each component of the scientific method is designed to compare whether predicted and actual outcomes are one and the same. Are people with sensation-seeking personality traits willing to try novel foods while high-anxiety individuals prefer bland foods? Happy is the researcher whose projected results come to fruition. However, this is not always the case. Hypothesis testing typically results in one of two findings and subsequent options.

1. Agreement between the predicted and the actual results suggests that you may rest on your laurels or expand your research to formulate and examine new hypotheses.
2. Lack of agreement between the projected and actual results suggests that a misstep occurred somewhere along the way. Your question, explanation, your hypothesis and or prediction might have been faulty. Alternatively, your hypothesis test might have been defective. Regardless of the reason, you may need to go back to the drawing board and revise your hypothesis, prediction, or methodology.

Regardless of the findings, the researcher circles back through the scientific process repeatedly. As the evaluation component of the scientific method suggests, the process is not a linear but a circular one. Observations lead to questioning, which leads to generating explanations, formulating a hypothesis, making a prediction, testing the hypothesis, evaluating the results, and back to hypothesizing again. With each cycle, investigators gain new knowledge that adds to the body of literature.

2.2.7 Reporting Results

An important role of the scientific method is making the process of developing knowledge accountable. This is accomplished by virtue of making the application of the scientific method a public process. Therefore, the entire research study is available for evaluation by others within the research community. In addition, interested scientists should be able to replicate the process that led to the original results. Replication or reproduction of the research procedure allows for accountability and verification. Making the study a public process permits others to examine the rigor of the procedures and the absence of methodological flaws that might place findings into question. Reproduction of a study also offers the opportunity to verify the findings and provide additional evidence of a study's contribution to the research literature.

Two vital routes to placing research in the public arena is publishing findings in peer-reviewed research journals or presenting results at research conferences and other professional meetings. It is primarily in these environments that research peers may have access to the details of the studies for purposes of evaluation. With the principles of the scientific model in place, peer evaluators have access to the details of the methodology needed to conduct critical reviews of research that keeps the field accountable.

2.3 The Research Administration Process

The scientific method provides a high-level view of the principles on which research is based. The research process is based on the scientific method. It is a fleshed-out version of the scientific method. This section will describe the research process and will focus on a critical element of this process that encompasses testing the selected hypothesis.

2.3.1 Research Question

As was discussed in Section 2.2, the scientific method is a set of general principles used to guide the systematic study of a phenomenon, and these principles include in the early stages, asking questions. Research questions are often one of three types:

- **Questions that seek a description**
  For example, a researcher may be interested in online learning at the university level, and questions may include: How many courses are online? How many courses are online across the academic disciplines? What are students' attitudes toward online learning?
- **Questions that seek to investigate relationship among variables**
  For example, a researcher may wish to investigate a relationship between college students' internet self-efficacy and student achievement. Specifically, the researcher may ask: Is there a relationship between college students' internet self-efficacy and student achievement? Does students' performance increase as their internet self-efficacy increases?
- **Questions designed to investigate differences among groups**
  For example, a researcher may be interested in investigating whether there are differences in achievement of students exposed to three different instructional methods. Is there a difference in the achievement of students in online, in-person, and blended learning instructional methods? Do students perform differentially across the online, in-person, and blended instructional methods?

Questions are the driving force for explanations or answers to questions. That brings the discourse to the need to become deeply acquainted with the work of researchers who answered the same or similar questions in the past.

2.3.2 Related Theory

In an interview with Burton (2008), Alison Gopnik, an American professor of developmental psychology, said, "Asking questions is what brains were born to do, at least when we were young children. For young children, quite literally, seeking explanations is as deeply rooted a drive as seeking food or water." And the two are linked even into adulthood and scholarly work.

Asking questions about a notable experience is typically accompanied by seeking explanations that answer the questions. The explanations are based in theory. For instance, the first explanation for the tinkin' toe example in Section 2.2.2 is based on personality trait theory, which says that personality traits (e.g., openness to new experiences or sensation-seeking) are related to food preferences (Spence, 2022). Specifically, those who are open to new experiences like a wider range of foods (like tinkin' toe) than those who are not.

The second explanation for the tinkin' toe example comes from food theory, which states that our "sociocultural context (and ultimately the cognitive context) has … a large a role in defining what is and is not food, and what should or should not be eaten" (Allen, 2012, p.4). Therefore, our sociocultural environment and associated food experiences are related to our food preferences.

Most often, conducting a search of studies related to the topic of interest is what leads to the discovery of explanatory theories. Searching for what investigators found in previous related studies leads to connected theories. Interaction with peer researchers, mentors and acquaintances may also result in the discovery of new theory that becomes the foundation of a study.

2.3.3 Formulating the Research Hypothesis

Section 2.2.3 established some general characteristics of hypotheses; for example, hypotheses must be refutable and measurable. In this section, we will examine additional qualities of this important element of research studies.

The hypothesis is an extension of the research question. Some might say that it is the research question reformulated as a statement. Moreover, the hypothesis is stated in terms of the target population, since the intent is to use the study outcome from a representative sample to make inferences about the target population.

Like research questions, hypotheses take different forms. Consider the questions and their associated hypotheses in Figure 2.2. The research hypotheses are what are called nondirectional or two-tailed hypotheses. Typically, these hypotheses state that there is a difference between two or more groups or a relationship between two variables, without specifying the direction of the differences or the relationship. More about hypotheses in Section 2.4.

Figure 2.2 Research Questions and Hypotheses

Question	Hypothesis
1. Is there a relationship between college students' internet self-efficacy and academic performance?	There is a relationship between **college students' internet self-efficacy** and **academic performance**.
2. Is there a difference in the willingness of tourists with different personality traits to try novel foods?	There is a difference in the **willingness** of tourists with different **personality traits** to **try novel foods**.

Although researchers need to make additional decisions about the details of a study, the statement of the research hypothesis is the basic framework of the study. It describes the variables (attributes or characteristics) in the study, and it points to the type of research strategy and design that will be used in a study.

2.3.4 Variables

A hypothesis incorporates the specific variables that researchers will investigate in a study. Variables are attributes or characteristics that can change or can be expressed as more than one value (Vogt, 1993) during the course of a research study. Consider the hypothesis for Question 1 in Figure 2.2. The hypothesis names each variable: internet self-efficacy and academic performance. In addition, the hypothesis specifies that there is a relationship between the variables, internet self-efficacy and performance. The condition that there is a relationship between the variables suggests that the variables are not constants, but rather that the values of the variables can change. Some study participants may have high self-efficacy scores. Others may have low self-efficacy scores. Scores that represent academic performance will also vary across students.

Therefore, variables may be defined by not only naming them, but also indicating whether their values can change. They may change among people (academic performance, gender, height), among places (schools, churches, parishes) or across other factors (mood, income, and expenses).

2.3.4.1 Types of Variables

In framing hypotheses, variables are assigned to broad implicit functions or types. Perhaps the most commonly used variables are independent and dependent variables. An independent variable (sometimes called a factor) is one that does not depend on any other variable and one that the researcher generally controls or manipulates (e.g., age ranges, education levels). Sometimes these variables fall into natural groupings (e.g., vehicle brand and color). Independent variables are sometimes, but not always, based on qualitative characteristics; for example, variables used to name, label or categorize an attribute like a food-related personality trait (namely, food neophilia—overt willingness to try novel foods, and food neophobia—reluctance to try novel foods). In these situations, values associated with independent variables have no intrinsic ordering. They are merely used for grouping purposes. However, in quasi- and true experiments, researchers can directly manipulate independent variables in order to see how they affect outcomes (e.g., the use of different drug dosages to compare the effects on hypertension). In these situations, independent variables are identified as having a causal role in research studies and are often referred to as treatment variables.

Sometimes independent variables are termed predictor or regressor variables. This term is often used in studies where one or more variables are used to predict a single outcome. For example, several research studies used food neophobia, personality traits, and culture to predict willingness to try unfamiliar foods (Otis, 1984; Spence, 2022). In some studies, internet self-efficacy might be used as a predictor variable to predict the outcome of academic achievement.

A dependent variable is one that depends on another variable; it is the variable that "depends on" the independent variable or that is predicted by one or more independent variables. Researchers expect to see some change in the dependent variable as the values of the independent variable changes. In a nonexperimental study, the dependent variable may be called the outcome variable or the criterion variable in that it may be related to the independent variable, but not necessarily caused by it. Therefore, Hypothesis 2 suggests that researchers expect that willingness to try new foods is dependent on food-related personality traits (independent variable).

2.3.4.2 Variable Measurement Level

Before a researcher tests a hypothesis, the investigator must ensure that the variables to be included are measurable. To determine if a variable is measurable, it is helpful to understand the scales of measurement used to define a variable as that governs research design alternatives.

The level of measurement or scale of measurement of a variable is a classification system that describes the type of information contained in variable values and, based on that, what kind of statistical techniques and analyses are applicable for them. American psychologist, Stevens (1946), developed the best-known classification system of variable values which describes the properties and applications of four levels of data: nominal (includes dichotomous), ordinal, interval, and ratio (Figure 2.3). Properties of the measurement levels are cumulative, in that within the hierarchy of measurement levels, each includes all the properties of the one before it as well as one or more additional features.

The scales of measurement fall into two broad classes: categorical and continuous variables. Categorical variables are also known as qualitative or discrete variables that name the groups that belong to a variable. Categorical variables might include types of trucks, fruit trees, languages or animal species. The items in each categorical variable are mutually exclusive. Therefore, they cannot be mathematically manipulated; it is nonsensical to add one half of a dog to half of a cat.

Figure 2.3 Characteristics and Applications of Scales of Measurement.

Class	Scale of Measurement	Property	Descriptive Statistics Applications
Categorical/Discrete/ Qualitative	Dichotomous	variable with two discrete categories; e.g., male/female; correct/incorrect; alive/dead.	mode, frequency, proportions.
	Nominal	two or more unordered categories; e.g., parishes of Barbados, political affiliation.	mode, frequency, proportions.
	Ordinal	two or more ordered or rank-based categories; e.g., letter grades, Likert scales, age group.	all above + median, percentiles, rank order correlation
Continuous/ Quantitative	Interval	numerical variables whose values are measured along a continuum; e.g., temperature in degrees Celsius or Fahrenheit.	all above + mean, standard deviation, range, variance
	Ratio	interval variable where zero means absence of the characteristic measured; e.g., height, distance.	all above + coefficient of variation.

Sometimes the categories of a categorical variable are denoted by numerical values. However, the values are solely for labeling and distinguishing between categories. According to the World Bank (2019) 560 indigenous languages are spoken in Latin America and the Caribbean. If we were to assign a number to each of those languages, we could not claim that the language labeled 560 is any better or worse than the language labeled 1. The numbers are assigned for purposes of convenience, and the categories are equivalent in value.

The broad class of categorical variables includes two types of such variables: dichotomous and nominal variables. A dichotomous variable is the most basic of categorical variables. Each dichotomous variable is limited to two options—male or female, yes or no, heads or tails. Because the groups do not overlap, a data value cannot occupy both categories: male or female, yes or no, heads or tails.

When a categorical variable consists of more than two groups, it is known as a nominal variable. Beyond having more than two categories (e.g., types of fruit drinks, or languages), nominal and dichotomous variables share the same characteristics:

- The categories of nominal data are mutually exclusive.
- Descriptive labels are used to group nominal data. These labels do not have any numerical value.
- Nominal data cannot be grouped in a meaningful hierarchy because all nominal data have equivalent value.

Categorical data are most usefully summarized using descriptive statistics like mode and frequencies. Accordingly, we might use the mode to identify which of the 560 indigenous languages most Caribbean and Latin American people speak. We might also use proportions or percentages to summarize language frequency.

A third type of categorical variable is known as ordinal variables. Like dichotomous and nominal variables, this variable type groups data into descriptive categories. However, instead of grouping data in categories with no intrinsic order, ordinal data are ordered or rank-ordered groups. Ordinal data may be placed in ordered groups; for example, measuring economic status using the rankings poor, middle income, and wealthy, or measuring educational level with the ranking of primary, high, and university level. Ordinal level variables place data in ordered categories. For example, educational levels tell us that university level is higher than high school level, which is higher than primary school level. However, the categories do not reveal precise differences between levels.

As an example, consider the Secondary Entrance Assessment (SEA) in Trinidad and Tobago, an assessment for 11-13-year-olds. For that assessment, students respond to either a narrative or report writing prompt over a 50-minute period to demonstrate their writing skills. Each writing sample is then independently judged by two writing experts (typically educators) based on an ordered scale of 0 5, where 0 means unsatisfactory, and 5 means exemplary. The ordered scale conveys that student with an overall score of 4 demonstrated better skills than those earning a score of 2. However, even though there is a difference of two between 4 and 2, it is not clear that the distance between 4 and 2 in skill is equivalent to the distance between 5 and 3 or 0 and 2.

While categories on an ordinal scale may be labeled numerically, as in the case of writing assessment scores, numeric values have no intrinsic mathematical value. They could be replaced by letters because they are purely descriptive. Descriptive statistics are used to summarize ordinal level data (Figure 2.3). In this text, statistical tests using ordinal data include nonparametric tests like Kruskal-Wallis and Spearman's rank correlation coefficient.

The next level of measurement is continuous or quantitative types of variables. Like categorical variables, continuous variables can be placed into groups and can be ranked. However, they are distinctively different in that they can assume an unlimited number of numerical values between the highest and lowest points in a range of values.

As Figure 2.3 illustrates, continuous variables include two types—interval and ratio level variables. Like ordinal-level variables, data for interval-level variables are in ordered categories. However, there are differences between the two:
- Interval level data are quantitative or numeric values rather than purely descriptive features of nominal and ordinal level variables.
- There are equal differences between adjacent numerical values on the interval scale. Equal differences represent equal intervals.

One example of an interval level variable, described in Figure 2.3, is the Celsius or Fahrenheit temperature scale where the differences between any two points on the scale are equidistant. Another example of interval-level data is the difference between scores on an assessment. Consider an assessment where the minimum score is 0 and the maximum is 100. On that scale, the difference between scores of 50 and 70 is same as the difference between 70 and 90 on the same test —20 points. Other examples of interval level measures include times of the day (e.g., 8 am, 10 am, 12 pm, 2 pm), income levels (20,000; 40,000; 60,000; 80,000) or years (e.g., 1940, 1965, 1980, 2005).

The equal interval and numeric characteristics of interval level data make it especially useful because it lends itself to arithmetical operations like finding consistent differences by adding or subtracting numerical values. Similarly, as interval measurements have equal intervals, mathematical operations like finding the average of a set of values are allowable. Therefore, the average amount of rainfall for one week during the rainy season may be calculated by adding the amount of rainfall each day and dividing by 7. These characteristics make interval level data more useful in carrying out descriptive (Figure 2.3) and inferential statistical analyses like *t*-tests and analysis of variance (ANOVA). Much of the statistical procedure in this textbook requires the use of interval-level data.

Slotting variables into a specific measurement level is sometimes quite challenging. In some instances, there is considerable and longstanding disagreement among scholars about whether the measurement level for some data is ordinal or interval. For example, consider Likert scales developed by Likert (1932) to measure attitudes. The Likert scale response formats in Figure 2.4 may be the response component of a customer service survey in which respondents rate their level of satisfaction. Some authors (e.g., Jamieson, 2004) consider item-level survey data based on a Likert response format to be ordinal level because respondents rank order their attitude toward each statement on the survey. Researchers argue that individual item responses do not have an even distribution as required for interval-level data. Adjacent anchors may not represent equal distance (e.g., the distance between "very satisfied" and "satisfied" may not be the same as the difference between "dissatisfied" and "very dissatisfied"). The researchers assert that while it is possible to perform arithmetic operations on the data, the outcome may not be interpretable. For example, it is challenging to interpret a mean dissatisfaction of 3.5 on the five-step response format in Figure 2.4b, even more challenging when there are only three scale points (Figure 2.4a) on a Likert-style response format.

Other scholars (e.g., Norman, 2010; Carifio & Perla, 2007) disagree with uniformly limiting Likert scale item response formats to the realm of ordinal level of measurement, and thereby, restricting statistical techniques that might be used with such data. They assert that their empirical evidence shows that Likert scale item types may be treated as interval-level data at the individual item level if the number of scale points for the response format "is sufficient (preferably 7)" (Carifio & Perla, 2007, p. 111), and even more so if they represent a symmetry of scale points about a midpoint as in Figure

2.4c. In addition, a variable measured by a set of survey items with Likert-style response formats is also an interval-level measure if the response format includes five to seven scale points, and there are five to eight items in the set (Carifio & Perla, 2007). The scale points may not be equidistant from each other, but if well-presented, they are close enough to approximate interval-level measurement and make outcomes interpretable (Meyers et al., 2016).

Figure 2.4 Satisfaction Likert Scales

(a) Three-Point Response Format	(b) Five-Point Response Format	(c) Seven-Point Response format
1. Satisfied 2. Neutral 3. Dissatisfied	1. Very satisfied 2. Satisfied 3. Neutral 4. Dissatisfied 5. Very dissatisfied	1. Very dissatisfied 2. Moderately dissatisfied 3. Slightly dissatisfied 4. Neutral 5. Slightly satisfied 6. Moderately satisfied 7. Very satisfied

The second level of measurement within the continuous category is the ratio level of measurement—the fourth and final of the four measurement levels. Like variable data at the interval level, ratio-level variables are quantitative with all the properties of the other three levels of measurement—nominal, ordinal and interval. The distinguishing characteristic of the ratio level is that it has a true zero point, so that a value of zero on the ratio scale means that the variable being measured is absent. For example, if you have a course with a registered student count of zero, that means there are no students registered for the course. Because ratio level measurements have a true zero point, there are no negative values. As a result of the distinctive properties of ratio-level measures, it is possible to calculate numerical ratios; hence the name of the measurement level, "ratio." Accordingly, if a driver is traveling at 50 miles per hour and the speed limit is 25 miles per hour, you can say that the driver's speed is twice the speed limit (50/25).

Examples of variables at the ratio level of measurement include body weight, height, size of a population, age in years, speed in miles per hour, and salaries. In theory, some of these variables can take on infinite values (e.g., income or the number pi or π), while others are finite (e.g., number of faculty members at a college).

With ratio level data at the top of the scales of measurement, almost all mathematical operations and statistical techniques may be used to analyze variables of this type. The same descriptive statistics used for interval level measurements may be used for ratio measurements as well. Nevertheless, when analyzing data, there is little practical difference in the statistical tools available for interval and ratio level measurements.

Defining variables by type helps researchers determine how best to measure the variables. Consider Hypothesis 1 regarding the relationship between internet self-efficacy and student achievement. Should the researcher measure internet self-efficacy via: observation or self-assessment? a questionnaire with a Likert-response item format? online or in-person? How should the researcher measure achievement: standardized test scores? marks in a course? Although these concerns are not stipulated in the hypothesis, how to accomplish the measuring of variables using an empirical method is an important part of crafting the hypothesis. In a study of the relationship among internet self-efficacy, self-regulation, and student performance, Kuo and Tseng (2020) used total scale-level scores from an internet self-efficacy scale with a 7-point response format. In this case, individual item responses could be classified as interval-level, but certainly the total scale scores which were used in the data analysis could be considered ratio-level data. Performance scores could be ordinal (e.g., grades) or interval level (test scores). Defining variables by type helps researchers to transform the research question and hypothesis into a well-defined research design. Typically, this happens under the guidance of similar previous research.

2.3.5 Research Approach and Design

The term research approach describes three broad classifications of research studies: qualitative, quantitative and mixed methods. Boffa et al. (2013) described the difference between qualitative and quantitative research this way: "Simply, the difference between qualitative and quantitative research is in the prefix: qualitative studies investigate the qualities of an experience, while quantitative studies measure and interpret numerical output" (Boffa et al., 2013, p. 103). Qualitative research uses primarily non-numeric data (e.g., data from interviews, observations and artifacts) to answer research questions. Quantitative research uses quantifiable numerical data (e.g., test scores, physical measurements, attitudinal survey responses) and associated statistics to answer research questions. Mixed methods research uses both quantitative and qualitative approaches to conduct research.

The term *research strategy* or *design* commonly refers to the framework used to implement the general approach and goals of a research study. This is usually determined by the kind of question you plan to address and the kind of outcome you hope to obtain. In quantitative research, which is our primary interest here, research designs are classified in two broad categories: experimental and nonexperimental. Some of the more commonly-used sub-types are presented in Figure 2.5, along with a description of each.

Nonexperimental and experimental research initially appear to be monumentally different approaches to conducting investigations. However, the characteristics of each type of design contribute to a wide selection of established defensible practices and processes for conducting quantitative research. Fundamental characteristics of nonexperimental and experimental research designs are presented in Figure 2.6. Exhaustive discussions of research designs may be found in the classic references of Campbell and Stanley (1963); and Shadish et al. (2002). Here we will briefly describe some of the most commonly-used designs.

2.3.5.1 Nonexperimental Research: Descriptive Designs

Descriptive research is a nonexperimental approach that seeks to describe the current characteristics of individuals, settings, conditions, or events as they exist naturally (Mertler, 2016). These research designs do not include any manipulation of individuals, conditions, or events. They ae meant to provide systematic descriptive information about a phenomenon. Frequently-used quantitative descriptive research designs include case studies, observational studies, and surveys.

Figure 2.5 Common Quantitative Research Approaches

Research Design Types	Description
Nonexperimental	a group of techniques where there is no manipulation of any variables in the study. Instead of manipulating variables, researchers measure the variables as they naturally occur.
Descriptive	describes the characteristics of a population or phenomenon. It answers the question what, when, where and how.
Correlational	measures the relationships between two or more variables. These include explanatory and predictive studies.
Causal-Comparative	compares differences (e.g., in performance or behaviors) between two or more groups based on group qualities that existed before the investigation; also known as ex-post-facto research.
Experimental	a group of techniques where the researcher randomly selects and/or assigns study participants to various treatments or conditions and then studies their effects on study participants.
Quasi-experimental	an investigation of the cause and effects of a specific intervention or condition that involves only the random selection (no random assignment) of study participants.
True experimental	a study that involves the random selection and random assignment of study participants. This approach controls for nearly all threats to the validity of a study.

Quantitative case studies are in-depth investigations to gain a better understanding of an individual, a group, an organization, and event or belief system. These studies may involve the use of self-administered questionnaires or achievement tests to collect case study data.

Quantitative observational research generally concentrates on certain aspects of behavior that researchers can quantify. For example, a researcher may conduct a study in which trained observers quantify disruptive behaviors among particular students or groups of students during particular activities or time of day. The researcher may develop an observational protocol to record frequency of disruption or to rate the intensity or mastery of behaviors of interest.

Survey research includes cross-sectional as well as longitudinal surveys with the central purpose of describing the attitudes, experiences, or other characteristics of a group (e.g., sample surveys) or an entire population (e.g., census surveys). Researchers may use surveys in a descriptive study, or to examine relationships between variables (e.g., income and age) within a population, or to compare groups of survey respondents (e.g., novice and experienced teachers).

Figure 2.6 Common Research Designs: Characteristics and Statistics

Research Design	What is Measured?	IV Manipulation	Random Assignment/ Selection	Measures Differences Between or Within	Parametric Statistics	Nonparametric Statistics
Non-experimental						
Descriptive	characteristics, trends, categories	no	no	no	central tendency, variability, dispersion	central tendency, ranks, frequency
Correlational	relationships	no	no	no	Pearson's R regression coefficients	Spearman Rank, Kendall's Tau
Causal-Comparative	potential cause/effect	no	no	between	t-test ANOVA ANCOVA	Mann-Whitney K-W test
Experimental						
Quasi-experimental	cause/effect	yes	random selection	between and within	t-test ANOVA repeated measures ANCOVA	Mann-Whitney K-W Test Wilcoxon's Test, Friedman's
True experimental	cause/effect	yes	random assignment and selection	between and within	t-test ANOVA repeated measures ANCOVA	Mann-Whitney K-W Test Wilcoxon's Test, Friedman's

Examples of descriptive research topics include:
- a description of the types of activities in which elementary students in Bridgetown, Barbados, engage while on summer vacation
- a description of how student-nurse educators perceive their roles as subject matter and pedagogical experts
- the frequency of student disruptive behavior in novice and experienced teachers' classrooms
- a longitudinal survey of student post-secondary employment outcomes

2.3.5.2 Nonexperimental Research: Correlational Designs

Correlational research investigates the extent of a relationship among two or more variables. This type of research falls into two categories: explanatory research studies and predictive correlational studies.

Explanatory correlational studies include investigations in which researchers measure two or more characteristics of a single group of study participants. They then analyze the data to measure and interpret the pairwise statistical relationships between the variables, one pair at a time.

Predictive correlational studies include inquiries in which researchers attempt to ascertain whether one or more variables (predictor(s) or independent variable(s)) are linearly related to a single outcome variable (or dependent variable). For example, height might be used to predict weight, or scores on various high school assessments may be used to predict college-level GPA.

Regardless of whether the studies are explanatory or predictive, they are intended to identify trends and patterns in data, but they do not reveal or imply causes for the observed patterns of relationships. Unlike experimental research, variables are not manipulated; they are only identified and are studied as they occur in a natural setting. Because no variables are manipulated in correlational research, sometimes it is considered a type of descriptive research, and not as its own type of research. Examples include:
- the relationship between Caribbean Secondary Education Certificate (CSEC) test scores and grade point average (GPA) in the first year of college
- the relationship between church attendance and disruptive behavior among elementary school students.
- the relationship among diet, exercise, smoking, and heart disease.

2.3.5.3 Nonexperimental Research: Causal-Comparative Design

Causal-comparative research incorporates investigations that compare two or more groups to explain differences between the groups on one or more dependent variables. This type of design is similar to experiments, but with some key differences. For example, in causal comparative designs, researchers do not equate groups at the beginning of a study using random selection and/or assignment. They do not randomly choose individuals to participate in research studies (random selection). Once chosen to participate in research studies, individuals are not randomly assigned to treatment and comparison groups. Instead, researchers form participant groups based on previously existent naturally-occurring characteristics (e.g., male or female; infant school, basic school, and private preschool education) that may not lend themselves to equated groups; hence this design is also known as "ex post facto," literally "after the fact" research.

In causal-comparative research, groups (independent variables) either cannot be manipulated by the researcher (e.g., gender) or should not be manipulated for ethical and other reasons (e.g., the number of cigarettes smoked each day). Researchers may attempt to strengthen a causal-comparative design using a variety of group composition strategies (e.g., participant matching) and statistical techniques (e.g., analysis of covariance or ANCOVA, Chapter 9). However, in spite of efforts to bolster the design, causal-comparative research does not approach the rigor of experimental research designs.

The intent of a causal comparative research design is to compare the effect of the independent variable groups (e.g., smokers/nonsmokers) on the dependent variable (e.g., the development of lung cancer) in an effort to explain any differences between the groups. Therefore, this nonexperimental design is also known as a between-subjects design.

Due to the lack of controls in causal comparative studies, researchers may identify variables as having potential cause-and-effect relationships following statistical data analysis results, as variables that are not in the research design, both known and unknown, might affect the outcome as well. As a result

of the absence of random selection and assignment, researchers exert caution in making cause/effect claims based on study results. Instead, causal comparative studies often form the foundation for more rigorous experimental studies. Examples of investigations using this type of research design include:
- the effect of nursery school attendance on social maturity at the end of first grade,
- the effect of gender and study methods on student math achievement,
- the effect of cigarette smoking on the development of cancer.

2.3.5.4 Experimental Research:

True experimental research includes studies that strictly adhere to the use of the scientific method to establish a cause-effect relationship among a group of variables in a study. The true experiment is often thought of as a laboratory study, but true experiments may also take place in a field setting that is a controlled environment. Like causal comparative research, one goal of true experiments is to investigate differences between groups. However, unlike causal comparative research, controls inserted in true experiments allow investigators to demonstrate cause and effect.

A true experiment is any study where an effort is made to identify and impose control over all other variables except one—the dependent variable(s). For example, in a true experiment, instead of using naturally-occurring groups, researchers randomly select participants for studies. Following selection, researchers randomly assign participants to one of two or more independent variable groups:
- an experimental group, which receives the treatment of interest,
- a control group that receives no treatment, and
- possibly a comparison group that receives an alternative treatment.

This manipulation of the independent variable helps to determine differential effects on the dependent variable. Researchers have at their disposal several true experimental designs. Three commonly-used designs (Figure 2.7) are known as the posttest only control group design, the pretest-posttest control group design, and the counterbalanced design. The posttest-only design (Figure 2.7a) includes at least two groups of randomly assigned participants (R). Neither group is pretested. One group receives the treatment and both complete a posttest. In the pretest-posttest control group design (Figure 2.7b), both groups are pretested (O), receive either a predetermined treatment (X) or no treatment at all, and complete a posttest (O) at the end of the study. In the final design, the counterbalanced design (Figure 2.7c), the randomized groups are each exposed to each treatment, but in a different, randomly determined, sequence.

Examples of true experiment studies include:
- a comparison of the efficacy and safety of different doses of dexamethasone (a corticosteroid drug) in the treatment of COVID-19
- a comparison of the effect of background music (no music, classical, and hard rock) on memory

2.3.5.5 Experimental Research: Quasi-Experimental Designs

In quasi-experimental research, "quasi" means "partly" or "almost." Accordingly, of all the research designs, quasi-experiment designs are the closest to true experiment designs. Quasi-experimental designs share most, but not all of the features of a true experiment. Perhaps the most distinctive difference is the absence of the element of random assignment to experimental and control conditions in a study.

Figure 2.7 True Experiment Designs

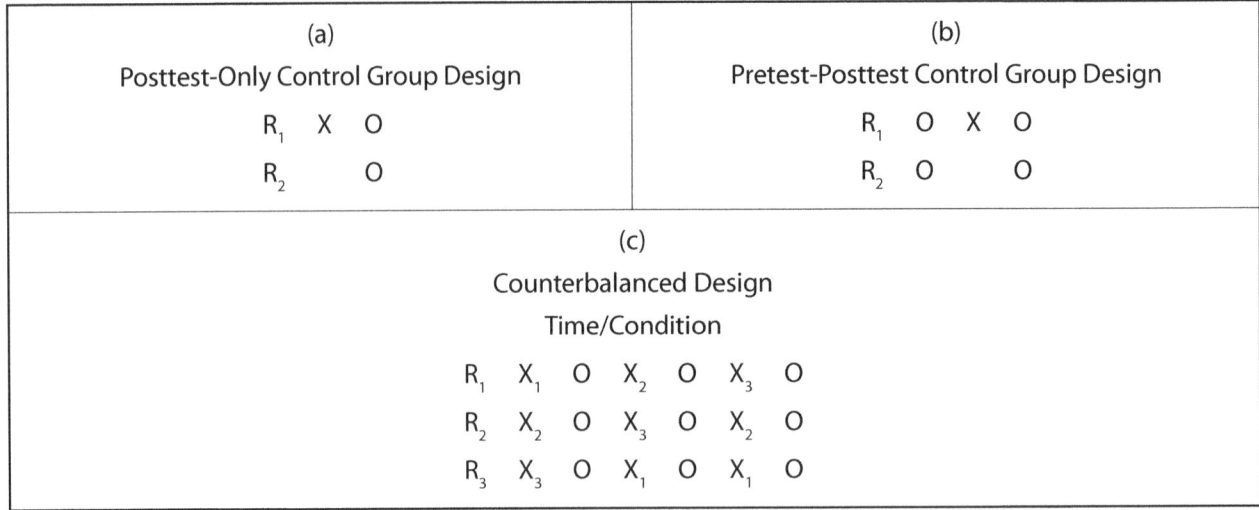

Unlike true experiments, quasi-experiments often take place in a field, non-laboratory setting where it is not possible to randomly assign participants to groups due to ethical concerns or the impracticality of group assignment. For example, arguably it would be ethically untenable to randomly withhold a drug treatment from patients or to randomly assign students in a school-based study where students are already assigned to intact groups.

Figure 2.8 Quasi Experimental Designs

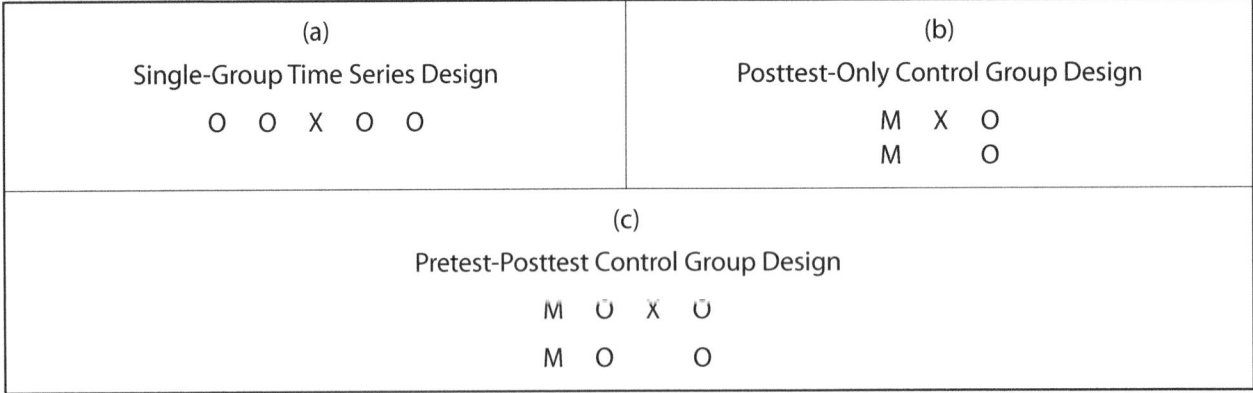

Researchers have a variety of research designs available to eliminate threats to the validity of quasi-experimental study results due to the absence of random assignment. Three commonly used designs are presented in Figure 2.8. The first is known as a time-series design, which is sometimes regarded as a pre-experimental, single-group design (e.g., Gall et al., 2007). The time-series design (Figure 2.8a), is a single-group design illustrated by the symbols that are in a single row. The design includes a single group of study participants who are pretested repeatedly before the treatment and posttested repeatedly after the treatment. The remaining designs in Figure 2.8 are the same as those in Figure 2.7, with the exception of the random assignment (R) component.

As Figure 2.8 illustrates, a matching component (M) replaces random assignment. Consider, for example, a study of the effects of a new instructional method on reading achievement in which the

posttest-only control group design in Figure 2.8b is implemented. In this case, researchers may randomly select two intact classrooms from the same school system in an attempt to control for extraneous influences. In addition, in the absence of random assignment, students may be matched on certain background factors to enhance pre-treatment equivalence of the groups. Researchers then expose one class to the new instructional method (X) and the other, the control group, continues to benefit from the existing method of instruction. Both classes complete the same reading assessment after a period of time (O), which is subsequently used to assess differences in student achievement. Figure 2.8c displays a design that is an improved version of the design in Figure 2.8b. Random selection remains in place, but also the design includes a pretest which serves as a method for matching students across the groups. Therefore, all the students take the pretest, and their results are used to statistically equate the groups before comparing their posttest performance. Examples of quasi-experimental studies include:
- a comparison of the effects of gender and age on texting speed
- a comparison of the effects of different parental styles (e.g., authoritarian, authoritative, permissive, and neglectful/uninvolved) on student achievement.

2.3.5.6 Between-Subjects and Within-Subjects Designs

Section 2.3.5.3 disclosed that causal-comparison research is the only nonexperimental research design that is a between-subjects design (Figure 2.6). Experimental research also utilizes between-subjects designs, two of which were explored in Section 2.3.5.4 and 2.3.5.5. In between-subjects designs, study participants are arranged so that each is restricted to a single group of an independent variable (e.g., educational level) that comprises two or more groups. Researchers then compare the groups or levels of the independent variable to learn whether there are significant differences between them.

A second method of arranging study participants or the data they provide is known as the within-group, or repeated measures design. In this type of design, there are two or more treatment groups or conditions. Researchers secure data from each participant for each of those treatment situations or conditions and statistically analyze the data to compare and identify any significant differences between the treatment conditions administered. As indicated in Figure 2.6, this type of design is commonly seen in true- and quasi-experimental studies. The designs in Figure 2.7c and 2.8a are two examples of repeated measures designs, Figure 2.7c being an example of both between and within components. This text will explore statistical techniques developed to analyze data for between- and within-subjects designs.

Regardless of the research design selected, research questions and hypotheses point to the nature of the design of the research study, and the research design points to the type of data collection and data analysis necessary.

2.4 The Research Plan

While a research approach provides the broad contours of a research inquiry, a research strategy goes one step further to provide the blueprint or set of procedures that a researcher uses in order to carry out a study and secure the most valid findings possible. It is also often called the research plan or research method that includes the following components for each type of design (some of which were discussed in previous sections on the research approach):
- stating the research question(s) (discussed in Section 2.3.1)
- describing the variables (discussed in Section 2.3.4)

- defining associated hypotheses
- setting the level of significance
- describing data collection methods
- selecting statistical techniques
- statistical analysis (conducting assumptions tests, selecting and calculating test statistic, interpreting results)

2.4.1 Two Hypothesis Statements

A measurable hypothesis is developed during the planning and design phase of a study, and one important feature is that it must be falsifiable; that is, it must be possible to prove the hypothesis wrong. To do that, researchers often begin with the assumption that a stated relationship (or difference or treatment effect) does not exist. The research goal is to provide convincing evidence to the contrary. Since a hypothesis can either be true or false, two statements are constructed to represent those outcomes: the null hypothesis and the alternative hypothesis.

The null hypothesis, signified by H_o, is the statement that represents falsification. Perhaps the most common type of null hypothesis is the nondirectional (two-tailed or two-sided) null that always includes the condition of equality in that it states that there is "no difference" or "no relationship" without specifying the direction of the differences. It is the hypothesis tested by the selected statistical procedure. For example, consider the null hypothesis for the questions in Figure 2.2:

Null Hypothesis for Question 1: There is no relationship between student internet self-efficacy and student performance.

Null Hypothesis for Question 2: There is no difference in the willingness of people with various personality traits to try novel foods.

The null hypothesis is often called the straw man, in that the objective of a research investigation is that the evidence (data) collected will result in convincingly rejecting the null hypothesis. If, however, there is insufficient evidence to reject the null hypothesis, the result is to "fail to reject" the null hypothesis. Failing to reject the null does not mean that a relationship or difference does not exist. It simply means that there is insufficient evidence to support that conclusion, and therefore the null hypothesis stands until opposing evidence is presented.

The alternative hypothesis, also called the research hypothesis, and symbolized H_a, is the statement that denotes the alternative to the null hypothesis—what the study was designed to investigate and what the researcher predicts will be the outcome of the investigation. In a nondirectional alternative hypothesis, the statement declares that there is a difference between groups, or there is a relationship between variables without specifying a direction. For example, the alternative hypothesis for the questions noted in Figure 2.2 is nondirectional as follows:

Alternative Hypothesis for Question 1: There is a relationship between college students' internet self-efficacy and academic performance.

Alternative Hypothesis for Question 2: There is a difference in the willingness of tourists with various personality traits to try novel foods.

When there is sufficient evidence to reject the null hypothesis, the alternative hypothesis is automatically retained. Therefore, if the evidence does not support the absence of a relationship between student self-efficacy and achievement, we automatically conclude that there is a relationship between the two variables. Again, this conclusion stands until, as a result of the scientific process, new research presents evidence challenging that conclusion.

There are occasions when a researcher wishes to test a hypothesis with a specific direction in mind. These are described as directional, one-tailed or one-sided hypotheses. For example, imagine that the researcher wants to investigate whether there is a positive relationship between internet self-efficacy and academic performance, the hypotheses might look like this:

Null Hypothesis: There is a negative relationship between internet self-efficacy and academic performance.
Alternative Hypothesis: There is a positive relationship between internet self-efficacy and academic performance.

If a researcher wishes to investigate differences among groups using directional hypotheses, the hypothesis statements might look like this:

Null Hypothesis: Of 20 novel foods offered, food neophobic tourists were willing to try a mean of 5 or fewer of them.
Alternative Hypothesis: Of 20 novel foods offered, food neophiliac tourists were willing to try a mean of more than 5 of them.

Figure 2.9 presents a summary of types of hypotheses. Note that it is important to ensure that the signs of the null and alternative hypotheses statements are the opposite of each other.

Figure 2.9 Types of Hypotheses

Hypothesis	Nondirectional (2-tailed)	Directional (1-tailed)	
Null (H_o)	=	≥	≤
Alternative (H_a)	≠	<	>

2.4.2 The Statistical Level of Significance

So far, this section has explained that a hypothesis includes two statements—the null hypothesis and alternative hypothesis. These statements identify the variables to be measured, the type of research strategy as well as the research design to be implemented. Most importantly, the statements signify a researcher's intention to investigate whether the research outcome happened by chance.

Imagine that a researcher conducts a study, using a representative sample, to answer Question 2 in Figure 2.2. The researcher finds what appears to be a large differential willingness to try novel foods. Suppose the evidence suggests that participants characterized as having a food neophiliac personality (willingness to try novel foods) were more willing to try novel foods than those characterized as food neophobic individuals (unwillingness to try novel foods). The large difference found in the study sample does not necessarily falsify the null hypothesis. In order to falsify the null hypothesis, the difference must be large enough so that it is unlikely that it happened by chance. These tests are also called tests of significance.

To determine whether the difference represents chance effects, the researcher conducts statistical tests (more on this in Section 2.4.5) which measure the likelihood that the difference found in the investigation would be observed in the target population if the null hypothesis (which states no difference) were true: in other words, the likelihood of saying there is a difference, when in fact, there is no difference (or the likelihood of wrongly rejecting a true null hypothesis). If the likelihood of wrongly rejecting a true null hypothesis is sufficiently small (e.g., a five percent likelihood), a researcher can assume that the difference does indeed exist. Therefore, differences in the willingness to try novel foods are statistically significant. However, if there is a high likelihood of incorrectly concluding a difference exists, any difference noted between groups may be due to random chance (error).

The hypothesis points to the intention to conduct statistical tests by including two hypotheses—the null and the alternative hypothesis. The inclusion of the phrase, "statistically significant" in the hypothesis statements is intended to indicate the intention of conducting statistical testing of the hypothesis statements. Therefore, the revised null hypothesis statement for Questions 1 and 2 may appear as follows:

Null Hypothesis for Question 1: There is no **statistically significant** relationship between student internet self-efficacy and student performance.

Null Hypothesis for Question 2: There is no **statistically significant** difference in the willingness of people with different personality traits to try novel foods.

A vital part of the planning involved in statistically testing hypothesis statements is to establish a standard (or cut-off) for rejecting the null hypothesis. In keeping with ethical research standards, the researcher sets the standard during the planning phase of a study, before collecting or analyzing any data. This a priori standard is called the statistical level of significance (also known as the probability level or alpha level). The statistical level of significance is an allowable range of probability levels that represents acceptable levels of claiming that there is a significant difference among groups or a relationship between variables when there is not. This is called a Type I error. More on this in Section 2.4.6.1.

Perhaps the most common statistical level of significance is $p \leq .05$. Once data has been collected and statistical testing of the hypothesis is completed, the value resulting from the hypothesis test is compared with a pre-established standard. If it falls within the a priori range, then the researcher may reject the null hypothesis and accept the alternative hypothesis. If the result from hypothesis testing falls above the range, then the researcher fails to reject the null hypothesis and cannot accept the alternative hypothesis.

A study of internet self-efficacy and student performance (Tseng & Kuo, 2020) provides some insight into the role of the statistical level of significance. Tseng and Kuo wished to investigate, among other questions, whether there was a significant difference in the internet self-efficacy of online college students at different performance levels. The hypothesis for this component of the study might have been:

Null Hypothesis: There is no statistically significant difference in the internet self-efficacy of students at five performance levels.

Alternative Hypothesis: There is a statistically significant difference in the internet self-efficacy of students at five performance levels.

Hypothesis testing sought to answer the question: If there is no difference among students in the five performance levels in the target population (assuming that the null hypothesis is true), how likely would it be to observe a difference in internet self-efficacy among the five performance levels in the population that is as large or larger than that seen in the study sample? Using statistics computer software, the researchers computed a statistic that is used to measure differences among groups (F statistic) which subsequently converted to a probability value (p value) of .013 ($p = .013$). The p value represents the likelihood that the result of the study would have been obtained by random chance alone. Therefore, the p value of .013 indicates that there was a 1.3% chance of finding differences as large as (or larger than) the results from the study sample in the target population, if the null hypothesis was true. In other words, there was a 1.3% chance that the null hypothesis was true.

Although the researchers obtained a p value, there was still an outstanding question: Does the p value mean there was a statistically significant difference among the performance level groups? To answer that question, the researchers compared their computed p value of .013 against the a priori standard—the statistical level of significance—of $p \leq .05$. This means that:

If there is a chance equal to or less than five percent (5 out of 100) that there would be no difference in the population, then researchers must reject the null hypothesis and accept the alternative hypothesis. Conversely, if the chance is greater than five percent, researchers must fail to reject the null and may not accept the alternative hypothesis.

Comparing the computed p value of .013 against the standard of $p \leq .05$ showed that the computed value was less than the a priori standard. This means that the difference observed in the study sample would occur far too often in the target population for the differences to be regarded as chance occurrences. Therefore, researchers rejected the null hypothesis and retained the alternative hypothesis.

While $p \leq .05$ is perhaps the most widely used statistical level of significance in academic research, other frequently used standards are 1% ($p \leq .01$) and 10% ($p \leq .10$). The lower the statistical level of significance, the more stringent the standard for wrongly rejecting a null hypothesis that is true. Therefore, a standard of $p \leq .01$ is more rigorous than $p \leq .05$, which in turn is a higher standard than $p \leq .10$.

2.4.3 Data collection Methods

Without data there is no hypothesis testing or, indeed, any way of answering the research question that triggered the research investigation. In this section we will explore a vital tool in collecting research data—the data collection instrument.

The choice of a measurement procedure involves a number of decisions. For example, can the researcher measure the variable directly or indirectly? While some variables may be easily observed and measured directly (e.g., height, visual acuity), many are not readily observable and therefore more complicated to measure directly. For example, in a doctor's office, a stadiometer (specialized ruler) may be used to measure a person's height. At home all that is needed is a tape measure. By contrast, measuring academic achievement cannot be accomplished by placing an achievement sensor next to a student's head.

2.4.3.1 Defining Constructs (Underlying Variables)

An essential component of research planning is to unambiguously define important terms associated with the goals of intended studies. Much of the data that researchers collect are intended to contribute to the measurement of concepts like intelligence, altruism, achievement, self-esteem, and anxiety. These are all concepts that we hear about, but cannot directly measure. These concepts are called theoretical constructs, and they are often the focal point of research studies. However, while these terms have dictionary or conceptual definitions, those definitions are not sufficiently precise for scientific investigations. The meanings of constructs that are used for scientific purposes must be clear and precise. They must be defined operationally.

Although constructs are not directly observable, their meanings are manifested in phenomena that are observable and measurable and that indicate the presence or absence of the underlying construct. One way of operationally defining constructs is to point to how they are measured.

For example, The American Psychological Association (VandenBos, 2015) partially defines *anxiety* as "an emotion characterized by apprehension and somatic symptoms of tension in which an individual anticipates impending danger, catastrophe, or misfortune." This is a conceptual definition of *anxiety*. An operational definition of the construct would be based on specific observable physiological symptoms, such as having sweaty palms or increased heart rate. The definition could also include a self-assessment, based on a self-rating measurement scale, such as the Generalized Anxiety Disorder 7-item scale (Spitzer et al., 2006). Once researchers identify the behaviors associated with a construct, they can then precisely point to the goals of their studies and use one or more of the behaviors as proxy measures to define the construct.

In their investigation of internet self-efficacy and student performance, Kuo and Tseng (2020) operationalized the construct of academic performance as it relates to a specific domain—the final letter grades students attained in a web-based research course the authors teach. They used an existing measure of internet self-efficacy (Eastin and La Rose, 2000) and the subsequent associated definition of internet self-efficacy, "the belief in one's capability to organize and execute Internet actions required to produce given attainments" (p. 1).

2.4.3.2 Identifying Methods of Measurement

Another decision related to data collection is the method to be used in measuring constructs. The methods may be direct or indirect. Examples of direct measures include taking a person's vital signs in the doctor's office (temperature, height and weight, pulse and respiration rate). In a school setting, direct measures might include documenting school attendance or disruptive behavior.

An indirect measure may include administering an academic test or quiz to measure student learning in a specific subject area, completing a course evaluation survey or a survey that measures depression. Of course, today, indirect data collection methods incorporate the use of a variety of different media, including in-person, telephone, computer, and social media.

Hypotheses 1 and 2 in Section 2.3, suggest a variety of methods for collecting data. Consider Hypothesis 1 that addresses internet self-efficacy and academic performance as the two variables of primary interest. We noted earlier that Kuo and Tseng (2020) in their study of the relationship between the two variables used two indirect measures of the variables: a self-reported, internet self-efficacy survey as well as the course grade to measure student performance.

For Hypothesis 2 which investigated whether differences existed between two groups (neophobics and neophiliacs) in their willingness to try novel foods, a combination of direct and indirect measures may be used. An indirect, self-report measure, such as the Food Neophobia Scale (Pliner & Hobden, 1992) may be employed to characterize study participants as food neophiliacs or food neophobics. Participants' willingness to try new foods could be recorded using a direct measure like one used in a food behavior study conducted by Otis (1984) who placed bite-sized pieces of 12 actual foods in front of participants: octopus, hearts of palm, seaweed, soya bean milk, blood sausage, Chinese sweet rice cake, pickled watermelon rind, raw fish, quail egg, star fruit, sheep milk cheese, and black beans. Study participants used a 7-point rating scale to record their willingness to try each.

There are a multitude of other data collection tools. Commonly-used types of data collection instruments include paper-based or computer-assisted interview protocols, protocols for in-person or computer-assisted observations, and performance rating scales.

Regardless of the type of data collection devices researchers employ, there are two widely used approaches for acquiring data collection instruments for a research investigation—constructing the instruments or acquiring existing instruments. The construction of data collection instruments is outside of the scope of this textbook. However, there are many excellent resources available on constructing data collection instruments (e.g., Dillman et al., 2014). Regardless of whether a researcher is designing an original data collection tool or acquiring an existing one, an important selection consideration is deciding which instrument is best.

A good starting point for selecting a measurement procedure is to review past research involving the variables or constructs to be examined. Investigating similar previous research provides invaluable insights into the most appropriate measures to use. In addition, standard references like Mental Measurement Yearbook provide descriptions and evaluations of thousands of assessments. Regardless of the method used for selecting a measurement procedure, paramount in making that decision is assessing the quality of data collection instruments. Two commonly-used procedures for doing that is to assess the reliability and the validity of the instruments.

2.4.3.3 Reliability and Validity

Even the most carefully constructed data collection instruments are imperfect. There is some error in the measurement of the construct it is intended to measure. To keep this error to a minimum, two general criteria are commonly used to evaluate the technical quality of any measurement procedure—empirical evidence of properties of reliability and validity. The first property, reliability, is whether an instrument can be interpreted consistently across different situations. Three types of commonly used evidence of reliability are: reliability over time (test-retest reliability), reliability across items (internal consistency), and reliability across different researchers (inter-rater reliability).

A reliable measure is not necessarily a valid measure. For example, imagine that a researcher wishes to measure students' reading comprehension. To do that the investigator administers an oral reading test. While the oral reading test is not a completely useless method of estimating reading comprehension, it is not as good as a test designed specifically to assess reading comprehension. If the reading comprehension test is not valid, then it is not reliable. Validity is defined as the extent to which a data collection procedure measures what it was designed to measure. Widely-used evidence of validity include: face validity, content validity, criterion-related validity, and construct validity.

Figure 2.8 presents methods available for evaluating data collection instruments for evidence of validity and reliability. A description of each type of validity and reliability follows.

Figure 2.8 Methods of Evaluating Validity and Reliability

Validity	Reliability
face validity	test-retest reliability
content validity	Internal consistency reliability
criterion-related validity	split-half reliability
concurrent validity	Cronbach alpha
predictive validity	inter-rater reliability
construct validity	parallel-forms reliability
convergent validity	
divergent validity	

Face validity is perhaps the simplest form of evidence of validity. It is the face value of a measurement procedure or the extent to which members of the target population believe a measure looks like what it is intended to measure; that is, does a survey respondent recognize in the survey what the investigator claims to measure.

Content validity is concerned with the extent to which experts in the area of interest judge the content of a measure to adequately sample the domain of content. While face validity is concerned with how the target population (e.g., survey respondents or test-takers) views the items on a test or attitudinal scale, content validity employs expert adjudicators to assess whether the content of a measure adequately represents the desired construct or test objectives. For example, an important component of the development of standardized assessments, like the Caribbean Secondary Education Certificate (CSEC) mathematics examination, involves the empaneling of mathematics subject-area experts (mostly teachers) to systematically evaluate whether algebra items adequately represent all facets of the domain of algebra concepts being tested.

Criterion-related validity is the investigation of the relationship between a test, survey or some other data collection instrument and a well-established external measure (criterion) that more directly measures the construct of interest. Depending upon the time frame within which the data is collected for both measures, the study may be concurrent or predictive.

Concurrent validity is the amount of agreement between two assessments administered at approximately the same time. Typically, one measurement technique is well-established with solid psychometric properties, and the other is an external (sometimes newer) procedure that purportedly represents the same construct (criterion). The measurement methods do not need to be the same. One could be a survey and another could be an observation protocol. A strong relationship between the measures implies high concurrent validity for the measure of interest. An example of concurrent validity is the development and validation of the Food Neophobia Scale (FNS), in which Pliner and Hobden (1992) found a significant relationship between FNS scores and those of the Experience-Seeking subscale of the Sensation-Seeking Scale, a scale designed to measure a person's willingness to seek out and take risks (physical or social) in order to have novel and intense experiences (Zuckerman, 1979).

Another example of concurrent validity is the amount of agreement between students' ratings of self-efficacy using a well-established instrument and instructors' judgments of the same students' self-efficacy completed at about the same time. Similarly, evidence of a relationship between scores on an established traditional IQ test and newer Web-based IQ tests administered shortly thereafter (e.g., Firmin et al., 2008) provides yet another example. As the correlation between the test scores increases, the evidence of concurrent validity increases as well.

Predictive validity is another type of criterion-related validity, which weighs evidence of the predictive relationship between two assessments—the assessment of interest and an external criterion. Alternatively, evidence could be assessed by investigating the predictive relationship between one assessment and future behavior (criterion). In investigating predictive validity of the FNS, Pliner and Hobden (1992) found that participants' scores on the survey scale were significantly predictive of their later rankings of unfamiliar foods.

Another example of evidence of predictive validity is the predictive relationship between high school admissions test scores in the Caribbean and later scores on high school exit examinations or college admissions test scores and college GPA. In one study, an investigator (Hall, 2015) found a moderately high relationship between high school admissions and high school exit exam scores.

Construct validity represents another source of evidence of a measure's psychometric quality as you evaluate candidates for data collection instruments. This type of validity is frequently described as the overarching concern of validity research that subsumes all types of evidence of validity. It concerns the collection of evidence that a survey, test, or other type of measurement can be considered to reflect the intended construct or trait. Does the measurement behave like the theory behind the measure says it should? There are two types of construct validity: convergent and discriminant (or divergent).

Convergent validity is evidence of a positive and strong relationship between scores from two different methods of measuring the same underlying variable. In this way, a convergent validity study shows that different measurement procedures "converge" on the same construct. One example of this type of evidence comes from the study leading to the development of the Food Neophobia Scale. In the convergent validity component of the study, scores from the Food Neophobia Scale and from another similar scale the General Neophobia Scale (GNS) were investigated. Research investigators found that the measures were moderately correlated—as the scores on FNS increased, so did the scores on the GNS (Pliner & Hobden, 1992). Another method of investigating convergent validity is to examine the relationship between an indirect measure like a survey and actual behavior that previous research records as reflecting the same construct. One such study examined the relationship between the FNS and a behavior-based measure called the Unknown Flavor Sampling Test (UFST) (recounted in Alley & Potter, 2011), in which college students were presented with eight small containers of unusually colored small candies in each. For each container, they were asked (1) whether they thought they knew what each flavor was ("yes" or "no") and (2) whether they would be willing to try it. Students were given the impression they would later eat the items in the containers. Each participant's score was the ratio of the number of unknown, rejected candies to the total number of unknown candies. Here again, researchers found a positive relationship between the UFST and the FNS scores, indicating evidence of convergent validity.

Discriminant validity is evidence of little to no relationship between the scores from two distinctively different constructs when they are measured using the same method. In contrast to convergent validity, the purpose of discriminant validity is to be able to discriminate between measures of dissimilar constructs. Accordingly, constructs that are different from each other should exhibit little or no relationship. For example, loneliness and narcissism are two different constructs. Therefore, there should be no relationship between measures of these two constructs when the same method (e.g., self-rating scale) is used. Imagine that researchers developed a scale measuring loneliness which they believed was distinct from narcissism. To establish discriminant validity, they must then show that the scores for narcissism and loneliness diverge when compared. To do that, researchers might do the following:

- For each trait (e.g., narcissism), investigate whether agreement exists between scores from the trait using two different methods (survey and observation scores). High agreement between the scores for the same trait across different methods suggests convergent validity.
- For each method, investigate whether scores for two different traits (narcissism and loneliness) measured using the same method (e.g., survey scores) result in low agreement between the scores. Low agreement between the two traits using the same method means high discriminant validity.

Test-retest reliability, also known as stability reliability, is precisely what its name suggests. It is designed to measure the stability of, say, test results, when a test-developer re-administers a test to the same group of study participants after a period of time. For example, a vision test administered to the same group of individuals three weeks apart should elicit similar results.

To calculate test-retest reliability, researchers correlate the results of the test and the retest. If the correlation of the scores is low, the test-retest reliability of the instrument is also low. The closer the results are, the higher the correlation of scores and the greater the test-retest reliability of the instrument.

Test-retest reliability is not always the best test of reliability to employ. It is particularly appropriate when used to assess stability of something you expect to remain constant over the period of the study. Assessing the stability of a reading test by administering the test before and after a new reading instruction is introduced is likely to result in inaccurate results due to increased student learning during the intervention. Other indications that test-retest is the wrong assessment of evidence of reliability include:

- when the time between test administrations is too short (e.g., a few days)
- when study participants have little in common with groups the test was designed for
- when internal consistency reliability is poor

Internal consistency reliability measures the intercorrelations of items on the same data collection instrument (e.g., academic test, questionnaire, or survey) that are intended to measure the same construct. For example, on Kim and Glassman's (2013) Internet Self-Efficacy Scale (ISS), individuals who agree that they have high internet self-efficacy should agree that they are good organizers of information on the internet. If responses to the items measuring information organization and self-efficacy are not similar, then you might be measuring different constructs with those items. Hence, the internal consistency of the group of items will be low. If responses to the items are similar, that suggests that they measure the same underlying construct, organizing self-efficacy. Similar responses lead to item intercorrelations that are high, and therefore, internal consistency of the group of items is also high.

There are several forms of internal consistency measures. One approach is called **split-half correlation**, which involves randomly splitting items for each construct in an assessment into two sets. The odd-numbered items might be placed in one set, while the even-numbered items are placed in the second set. Alternatively, the first half of the items measuring the construct may be placed in Set 1 while the second half of the items is placed in Set 2. Once the items have been split into two groups, researchers find the total score for each set of items for each study participant and then calculate the correlation between the scores of the two groups of items. A split-half correlation greater than 0.8 is regarded as an acceptable level of correlation.

Perhaps the most common type of internal consistency reliability measure is **Cronbach's alpha**. Mathematically speaking Cronbach's alpha is the mean of all possible split-half correlations for items on a measure for a single sample of study participants. Manually calculating this statistic would be highly labor-intensive and subject to error. Fortunately, most statistics software packages routinely calculate Cronbach's alpha in a fraction of the time it takes to accomplish the feat by hand calculations and with much greater accuracy.

Other measures of internal consistency include:
- average inter-item correlation
- average item-total correlation
- Kuder-Richardson's formulas used for measure with dichotomous choices

Parallel forms reliability measures the correlation between two equivalent forms of an assessment. Imagine that a classroom teacher wishes to introduce a new instructional method in reading and desires to investigate whether student learning improves as a result of the new method. The teacher decides to administer a test before and after introducing the new method and compare student performance on each test for signs of improvement. In this situation, using the same test before and after the intervention might result in an increase in achievement due to students' memory of what appeared in the test the first time around rather than true learning (a threat to internal validity). To address this problem, a different form of the test may be used as a posttest after the intervention: one that is pulled from the same bank of items and that is, therefore, similar in content and constructs covered. This form is described as a parallel form of the test. Evidence that the test forms may be used interchangeably comes from the assessment of the consistency (correlation) of the two parallel forms of the test.

Inter-rater reliability (also known as inter-rater agreement) is the extent to which two or more independent experts (e.g., examiners, assessors, observers, reviewers, inspectors) are consistent in their assessment of the same phenomenon. The use of inter-rater reliability is wide and varied. Depending on their field of expertise, the judges may be engaged in assessing teacher performance in the classroom, student performance on a written communication test, performance of ice-skaters in figure skating competitions, candidates seeking their driving license, or psychologists rating the level of aggression exhibited by young patients. The common element among these situations is that two or more judges independently evaluate the same set of items, people, or conditions.

In general, judgments of behavior or performance are conducted based on a scoring rubric, rating scale or checklist. In writing assessment, rubrics describe characteristics of levels of writing ability that are the basis for judge's ratings. Consider papers for the creative writing component of the Secondary School Entrance Assessment of Trinidad and Tobago. These papers are scored using a scoring rubric that includes all possible performance categories assessed (e.g., content, language use, organization) as well as annotated performance levels for each category. For example, the scoring rubric for the report-writing component of the assessment describes the "Superior" performance level of Language Use (score 9-10) as a student who "uses formal language that conveys precise meaning relevant to report." By contrast, a student at the "Makes an Attempt" performance level of Language Use (score 1-2) is described as having an "inability to use language accurately" (Curriculum Planning and Development Division, 2015).

Once scores are independently assigned to each paper by each of the judges, scores are examined for their consistency. In some situations, if two judges scoring a paper assigns scores that are greater than two points apart, the paper is read by a third scorer, called an adjudicator, who makes the final call. Score consistency may also be examined by investigating the overall correlation between the scores assigned by first and second raters or by the percentage exact agreement between the judges' ratings. Other measures of inter-rater reliability include Cohen's kappa (Cohen, 1960) and intra-class correlation (Bartko, 1966).

High inter-rater reliability means that judges' ratings for the same phenomenon are highly consistent. Low inter-rater reliability means that judges' ratings are inconsistent with each other: the greater the difference in ratings, the less trustworthy the results of an assessment.

2.4.4 Data Collection Procedures: The Study Sample

Data collection is the systematic gathering of information needed to measure the variables in a study. Using defensible methods to collect data for a study enables the researcher to test hypotheses, answer associated research question(s), and draw accurate conclusions from research results. With the increasing use of technology to communicate, methods of data collection have grown beyond the use of snail mail distribution of paper surveys and in-person interviews to a wide variety of data collection instruments (e.g., tests, surveys, checklists, interviews, observations, and online tracking or monitoring protocols), and myriad ways of administering these tools in-person and remotely (e.g., mail, phone, audio recording, computer, internet, social media and satellite technology).

Irrespective of the type of data collection method used, accurate and valid data collection methods are paramount to preventing faulty data collection practices which can threaten the integrity of a research study. Consequences from improper data collection procedures incorporate:
- inability to answer research questions and accurately test hypotheses
- reduced ability to replicate studies
- distorted findings, and
- possible harm to study participants

Detailed discussion of data collection procedures is beyond the scope of this book. Nevertheless, one critical area of data collection that we will touch on is sampling strategies and procedures.

2.4.4.1 Populations and Samples

In addition to crafting your hypothesis, defining the variables to measure, and deciding how to measure those variables, a vital component of operationalizing your research study is selecting study participants. For example, consider the studies associated with the hypotheses in Figure 2.2. In the study of the relationship between college students' internet self-efficacy and grades, who should be recruited for that study? The research question has already restricted the study to college students rather than all students. Nevertheless, even with the specified boundary of the research, should the study target all college students worldwide? Similarly, who should participate in the study of tourists' willingness to try novel foods? Should the study include all tourists?

In the world of research design, all entities within the boundaries of the research study are called the population (also called the target population). This group possesses common characteristics described by the research criteria. Therefore, for Research Question 1, the population includes all college students.

For Question 2, the population includes all tourists. In other studies, the population may be all students in a specific school, all sports events, all insects, or all countries. In effect, populations may be human or non-human species or even inanimate objects.

Populations may range from a relatively small group of members (students in a high school) to an infinitely large group of members. However, due to fiscal and time and other resource-related practical constraints, in most cases a population consists of too many members, often in far-flung places, to include all of them in a research study. If, however, all members of a population are included in a study, the study is described as a census study, and the outcome of the study are called parameters. For example, imagine that two researchers at a university want to know which undergraduate majors were most popular among students in a specific year. Assume the researchers selected incoming undergraduates and derived the percentage of students declaring each available major. In this case, the population would be incoming undergraduates in the year of interest. The data (percentages) from this census study would be called parameters.

It is frequently the case that all members of a population cannot participate in a research study. Nevertheless, for researchers, an essential component of many studies is to secure results that provide information about the population of interest. In order to accomplish that goal of generalizability as well as address the research constraints imposed by a population, researchers rely on a small, but representative subset of the population to conduct the research study. That group is called a sample. Researchers use data (often scores) from the representative sample to conduct the research study, and then use the results to infer (generalize) something about the characteristics of the population from which the sample was drawn. However, selecting the sample is only part of the process for generalizing results to the population.

In order to generalize from the sample to the target population, researchers follow specific statistical procedures (e.g., hypothesis testing) to accomplish this objective. Most of the statistical procedures described in this textbook include material on appropriate tests of hypotheses that may be employed when using study samples to answer questions about their respective target populations. For example, consider the study of the relationship between college student internet self-efficacy and student achievement. The purpose of that study was to test a hypothesis that concerned the target population; specifically, "There is no relationship between college student internet self-efficacy and student performance."

With the inability to access the target population of all college students, a sample was drawn from a narrowly-defined, accessible population of college students: students attending the university where the research investigators are faculty members. Using hypothesis testing statistical procedures that will be described in greater detail in the remainder of this textbook, researchers analyzed data collected from the study sample in such a way that they could make inferences about the relationship between internet self-efficacy and achievement in their target population—students at their college. These are called statistical inferences or statistics. Statistics from the representative sample produce estimates of the probability that what occurs in the study sample also takes place in the narrowly defined target population.

2.4.4.2 Sampling Techniques

In the previous section, we noted that a sample is a group of members (or data) that is representative of the larger target population (data). The degree to which the sample is representative is the extent to which it mirrors population characteristics of interest. Selecting a representative sample is an important precursor to generating accurate statistics that is generalizable to a target population. The representativeness of the sample used in a research study depends on the method used to select the study participants.

Over the years, researchers have developed a variety of sampling techniques (also called sampling methods or sampling procedures). These techniques belong to two major categories—probability and nonprobability sampling. Probability sampling is a procedure in which the researcher can calculate the probability that each study participant will be included in the study from a known population. For example, in a target population of 100, if each person is equally likely to be selected, the probability of selection for each study participant is 1/100. With nonprobability sampling, the odds of selection are not known because the researcher does not know much about the population size or its membership.

Probability sampling is the preferred method for selecting study participants when researchers wish to accurately generalize findings to a target population. Among probability sampling techniques, perhaps the ideal representative sample is one that is truly a random sample, which means that every member of the target population has an equal chance of inclusion in the sample. Besides simple random sampling, probability samples include systematic sampling, stratified random sampling, and cluster sampling. However, probability sampling is also more challenging and expensive to implement. For example, when selecting a probability sample, one common source of identifying the members of a target population is directories (e.g., telephone, voter registration, or professional organizations). Often, however, it is hard to find directories with exhaustive lists of current members of the population, so generalizations from study findings end up being limited to population members who are in the directories.

Probability sampling uses selection techniques designed to produce unbiased samples. Nonetheless, even when ready-made directories are complete and up-to-date, accurate contact information may be missing. Therefore, although it is feasible to select an unbiased sample, inability to contact participants results in missing subjects and subsequently an increased risk of systematic bias in sampling. A biased sample leads to results that may not be trustworthy, which means researchers are limited to making cautious generalizations about the population.

Nonprobability sampling does not presume to use unbiased sampling techniques for participant selection. While there is an effort to maintain representativeness, the focus is on convenience and cost-effectiveness. One of the hallmarks of nonprobability sampling is its practicality. Social science researchers are often unable to secure the information required by probability sampling procedures to produce unbiased samples. Their alternative in those instances is nonprobability sampling. In the social sciences, one of the most commonly-used nonprobability sampling techniques is **convenience sampling**, which is the selection of study subjects based on convenience (e.g., a college professor using his own students in a research study) rather than representativeness of the population. **Quota sampling**, another nonprobability sampling technique, combines the ease of convenience sampling together with some of the techniques used in probability sampling.

Providing an exhaustive discussion of sampling methodology is beyond the scope of this textbook. Nevertheless, understanding sampling and sampling techniques is crucial to the selection of the most appropriate research design and the evaluation of research designs as a consumer of research products. More detailed information is widely available in research methods textbooks (e.g., Lohr, 2022) that focus on sampling and sampling techniques exclusively. Figure 2.9 presents a summary of the most common sampling methods.

Figure 2.9 Sampling Techniques

Sampling Technique	Method
Probability Sampling	
Simple Random Sampling	Select a group of subjects from the target population so that each subject is chosen entirely by chance. Methods to accomplish this include (a) defining your target population and acquiring a list of its members, (b) deciding on your sample size, (c) putting the list in a computer database, and (d) running a computer program to randomly select the subjects you want in your study sample from the list in your computer database.
Stratified Random Sampling	Classify or separate target population into mutually exclusive groups based on specified characteristics (e.g., gender, income level) and then using simple random sampling to choose members from groups.
Systematic Sampling	Researchers select or instruct computer software to select members from the target population database at regular intervals (e.g., every 12th member on the population list) after randomly selecting a starting point.
Cluster Random Sampling	For populations that are already clustered into groups, randomly select groups (clusters) rather than individual members of the population from a list of all the pre-existing subgroups in the population. For example, in a study of 6th grade performance in a school region, randomly select schools, rather than individual students.
Nonprobability Sampling	
Convenience Sampling	Researcher selects study participants from sources that are accessible and easily available (e.g., students in a course the researcher teaches, phone surveys, social media surveys).
Quota Sampling	The researcher separates a target population into mutually exclusive groups (e.g., gender). Then the researcher selects study participants from each group based on a pre-specified quota (e.g., 10 males, 10 females). The researcher stops selecting study participants from a subgroup after meeting the quota for that group.

2.4.4.3 Sample Size

In some studies, like a population census study, researchers collect data from every member of a population. For most studies, a census is impractical due to the time and financial resources associated with such studies. Although censuses are infeasible in many situations, most researchers' primary goal remains to draw precise inferences about a population. To accomplish their objective, researchers most often are limited to selecting a relatively small study sample to represent a population. In doing that, one fundamental question researchers face is the size of the sample that is needed to adequately represent a target population.

The size of a study sample that a researcher needs for a study depends on a range of issues. These include the precision needed in the results, the diversity of the target population, the financial costs associated with the study, and the demands of the statistical procedure involved.

In the previous section, we noted that the desire to control financial costs is one contributor to researchers' use of sampling over census studies. For example, Caribbean countries routinely conduct a population and housing census every ten years. In Jamaica, the most recent census planned for 2022 (Ministry of Finance and the Public Service, July 2022) had an estimated cost of $2.4 billion. Much of this cost is a result of the time needed to make contact with each individual at a time when they are available to provide extensive, time-consuming data, which must be accomplished in-person. The cost of research studies that are based on representative samples of a target population are miniscule compared to censuses. However, beyond the cost savings, samples can result in sacrificing accuracy.

All things being equal, studies with larger samples (e.g., correlational studies) have the capacity to detect the smallest of differences between groups. By contrast, research studies with smaller sample sizes are less likely to detect small group differences in the population. Assuming that a sample is representative of the target population, the larger the sample, the more confidence we can have that sample values approximate population values: the greater the importance of the study, the greater the need for precision and accuracy. Often this implies larger representative sample size.

Perhaps one of the largest, most important studies on record is one conducted as a result of the coronavirus pandemic in 2020. The study led by the University of Edinburgh and the University of Birmingham in the United Kingdom investigated the impact of COVID-19 on surgical patients and was awarded the Guinness World Record title for the world's largest scientific collaboration, involving over 140,000 patients in 116 countries. One important finding of the study was that patients around the world waiting for elective surgery should have priority access to COVID-19 vaccines to avoid post-operative deaths due to the virus.

Another driver of sample size decisions is the diversity of the target population. As the diversity of a target population increases, so does the need to increase sample size to adequately mirror the population. If a researcher wishes to draw a sample that is representative of the language groups in a population, and everyone speaks English, then a very small sample of one may be sufficient representation. If, however, there are many language groups in the population, a small sample of a single language group is likely to misrepresent the population. A larger sample of greater language diversity would be needed to avoid a biased sample.

There is no question that assembling large representative samples are typically more accurate than smaller samples. However, small representative samples are less expensive and more time-efficient than larger ones. With all of these concerns in play, researchers must carefully consider what sample size represents a good balance of cost, efficiency, and accuracy. Some researchers suggest that studies requiring minimal error need large representative samples, while small samples will suffice in studies where more error can be tolerated.

In general, there is no simple solution to determining how many individuals should be in a sample. In complicated studies, there may be different sample sizes in the same study. For example, in a study of a diverse population, stratified sampling may be used, with some smaller groups over-represented in some strata to meet sample size assumptions in statistical analysis. In experimental design studies, where participants are divided into different treatment groups, each group may have different sample sizes.

One helpful guide in selecting study samples is to review peer-reviewed, published reports of similar research studies to see how many participants were used. Another strategy is to use one of many widely-available software programs during the design component of your study to help determine the sample size needed to enhance the statistical power of your study (more about this later) and increase your chances of detecting a true effect, that is having a large enough sample size that provides the statistical power to reject the null when it is false. Many of these sample size calculators are available on the internet and are free of charge.

2.4.5 Data Analysis: Types of Statistics

Thus far, Section 2.4 has included discussions of the initial research problem and subsequent considerations that lead to a research study. The discourse has included hypotheses and predictions associated with research questions, overall research strategy and design as well as variables that are measured in the study. We have also discussed how participants in a study are selected and methods for collecting data to answer research question. Once a researcher has the data in hand and the data have been entered into a computer database, it is time to commence with executing analysis of the data or the information collected, in a way that accurately implements the research strategy and design. However, data analysis implementation does not happen spontaneously. Rather, data analysis decisions are an important part of the research planning process whereby researchers specify the statistical techniques to be employed for two fundamental objectives:
- to summarize a large amount of data collected from study participants
- to answer the research question(s) that initiated the study

2.4.5.1 Descriptive Statistics

Research data may be qualitative or quantitative. Qualitative research data may include a researcher's observations of student behavior in a classroom, responses to interview questions, archival historical data, photographs, videos, or social media postings. Qualitative data also include a wide variety of non-numeric information; they may be words, sounds, or images. Our primary interest in this text, though, is quantitative research and therefore our data are primarily numerical, such as scores on a test (e.g., Caribbean high school student scores on the math CSEC exam), or responses to a survey (e.g., the Internet Self-Efficacy Scale). The statistical methods used to analyze such numerical data are specifically designed for that purpose.

Typically, the purpose of summarizing data is to use statistics to describe and organize quantitative information. Researchers use descriptive statistics to summarize large amounts of data into a few meaningful numerical values that are easier to understand and interpret than the original data. Examples of summary descriptive statistics include the average CSEC math score of a sample of 200 incoming university freshmen or the median income of a sample of tellers at a local bank. Note in the two examples the summary values (mean and median) are statistics. When data come from a sample of study participants, summary values used to describe those samples are called statistics. Figure 2.10 shows statistics for responses to one question on the 2013 World Values survey conducted in the nation of Haiti, and the nation of Trinidad and Tobago. The statistic reported is the percentage of survey respondents reporting differing levels of satisfaction with life in their countries.

Researchers may also apply descriptive statistics to a population. For example, if a researcher's sole interest was in the mathematics performance of a group of 200 incoming freshmen, then that group would be a population, and the average CSEC mathematics score would be a parameter. A population may be small, like the group of incoming freshmen, or large, like all young adults or nursery school-age

Figure 2.10 Satisfaction with Life

[Bar graph comparing Haiti and Trinidad & Tobago responses from "Completely Dissatisfied" to "Completely Satisfied"]

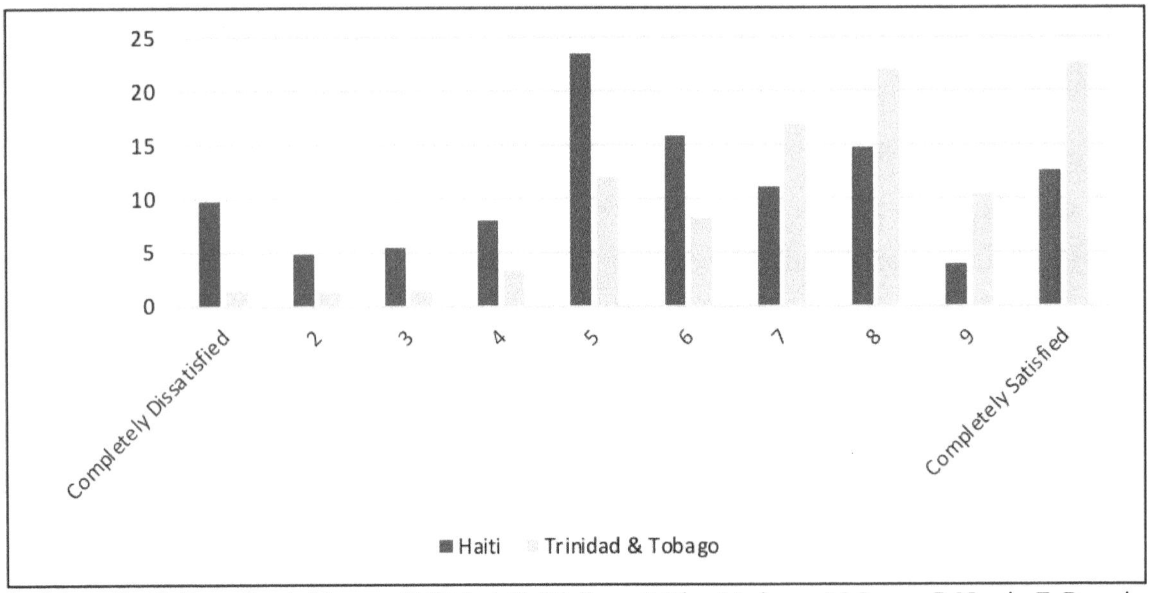

Inglehart, R., C. Haerpfer, A. Moreno, C. Welzel, K. Kizilova, J. Diez-Medrano, M. Lagos, P. Norris, E. Ponarin & B. Puranen et al. (eds.). 2014. World Values Survey: Round Six - Country-Pooled Datafile Version: http://www.worldvaluessurvey.org/WVSDocumentationWV6.jsp. Madrid: JD Systems Institute.

children in a country. Therefore, youth unemployment rates collected by the Statistical Institute of Jamaica and school enrollment rates reported by the Ministry of Education, Youth and Information in Jamaica are two examples of descriptive parameters (the summary values describing a population). Parameters are the population equivalent of sample statistics.

Descriptive statistics may also be used as the basis of graphs or plots that visually organize and summarize information about study participants. For example, researchers might use a bar graph to report the comparative percentage of levels of satisfaction with life expressed by respondents from Haiti, and Trinidad and Tobago on the 2013 World Values Survey (Figure 2.10).

The mean, median, mode, and frequency are called measures of central tendency or values that show the typical point of a set of data. Measures of variability, or the spread of the data are another set of descriptive statistics. Imagine that 30 students received a mean score of 55 out of 100 on a biology examination. Although the average is 55, not all students received a score of 55. Some received higher and others received lower marks. In other words, the scores are spread out. Measures of variability are statistics that describe the spread of score values. Measures of variability include statistics such as range, quartiles, variance, and standard deviation.

Descriptive statistics is not the central focus of this text. However, it is an important preliminary step for most statistical techniques that should be incorporated in all data analyses and will be used throughout the remainder of this book.

2.4.5.2 Inferential Statistics

Another objective of statistical analyses of research data is to answer the research question(s) that initiated a study. Sections 2.4.1 and 2.4.2, focused on researchers' general procedures for testing null hypotheses (also called Null Hypothesis Significance Testing or NHST).

Another important component for accurate generalizations is the use of inferential statistics in conjunction with hypothesis testing (e.g., Section 2.4.1 and 2.4.2). Employing inferential statistical techniques facilitates moving beyond findings from a sample of subjects to using the findings to draw conclusions about target populations from which the samples were drawn.

Inferential statistics are needed because samples do not perfectly represent target populations. Not only is a sample frequently much smaller than the target population, analyses of two samples of data drawn from the same population may end up with different results; sample statistics generated from a study are not exactly the same as population parameters. The difference between sample statistics and population parameters is known as sampling error. For example, imagine that the population parameters entering freshmen in a private university equaled a total of 1,200 students and of that 53% were female and 47% were male, with a minimum age of 15 and a maximum age of 21. Two random samples taken from the target population revealed the following sample statistics.

Demographic Variable	Sample 1	Sample 2	Population
Total	10	15	1,200
% Female	49	55	53
% Male	51	45	47
Minimum Age	16	18	15
Maximum Age	17	19	21

Note that neither sample statistics match the population parameters exactly. The difference between the population parameters and the sample statistics is sampling error.

In order to link results from samples to the target population, inferential statistics acknowledge the presence of sampling error which is a natural discrepancy between samples and the actual values in the population from which they are drawn. Inferential statistics help researchers to determine whether findings of differences between groups or similarities between variables (relationships) are evidence of chance occurrences due to sampling error as opposed to true differences or relationships in the population.

Much of this textbook is dedicated to different types of inferential statistical techniques in different research situations. Consult Chapter 1 for a description of the statistical techniques to be covered in this text. The associated statistics are designed to evaluate differences between two groups (t-statistic), and among more than two groups (e.g., F-ratio or Kruskal Wallis); the correlation between two or more variables (e.g., Pearson's r or Spearman's rho); and the multiple correlation (Multiple R or R) between an outcome variable and a set of one or more independent variables.

2.4.6 Data Analysis: Null Hypothesis Significance Testing

An important portion of conducting inferential statistical procedures is null hypothesis significance testing. Broadly speaking, null hypothesis significance testing, or NHST, involves preliminary activities like the acquisition of a data collection tool, selecting a representative sample, and collecting and screening data. However, the central component is all about analyzing the data, often by using inferential statistics, to rule out chance as a plausible explanation for study findings. To accomplish this, researchers engage in an established procedure for deciding between the null and alternative hypothesis statements discussed in Section 2.4.1.

Making the decision between the null and alternative hypothesis involves several phases:
- computing a sample statistic, which represents the difference between groups or the relationship among variables
- computing the test statistic, which is a number that quantifies expected levels of a characteristic (e.g. difference between groups) based on the null hypothesis
- comparing your sample statistic against the test statistic for your prespecified significance level; the larger your sample statistic, the more it represents differences between sample data, increasing the likelihood of rejecting the null

In bygone years, researchers generated all test statistics solely by hand calculations of associated formulae. Today, it is still helpful to have a conceptual understanding of the derivation of test statistics, hence the inclusion throughout this text of the derivation of selected statistics (e.g., the Pearson correlation coefficient in Chapter 4 and the F-ratio for ANOVA in Chapter 7). However, with widely-available statistical software applications (e.g., IBM SPSS, SAS, Minitab and R), researchers can quickly generate needed study sample test statistics for research studies.

Once sample test statistics have been computed, the next step in the hypothesis testing process is to complete the researcher's primary objective—to determine the likelihood (probability or p value) of finding the study sample results in the target population if the null hypothesis is true. In essence, the researcher sets about making inferences about the target population based on sample test statistics and the null hypothesis. If the sample results are highly unlikely to be found in the population, then the researcher is unable to reject the null hypothesis. With statistical software applications commonly available, we can generate the test statistic and the p values to choose between the null and alternative hypotheses.

Researchers can manually calculate whether a statistic is statistically significant or use statistical software to do the same task. For example, imagine that a researcher finds a mean difference in math performance of 26.2 in a sample of students from two different classes. To determine if the difference is statistically significant, the researcher can use William Gossett's t-test formula to hand calculate the t statistic (e.g., $t = 2.528$) that represents the observed difference between the groups. With the test statistic in hand, the researcher may use a table of critical values (found in the back of most statistics text books) to determine whether to reject the null hypothesis.

Critical values are values of a test statistic at or beyond which we can reject the null. The values are based on the number of subjects in a study, the number of groups of subjects, and the selected level of probability (p value) that sample results will be found in the population.

Imagine that the researcher uses the most commonly used alpha level (statistical significance level), which is $p \leq .05$, meaning that there is a probability less than or equal to five percent that the study sample results occurred by chance. In other words, there is a 95% or greater likelihood that results seen in the sample occur in the population. If the test statistic exceeds the critical value at the preset alpha level, the researcher can reject the null hypothesis. If the test statistic does not exceed the critical value, the researcher must fail to reject the null hypothesis and conclude that the findings likely occurred purely by chance.

In the case of the difference in mathematics performance, the critical value for the statistical significance level of $p \leq .05$ is 2.306. Therefore, the test statistic ($t = 2.528$) exceeds the critical value, and the researcher rejects the null hypothesis that the mean difference is math performance is not significantly different.

Although researchers occasionally manually compute statistical significance today, the calculation of most inferential statistics need not be as labor-intensive as they formerly were. Computer programs routinely compute statistical significance in seconds and report exact probability levels; for example, in the math example, computer software yields a p value of .035, which means that there is a 3.5% probability that the results occurred purely by chance. The publication manual published by American Psychological Association (APA, 2020), and used by most journals in the social sciences, now requires exact probabilities (e.g., $p = .035$) rather than standard alpha levels (e.g., $p \leq .05$) when reporting results of null hypothesis significance testing. This level of precision contributes to a more powerful test.

2.4.6.1 Errors in Null Hypothesis Significance Testing

Ruling out chance as a likely cause of study findings is the primary objective of hypothesis testing. The extent to which researchers can rule out chance as the source of their results helps them determine how reliably they can extrapolate findings from their study sample to the population from which it was drawn. Unfortunately, the conclusions are not always correct (Figure 2.11).

Figure 2.11 Correct Decisions and Errors in Null Hypothesis Significance Testing

Decision Based on Study Sample	Truth About the Target Population	
	Null hypothesis is true (The alternative is false.)	Null hypothesis is false (The alternative is true.)
Reject the Null (The alternative is true.)	Type I error	correct decision
Fail to Reject the Null (The alternative is false.)	correct decision	Type II error

There are four possible outcomes of a null hypothesis test, summarized in Figure 2.11:
✓ rejecting the null hypothesis when it is false (or rejecting a false null),
✓ retaining the null hypothesis when it is true (or retaining a true null),
✗ rejecting the null hypothesis when it is true (or rejecting a true null), and
✗ retaining the null hypothesis when it is false (or retaining a false null).

The first two conclusions are correct. The second two are incorrect decisions. Incorrect decisions occur under two conditions—rejecting a null hypothesis when it true (called a Type I error), and retaining a null hypothesis when it is false (called a Type II error).

Type I errors are also known as false positives because the researcher concludes there **is** a statistically significant difference (or relationship) when there is no corresponding occurrence in the target population. Imagine, for example, that a researcher is investigating whether there is a statistically significant difference in the effectiveness of Drug A compared to Drug B in lowering symptoms of a disease. Based on results from the research sample, the investigator concludes that the difference observed in the effectiveness of the drugs probably did not occur by chance alone and rejects the null hypothesis. However, if there is no corresponding difference in the target population, the findings represent a Type I error. Some refer to this as a false alarm.

A Type II error is the reverse of a Type I error. In this situation, a researcher retains a null hypothesis that is not true, producing a false negative. In this situation, a researcher concludes that there **is no** statistically significant difference (or relationship) when there is a difference (or relationship) in the target population. Hence the researcher fails to reject a false null hypothesis when findings did not occur by chance alone. For example, imagine that a researcher wants to study whether there is a significant relationship between students' grades in high school and in college. Based on results of the study, the researcher concludes that the findings observed in the research sample occurred by chance alone, and there is not a statistically significant relationship between high school and college grades. The researcher therefore fails to reject the null hypothesis. However, there is a relationship in the target population. In other words, the researcher failed to find a relationship that actually existed.

2.4.6.2 Shifting Significance Testing Standards

Ideally, there should be no error in research. However, there is a risk of making each type of error in every data analysis, and neither can be circumvented completely. A Type I error could result in finding an innocent person guilty of murder. A Type II error could result in failing to detect the effectiveness of a new antiviral drug to treat a deadly disease. Hence, it is important to consider the consequences of errors during the design phase of every research study and to incorporate safeguards to reduce the likelihood of error and the associated consequences when implementing research. In 2009, the American Psychological Association Publication Manual instituted new standards for conducting and reporting research studies to boost their statistical power. These standards include:

- exact probabilities associated with hypothesis testing rather than the exclusive use of a priori alpha levels
- effect size, which is a measure of the practical significance of research findings

Both standards, as well as sample sizes and confidence intervals, are methods of increasing statistical power, which will be examined in Section 2.4.7.

2.4.7 Statistical Power

The primary objective of null hypothesis significance testing is to learn whether there is enough evidence from a study sample to allow a researcher to reject the null hypothesis when it is false and conclude that there is evidence that an effect exists. The null hypothesis may be that high- and low-performing students do not significantly differ in average hours of TV viewing. Alternatively, the null hypothesis in a correlational study may state that the relationship between body fat and fitness is not significantly different from zero. The probability of statistically detecting a difference between two or more groups when such a difference exists, or detecting a correlation between two variables when the correlation is significantly different from zero is called statistical power.

Statistical power is defined statistically as 1 – beta, where beta is the probability of committing a Type II error. More power increases the likelihood of correctly rejecting the null when it is false and avoiding a Type II error (failing to detect an effect when it exists, or a false negative). High power maximizes the ability to use sample data to distinguish a true effect in the target population. Low power means a study has a small chance of detecting a true effect. Then again, if there is too much power, hypothesis testing will be sensitive to very small true effects. Therefore, although the effects are statistically significant, the results are not meaningful in the real world.

To strike the right balance—power that reveals results that are useful in the real world—it is best to conduct a power analysis during the design phase of a study. While statistical power can be complicated to calculate manually, there are many software applications (e.g., G*Power) available to estimate statistical power for research studies. These software applications are used for power analyses that identify how to boost the power of null hypothesis statistical significance testing.

Power analyses mainly take three factors into account: sample size, the effect size, and the significance or alpha level. To reduce Type II errors, power analyses are often calculated at the outset of a study to identify the minimum sample size for a specified power level, alpha level, and effect size. Inputs for the power analysis to calculate minimum sample size include the desired alpha level (typically $p \leq .05$), the desired power level (usually .80 or 80 %), and the expected effect size, which is a measure of the magnitude or importance of study results, which varies according to statistical applications.

2.4.7.1 Alpha Level

The alpha or significance level of a research study specifies the probability of making a Type I error in rejecting the null hypothesis. The alpha level of $p < .05$ means that the researcher is willing to incorrectly reject the null five percent of the time. Therefore, there is a five percent chance of committing a Type I error by rejecting a true null, which means a five percent chance that findings occurred by chance.

More stringent alpha levels (e.g., $p < .01$) represent lower chances of committing a Type I error. However, decreasing alpha levels also lead to more conservative tests that are less sensitive to detecting true effects and more likely to lead to committing a Type II error. A researcher who sets a conservative alpha level to be sure that there is a significant relationship between two variables may very well miss significant relationships because those relationships do not meet the more rigorous standards. Reducing the likelihood of Type I errors may result in increasing the likelihood of Type II errors.

Researchers can reduce the likelihood of committing a Type II error by increasing the alpha level at the outset of a study. For example, instead of setting an alpha level of $p < .05$, a researcher may decide on an alpha level of $p < .15$ instead, which means there is a 15% chance of wrongly rejecting a true null hypothesis. This approach could lead to the identification of more statistically significant group differences or relationships between variables, reducing the likelihood of committing a Type II error. But in reducing the chances of a Type II error (false negative or failing to detect a true effect), the researcher increases the likelihood of a Type I error (false positive or failing to reject a null that is true).

In settling on an alpha level, it is important to weigh the severity of the consequences of committing a Type I or Type II error and the amount of risk researchers are willing to take. Minor consequences may free researchers to take greater risks of falsely rejecting the null.

2.4.7.2 Effect Size

Effect size is a measure of the magnitude of the average difference between groups or the relationship between variables. Some describe effect size as an index of the practical significance of the findings of a study. Two widely-used effect sizes are Cohen's d and the coefficient of determination.

- Cohen's *d* (or *d*) is the effect size used in studies of group differences. Computationally, it is the difference between two group means divided by the standard deviation of the target population.
- The coefficient of determination (r^2) is used in correlation and regression studies. This statistic represents the proportion of shared variation.

Effect sizes greatly influence statistical power of a study. Large effect sizes are related to greater levels of statistical power. Studies with high statistical power are likely associated with the detection of medium and large effect sizes. Studies with low statistical power are limited to finding large effect sizes. In these studies, effect sizes are more likely to overstate true effects.

Researchers frequently use previous studies and pilots to assist them in determining appropriate effect sizes. In addition, Cohen (1988), the inventor of effect sizes, provides guidelines for evaluating the magnitude of effect sizes in various applications. For example, for Pearson *r*, effect sizes of .10, .30 and .50 are considered to be small, medium, and large, respectively. For the eta-square statistic (η^2) used in the analysis of variance, effect sizes of .01, .06, and .14 are considered to be small, medium, and large.

2.4.7.3 Sample Size

Sample size greatly influences statistical power. In general, as sample sizes increase so does the statistical power of a research study. Larger sample sizes have more stable estimates of population parameters. In addition, smaller values of statistics (e.g., *F* statistic) are needed to reject the null hypothesis. All of these lead to greater statistical power. However, there are also drawbacks to large sample sizes.

When samples are large, adding a few more study participants are unlikely to measurably increase power. In addition, very large sample sizes can result in statistically significant findings even when the effects are weak. Effect sizes help to provide practical interpretations in these situations.

2.4.8 Ways of Increasing Power

Having sufficient statistical power in an investigation is a central concern of researchers. However, under-powered research studies remain an enduring challenge. Fortunately, researchers can take measures to increase power. Below are some strategies for doing so.
- Raise the significance level.
- Switch from a two-tailed test to a one-tailed (directional hypothesis).
- Increase sample size.
- Increase the effect size.
- Use a repeated measures (within-group) design.
- Decrease measurement error by increasing the precision of data collection measures and procedures.

2.5 Research Reporting

Once data analysis has been conducted and interpreted, the next important step in the research process is to disseminate the results. Sharing the details of a research study is important:
- They add to the knowledge base that other researchers may use to answer their own questions or to generate ideas for other investigations.
- They enable the research community to critically review the work that has been done.
- They provide information that may be used to replicate studies. Replication may lead to confirmation, refutation, or to expansion of the findings.

The research report is a written description of the investigation, which includes
- the purpose of the study,
- a review of the relevant research literature that led to the investigation,
- a description of the methods used to conduct the study,
- the results of the investigation,
- discussion and interpretation of the results, and
- recommendations based on the results.

Several writing styles exist for research reports in the various scholarly disciplines. However, within the social sciences the guidelines commonly-required by publishers and institutions is the Publication Manual of the American Psychological Association.

Throughout this textbook, case studies will be used to introduce and (using SPSS) to demonstrate the application of each statistical method. Including complete research reports for the associated cases is outside the scope of this textbook. However, brief summary reports of the cases and results of the data analysis are designed to underscore the importance of research reports.

2.6 Key Terms

Alpha level
Alternative hypothesis
Between-subjects design
Concurrent validity
Content validity
Continuous variable
Correlational research
Counterbalancing
Critical value
Criterion-related validity
Criterion variable
Dichotomous variable

Discrete variable
Falsification
Null hypothesis
Interval-level variable
Level of measurement
Longitudinal survey
Measurement level
Nominal variable
Null Hypothesis significance testing
Ordinal variable
Population parameter

Predictive validity
Predictor variable
Probability
Random selection
Random assignment
Ratio variable
Reliability
Repeated measures
Test-retest reliability
Validity
Variance
Within-subjects design

3

Data Preparation: Cleansing and Screening

Key concepts from previous chapters or courses:
    independent variable: the variable that a researcher manipulates or is studying.
    dependent variable: the variable that a researcher measures using its data values.
    levels of measurement: also called scales of measurement or measurement levels. A system that describes the properties or types of information contained in variable values.
    categorical variables: also known as qualitative and discrete variables. These variables distinguish among study participants by placing them into a limited number of mutually exclusive labeled groups that may be unordered (e.g., sex) or ordered (e.g., age group).
    continuous variables: also known as quantitative variables, may assume an unlimited number of numerical values between the highest and lowest points in a range of values.
    descriptive statistics: procedures for describing quantitative information by summarizing, organizing, or graphing that information.
    inferential statistics: statistics that allow researchers to draw conclusions or inferences from data about a target population based on data from a representative sample of that population.
    significance tests: statistical tests that determine the probability that observed characteristics of samples drawn from a target population occurred by chance alone. If characteristics are unlikely to have occurred by chance alone, they are regarded as statistically significant.
    linear relationship: also known as linear association or linearity describes a relationship between two variables, that when plotted on a graph forms a straight line.
    normal distribution: a probability distribution of variable values that is symmetric about the mean. Values are often clustered around the mean in a pattern known as a bell-shaped curve.

THIS CHAPTER INTRODUCES:

- coding data
- methods of cleansing and screening data
- methods of evaluating data assumptions of statistical techniques

3.1 Introduction

Researchers today have a wide variety of data collection instruments and administration methods available to them. Here are some of the most common data collection methods. Examples of modes of administration are noted in parentheses.

1. Experiments (in-person, internet-based)
2. Tests (in-person, computer-based)
3. Questionnaires (face-to-face, telephone, mail, internet-based)
4. Interviews (telephone, face-to-face, internet-based)
5. Observations (in-person, internet-based)

Selecting a data collection instrument is a key step in the research process. Conducting a literature search is a common method of identifying and selecting a data collection tool.

In the event that a relevant data collection instrument related to a researcher's topic does not exist, then the researcher typically has two options:

- Modify and cross-validate an existing instrument
- Develop an entirely new instrument.

One advantage of using an existing instrument is the strong likelihood that information about the technical quality of the instrument (validity and reliability) already exists. Therefore, the instrument will be ready for immediate use, or at the very least, ready for adaptation for the specific unique needs of the research study.

Developing an entirely new instrument has significant drawbacks. This includes the reality that instrument development is an additional research project to complete that will require substantial time and effort and, possibly, added coursework or self-study. Furthermore, the final instrument may have as many flaws as existing instruments with the extra shortcoming that the results of the research may lack comparability with previous similar research.

Development of valid and reliable data collection instruments are beyond the scope of this textbook. For more information about instrument development, Dillman et al. (2014), and DeVellis and Thorpe (2022) deliver extensive coverage of survey development. Also, Lane et al. (2015) offers a comprehensive handbook on the latest developments in the field of test development.

Regardless of the method used in gathering research data, preserving data integrity is critical. Data integrity is the development and implementation of procedures to ensure quality assurance and quality control of the data collected. Quality assurance (QA) is a set of activities that ideally takes place before actual data collection begins. Large worldwide organizations like the World Health Organization (WHO) and the International Association for the Evaluation of Educational Achievement (IEA) routinely produce extensive quality assurance guidelines for data collection and research activities (e.g., Ustün et al., 2005; Musu et al., 2020) that are designed to minimize data collection errors by:

- using the appropriate data collection tools (questionnaires, tests, or interviews) accompanied by clear instructions to standardize the administration process,
- pilot-testing data collection tools and administration procedures to identify and correct potential issues of validity,
- recruiting data collection teams equipped with the necessary background knowledge related to the research study,

- training data collection crews before and during the data gathering task to support uniformity in the administration process. This may involve the training of those directly administering the data collection tools (e.g., interviews and questionnaires) or the training of monitors of the data collection process (e.g., observers of the process at research sites).

The primary focus of quality assurance is prevention by forestalling the occurrence of data collection problems likely to undercut the validity of the research study. Hence the focus is on the development of standardized administration protocols and training of data collectors to minimize administrators' unintentional deviation from the protocols; in other words, these protocols are put in place to minimize drift. QA might also include measures of social desirability in a protocol designed to reveal the honesty of respondents.

According to Musu et al. (2020), while QA, like those in international large-scale student assessments, occurs primarily before data collection, quality control (QC) is a component of QA that occurs during and after data collection. It is intended to capture errors in processing the data through exploratory data analysis.

Quality control may include the monitoring of data collection implementation and administration as well as a clearly defined system of communication between the principal investigator and research assistants should errors in data collection surface. Interviewers and observers of research participants may use field notes to document exceptional occurrences, like incidences of study participants exhibiting socially desirable responses. Research assistants may conduct site visits to determine whether the data are collected as intended and address shortcomings (e.g., errors in questionnaires or test items, test administration security violations) where necessary.

Quality control also includes what happens once data have been collected. Typically, coding and entering the responses of study participants into a database for analysis follow the collection of research data. Our discourse on data preparation will begin here.

3.2 Coding and Data Entry

Methods for preparing data for quantitative analysis have changed considerably over the years. In bygone years, manually coding collected data, developing codebooks, and hand-entering data were part and parcel of the quality control process of collecting and assembling data for analysis. Coding is the process of converting study participants' responses to numeric or alphanumeric values or symbols for each variable. Codebooks for research studies are documents that provide information about the structure and content of each of the variables in a dataset. As Figure 3.1 illustrates, for each variable, the codebook is likely to include its name (Name), a descriptive label (Label), numeric or string variable type (Type), the character width (Width) of variable values, the number of decimal places for variable values (Decimals), the measurement level (Measure), and allowable values (Values) and value labels (Value Labels) for each variable.

Like data collection administration manuals, codebooks were designed to help to control the quality of the data entered for analysis. Coding allows the researcher to reduce large quantities of information into a form than can be more easily handled, especially by statistical software. In spite of the use of codebooks, manually entering data leaves much room for human error.

Figure 3.1 Example of Codebook Entries

Name	Label	Type	Width	Decimals	Measure	Values	Value Labels
ID	ID Code	String	3	0	Nominal	001-350	
Gen	Gender	String	1	0	Nominal	0 - 1	0 = Female, 1 = Male
Age	Age	Numeric	2	0	Scale	17 - 55	
Age_Grp	Age Group	Numeric	1	0	Ordinal	1 - 6	1 = 18 & under, 2 = 19 – 25, 3 = 26 – 30, 4 = 31 – 35, 5 = 36 – 40, 6 = More than 40
HSGPA	High School GPA	Numeric	3	1	Ordinal	0.0 – 4.0	0.0 = E/F, 1.0 = D, 1.3 = D+, 1.7 = C-, 2.0 = C, 2.3 = C+, 2.7 = B-, 3.0 = B, 3.3 = B+, 3.7 = A-, 4.0 = A
UNIVGPA	University GPA	Numeric	3	1	Ordinal	0.0 – 4.0	0.0 = E/F, 1.0 = D, 1.3 = D+, 1.7 = C-, 2.0 = C, 2.3 = C+, 2.7 = B-, 3.0 = B, 3.3 = B+, 3.7 = A-, 4.0 = A
Univ_Sched	College Schedule	Numeric	1	0	Nominal	1 - 2	1 = Part-Time, 2 = Full-Time

Today, web- and cloud-based data collection methods reduce the occurrence of human error. Contemporary data collection applications allow researchers to embed data constraints that minimize the entry of invalid information. For example, data-type constraints may be embedded to prevent study participants from entering text rather than numeric responses (e.g., "F" rather than "1" to indicate female). Data range constraints prohibit responses that are beyond a stated range (e.g., preventing the insertion of age 35, which is beyond a stated 18 – 30 age range).

Modern data collection applications also merge the steps of collecting data with those of coding and data entry by incorporating coding and data entry into the data collection process. For example, several online survey tools assist researchers in designing and distributing data collection instruments as well as coding and exporting data to data analysis software packages (e.g., Statistical Package for the Social Sciences or SPSS, Excel). Most also include some basic data analytics that may be used for monitoring data errors or for describing study participants before moving on to more complex analyses. Examples of free software packages for small data collection projects are listed in Figure 3.2.

Figure 3.2 Common Software Packages for Internet Surveys

Software Packages for Internet Surveys	Websites
Google Forms	https://www.google.com/forms/about/
SurveyMonkey	https://www.surveymonkey.com/
SurveyPlanet	https://surveyplanet.com/
Qualtrics	https://www.qualtrics.com/free-account/

In quantitative research, whether researchers code and enter data manually or with the help of web-based applications, coding guides data entry experts to build datasets that are ready for analysis. For most studies, this means that data is entered into a table or spreadsheet that is housed in the data editor of a software application (e.g., SPSS, SAS, R). In most packages, each row of this spreadsheet contains information for a single study participant. Each column includes information for a single variable. The finished product is a row by column matrix of mostly numbers, where the name for each variable is at the top of its respective column (Figure 3.3). In SPSS nominal (string) variables are represented by a trio of circles, ordinal level variables by a bar chart, and scale level (interal of ratio level) variables by a ruler.

Figure 3.3 Example of Data Spreadsheet in SPSS Data Editor

	ID	Gender	Gen_Alpha	Age	Age_Group	HSGPA	UNIVGPA	Univ_Sched
1	001	0	F	43	6	3.0	3.3	1
2	002	0	F	27	3	3.7	3.3	1
3	003	1	M	38	5	3.3	3.3	1
4	004	0	F	33	4	3.0	3.7	1
5	005	0	F	33	4	4.0	3.3	1
6	006	1	M	44	6	3.7	3.7	1
7	007	1	M	26	3	3.0	3.0	1
8	008	1	M	19	2	3.0	3.3	2
9	009	1	M	34	4	3.3	3.7	1
10	010	0	F	30	3	3.7	3.3	1

Regardless of the data entry approach researchers use, there are some essential ground rules that researchers must follow in order to maintain data integrity. Although all datasets for exercises in this textbook are provided and no coding or data entry is required, it is essential to become acquainted with coding and data entry rules in preparation for future research studies; if the data is not set up correctly, statistical software cannot conduct the desired analyses.

Detailed coding rules may be found in most statistical software users' guides as well as statistics textbooks (e.g., Leech et al., 2015). Below are some examples of coding basics.

3.2.1 Rows and Columns of the Dataset

One case, one row. Most statistical analyses, with the exception of analyses like repeated measures or multi-level modeling, require a single row of data for each unique case about a person, place (e.g., school, town, church) or thing (e.g., years, foods, vehicles) being measured. This is called the wide format and is the only format used in this textbook. Coders should ensure that data entry is designed to accommodate all data for a case on a single row.

System-numbered rows. Each row automatically includes a prenumbered row number. However, these should not be substituted for the researcher-assigned identification code for each study participant. These pre-numbered row numbers are visual guides and are not attached to specific cases. Should a researcher rearrange the data the prenumbered rows will be associated with different cases.

One variable, one column. For most statistical procedures, each variable (e.g., age) occupies a single column (Figure 3.3), and each variable name is listed at the top of its respective column. All data for each variable are entered in the same column. For example, in Figure 3.3, the age for each study participant may be found in the column of the variable, Age.

3.2.2 Numbers or Letters

SPSS and other statistical software packages accommodate different types of variables. The most common are numeric and string. Numeric variables are numbers and are most frequently used for numeric calculations. String are nominal variables which may include letters, numbers and other characters. These are sometimes referred to as alphanumeric or character variables, with large numbers of categories, like email addresses, names, or phone numbers. These are not useful for calculations.

Other string variables have very few categories (e.g., gender coded as "M" for male and "F" for female) and may be used as grouping variables in some statistical analyses (e.g., analysis of variance). However, the best rule of thumb is to use numeric data that can be counted or measured (quantifiable) in quantitative research.

3.2.3 Creating Mutually-Exclusive Groups

Some questionnaire or test items may allow respondents to select more than one response. These items present an item stem along with a list of response options from which a respondent may select all options that are true for them. For example, imagine that a researcher wishes to know more about social media use, using the following question:

Which of the following social media sites do you use? Check all that apply.

- ❑ Facebook
- ❑ WhatsApp
- ❑ YouTube
- ❑ Instagram
- ❑ TikTok
- ❑ Snapchat

Respondents will have different patterns of responses in this situation. However, responses for these and all other items must be coded so that all values are mutually exclusive. To accomplish this for a multiple-response question, the item must be divided into a separate binary variable for each possible answer choice. In effect, each possible answer choice must be coded as if it were a separate question with a binary option (0=no or 1=yes), thereby, resulting in breaking one question into six different items.

3.2.4 Coding to Get the Most Information

Avoid Collapsed Categories. To get the most information possible, it is best to avoid including collapsed categories on a questionnaire. For example, in Figure 3.3, study participants are asked to insert their age. Notice that there is also a variable, Age Group. The variable age, provides the most information possible for the researcher, who may analyze the ages as entered or use computer software applications to collapse the ages into specific age groups.

In contrast to Age, the variables HSGPA and UNIVGPA provide more limited information. Ideally, asking respondents to insert their actual GPAs rather than letter grades converted to GPA categories would have provided more information.

A value for every response. Every variable must be assigned a code, preferably a numeric code. Missing responses must also have codes. Missing data may be due to items that respondents found inapplicable to their situation. Alternatively, they may be due to respondents' refusal to answer items, or the provision of invalid responses. These are user-missing values that a coder typically assigns values of 9 for single-column field, or 99 for a two-column field. Whatever code is used for a user-missing value, it should not be the same as a code for a valid response.

To avoid the inclusion of user-missing codes in statistical calculations, the coder must also programmatically identify these values as user-missing codes. Conversely when statistical software is unable to determine the value of a variable, it automatically assigns a system-missing value which is typically denoted by a dot in SPSS. One source of system-missing values is conditional branching in surveys in which some survey questions are not presented to respondents based on their response to the current item.

Code Likert response formats according to the preferred attitude measured. To avoid confusion in interpreting results, it is best if high values are associated with the attitude or characteristic being measured. Consider the Likert response formats in Figure 3.4 for two characteristics: TV-watching frequency, and exercise frequency.

Figure 3.4 Likert Response Formats

TV Watching Frequency	Exercise Frequency
1. Very Frequently	1. Never
2. Frequently	2. Very Rarely
3. Occasionally	3. Rarely
4. Rarely	4. Occasionally
5. Very Rarely	5. Frequently
6. Never	6. Very Frequently

In both cases the response formats are the same, but with different implications. In the first, higher values are associated with low frequency of a negative behavior. In the second, higher values are associated with high frequency of a positive behavior.

3.2.5 Create a Codebook

Do not wait until data collection is completed before constructing the codebook. In fact, making decisions about how to code the data is an integral part of the creation of the data collection tool. This is especially pertinent when the information collected is not in a form that can be directly entered into the database.

Creating the codebook while the data collection tool is under development also provides the opportunity to troubleshoot and fix problems with questionnaires. For example, in preparing a coding sheet, researchers may discover that they can more accurately derive respondents' ages if they collect dates-of-birth rather than ages or age-ranges.

3.2.6 Coder Training and Monitoring

For manual data entry, once coding is completed, researchers must train data entry assistants to make the coding rules clear. In the process of training and throughout the course of data entry, researchers may discover the need to integrate additional coding rules and make decisions about how to incorporate them. In addition to adjusting the code book, researchers will need to monitor data entry and retrain data entry specialists to enhance accuracy.

3.3 Before Cleaning and Screening Data

Even with careful coding and close monitoring of data entry for quality control, datasets often contain errors or other problems. Hence, it is important to clean and screen the entered data. Before that point, check the dataset to confirm the following:
- Variables are in columns, and cases are in rows.
- All the variables needed are in the dataset.
- In addition to the prenumbered row number, there is at least one unique identifier (ID code) for each case.
- There is a back-up copy of the dataset.
- The codebook is updated and complete.
- Syntax that document changes to the dataset has been saved to a separate data file. Syntax might include those for variable recodes, transformations, and computations of new variables. In IBM SPSS, click on the Paste button in dialog boxes to save syntax.

3.4 Data Cleaning and Screening

According to Meyers et al. (2016), cleaning the data, or value cleaning, is the process of ensuring that values are within the limits of reasonable expectation. Data screening is defined by the American Psychological Association dictionary (VandenBos, 2015) as a preliminary review of data that incorporates basic checks for data accuracy and more advanced screening activities, including missing data and unusual data patterns (e.g., outliers), to determine whether data assumptions (e.g., normality) have been met. This section will examine data cleaning and screening in two subsections:
- 3.4.1: basic data review and adjustments to maximize accuracy, consistency and easier use;
- 3.4.2: advanced evaluations of data properties and assumptions that might affect planned, central data analysis and interpretation as well as corrective options.

3.4.1 Basic Review

Some of the most common concerns in review for data accuracy and consistency include:
- implausible or unlikely responses, like out-of-range data values;
- duplicate cases or responses (e.g., two rows of data with identical information);
- lack of uniformity in variable units of measure throughout the dataset (e.g., metric and English system weight values);
- unexpected value formats (e.g., numerical values treated as texts);

- inconsistent data across variables (e.g., a study participant enters "8" for age and educational level as "college sophomore");
- commingled user-defined missing and actual values (e.g., use of user-defined missing value "9" in item requesting children's ages that include age "9").

3.4.1.1 Implausible or Unlikely Responses

These are values that are highly unlikely in the dataset and may be user-based errors or mistakes in data entry. For example, in response to a health survey question that asks the amount of water consumed each day, a respondent answers 10 liters. With 10 liters being the equivalent of 42 glasses of water, this response is exceedingly unrealistic and thus a good candidate for deletion. In some instances, implausible answers are not user-based but data entry errors. For a question in which gender is coded "0" (male) and "1" (female), a data entry clerk may unwittingly enter the number 2 instead.

These inaccuracies are more likely to occur in open-ended paper-and-pencil questionnaire entries than computer-based questionnaires where value constraints prevent respondents from entering irrelevant information. A value constraint is a data validation technique in which a condition is attached to a survey item, and users must satisfy the condition before their entry or selection is accepted. The survey participant cannot move on to another item until the constraint is satisfied. One common constraint is to exclude entries that are outside of a specified range of response alternatives, such as 0 and 1 for gender. If a survey respondent attempts to enter a different code, an invalid response error message appears.

Exploratory data cleaning analyses (detailed later in this section) may be used to detect these inaccuracies for further data validation by comparing minimum and maximum values against allowable ranges. Unless the correct responses can be identified by consulting the original data source (original survey documents or survey participants), they are removed from the working data file prior to analysis.

3.4.1.2 Duplicate Cases and Responses

For most statistical procedures, it is critical that each record in a dataset is unique. Leaving duplicates in datasets will result in skewed survey results, weighting some survey respondents' submissions more than others. Therefore, when planned statistical procedures are limited to unique cases, knowing how to handle duplicated records in a data file is a vital element in the data cleansing component of data analysis.

Duplicate cases happen as a result of data entry errors or respondent missteps:
- A data entry specialist may record survey responses from the same participant more than once.
- Duplicates may occur when multiple data entry specialists or data collection agents unwittingly enter data for the same respondents. Duplicates end up in the dataset when files from different sources are merged into a single master file.
- A study participant may submit multiple copies of an online questionnaire when they unintentionally hit the survey submission button twice.

Duplicate ID codes or email addresses are two common indicators that there may be duplicated survey submissions in your data base. Alternatively, an entire row in a dataset may be duplicated.

One strategy for detecting duplicate entries is to generate a table of frequencies of the ID codes or email addresses. The table will reveal which cases include duplicate entries. Duplicates can be selected and filtered out or removed prior to analysis.

Another method of detecting duplicated data is to sort the data by the variable of concern (e.g., ID), which will result in rows with identical responses located next to each other. Use of the lag function can then confirm which cases are duplicates.

A third option in IBM SPSS is to use the Identify Duplicate Cases wizard to locate rows with identical data. You may then select and filter out or delete the unwanted cases.

3.4.1.3 Lack of Uniformity: Same Variable—Different Units of Measure

Imprecise instructions can sometimes result in a lack of uniformity in study participants' responses. Consider, for example, an open-ended item in a health survey that asks participants to enter their weight. However, the item does not specify the unit of measure to be used. Hence, some respondents report their weight in kilograms (a metric system of measurement), while others report their weight in pounds and ounces (the avoirdupois pound or English system of measurement). The units of measure lack uniformity across study participants. Therefore, researchers are unable to identify whether participants use the metric system or the English system of measurement to report their weights. Survey developers who fail to specify the desired unit of measurement are likely to end up with uninterpretable data.

Researchers need uniform data for accurate analyses. The best cure for nonuniform data is prevention—careful specification of data constraints in data collection instructions and code books for manual data entry. Data constraints, like limiting income reports to gross (rather than net) annual income, minimize the likelihood of uninterpretable data.

3.4.1.4 Incorrect/Unexpected Value Formats

Specifying the data type of each variable in a research study is essential for accurate data entry and analysis. Formatting variables in accordance with the type of data collected is a necessary constraint that should be incorporated into computer-based data collection tools and in data entry codebooks to ensure valid and trustworthy data.

Of the several types of data that may be used in SPSS, the two most common are numeric and string data types. Numeric types of data are essentially data that can be used in statistical or arithmetic calculations. However, not all numeric values are used for calculations. Some may be used to denote categorical variables. These types of numeric values (e.g., 0 and 1 designating male or female for the variable gender) are not designed for use in mathematical or statistical procedures.

Typically, numerical data to be used in statistical and mathematical calculations are formatted to denote the number of decimal places for the values. By contrast, categorical data, which may be ordinal or nominal level, are formatted as integers. Dates are another type of numeric variable that utilize special formatting. Date formats include the use of commas, blank spaces, hyphens, periods, or slashes as space delimiters. In SPSS, data using a date format must be assigned the date variable type in order to be correctly processed.

String variables, also called alphanumeric or character variables, may be numbers, letters, symbols or a combination of all three and are always treated as text. This means that even if string variables are made up of numbers, their values cannot be meaningfully used in mathematical calculations. Models of a type of automobile, names of players on a soccer team, and phone numbers are all examples of string variables.

Most statistical software applications are restricted in the types of data they can use for statistical analysis. The consequence of failure to honor those constraints is often nonsensical results.

- Attempting to find the mean of alphanumeric string variables, like the names of students in a class, will result in an error message.
- String variables, like phone numbers, that are mistakenly formatted as numeric variables can be used in numeric calculations, but the results will be uninterpretable.
- Unexpected value formats, like numeric variables formatted as string (text) or alphanumeric variables, means that those variables cannot be entered as decimals and cannot be used in calculations.

Careful pre-specification of data types and formats as well as calculations that adhere to data type constraints all contribute to clean and accurate datasets.

3.4.1.5 Inconsistent Responses Across Variables

Consistent data may be described as how closely a data value for one variable aligns with a related data value for another variable. If a person chooses the survey response, "I watch 30 hours of TV per week," but in answer to a later question, states "I don't watch TV," those responses are inconsistent. Alternatively, a reported age of 13, along with a birth year of 1940 also constitutes an impossible or nonsensical combination. Examining the consistency of responses across questions is one way of evaluating the trustworthiness of research data. Contradictory responses across items in a dataset suggest that the data may be undependable and unsuitable for use.

Inconsistencies may be identified using basic descriptive statistic procedures, such as a crosstabs analysis. In the case of the age inconsistency, you might accomplish this by recoding date-of-birth values into values for a new variable, called dob_age, and then conducting a crosstabs analysis with the variable age and the new variable dob_age to expose inconsistencies.

Another method of identifying inconsistent data is to use filters. In the case of TV watching, first filter for all respondents who stated that they watched TV (any number of hours). Second, filter for those respondents who selected "I don't have a TV." All cases in the file with both of these responses have conflicting data values.

According to Warner (2008), records with inconsistencies may be handled in a variety of ways.

- If incorrect data were entered manually, go back to original documents to locate correct information, then use the **Recode into Different Variable** procedure to replace the implausible values with accurate ones. Using the Recode procedure corrects the data error while preserving the ability to keep track of modifications.
- Remove the improbable value from analysis by replacing the flawed value with a blank cell or code representing missing data.
- If a record has multiple inconsistencies, consider temporarily or permanently removing the record from the file.

3.4.1.6 Missing Data

Ideally, a datafile should be complete with valid values for every participant in the study. Unfortunately, that rarely occurs. Missing data are among the most challenging problems that social science data analysts face. It happens when there are no valid values for one or more variables for some subjects in a study. For example, in a study of whether there is a difference in the average texting speed of males and females, texting speed or gender may be missing for some study participants.

3.4.1.6.1 Reasons for Missing Data

Data go missing for myriad reasons. In some cases, data should be missing. For example, survey branching is a survey method in which the questions respondents see are based on their answers to previous items. Therefore, a single person who has never married will not be presented with a question that asks how long they have been married. Consequently, all single respondents will have missing responses for the length of marriage question.

Beyond missing data that is attributable to instrument methodology, there are those that are due to technical failures, human error, or respondent refusals. For example, technical failures may occur due to research equipment malfunction, errors in questionnaire items, or malfunctions in computer delivery systems. Researchers may lose student test responses due to faulty security measures, or violation of protocol. Data entry specialists may make data entry errors. Study participant refusals may include participants' decisions to skip very personal questions. Longitudinal studies where measures (e.g., blood pressure) are collected repeatedly over a period of time are especially prone to missing data as a result of participant attrition.

3.4.1.6.2 Consequences of Missing Data

Missing data can bring about a series of negative consequences. Researchers may be obliged to rely on smaller sample sizes for statistical analysis, and smaller sample sizes may result in a reduction in statistical power, bias in statistical estimations, and subsequently misrepresentation of target populations. Inaccurate generalizations about populations may ultimately contribute to ineffective or harmful interventions and policies.

Flawed statistical inferences may not only be attributable to the quantity of missing data, but also the pattern of missing sample data is often a key contributor. Drawing valid statistical inferences require a research sample that is representative of the population. If data values are missing at random, it is likely that there is no significant difference between available and missing data. For example, in a study of self-esteem, researchers may construct a dummy variable with two groups, subjects with missing (0) and non-missing values (1) of weight. Researchers may then compare the frequency of missing values against other variables of interest, like gender, to determine whether missing values are related to other variables in the dataset. If they are not related to other variables, the missing values are likely to be missing at random. If there are identifiable patterns in the missing data, statistical results based on such data are likely to be questionable. Another assessment may involve examining whether there is a significant difference in self-esteem between the groups. A nonsignificant difference suggests equivalent groups with missing values randomly distributed.

3.4.1.6.3 Handling Missing Values

Arguably, the best approach for handling missing values is data collection design strategies that reduce the quantity of missing data. For example, data collection instruments may include embedded data validation techniques or checks that ensure the responses do not exceed the expected range, or data type checks that prohibit the entry of letters in a field meant for numeric values. Additionally, researchers may provide incentives to reduce the dropout rate in a longitudinal study. Once data collection is complete, researchers turn to corrective measures intended to enhance the quantity and the integrity of the data. The first step is typically to identify which values are missing and to determine the extent of missing values. Following that preliminary examination, there are three common options for handling missing data: ignoring, deleting, or replacing missing values.

In SPSS, missing values are easily identifiable. They are either system- or user-defined missing values. SPSS assigns a dot as a default (automatic) system-missing placeholder for missing values of a quantitative variable; SPSS automatically assigns the system-missing symbol when there is no value for a case in the dataset. SPSS assigns a blank space when a value for a string (categorical) variable is missing. A user-missing value is an actual value that users of the dataset assign to signify a nonresponse. Special numeric codes signify specific types of missing observations or reasons for a missing value (e.g., refusal to respond, or a sudden emergency). These user-missing codes may also be used as a monitoring device to distinguish between valid missing values and inadvertent data entry errors.

Whatever the reason for specifying user-missing values, it is essential to select user-missing codes that are unlikely to be confused with other codes in the dataset. For example, consider the 9-point Likert scale rating satisfaction with service at a new restaurant:

1	2	3	4	5	6	7	8	9
Extremely Dissatisfied	Very Dissatisfied	Moderately Dissatisfied	Slightly Dissatisfied	Neither Satisfied nor Dissatisfied	Slightly Satisfied	Moderately Satisfied	Very Satisfied	Extremely Satisfied

A codebook that specifies the number "9" as a nonresponse for data entry could result in confusion with the existent code for scale point 9 on the Likert scale. To avoid confusion, some researchers go beyond the typical user-defined missing codes of "9" and "99" to use negative numbers (e.g., -1, -9, -99) or very large numbers (e.g., 888 or 999). For proper processing, these special codes are prespecified in SPSS's Missing Value column in the Variable View window, and the number of special missing value codes is limited to three per numeric variable. Consider a survey in Trinidad and Tobago with the following instruction:

Circle the number next to your political party affiliation:
1   People's National Movement
2   United National Congress
3   Progressive Democratic Patriots
4   Other Political Party (registered with Elections and Boundaries Commission)
5   Other Political Party (**not** registered with Elections and Boundaries Commission)
6   Unaffiliated with Political Party
9   Prefer not to say

The response code "9" is a user-defined missing value recorded in the Missing Value column in Variable View. In preliminary data analysis, it is important to detect which variables have missing values. One way of doing that is to generate a frequency count of all values for each variable. The result is a table that should label the frequency of values that are nonmissing separately from those that are missing (e.g., Figure 3.4 and 3.8).

Quantifying missing data allows researchers to estimate the severity of missing data values in a dataset. According to Tabachnick and Fidell (2018), "If only a few data points—say, 5% or less— are missing in a random pattern from a large dataset, the problems are less serious and almost any procedure for handling missing values yields similar results" (p. 97). Therefore, calculating the overall percentage of missing data values as well as the percentage within each variable is an important preliminary task when considering options for handling missing data values. With a small percentage of data points missing at random, ignoring the affected cases by leaving the default system or user-missing values is an option. Deleting the cases with missing values from an analysis, the default for many statistical applications, is another option. SPSS offers two methods for deleting cases: listwise deletions and pairwise deletions. Listwise deletion means that all the values for study participants are removed from an analysis when they have missing values for one or more variables used in the analysis. Therefore, if a researcher is conducting a correlation of weight and height and the value for weight is missing from a case, listwise deletion would completely remove that record from the correlation analysis. Only complete cases would be used.

In pairwise deletion, a record which has missing values on some variables, may be excluded from statistical procedures that involves those variables. However, the same record may be included in other procedures when analyzing variables with nonmissing values. Pairwise deletion is the default for correlation, descriptive statistics, and other fundamental procedures in SPSS. Some methodologists (e.g., Graham, 2009) do not recommend pairwise deletion due to the likelihood that parameter estimates will be biased, and there is no way of estimating the degree of uncertainty (standard errors) about the parameter estimates.

If missing values are not randomly distributed, replacing (imputing) the missing values with estimates is another method of securing a complete dataset for analysis. Methods of imputation include using the means of available data to estimate missing values (mean substitution) or using values from similar cases to replace missing values (hot deck imputation). For a comprehensive treatment of missing data, consult Graham (2012).

3.4.1.7 SPSS Specifications for Basic Value Cleansing: Checking for Errors

Throughout the design, data collection, and data entry phases of a research study, errors are likely to occur. Data collection agents or respondents themselves may make errors in recording responses to questions. Variable values may be incorrectly coded, or data entry specialists may have lapses in interpreting or transferring codes to the dataset. This section demonstrates common methods of checking for errors in a dataset.

- Screen for appropriate data types and formats of variables in the dataset.
- Use sorting, frequency distribution or the Identify Duplicate Cases wizard to handle duplicate entries.
- Use frequency distribution tables to summarize the values of variables. The frequency procedure may be used for quantitative or qualitative variables (nominal or ordinal).
- Generate descriptive statistics to identify out-of-range values for quantitative variables.
- Generate means and standard deviations to evaluate reasonableness of variable values.
- Use the crosstabulation procedure to fix inconsistent responses across variables.
- Review the quantity, types, and extent of missing data.

68 Chapter 3: Data Preparation: Cleansing and Screening

Section 3.4.1.7.1 SPSS Screening to Review Data Types and Formats

This data cleansing demonstration will use IBM SPSS to review data from a small survey of the heights and weights of 40 undergraduate and graduate university students from across the Caribbean. The survey data were collected manually by graduate students. Some of the data values have been fictionalized to accommodate this exercise. The dataset may be found in the datafile, Carib_Height_Weight_Dups.sav, and in Appendix B. The variable names for each variable in the dataset as well as their variable labels appear in Figure 3.5.

Figure 3.5 Variables in Carib_Height_Weight_Dups.sav

Variable Name	Variable Label	Variable Name	Variable Label
ID	Identification code	Gender	Gender
Country	Country of Residence UN Country Code	Age	Age
UNCode		Heightcm	Height (cm)
Edlvl	Educational Level	Weightkg	Weight (kg)

The data dictionary and variable value tables for Carib_Height_Weight_Dups.sav are displayed in Figures 3.6 and 3.7.

Figure 3.6 Data Dictionary Table

Variable Information									
Variable	Position	Label	Measurement Level	Role	Column Width	Alignment	Print Format	Write	Missing Values
ID	1	Identification Code	Nominal	Input	12	Left	A5	A5	
Country	2	Country of Residence	Nominal	Input	24	Left	A24	A24	
UNCode	3	UN Country Code	Nominal	Input	8	Left	A8	A8	
Edlvl	4	Educational Level	Ordinal	Input	8	Right	F8	F8	99
Gender	5	Gender	Nominal	Input	8	Right	F8	F8	
Age	6	Age	Scale	Input	8	Right	F8	F8	999
Heightcm	7	Height (cm)	Scale	Input	8	Right	F8	F8	
Weightkg	8	Weight (kg)	Scale	Input	8	Right	F8	F8	
Variables in working file									

The data dictionary in Figure 3.6 presents all of the variables in the dataset, along with characteristics of each variable. The first two are alphanumeric or string (text) variables. Variable 3 is the numeric, string version of the variable, Country, as indicated by the print and write format codes beginning with "A." The remaining variables are numeric with values that are whole numbers as indicated by the default print and write format codes, beginning with "F."

The values for some of the variables in the dataset are presented in Figure 3.7. The value labels will appear in all SPSS output to make it easier to understand and interpret the results of analyses. The value codes, 99 and 999, are user-defined missing variable values.

Figure 3.7 Variable Values Table

Variable Values		
Value		Label
UNCode	044	The Bahamas
	052	Barbados
	136	The Cayman Islands
	212	Dominica
	308	Grenada
	388	Jamaica
	530	Sint Maarten
	659	Saint Kitts and Nevis
	660	Anguilla
	662	Saint Lucia
	780	Trinidad and Tobago
	796	The Turks and Caicos Islands

Variable Values		
Value		Label
Edlvl	1	Undergraduate
	2	Graduate
	99[a]	Prefer Not to Say
Gender	0	Male
	1	Female
Age	999[a]	Prefer Not to Say
a. Missing value		

3.4.1.7.2 SPSS Screening for Duplicate Cases

One method of identifying duplicates in a dataset is to use the **Identify Duplicate Cases** wizard to locate duplicates. Once identified, one of two approaches may be used to handle the duplicates: (a) Retain them in the dataset, but filter them out so that they will be excluded from any further analysis; (b) remove them from the dataset completely, leaving only unique cases in the file. To use the wizard, locate the dataset, **Carib_Height_Weight_Dups.sav** and double-click on the filename to bring the dataset into SPSS.

Select **Data → Identify Duplicate Cases** from the menu options at the top of the IBM SPSS Statistics Data Editor worksheet (Figure 3.8).

In the **Identify Duplicate Cases** dialog window (Figure 3.9), select the variable(s) in the panel on the left to be examined for duplicate entries, and move them to the **Define matching case by** panel on the right. In this case, drag or click over the variables, **Country (Country of Residence)**, **Edlvl (Educational Level)**, **Gender**, **Age**, **Heightcm (Height (cm))**, and **Weightkg (Weight (kg))**. Under **Variables to Create**, confirm the following default settings have been selected.

- **Last case in each group is primary** (the last case marks the one that will be retained).
- **Move matching cases to the top of the file** (duplicate cases will be at the top of the file).
- **Display frequencies for created variables** (generate a frequency table with unique and duplicate cases).

Figure 3.8 IBM SPSS Data Editor

		File Edit View					Extensions Window Help				
						UNCode	Edlvl	Gender	Age	Heightcm	Weightkg
1	1001					136	2	0	999	185	90
2	1002					388	2	0	.	185	109
3	1003					780	2	1	49	180	90
4	1004					388	2	1	39	173	68
5	1005					136	1	1	32	170	65
6	1006					308	1	0	29	175	84
7	1007					052	1	1	35	168	58
8	1008					780	2	1	50	185	79
9	1009					388	2	1	47	183	82
10	1010					388	2	0	.	191	79

Figure 3.9 Identify Duplicate Cases Dialog Menu Window

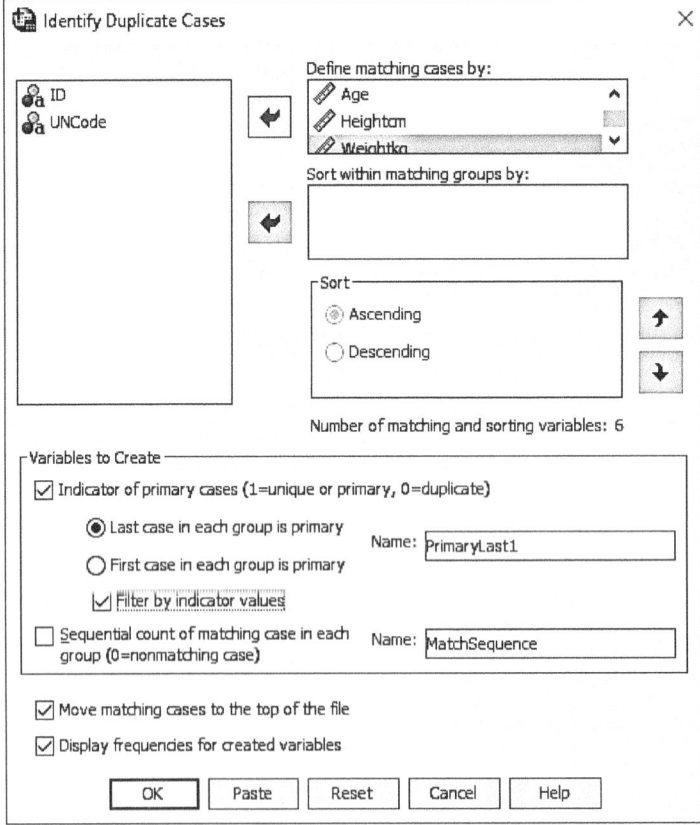

In addition, select **Filter by indicator values**. This specification means that a new variable (**PrimaryLast**) will be created in the active dataset with each case assigned the value 0 (duplicate, unselected) or 1 (unique, selected). Click **OK** to run the procedure.

Figure 3.10 displays the frequency table for the created variables. As the Statistics table reports, a total of 40 cases occupies the dataset. The second table (Indicator of each last matching case as Primary) records that of the 40 cases in the data file, seven were duplicates. Figure 3.11 displays all the matching cases and the duplicates in the file (e.g., ID 1007 and ID 2007).

With the duplicate cases identified, the next step is to create a new dataset with only unique cases in it. To create the new dataset, delete the duplicate cases (those identified with values equal to 0 for the variable **PrimaryLast**). Save the new dataset with a new filename. In this way, the original dataset is preserved, and a dataset without duplicates is available for analysis.

Figure 3.10 Frequency table of created variables

(a)

Statistics		
Indicator of each last matching case as Primary		
N	Valid	40
	Missing	0

(b)

Indicator of each last matching case as Primary					
		Frequency	Percent	Valid Percent	Cumulative Percent
Valid	Duplicate Case	7	17.5	17.5	17.5
	Primary Case	33	82.5	82.5	100.0
	Total	40	100.0	100.0	

Figure 3.11 Matching Cases that Include Duplicates

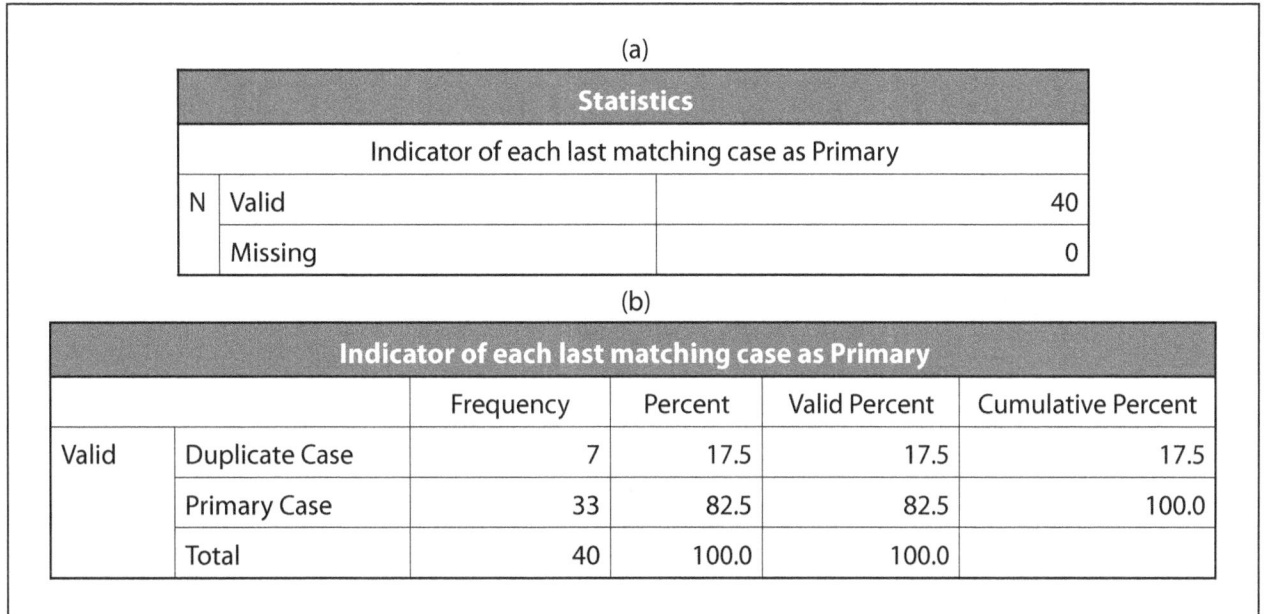

3.4.1.7.3 SPSS Specifications to Identify Out-of-Range Categorical Variables

Generating frequency counts of variable values is a common method of initiating data cleansing, particularly when variables have a small number of values. To generate frequency counts for the background categorical variables, open the new dataset **Carib_Height_Weight_Nodups.sav**, and from the main menu, select **Analyze → Descriptive Statistics → Frequencies** to open the main Frequencies dialog menu.

In the **Frequencies** main dialog screen, select the variables, **Edlvl** and **Gender** in the left panel, and click them over to the **Variable(s)** panel on the right (Figure 3.12). By default, the **Display frequency tables** checkbox is prechecked. This default specification will disclose whether there are more than two valid categories for **Gender** and **Edlvl** as itemized in the **Variable Values** table in Figure 3.7. In addition, the frequency analysis will identify whether any assigned user-defined missing values were used for the variables. Click **OK** to run the analysis.

Figure 3.12 Frequencies Main Dialog Screen

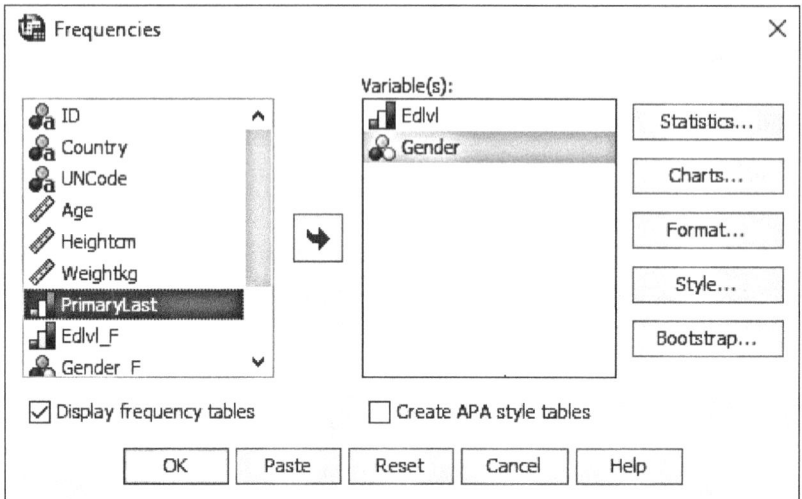

Results of the frequency procedure are displayed in Figure 3.13. The **Statistics** table (Figure 3.13a) shows that of the 33 unduplicated survey respondents, one (3.0%) has a missing value for the variable **Educational Level**. Further assessment of the results for Educational Level in the second table (Figure 3.13b) confirms that the missing value is a user-defined missing value (99) for respondents who stated that they "Prefer not to say" their educational level.

Although the values 1 and 2 represent valid responses for **Educational Level**, the value 3, entered for one individual, is possibly a data entry error. Checking the data entered against the original survey document will likely confirm whether the value is a data entry error or one committed by survey administrators who might have inadvertently entered incorrect information. Should the latter be the source of the error, a follow-up with the original data source could provide accurate information.

Figure 3.13c presents frequency results for **Gender**. In the dataset, values for males and females are coded 0 and 1, respectively. However, a value of 2 with a frequency of 2 is in Figure 3.13c. This means that the value, 2, appears in the dataset twice even though in the code book **Gender** has not been assigned a variable value of 2. Checking original data collection documents confirmed that the value, 2, is a data entry error for ID 3860 and 3990 and should have been 0 for both cases.

3.4.1.7.4 SPSS Specifications to Detect Out-of-Range Quantitative Variables

The frequency procedure identified out-of-range values and user-missing values for two categorical variables—**Edlvl** and **Gender**. Frequency tables do the same for quantitative variables. Unlike categorical variables that are often limited to a few values that are easily summarized using the frequency procedure, quantitative variables may have more values to summarize. Nevertheless, a frequency table of quantitative values can be useful in finding out-of-range values, as well as user-defined missing values and system-missing placeholders.

Figure 3.13 Frequency Table for Gender and Educational Level

(a)

		Statistics	
		Educational Level	Gender
N	Valid	32	33
	Missing	1	0

(b)

		Educational Level			
		Frequency	Percent	Valid Percent	Cumulative Percent
Valid	Undergraduate	14	42.4	43.8	43.8
	Graduate	17	51.5	53.1	96.9
	3	1	3.0	3.1	100.0
	Total	32	97.0	100.0	
Missing	Prefer Not to Say	1	3.0		
Total		33	100.0		

(c)

		Gender			
		Frequency	Percent	Valid Percent	Cumulative Percent
Valid	Male	15	45.5	45.5	45.5
	Female	16	48.5	48.5	93.9
	2	2	6.1	6.1	100.0
	Total	33	100.0	100.0	

The frequency procedure also allows specification of basic descriptive statistics that provide summaries of the distribution of quantitative values. Available descriptive statistics include measures of central tendency (e.g., mean), and measures of dispersion (e.g., standard deviation and range), and distribution (e.g., skewness and kurtosis). Examining the mean (or average score) together with the standard deviation provides researchers with an indication of the variation or dispersion of a set of values related to the mean. A low standard deviation means that values are clustered around the mean, while a high standard deviation means that the values are more spread out. Results that show an unusual mean or standard deviation (e.g., low dispersion when high is expected) could suggest the need to examine the data more closely for out-of-range scores and other errors. The range, the difference between the maximum and minimum score, is another indicator of the spread of scores.

This demonstration is limited to the numeric variable, **Age,** and will include the descriptive statistics of central tendency and dispersion. Other descriptive statistics will be examined in later sections. To screen for out-of-range and other unusual values as well as missing values, select **Analyze → Descriptive**

74 Chapter 3: Data Preparation: Cleansing and Screening

Statistics → **Frequencies** in SPSS to open the **Frequencies** dialog menu (Figure 3.14a). Select **Age** from the left panel and click the variable over to the **Variable(s)** panel on the right. Confirm that the default selection labeled "**Display frequency tables**" is already checked, to produce frequency tables.

Figure 3.14 Frequencies Main Dialog and Frequencies Statistics Windows

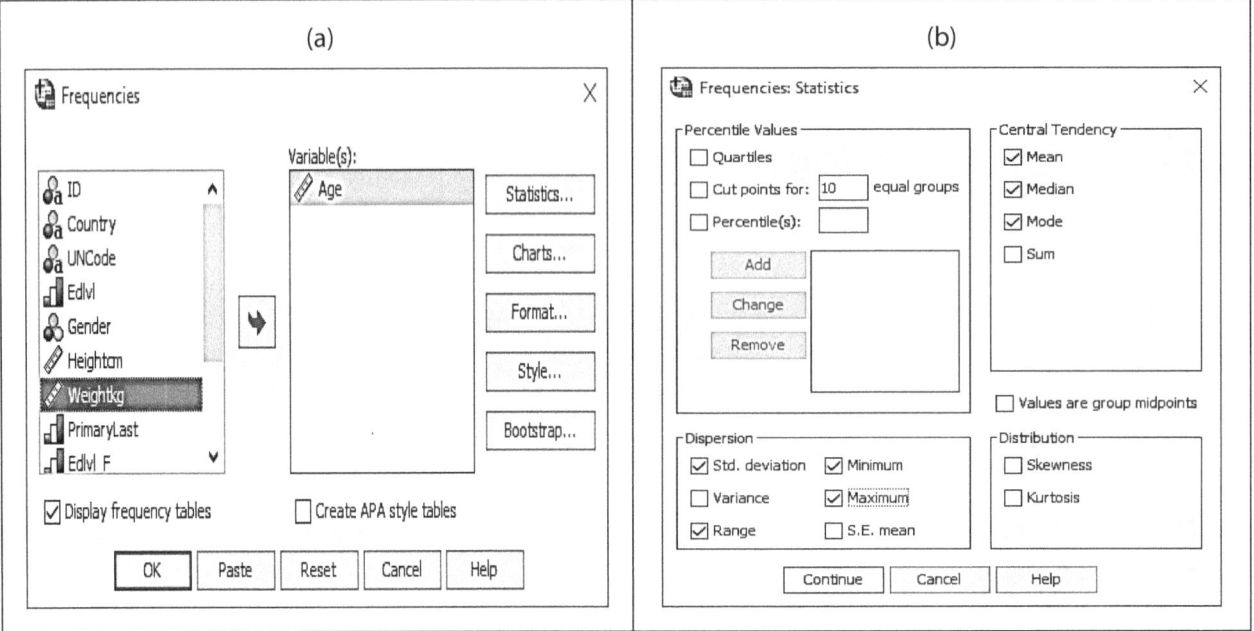

Next, click on the **Statistics** pushbutton in the top right corner of the **Frequencies** dialog window to produce the **Frequencies Statistics** window (Figure 3.14b). This window includes four sections: Percentile Values, Dispersion, Central Tendency, and Distribution.

As illustrated in Figure 3.14b, request the **Mean**, **Median** and **Mode** under **Central Tendency**. Additionally, select **Std. deviation** (standard deviation), **Range**, **Minimum** and **Maximum** under **Dispersion**. Click **Continue** to return to the **Frequencies** main dialog window. Then click **OK** to run the analysis. The **Frequency** table for **Age**, is presented in Figure 3.15.

Analysis results prove that student ages range from age 17 to 50, which is reasonable given that the study sample includes undergraduate and graduate students. Reported ages are quite variable, with most occurring once in the dataset.

As the **Missing** row in Figure 3.15 reports, three (10%) of the 33 survey respondents did not provide their age. Of the three missing values, one selected the user-defined missing value (**Prefer Not to Say**). The remaining two individuals (6.1%) had system-missing placeholders, meaning that they did not respond to the item. **Age** is the only variable with missing values that exceed the 5% rule of thumb commonly used for missing value intervention.

Figure 3.16 presents descriptive statistics results for the variable, **Age**. The table confirms the number of missing values—3 out of 33. The mean age is just short of age 34, with a standard deviation of 10.78, which indicates a fair amount of variability. The mode of age is 39, which indicates that 39 is the most frequently occurring age (occurring three times). The range of values from age 17 to age 50, appears reasonable for undergraduate and graduate students.

Figure 3.15 Frequency Table for Age

		Age			
		Frequency	Percent	Valid Percent	Cumulative Percent
Valid	17	1	3.0	3.3	3.3
	18	2	6.1	6.7	10.0
	19	1	3.0	3.3	13.3
	20	1	3.0	3.3	16.7
	22	1	3.0	3.3	20.0
	23	1	3.0	3.3	23.3
	24	1	3.0	3.3	26.7
	25	1	3.0	3.3	30.0
	26	1	3.0	3.3	33.3
	28	1	3.0	3.3	36.7
	29	1	3.0	3.3	40.0
	32	1	3.0	3.3	43.3
	33	1	3.0	3.3	46.7
	35	2	6.1	6.7	53.3
	38	1	3.0	3.3	56.7
	39	3	9.1	10.0	66.7
	40	2	6.1	6.7	73.3
	42	1	3.0	3.3	76.7
	45	2	6.1	6.7	83.3
	47	1	3.0	3.3	86.7
	48	1	3.0	3.3	90.0
	49	1	3.0	3.3	93.3
	50	2	6.1	6.7	100.0
	Total	30	90.9	100.0	
Missing	Prefer Not to Say	1	3.0		
	System	2	6.1		
	Total	3	9.1		
Total		33	100.0		

Figure 3.16 Descriptive Statistics for the Variable Age

Statistics		
Age		
N	Valid	30
	Missing	3
Mean		33.83
Median		35.00
Mode		39
Std. Deviation		10.780
Range		33
Minimum		17
Maximum		50

3.4.1.7.5 SPSS Specifications to Recode Variables to Correct Out-of-Range Values

As the **Percent** column in the **Educational Level** table (Figure 3.11b) illustrates, SPSS considers out-of-range values to be valid entries. Only user-defined missing values (e.g., Prefer Not to Say) are considered invalid data. Therefore, out-of-range values are included in data analyses, unless the errors are corrected. In this case, following up with the respondent led to the correct value, which is 2, rather than 3 for ID=3840.

To correct the value in the dataset while keeping track of the modification, use the **Recode** procedure (Figure 3.17) to create a new variable with the corrected value. From the top of the **Data Editor** window, select **Transform → Recode into Different Variables** to open the **Recode into Different Variables** dialog menu window.

Figure 3.17 IBM SPSS Main Data Editor Window

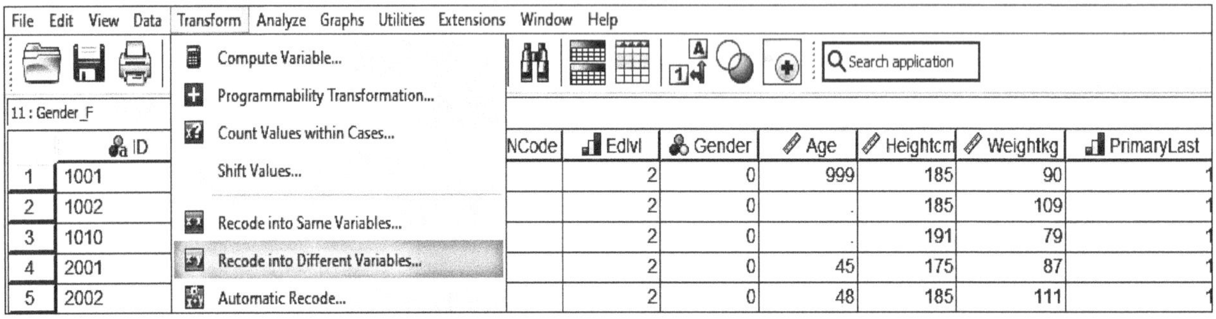

In the **Recode into Different Variables** dialog window (Figure 3.18), select **Edlvl** from the left panel, and click it over to the **Numeric Variable → Output Variables** panel in the middle of the window. To create a new variable for the corrected values, under **Output Variables** on the right, enter **Edlvl_F** in the box below **Name** (Variable Name). Under **Label** (Variable Label), enter **Corrected Educational Level**. Click on **Change** to execute the creation of the new variable.

Figure 3.18 Recode into Different Variables Dialog Window

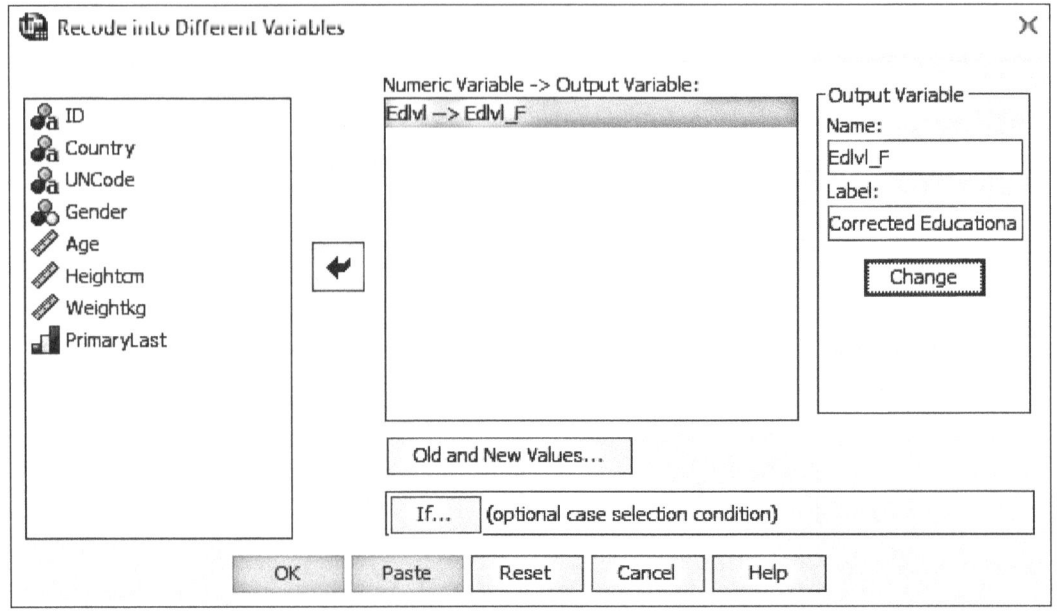

To correct the variable value for ID 3840, click on **Old and New Values** to open the **Recode into Different Variables Old and New Values** dialog window (Figure 3.19). Under **Old Value** on the left, enter the number 3, and under **New Value** on the right, enter the number 2. Under **Old → New**, click **Add** to execute the specified change. To add all the unchanged remaining values to the new variable, click on the button next to **All other values** on the bottom left of the dialog menu. On the right, click on **Copy old value(s)**, and click on **Add** to move the new specification to the **Old → New** panel (Figure 3.19). Click **Continue** to return to the **Recode into Different Variables** main dialog window. Click **OK** to run the procedure.

After the analysis ends, check the **Data View** window to ensure that all value changes were correctly assigned. In the **Variable View** window, reduce the decimal places in the **Decimals** column to 0.

Figure 3.13a reported that there were no missing values for **Gender**. However, closer examination of the **Frequency** table in Figure 3.13c disclosed two records with out-of-range values (coded 2). Both cases were male subjects and should have been coded 0. As a result of the **Recode into Different Variables** procedure, the new variable with the corrected values for Gender is **Gender_F (Corrected Gender)**. The new variable with corrected values for **Edlvl** is **Edlvl_F (Corrected Educational Level)**.

Figure 3.19 Recode into Different Variables Old and New Values

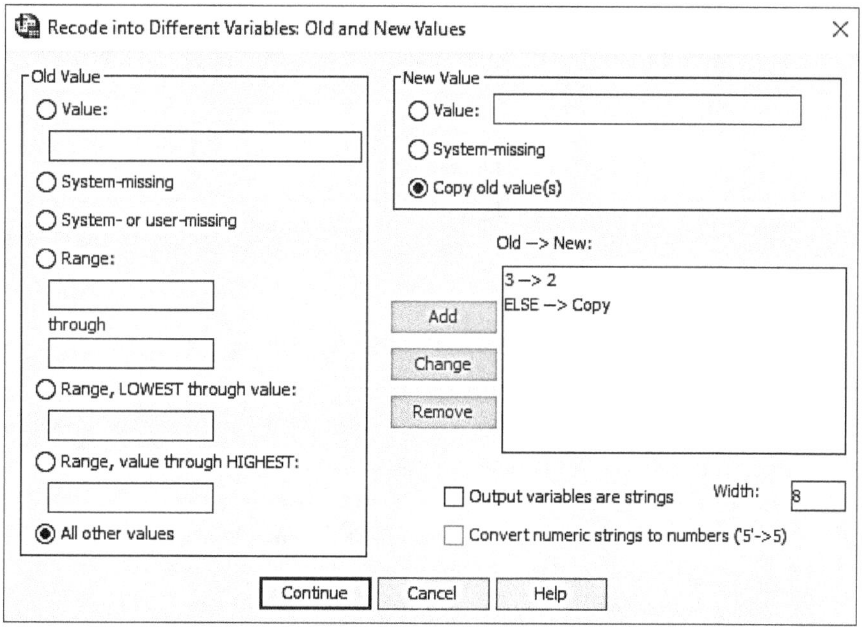

3.4.1.7.6 SPSS Specifications to Screen for Inconsistent Responses

Section 3.4.1.5 explained that one method of dealing with inconsistent responses is the removal or replacement of questionable values. Before the treatment can be considered, it is important to determine whether there is evidence of inconsistency. The use of cross-tabulation analysis is one way of screening the data for inconsistent or incongruent responses. Cross-tabulations (crosstabs) are an extension of the frequency procedure that generates two-way or multi-way contingency tables to compare one or more variables with another variable. This exercise will use that approach to screen for potential inconsistencies between age and educational level. For example, are there any cases of remarkably young students (e.g., age 16) at the graduate level? Figure 3.15 provides some information about respondents' ages.

78 Chapter 3: Data Preparation: Cleansing and Screening

Results show that respondents' ages range from 17 through 50. However, the frequency table does not allow for a comparison with another variable.

To generate a two-way contingency table for **Age** and **Educational Level**, select **Analyze** → **Descriptive Statistics** → **Crosstabs** to open the **Crosstabs** dialog menu window (Figure 3.20a). In the **Crosstabs** dialog window, select the variable **Age** from the left panel, and move it to the panel below **Row(s)**. Next, move **Edlvl_F** from the panel on the left to the panel on the right, below **Column(s)**.

To specify how the cells in the contingency table should be displayed, click on **Cell(s)** to open the **Crosstab(s): Cell Display** dialog menu (Figure 3.20b). Use all of the prespecified selections. Under **Counts**, confirm that the option, **Observed**, has been selected, indicating that only observed counts will be displayed in the results, and under **Noninteger Weights**, confirm that **Round cell counts** has been selected. Click on **Continue** to return to the **Crosstabs** main menu, and select **OK** to run the **Crosstabs** analysis.

Figure 3.20 Crosstabs Dialog Menu Window

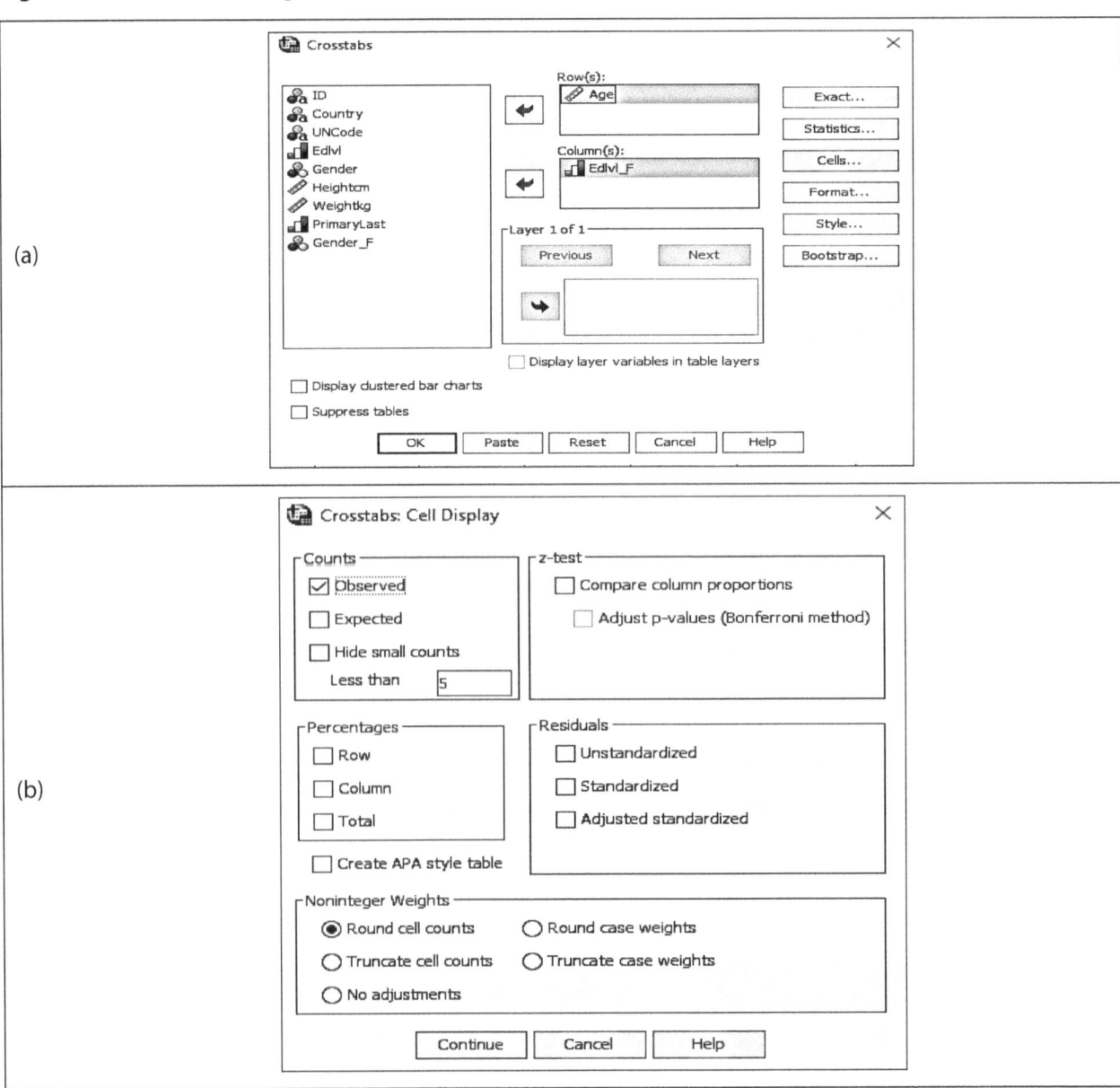

Figure 3.21 displays results of the Crosstabs analysis. As the figure illustrates, no inconsistencies or notable incongruencies are evident across the two variables. While some nontraditionally older students described themselves as undergraduate students, the youngest students (e.g., 17- and 18-year-olds) selected undergraduate level, as anticipated. The youngest graduate student is 25 years-old—well within the expected range. Therefore, the dataset may be retained without further adjustments.

Figure 3.21 Results of Crosstabs Analysis

		Age * Corrected Educational Level Crosstabulation			
		Count			
		Corrected Educational Level			
		1	2	99	Total
Age	17	1	0	0	1
	18	2	0	0	2
	19	1	0	0	1
	20	1	0	0	1
	22	1	0	0	1
	23	1	0	0	1
	24	1	0	0	1
	25	0	1	0	1
	26	1	0	0	1
	28	1	0	0	1
	29	1	0	0	1
	32	1	0	0	1
	33	0	1	0	1
	35	1	1	0	2
	38	0	0	1	1
	39	1	2	0	3
	40	0	2	0	2
	42	0	1	0	1
	45	0	2	0	2
	47	0	1	0	1
	48	0	1	0	1
	49	0	1	0	1
	50	0	2	0	2
Total		14	15	1	30

3.4.1.7.7 SPSS Specifications to Identify Missing Values

Two sources of missing data have been examined so far (e.g., user- versus system-missing data). Figure 3.22 presents one way of summarizing the findings based on the height and weight demonstration dataset.

Missing Data Total for each Variable. Figure 3.22 reports that a total of two of the five variables had missing data. Educational Level recorded one case with a user-missing value (3.0%), and Age had three cases with missing values (10%), one of which was a user-missing value while two (6.0%) had system-missing values.

Figure 3.22 Types and Quantity of Missing Data in Height and Weight of Caribbean Students

	Educational Level	Gender	Age	Height (cm)	Weight (kg)
Valid (N)	31	31	30	33	33
Out-of-Range	1	2	0	0	0
User-Missing	1	0	1	0	0
System-Missing	0	0	2	0	0

Missing Data Total for the Dataset. Figure 3.22 reports two (1.2%) user-missing values in the entire dataset and two system-missing, for a total of four (2.4%) missing values.

Missing Data for Each Case. Due to the small size of the dataset (33 cases) and the small number of missing values (four), manually count the number of missing values for each case. Using that method, it is evident that of the 33 study participants, four (12.1%) failed to respond to one of the two variables with missing values. The remainder responded to all items and had no missing values. With a larger dataset and more numerous missing data, it is often more efficient to evaluate the number of missing values in a dataset by using the **Recode into Different Variables** procedure described in Section 3.4.1.7.5 as well as the **Compute** procedure.

To specify the recode procedure in this instance, create a dummy variable for the two variables with missing values (M**iss_Edlvl** and M**iss_Age**). To do that, click on **Transform** in SPSS Data editor to reveal the pull-down menu, then click on **Recode into Different Variables** to open the associated dialog menu (Figure 3.23).

In the dialog menu, create the dummy variable M**iss_Edlvl** by highlighting the variable, **Edlvl_F**, in the left panel and clicking on the arrow to the right of the panel to place **Edlvl_F** in the **Numeric Variable → Output Variable** panel in the middle. Next, enter the output variable name (**Miss_Edlvl**) in the box below **Name** in the **Output Variable** panel on the right, and, optionally, enter a variable label (e.g., **Missing Ed. Level Values**) in the box below **Label**. Click on the pushbutton **Change** to complete the specification. Repeat the procedure for the second variable, **Age**. In this case, name the output variable **Miss_Age** and the variable label, **Missing Age Values**.

The next step in the recode process is to recode each system- or user-missing value (99, or 999) to the value of 1 and each nonmissing value to 0. To do that, click on **Old and New Values** below the middle panel to open the **Recode into Different Variables Old and New Values** dialog menu window (Figure 3.24).

In the **Recode into Different Variables: Old and New Values** dialog window, under **Old Value** on the left, select the button next to the **System- or user-missing** option. On the right, enter the number "1" in the box below **New Value**. Next, click on **Add** next to the panel below **Old →** to add the recode specification. To complete the recode, click on **All other values** on the bottom left of the dialog menu. Enter "0" in the box below **New Value** on the right, and click on **Add** to move the second specification to the **Old → New** panel.

Chapter 3: Data Preparation: Cleansing and Screening 81

Figure 3.23 Recode into Different Variables Dialog Menu

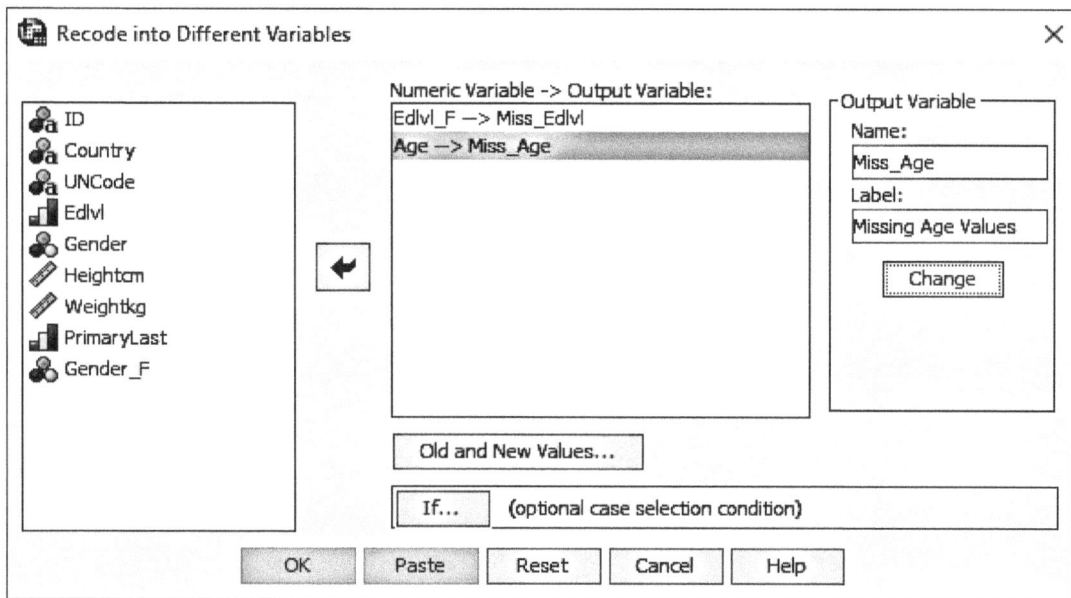

Figure 3.24 Recode into Different Variables Old and New Values Dialog Menu

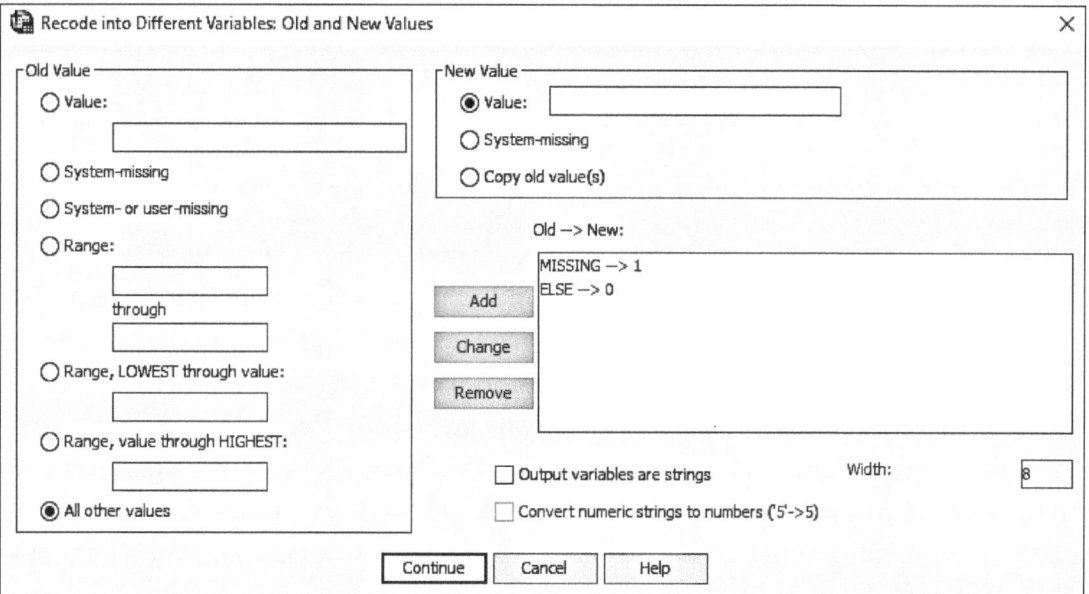

Figure 3.24 displays the completed specification. Click on **Continue** to return to the **Recode into Different Variables** menu, and click on **OK** to run the procedure that will add two new variables to the dataset: **Miss_Edlvl** and **Miss_Age**.

To compute the total number of variables with missing values for each study participant, create a new variable (e.g., **Miss_Total**) whose values will be the sum of the values of the dummy variables for each record. The **Compute Variable** procedure facilitates that exercise.

To use the **Compute Variable** procedure, select **Transform** from the top of the **Data Editor** worksheet. Select **Compute Variable** from the pull-down menu to open the **Compute Variable** dialog menu (Figure 3.25). There, enter **Miss_Total** in the box below **Target Variable** on the left. Optionally, click on **Type and Label...** to enter a label for the variable **Miss_Total**.

Figure 3.25 Compute Variable Procedure Dialog Menu Window

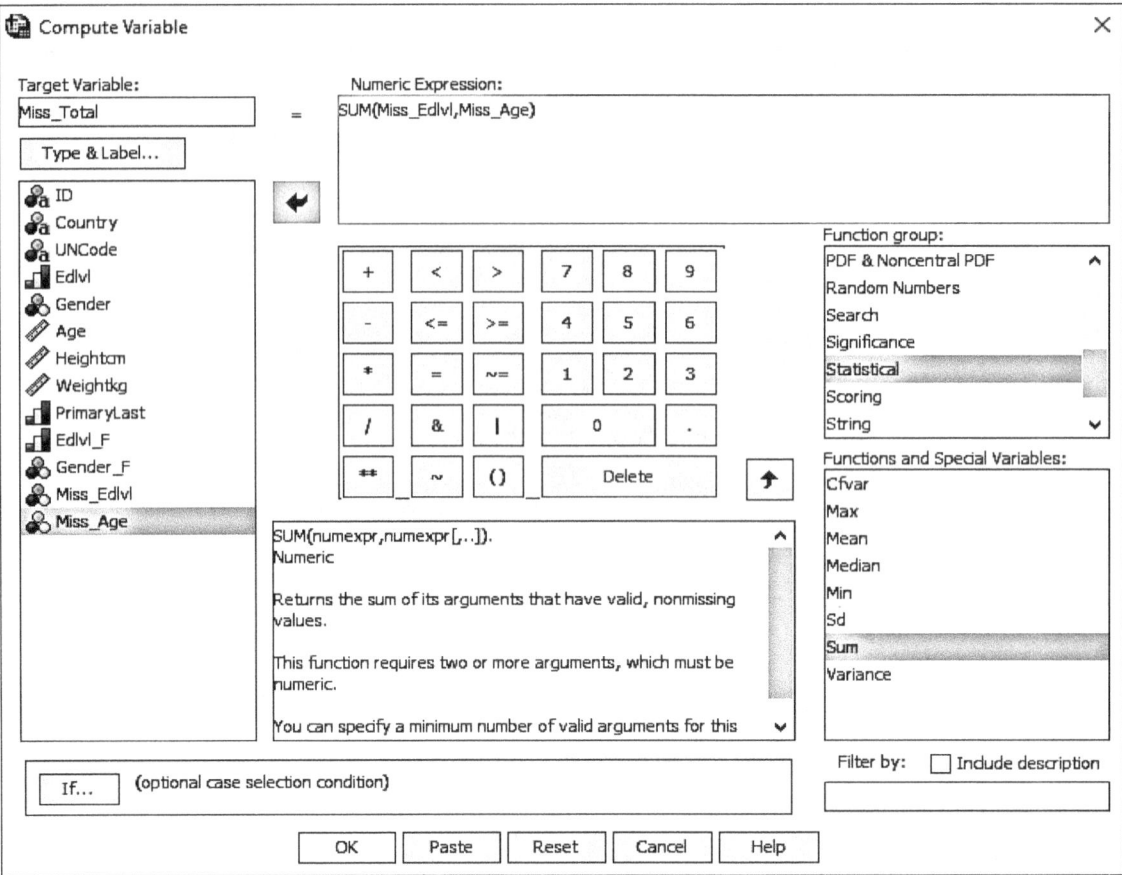

At the top right, under **Function Group**, scroll down and highlight **Statistical**, which will activate options under **Functions and Special Variables** in the panel below **Function Group**. Select **Sum** from the **Functions and Special Variables** panel, and click on the arrow left of the panel to move the **Sum** function (**SUM(?, ?)**) to the **Numeric Expression** panel at the top of the menu. Next, highlight **Miss_Edlvl** on the left, and double click it to replace the first question mark in the **Numeric Expression** box. Highlight the second question mark, then double click **Miss_Age** on the left to replace the second question mark in the **Numeric Expression** panel. Click **OK** to run the procedure.

To ascertain the number of participants with no missing values, versus one or more missing values, manually count the number of missing values in the **Miss_Total** column, or generate a frequency table of missing value counts. Figure 3.26 displays the frequency table for the number of missing values for age and educational level, which reports a total of four missing values, the same total reported in Figure 3.22.

Figure 3.26 Frequency Table of Missing Values

		Total Missing Values			
		Frequency	Percent	Valid Percent	Cumulative Percent
Valid	0	29	87.9	87.9	87.9
	1	4	12.1	12.1	100.0
	Total	33	100.0	100.0	

3.4.2 Advanced Data Screening

After generally cleansing a dataset of common errors and becoming acquainted with the data (e.g., the type, frequency, and range of data values), the next phase of preparing a dataset is to examine whether there are unusual data patterns that may affect the interpretability of data analysis results. Common methods of advanced data screening include checking for:

- outliers (extreme or unusual values) on a single or multiple variables
- violations of the data assumptions of statistical techniques (e.g., normality, linearity, homogeneity of variance, independence of observations)

3.4.2.1 Outliers

Outliers are cases in a dataset with extreme or unusual data values of quantitative variables. A univariate outlier is a case with a very high or very low value on a single variable. For example, the distances for the Olympic triathlon swimming, biking, and running courses are 1.5km, 40km and 10km, respectively. If the average biking time for the Olympic triathlon male athlete is 1 hour, 17 minutes, and the slowest athlete's time is 3 hours, the slowest athlete is an outlier in the set of biking times. While univariate outliers are restricted to a single variable, a multivariate outlier is a case with a combination of atypical values on two or more variables. For example, a 15-year-old patient with a low-normal weight of 45 kilograms, but an unusually high blood pressure reading of 160/90 mm Hg is likely to be a multivariate outlier.

Outliers send mixed messages. They could signal new, unexpected and exciting trends in a dataset, but they frequently reflect anomalies in a dataset that must be addressed before conducting central analyses. If outliers are not handled, they could distort the results of data analyses due to the sensitivity of commonly-used statistics (e.g., mean and standard deviation, and the F statistic) to outlier values. Inclusion of extreme values violates the assumptions associated with common parametric statistical procedures (e.g., analysis of variance, linear regression) leading to Type 1 or Type 2 errors.

Tabachnick and Fidell (2018) offer four reasons for the presence of outliers in a dataset:

- data errors due to incorrect coding or incorrect entry. Comprehensive data screening is one way of identifying these errors.
- failure to specify missing value codes as such in data analyses. This results in missing value codes that are analyzed as if they are real data.
- inclusion of a subject that is not from the target population. In these situations, the outlier case may be deleted.
- a case that is from the intended target population, but with variable values that have more atypical values than a normal distribution of scores. For example, a student may perform poorly in an exam due to illness. In situations with legitimate, though unusual data points, researchers may retain the case but transform values to reduce their influence on the outcome of the study.

Commonly-used methods for detecting unusual values fall into two broad categories:

- visualization or graphical methods
- statistical methods

3.4.2.1.1 Graphical Methods for Detecting Outliers

Applied statistics researchers frequently rely on histograms, boxplots (box-and-whisker plots) and scatterplots to uncover outliers in their datasets. All graphical methods provide an at-a-glance approach to identifying outliers. Isolated data points immediately pinpoint outlier values.

Sorting and Visual Inspections. A range of rules and statistical strategies exists that reliably detects outlier values in a dataset. In a small dataset, a quick and easy method is to sort data values for each variable and then visually inspect them for unusually high or low values, as illustrated in Figure 3.27. Although this method does not provide the degree of the extremity of the values deemed unusual, it quickly identifies the high and low values that are outlier candidates. In a large dataset, however, this task becomes less manageable.

Figure 3.27 Selected Heights and Weights of Caribbean College Students (Metric Measurement)

Height	185	139	142	145	148	150	155	158	158	160	163	164	166	167	168	170	(203)
Weight	(111)	40	43	47	50	55	55	60	60	60	65	58	68	54	58	70	79

Histograms. Typically, histograms are used when datasets include a large pool of quantitative values of continuous variables (e.g., test scores, blood pressure or heights and weights). In histograms, rectangular bars represent the frequency distribution of data values. Values for the variables are placed on the x-axis and the frequency of the values appear on the y-axis. The height of each bar is a graphic representation of the frequency of each value in the dataset. Isolated bars signify possible outliers.

Boxplots. Mathematician, John Tukey (1977) introduced the boxplot (also called a box and whiskers plot) as a compact way of displaying a summary of the distribution of scores in exploratory data analysis. This graphic summary helps researchers to identify outliers as well as other important characteristics (e.g., skewness and spread) of values in a dataset. To assist in the interpretation of information in this graphic, it is helpful to have a basic understanding of the fundamentals of a boxplot. For an in-depth description and explanation of boxplots, consult Cohen (1996).

The general form of Tukey's boxplot, typically displayed vertically in the research literature, is presented horizontally in Figure 3.28 for ease of understanding. The boxplot is used to display data based on a five-number summary that includes the sample median, the first (Q1) and third quartiles (Q3) of the dataset as well as the minimum and maximum values. Other critical elements of the composition of the boxplot are the interquartile range as well as the values representing the lower and upper whiskers of the dataset.

The **median** (Q2) is the middle value in the ordered set of values in the dataset denoted by the vertical line in the middle of the box in Figure 3.28. Also called the second quartile or the 50th percentile, it divides the box into two parts. The median is not always at the center of the box. A higher or lower location of the median line in the box denotes a negative or positive skew of the distribution of scores in the dataset.

The **first quartile** (Q1) is the left border of the box, which is also known as the lower quartile or the 25th percentile. Twenty-five percent of all values fall below this point in the dataset and 75% are above it. The Q1 data point is the middle score of the lower half of the dataset.

Figure 3.28 Vertical Display of the General Form of the Boxplot.

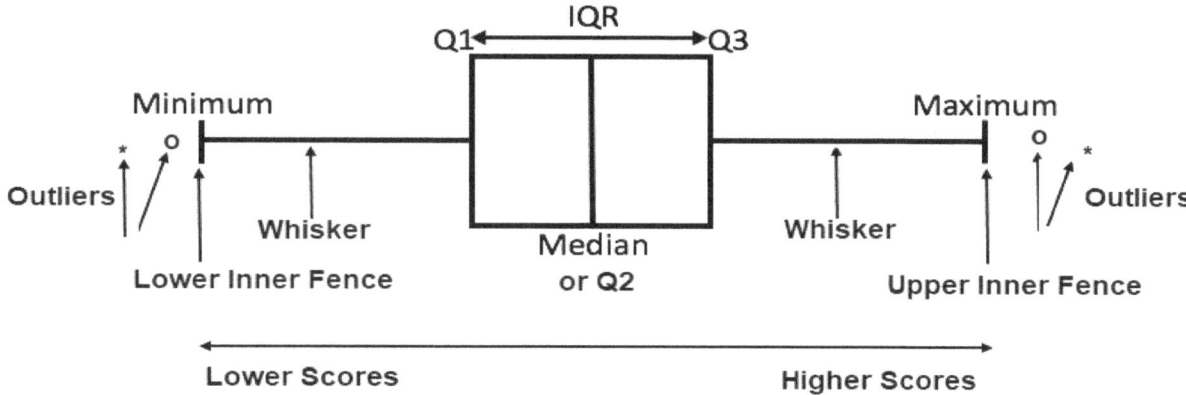

The **third quartile** (Q3) in Figure 3.28 is the right border of the box. It represents the 75th percentile, and is also known as the upper quartile. Seventy-five percent of all data points fall below this point and 25% are above it. The Q3 data point is the middle score of the upper half of the dataset.

The **interquartile range** (IQR) is the distance between the borders of the first and third quartiles (Q3 – Q1) and represents the middle 50% of the set of data values. This area captures the majority of the data points in a set.

The **lower whisker** is represented by the horizontal line that extends from the Q1 border to the lower inner fence. It is -1.5 of the IQR below the first quartile. All scores that fall within this range are not considered outliers. For example, consider a data collection that comprises the heights of 33 college students where the data point for Q1 is 158 centimeters and the corresponding data point for Q3 is 185.42. The IQR is Q3 – Q1 = 185.42 – 158 = 27.42. Subsequently, the lower inner fence comprises an area below Q1 that is 1.5 of the IQR or 1.5*27.42 = 41.13. To find the endpoint location of the lower whisker, subtract 41.13 from Q1, which is 158 – 41.13 or the lowest value closest to 116.87. Data points beyond the lower whisker are outliers. In SPSS, stars in the boxplot symbolize extreme data points that exceed -3.0 of the IQR.

The **minimum** is the lowest value in the dataset within the lower inner fence, excluding outliers. This is sometimes calculated as the lowest value between Q1 and the lower whiskers.

The **upper whisker** is the horizontal line that extends from the border that depicts Q3 to the upper inner fence is 1.5 of the IQR above the third quartile. Scores falling between the endpoints of the upper inner fence and the IQR are not outliers. However, higher scores beyond this point are outliers. For example, using the heights example, the endpoint location of the upper whisker is Q3 + 41.13 or 185.42 + 41.13 = 226.55. Values beyond this point are also extreme outliers. These values surpass 3 times the value of the IQR.

The **maximum** is the highest value in a dataset, excluding outliers or the highest value between Quartile 3 and the Upper whiskers.

Like histograms, boxplots may be used to spot outlier values:
- in a single boxplot for a single variable with a single group of cases, or conversely
- boxplots for multiple variables for a single group, or
- multiple boxplots for a single variable with several groups of cases.

Scatterplots. So far, the discourse on methods of detecting outliers has been limited to univariate data; that is, outliers in a single set of scores for a single variable. However, outliers may be found in

86 Chapter 3: Data Preparation: Cleansing and Screening

bivariate (two-variable) and multivariate (more than two variables) data as well. Scatterplots are a useful tool to detect outliers in these situations.

Consider the variables, mathematics scores and social studies scores. A histogram may identify cases of outlier scores for mathematics. However, it becomes challenging to use a histogram to identify all outliers when there is corresponding information about social studies (Figure 3.29).

A scatterplot of mathematics and social studies scores (a bivariate analysis) is one way of identifying additional outliers, like the data point circled in Figure 3.29, as well as corroborating existing evidence from a univariate analysis, like a histogram.

The rules for identifying outliers in a scatterplot are similar to those for other graphical approaches. In a scatterplot, each record is represented as a point on the x and y axes. Most cases fall within a swarm of data points. However, isolated data points that stand apart from the swarm of data values, like the data point circled in Figure 3.29, are likely to be outliers.

Figure 3.29 Comparative Methods of Identifying Outliers

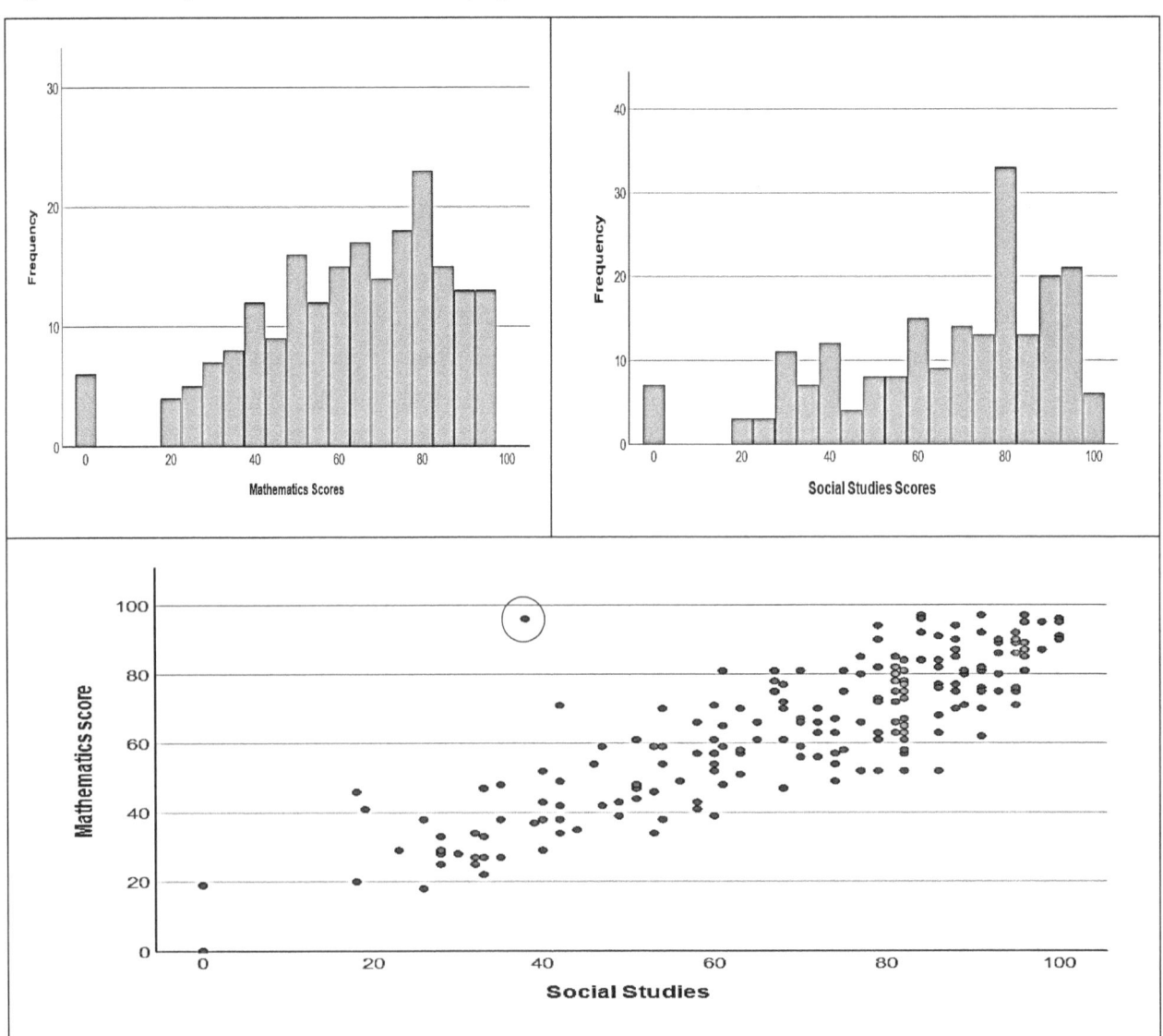

3.4.2.1.2 SPSS Specifications for Detecting Outliers

Histograms. To create a histogram, locate and open **Carib_Height_Weight_2022.sav** in IBM SPSS by double-clicking on the data file. Select **Graphs** to open the **Graphs** pull-down menu in the **Data View** screen of the **SPSS Data Editor** (Figure 3.30). Two dialog menus are available for building histograms:
- the legacy dialog menu designed for histograms, and
- Chart Builder, which consolidates functions, including histograms, previously available through stand-alone legacy dialogs.

The **Chart Builder** procedure may be used to create a single, multiple, or grouped histograms (e.g., by gender, academic discipline). This exercise will specify procedures to build a single simple histogram.

Click **Chart Builder** on the drop-down menu (Figure 3.30a) which opens the **Chart Builder** warning (Figure 3.30b). After confirming that variables have been assigned appropriate measurement levels, press **OK** to define the histogram.

Figure 3.30 IBM SPSS Main Menu and Chart Builder Reminder

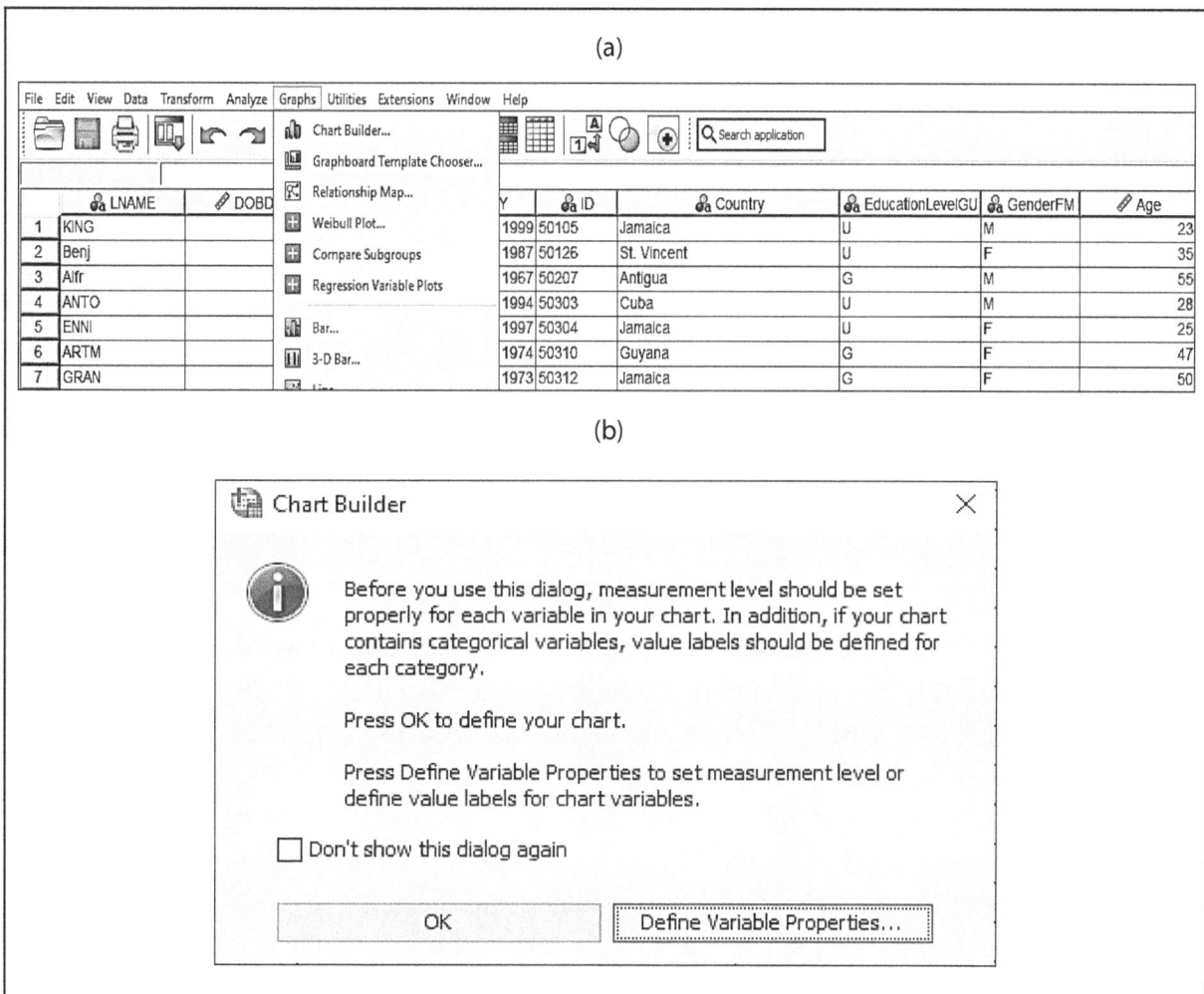

In the **Chart Builder** dialog menu (Figure 3.31), click on the **Gallery** tab in the bottom half of the dialog menu. Click on **Histogram** below the **Choose from:** panel to reveal four types of histograms as displayed in Figure 3.31. Click on the first histogram (Simple Histogram) next to the **Choose From:**

panel. Drag and drop the histogram selection in the Chart Preview panel and a crude chart will display based on SPSS example data.

Next, highlight the variable, **Weightkg**, in the **Variables** panel in the upper left portion of the dialog window. The **Weightkg** values will be used to build the bars in the histogram. Drag and drop the variable name onto **X-axis** in the **Chart Preview** panel. A crude chart will appear that is based on SPSS sample data and the **Element Properties** window on the right will open.

Figure 3.31 IBM SPSS Statistics Chart Builder Dialog Menu

Select **Bar 1** from the **Element Properties** panel below **Edit Properties of**. Under **Statistics**, confirm that **Weight (kg)** is next to **Variable:** denoting the variable to be used in the analysis and that **Histogram** appears below **Statistic**, signifying the statistical procedure to be used.

Select **X-Axis 1 (Bar 1),** confirm the label, **Weight(kg),** in the **Axis Label** field, and allow SPSS to select the scale ranges by retaining the checked boxes below **Automatic**. Customization of the scale range is available by unchecking one or more of the boxes under **Automatic** (**Minimum**, **Maximum**, **Major Increment** and **Origin**) and inserting the desired replacements.

Highlight the **Y-Axis 1 (Bar1)**; confirm the label for **Axis Label** (**Frequency**), and leave the **Scale Range** specifications at automatic. Scroll down to select **Title1.** Replace the SPSS title by selecting **None**. A title may be added later. Click **OK** to run the analysis.

The completed histogram is presented in Figure 3.32. The histogram displays a non-normal, highly skewed and kurtotic distribution of weights of college students. Most important, the histogram highlights bars representing data points that are isolated from the mass of data values. In Figure 3.32, the last rectangular bar displaying a high weight of 320 kilograms at the far right of the histogram is a likely outlier data point that should be further investigated as a data entry or data collection error or some other contributing factor to the atypical value.

Figure 3.32 Frequency Histogram of Weights of Selected Caribbean College Students

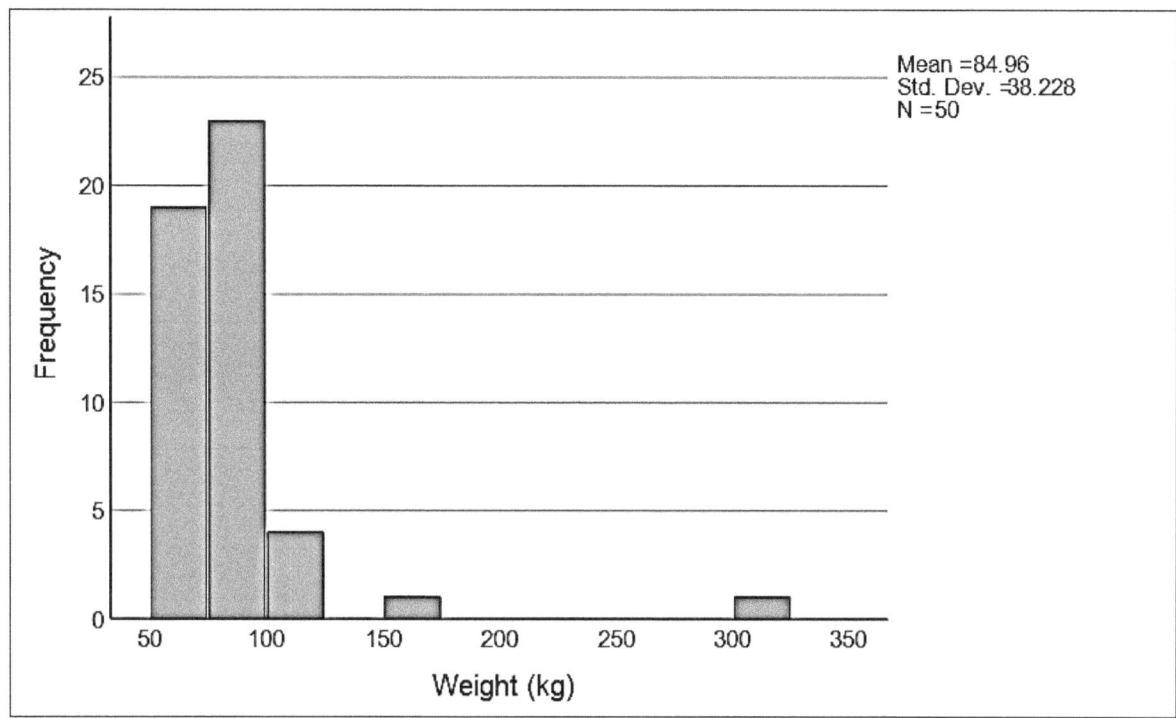

Boxplots. To create a boxplot to detect outliers, open **Carib_Height_Weight_2022.sav**. Both **Chart Builder** and the **Legacy** dialog menu may be used to create a boxplot. For this exercise, use the legacy dialog menu by selecting **Boxplot** from the **Graphs** drop-down menu (Figure 3.33) in **SPSS Statistics Data Editor**.

Figure 3.33 SPSS Statistics Main Menu

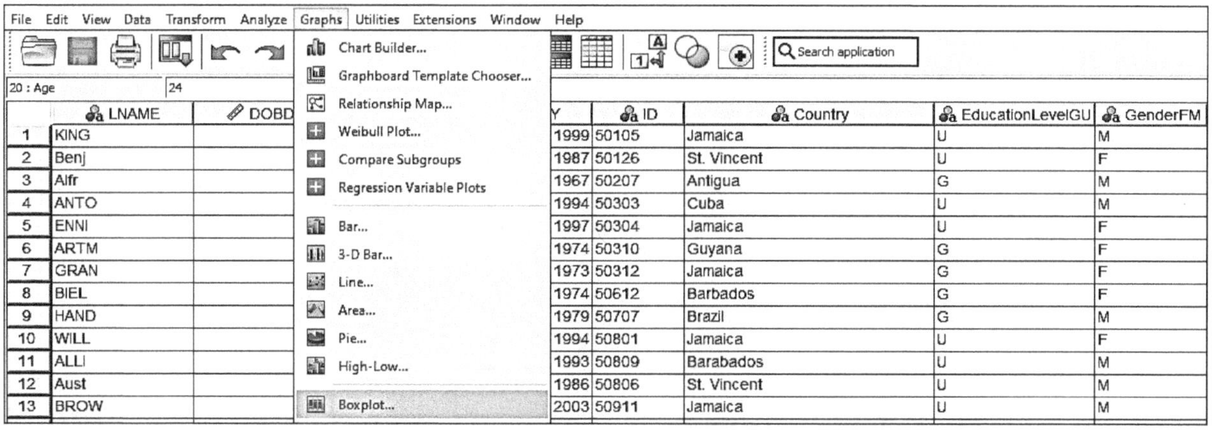

In the **Boxplot** dialog window, select the **Boxplot** icon labeled **Simple** (Figure 3.34a) and **Summaries of Separate Variables** under **Data in Chart Are**. Click **Define** to open the **Define Simple Boxplot:**

90 Chapter 3: Data Preparation: Cleansing and Screening

Summaries of Separate Variables dialog window. Click on the variable, **Weightkg,** on the left, then move it to the **Boxes Represent** panel by clicking on the arrow next to it (Figure 3.34b). Click **OK** to run the analysis.

Figure 3.34 SPSS Statistics Chart Builder Dialog Window

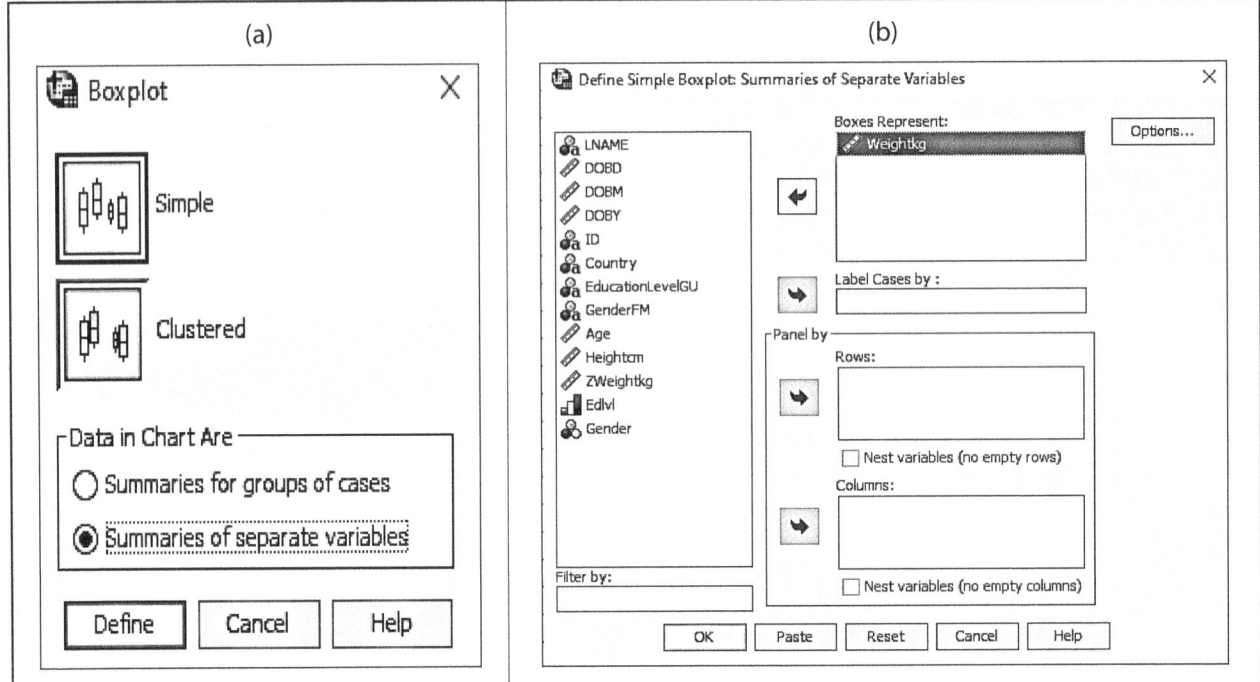

The Boxplot chart is displayed in Figure 3.35. As the chart confirms, the distribution of the variable values is very compact and bunched together in the lower ranges of the distribution, much like the distribution presented in the histogram in Figure 3.32.

Figure 3.35 Boxplot of the Weight of Selection of Caribbean College Students

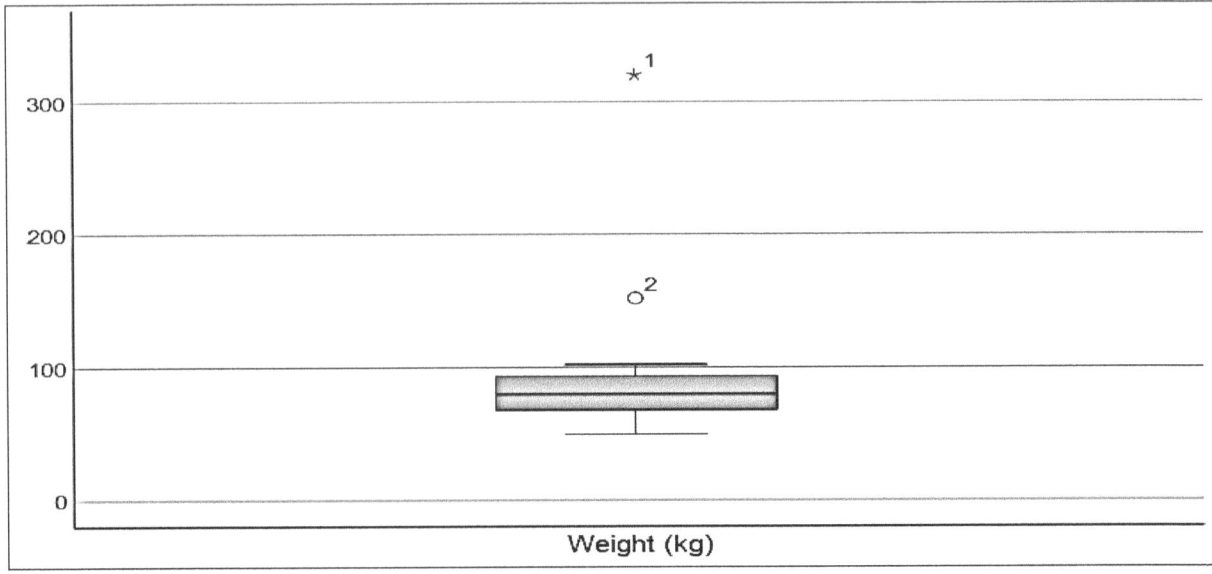

Two exceptions are beyond the inner fences. One weight value with SPSS label, o, is designated an outlier—but not an extreme one— in the region between 1.5 and 3.0 IQR. Of greater note is the highest weight in the dataset, and an extreme outlier (*), located beyond +3.0 IQR.

Scatterplots. To create a scatterplot to identify potential outliers, locate and open the dataset, **Carib_Height_Weight_2022.sav**. In the **Data Editor** window, select **Chart Builder** from the **Graphs** main menu to open the corresponding dialog window (Figure 3.36).

Figure 3.36 SPSS Statistics Chart Builder for Scatterplots

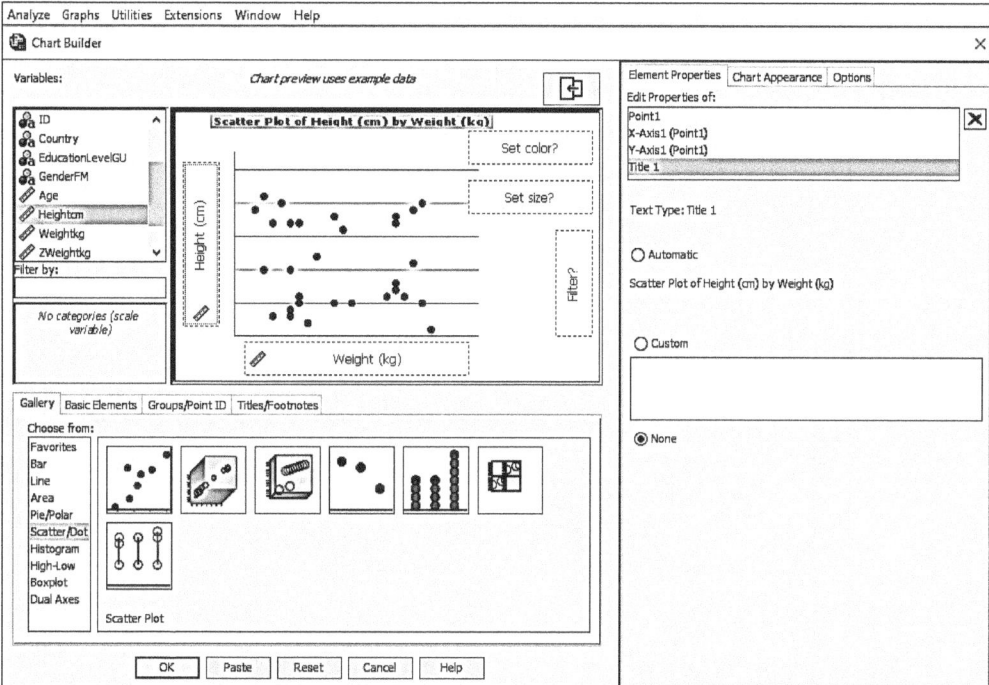

Select **Scatter/Dot** from the **Gallery** in the bottom half of the window in the **Choose From**: panel to reveal **Scatter/Dot** options. Select the first icon, then drag and drop it in the **Chart Preview** panel. From the **Variables** list in the upper left of the window, drag and drop the variable, **Heightcm** on the **Y-axis**. Then drag and drop **Weightkg** on the **X-axis**. In the **Element Properties** panel on the right, scroll down to **Edit Properties of: Title 1**. Click on the option **None** to exclude a title from the scatterplot. Click **OK** to run the analysis.

Mild and extreme outliers noted previously in the histogram and boxplots are also evident in the scatterplot of height and weight (Figure 3.37). The strong influence of extreme outliers extends beyond univariate statistical analyses to bivariate and multivariate situations. Outlier values have greater impact on correlation and regression coefficients than data values within the swarm of data points.

In multivariate research studies, an initial stage in detecting outliers is to generate a scatterplot matrix, which is a grid of bivariate scatterplots with every possible combination of the variables of interest. **SPSS Chart Builder** or **Legacy Scatterplots** may be used to create scatterplot matrices. An example of a scatterplot matrix is presented in Figure 3.38, where every possible bivariate combination of three properties of a sample of roller coasters (**height**, **length,** and **speed**) produces scatterplots for each grouping.

Figure 3.37 Scatterplot of Height and Weight of Selection of Caribbean College Students

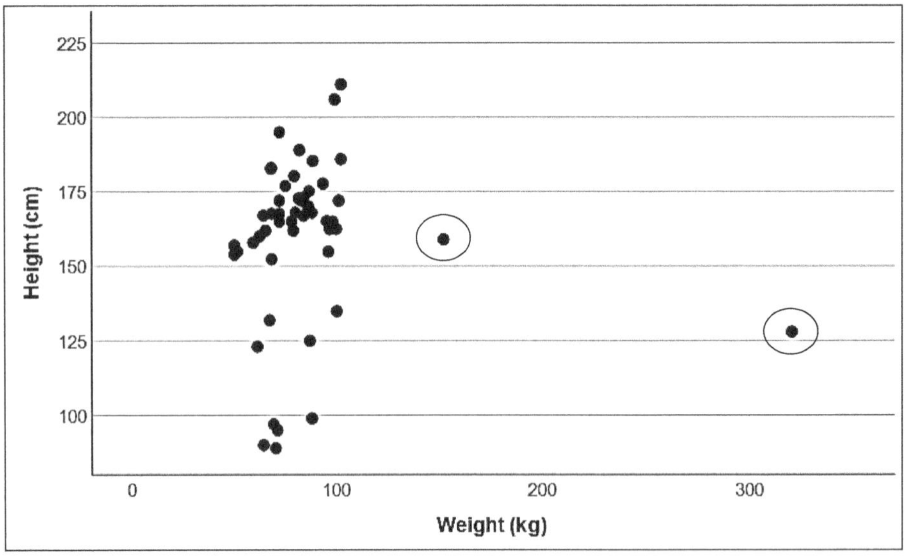

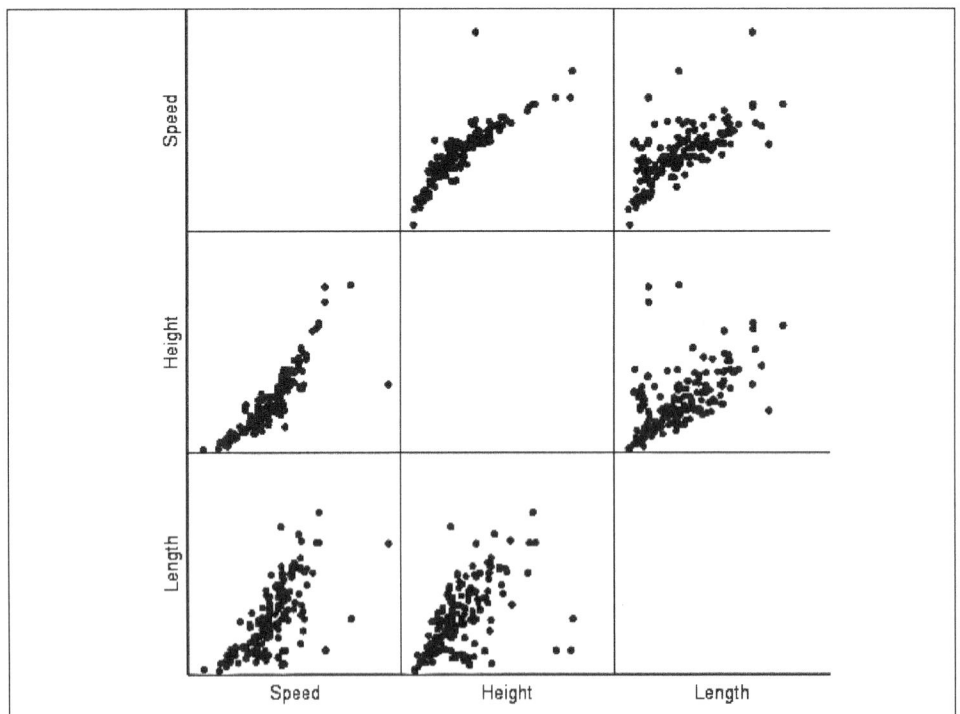

Figure 3.38 Scatterplot Matrix of Three Properties of a Roller Coaster*

*Height and length measured in feet. Speed measured in mph. Data retrieved from https://dasl.datadescription.com/datafile/coasters-2015/?_sfm_methods=Multiple+Regression&_sfm_cases=4+59943,.

Each image is a scatterplot of two variables, and the three plots on either side of the empty cells of the matrix are mirror images of each other. For example, the first plot on the first row depicts a plot of the **speed** and **height** of a selection of roller coasters. The mirror image of that display is the first image in the second row.

The scatterplots depict differing directions and strengths of the relationships among the three variables (or properties of roller coasters). The relationship between the **speed** and **height** of roller coasters appears to be somewhat curvilinear—as **height** increases **speed** tapers off. However, the relationship between **length** and **height** appears to be more linear. As we saw in the scatterplot of **height(cm)** and **weight(kg)**, there are some outlying cases in the plots (e.g., between **speed** and **height** as well as **speed** and **length**).

3.4.2.1.3 Statistical Methods for Identifying Outliers

While graphical methods of identifying outliers provide visual evidence of outlier data, quantitative statistics are also a useful tool in revealing anomalies in a dataset. Among the devices available to applied researchers are: descriptive statistics like frequency tables, z scores, and Mahalanobis distance.

Descriptive Statistics. Frequency tables are possibly the most prevalent method of identifying outliers. Researchers can readily identify univariate outliers among categorical variables by inspecting the frequency distribution of variable values. For example, extremely uneven splits in the values (e.g., 90% versus 10%) of dichotomous variables are very likely to be due to univariate outliers.

Standard scores also known as z scores may be used to detect univariate outlier scores among continuous variables. To use this method, methodological researchers (e.g., Hair et al., 2010; Tabachnick & Fidell, 2018) suggest converting scores to z scores, with a mean of 0 and a standard deviation of 1. Hair and others (2010) suggest that z scores exceeding +/- 2.5 should be regarded as potential outliers. Tabachnick and Fidell (2018) suggest a standard of z scores exceeding +/-3.29, but add that with a very large sample size, a few outliers may be expected.

Mahalanobis Distance and Cook's D. Mahalanobis distance and Cook's D are frequently-used indices that detect bivariate and multivariate outliers, cases with unusual combinations of values for two or more independent, continuous variables (e.g., a woman whose height is 150 cm and whose weight is 110 kg). An important assumption of multivariate statistical techniques, like multiple regression, is that there are no multivariate outliers. Mahalanobis distance and Cook's D are used to identify multivariate outliers through the multiple regression procedure of many statistical software programs, including SPSS. The statistics may be added to the dataset for each case for further review and assessment.

Mahalanobis distance is a measure of the multivariate "distance" between a subject's values on specific independent variables and the centroid for all other records. The centroid is a point in a multivariate space where the means of all variables intersect. The larger the multivariate distance, the farther away the data point is from the centroid. Using the Mahalanobis distance statistic (D^2), a large D^2 points to the existence of extreme values on one or more of the variables of interest. A rigorous alpha level of $p < 0.001$ is typical when evaluating Mahalanobis distance. Cases with D^2 that exceeds the critical value associated with the alpha level are potential multivariate outliers.

Cook's distance (also known as Cook's D or D_i) is used in regression analysis to find influential outlier cases in a set of predictor variables. Cook's distance is an estimate of how much a regression model changes when a record is removed from analysis. The higher the Cook's D, the more influential the associated record. According to Tabachnick and Fidell (2018), cases with influence scores greater than 1 should be investigated as potential multivariate outliers. The minimum value of D is zero. Procedures for using Mahalanobis distance and Cook's D to identify outliers may be found in Chapter 5. For a more detailed treatment of Mahalanobis and Cook's distance, including how to use SPSS to conduct these procedures, consult Chapter 4.

3.4.2.1.4 SPSS Specifications to Identify Outliers Using Statistical Methods

SPSS Descriptives or **SPSS Explore** may be used to calculate z scores and save them in a file for review. To calculate z scores using **SPSS Descriptives**, open the dataset **Carib_Height_Weight_2022.sav**. Sort the values of **Weightkg** from highest to lowest value. From the **SPSS Data Editor** main menu, select **Analyze → Descriptive Statistics → Descriptives** to reach the **Descriptives** window (Figure 3.39).

Highlight **Weightkg** in the left panel, and click it over to the variables panel on the right. Check the **Save standardized values as variables** option at the bottom of the window. Then click **OK** to run the procedure.

Figure 3.39 SPSS Descriptives Dialog Window

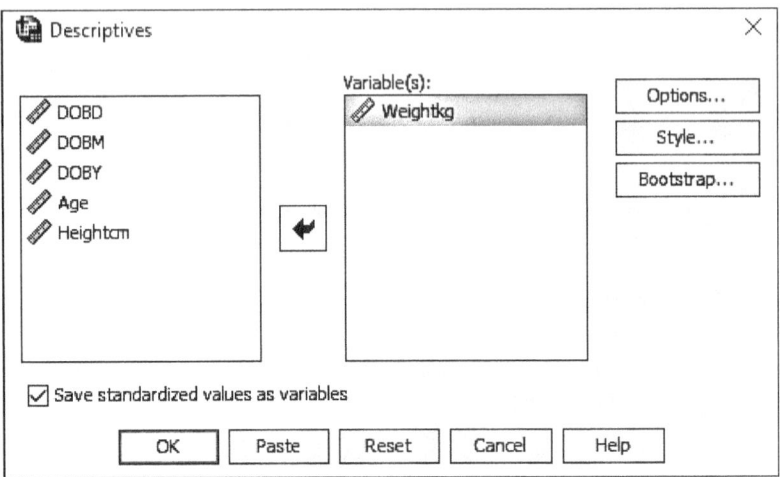

Return to the Data View window and scroll across to the last column to find the z scores for **Weightkg** that were added to the dataset (Figure 3.40). The name of the new variable is **ZWeightkg**. In addition, as a result of sorting the values for **Weightkg** in descending order, the highest weights are at the top of the column and the lowest at the bottom of the column.

Figure 3.40 Z Scores for Weightkg Added to Dataset

	ID	Country	EdIvl	Age	Gender	Heightcm	Weightkg	ZWeightkg
1	5	Jamaica	1	23	0	183	68	-.44261
2	6	St. Vincent	1	35	1	152	68	-.44261
3	7	Antigua	2	55	0	180	79	-.14597
4	3	Cuba	1	28	0	172	101	.41962
5	4	Jamaica	1	25	1	157	50	-.91723
6	0	Guyana	2	47	1	165	98	.34114
7	2	Jamaica	2	50	1	168	88	.07955
8	2	Barbados	2	47	1	168	80	-.12972
9	7	Brazil	2	43	0	135	100	.39346
10	1	Jamaica	1	28	1	163	97	.30491
11	9	Barabados	1	28	0	172	72	-.33899
12	6	St. Vincent	1	36	0	173	84	-.02732
13	1	Jamaica	1	18	0	155	51	-.88164
14	6	Saint Lucia	1	30	0	123	61	-.62674
15	5	Jamaica	2	27	1	165	72	-.33899
16	5	Jamaica	1	22	0	185	88	.09133

Bear in mind that z scores with a minus sign represent observations that are *below* the mean for the sample of values, while *positive z* scores represent observations *above* the mean. The weight for the first record is 6.14842 standard deviations above the sample mean. The weight for the second record is 1.75372 standard deviations above the sample mean. Using the two suggested standards indicating potential outliers (+/- 2.75 and +/- 3.29), we find that the first record exceeds the standard of +/- 3.29 and is therefore a likely extreme outlier.

3.4.2.1.5 Handling Outliers

Researchers have choices for handling outlier data points once they have been detected. Here are some common methods of dealing with outlier values.

- Do nothing. Conduct the analysis with the outlier included if it is representative of the target population.
- If outliers are the results of measurement, data collection, or data entry errors, replace the erroneous information, if possible; otherwise, remove the relevant outliers.
- Use statistical estimation procedures that are robust to the inclusion of outliers (e.g., nonparametric statistics).
- Run the analysis with and without the outlier. Then report the outcomes for both. In this way, consumers of your report have all the information available.
- If outliers accurately reflect the target population, removing them may lead to the exclusion of important information. In addition, for datasets that started out small, summarily removing outliers may lead to results that are less generalizable to the target population than results with the original dataset. Therefore, reduce the influence of extreme data values on statistical results by winsorizing or transforming variable values.

To winsorize outlier variable data values, assign them the next highest data values that are not outliers. For example, if an outlier score for a variable is 98 and the next highest score is 92, replace 98 with the score of 92. Researchers transform the values of a variable by applying a mathematical function to the original values so that the resultant data meet the assumptions (e.g., normal distribution) of the statistical procedure to be used. More on data transformations later in this chapter.

Regardless of the option selected to address outliers, research reports must document the outliers found and the guidelines used to handle the outliers. Avoid cherry-picking guidelines in order to present preferred results. Rather, guidelines for handling outliers should be in keeping with the research literature that undergirds a study.

3.4.2.2 Screening for Violations of Statistical Assumptions

Most statistical techniques used in the social sciences (and certainly in this text) are based on statistical assumptions about the data to be used. Testing whether the data meet particular statistical assumptions is a prerequisite for use of associated statistical techniques. For example, the *t*-test is widely used to reveal whether the differences between the means of two groups are significant. However, accurate interpretation of the outcome of the test relies on several data conditions. Two of those are that the data must be continuous scores and they must be normally distributed. To ensure that statistical tests perform as designed and that results may be interpreted with confidence, most statistical techniques require prior examination of data to determine whether they meet the necessary assumptions.

Fortunately, the data assumptions of many statistical analyses are quite similar. For example, a frequent assumption of most parametric tests (e.g., analysis of variance, Pearson correlation and linear regression) is that data values are normally distributed. This means that the frequency of data values of each variable involved should roughly resemble a bell-shaped curve. Assessment for violation of that data assumption is a prerequisite for many parametric statistical analyses.

Data assumptions for each statistical technique in this text and relevant methods of assessment of the statistical assumptions will be included throughout this text. However, this section will introduce methods of assessment of four of the most common data assumptions and how to handle the violations of the assumptions. The assumptions to be addressed in this section include the assumption of

- normality,
- linearity,
- homoscedasticity and homogeneity of variance, and
- independence of observations (also known as independence of errors).

SPSS procedures for assessing the assumption of independence of observations (also known as independence of errors) may be found in Chapter 4. SPSS procedures for testing the assumptions of homoscedasticity and homogeneity of variance may be found in Chapters 5 and 7, respectively. Additional assumptions for statistical techniques will be discussed as the techniques are introduced in the remainder of the text.

3.4.2.2.1 Normality

Section 3.4.1 discussed how a researcher might employ descriptive statistics to conduct basic error checks (e.g., out-of-range data, reasonableness of scores). Section 3.4.2.1 examined the use of graphical methods, including histograms, boxplots, and scatterplots in conjunction with descriptive statistics to assess the existence of outliers, which are often indicators of a lack of normal distribution in the data. This section will examine methods of assessing the assumption of normality, using some of the procedures we have already encountered, as well as additional procedures.

Properties of a Normal Distribution. In a perfectly normal distribution of scores, the shape of the distribution is symmetric so that the mean, median, and mode are all equal (Figure 3.41). In a normal distribution, scores are clustered around the mean and tail off on either side of the mean at about the same rate so that the curve of scores approach, but never touch, the horizontal axis. Scores may range from negative to positive infinity, but the mass of scores are between -/+ three standard deviations from the mean.

Normality Assumption. The assumption of normality is that research data are drawn from a normally distributed population of values. In addition, dependent variable data from the study sample drawn from the target population is also normally distributed. Furthermore, when a researcher is comparing two or more independent samples of subjects (independent variable), it is assumed that the scores (from the dependent variable) of each group is normally distributed. Non-normal distribution of data in one of the groups means that the normality assumption has been violated.

Normal distribution of scores is a vital assumption for parametric tests (e.g., independent *t* test, analysis of variance). Increasing departure from normality of scores in a single independent sample leads to the likelihood of a Type I error and subsequently degraded results. However, according to the central limit theorem, with large sample sizes, statistical techniques are robust to departures from normality.

Figure 3.41 Normal Distribution of Scores (Normal Curve)

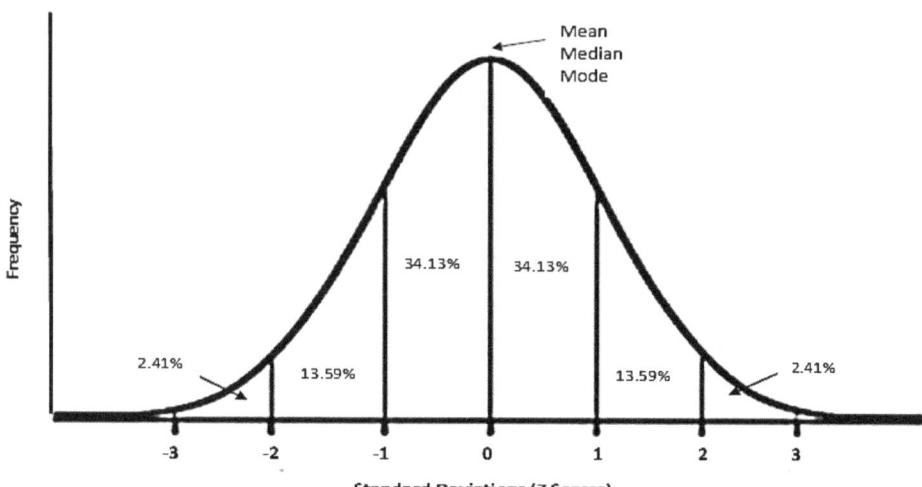

For example, Tabachnick and Fidell (2018) state that the F test is robust to violations of normality when there are at least 20 degrees of freedom in univariate ANOVA.

A variety of statistical gauges are available to detect departure from normal distribution, including the Shapiro-Wilk test and the Kolmogorov-Smirnov test. Perhaps the most frequently used statistics to assess data for normal distribution are skewness and kurtosis. These descriptive statistics measure the extent to which scores depart from the spread or height of a normal curve.

Skewness and Kurtosis. Skewness is the extent to which a set of scores departs from a symmetrical distribution. When scores are normally distributed, skewness has a value of zero. Skewed distributions of scores have skewness values markedly greater than or less than zero.

When most values in a dataset are at the lower end of the distribution of scores (a greater number of smaller values), skewness values are positive (Figure 3.42) because the scores tail off toward the right in the positive range of scores (positive skew). When most values are at the higher end of the curve (a greater number of larger values), skewness values are negative because the distribution of scores tail off to the left in the negative end of the range of scores (negative skew).

Figure 3.42 Skewed Distributions

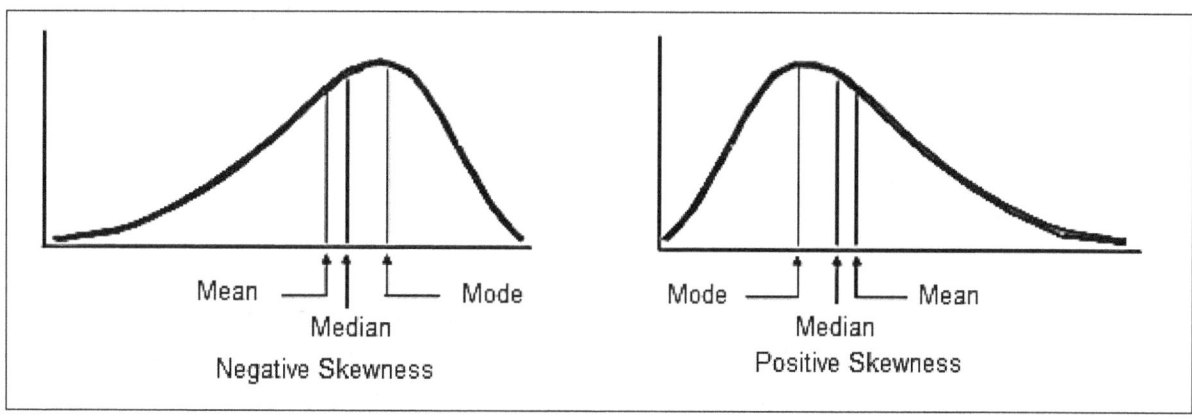

Kurtosis assesses the extent of the peak of the distribution of scores around the mean or center of the distribution. The measure describes the extent to which the distribution is either pointier (leptokurtic) or flatter (platykurtic) than the normal curve (Figure 3.43).

Figure 3.43 Kurtotic Distributions

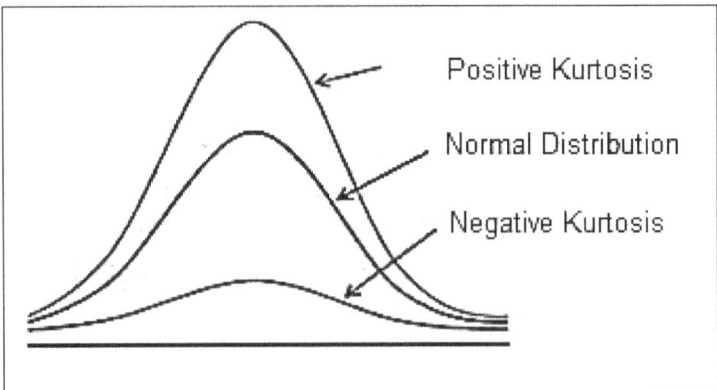

A normal distribution with a kurtosis value of zero is described as mesokurtic. In a leptokurtic distribution, most scores are in the middle of the distribution and kurtosis values are positive. In a platykurtic distribution, scores are more equally distributed across the range of the distribution, and kurtosis values are negative.

Methodology researchers continue to debate guidelines for labeling distributions as markedly skewed or kurtotic. Tabachnick and Fidell (2018) and Hair et al. (2010) suggest detecting non-normality by setting up a statistical significance test, dividing skewness or kurtosis by its standard error and comparing the result against the z distribution values that you might expect to get if the distribution were not different from zero. Findings of z scores greater than 1.96 ($p =< .05$) indicate significant departure from normality in small to moderate samples. In large sample sizes (e.g., greater than 200 cases for ungrouped analysis), the standard of z scores greater than 3.29 ($p =< 0.001$) may be substituted. SPSS's documentation for SPSS Statistics version 25 suggests that kurtosis or skewness greater than twice its standard error may be used as evidence of a significant departure from normality. Guidelines provided by George et al. (2019) suggest that skewness or kurtosis values between +/- 1.00 are excellent for most psychometric purposes and that values between +/- 2.00 may also be acceptable, depending on the statistical application.

Kolmogorov-Smirnov and Shapiro-Wilk. Two significance tests that are commonly used to test normality are the Kolmogorov-Smirnov (K-S) and Shapiro-Wilk (S-W) tests. Both tests are widely available in statistical applications.

K-S and S-W are goodness-of-fit nonparametric significance tests that compare empirical data (scores) against a known theoretical distribution of scores. The tests assume that score distributions are normal (null hypothesis). Therefore, the desired outcome of either test is a test statistic that indicates a failure to reject the null hypothesis ($p > .05$). Results that are significant ($p < .05$), means a rejection of the null hypothesis due to evidence that the score distribution is significantly different from a normal distribution.

Both tests have disadvantages. The S-W test has been found to be more powerful than the K-S test (e.g., Razali et al., 2011). Both tests are sensitive to sample size, with power increasing as sample sizes increase, and neither test performs well in small samples (30 and below). Accordingly, minor departures from normality may be labeled significant when sample sizes are large, and departures from normality in small sample sizes may go undetected. In sum, conclusions about normality should take place in conjunction with other tests of normal distribution.

Graphical Measures of Normality. Graphical methods for the detection of outliers are also an important component of assessing univariate normality. Assessment of normality frequently begins with visual inspections of frequency histograms, stem-and-leaf plots, and boxplots, to roughly judge whether the shapes are reasonably close to a normal distribution.

Frequency histograms are particularly useful tools, especially when they are fitted with a normal curve (Figure 3.44). An approximate symmetrical, bell-shaped curve suggests a normal distribution of scores. If the curve is asymmetrical, with a longer tail on one side or the other, then the distribution is skewed. Figure 3.44 redisplays Figure 3.32 with a normal curve superimposed on the frequency histogram, emphasizing the positive skewness of the distribution of the variable, Weight. The skewness of the distribution is 4.976 with a standard error of .337. Therefore, the skewness index is 4.976/.337 = 14.766, well above the standard, indicating a significantly skewed distribution of scores.

Figure 3.44 Frequency Histogram of Weight with a Normal Curve Overlay

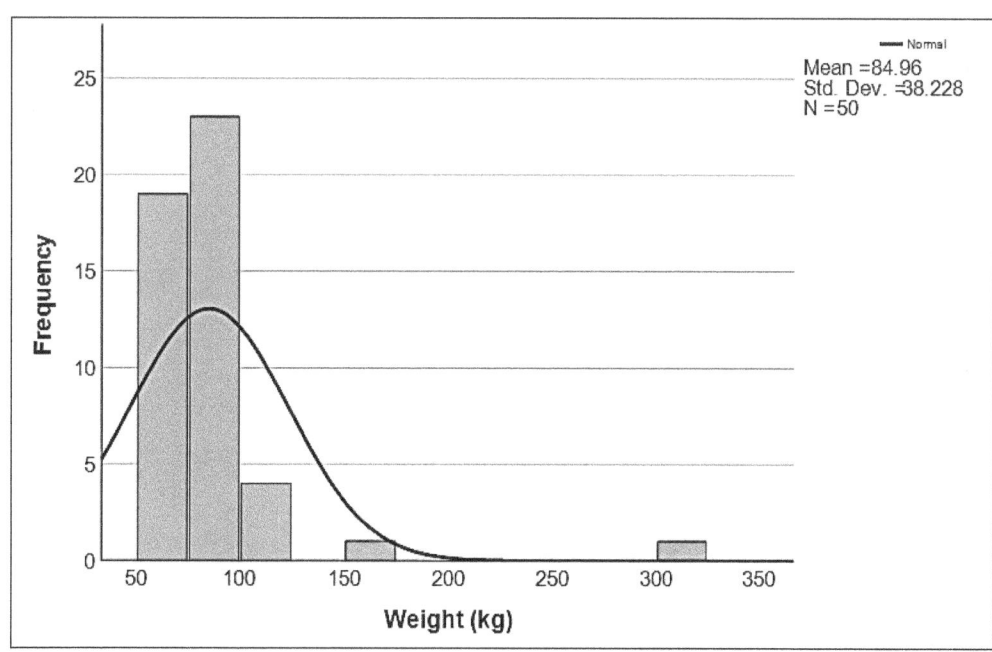

Stem-and-Leaf plots are similar to frequency histograms in that both summarize how often a score appears in a dataset, and both are graphical displays of the shape of the score distribution. However, one advantage of stem-and-leaf plots is that they preserve the actual values in the display of the shape of the distribution, and they identify extreme scores.

Figure 3.45 presents an example of a stem-and-leaf plot from a random sample of scores of a 5-item subscale of the short version of the Portrait Values Questionnaire (PVQ), included in the 2016 World Values Survey (WVS) (Englehart et al., 2014). The WVS is an international survey, conducted in more than 120 world societies every five years, of people 18 years or older. It is designed to study the social, political, economic, religious, and cultural values of people in the world and the impact they have on the social, political, and economic development of countries and societies over time.

The PVQ was developed based on Schwartz's (2006) theory that across cultures there are 10 dominant individual values. These values coalesce into two major categories: self-protective values and growth or self-expansive values. Responses to the PVQ scale are designed to present portraits of respondents' values. For example, people who avoid conflict, maintain tradition, and actively control threat are said to display self-protective values. The plot is based on responses to items representing self-protective values, collected in Haiti in Wave 6 of the WVS in 2016.

Stem-and-leaf plots encompass three components:
- the Frequency (Column 1), is the rate at which a response value occurred.
- the Stem (Column 2), is the unit digit of each data value. Each stem has a width equal to 1 unit (note at the bottom of the figure).
- the Leaf (last column), represents the decimal value of a score. Each leaf denotes two cases (note at the bottom of the figure).

Figure 3.45 Stem-and-Leaf Plot of Self-Protective Values Scores from the World Values Survey

```
Frequency      Stem   &     Leaf
15.00 Extremes (=<2.8)

 14.00          3     .     0000000
 21.00          3     .     2222222222
 33.00          3     .     4444444444444444
 56.00          3     .     6666666666666666666666666666
 47.00          3     .     888888888888888888888888
 77.00          4     .     00000000000000000000000000000000000000
 81.00          4     .     222222222222222222222222222222222222222222
 87.00          4     .     4444444444444444444444444444444444444444444
107.00          4     .     6666666666666666666666666666666666666666666666666666666&
 56.00          4     .     8888888888888888888888888888
 66.00          5     .     000000000000000000000000000000000
 36.00          5     .     222222222222222222
 28.00          5     .     44444444444444
 15.00          5     .     6666666
  7.00          5     .     888
  3.00          6     .     0

Stem width: 1.00
Each leaf: 2 case(s)
& denotes fractional leaves.
```

The stems are the integer values that combine with the leaf, or decimal values, to form the complete data values of participants' responses. For example, in the first row, the lowest score (3) has a frequency of 14. Each leaf has a value of 0, and there are 7 leaves in the first row. Since each leaf represents two cases, find the product of 7 and 2, which results in a frequency of 14 values of 3 in the first row.

The stem-and-leaf plot records a total of 15 cases (noted at the top of the figure) with extreme scores, which are values ≤ 2.8. The minimum and maximum PVQ scores are 1 and 6, respectively.

The distribution of "leaves" allows researchers to quickly assess whether the values are normally distributed, negatively or positively skewed. It is also possible to get a general impression of whether the values exhibit positive or negative kurtosis.

Boxplots are a common method used to detect outliers (Section 3.4.2.1.2), but researchers may also employ them to assess normality. The features of the segments of a boxplot reveal whether a dataset is normally distributed.

- A distribution is symmetric when the line representing the median is in the middle of the box, and the whiskers on either side are approximately the same.
- A distribution is positively skewed when the median is closer to the bottom of the box (or the left side if the boxplot is horizontal), and the whisker on the lower end (or left side) is shorter than the whisker on the upper end.
- A distribution is negatively skewed when the median is closer to the top of the box (or the right side of a horizontal boxplot), and the whisker on the upper end (or right side) is shorter than the whisker on the lower end.

Normal probability plots are frequently described as a more precise measure of normality than other graphical measures (e.g., Meyers et al., 2016; Tabachnick & Fidell, 2018). In these plots, variable values are rank-ordered and plotted against the expected normal distribution of the values. In the plot an expected normal distribution generates a straight diagonal line, and the actual values (displayed as circles or dots) are plotted as well. To assess normality, compare the plotted values against the expected values (straight line). If the data are approximately normally distributed, the plotted values will be on or close to the line. Deviations from normality shift the plotted values away from the line. Therefore, the farther the points are from the line, the farther they are from where they should be in a normal distribution.

Figure 3.46 displays the probability plot for the summative variable, Self-Protective Values, that was depicted in Figure 3.45. The plotted points are clustered close to the line, demonstrating that the values are approximately normally distributed.

3.4.2.2.2 Linearity

Section 3.4.2.1.1 discussed the use of scatterplots to detect outlier data points. Researchers also use scatterplots to test a common assumption of bivariate and multivariate statistical techniques: There is a linear relationship between pairs of variables in an analysis. When scores of two variables that are normally distributed are in compliance with the linearity assumption, the scores of the variables tend to change at the same rate. Therefore, for example, as the values of one variable decrease, the scores of the other variable increase. When the scores are plotted, the result is an oval-shaped or elliptical scatterplot.

In some situations, scatterplots reflect a mix of linear and curvilinear or completely curvilinear relationships. In a curvilinear relationship, the pattern of association does not follow a monotonic increasing or decreasing pattern. Rather the pattern changes direction at a certain point. For example, as the values of one variable gets larger, the other gets larger as well (positive correlation), up to the

Figure 3.46 Normal Probability Plot of Self-Protective Values

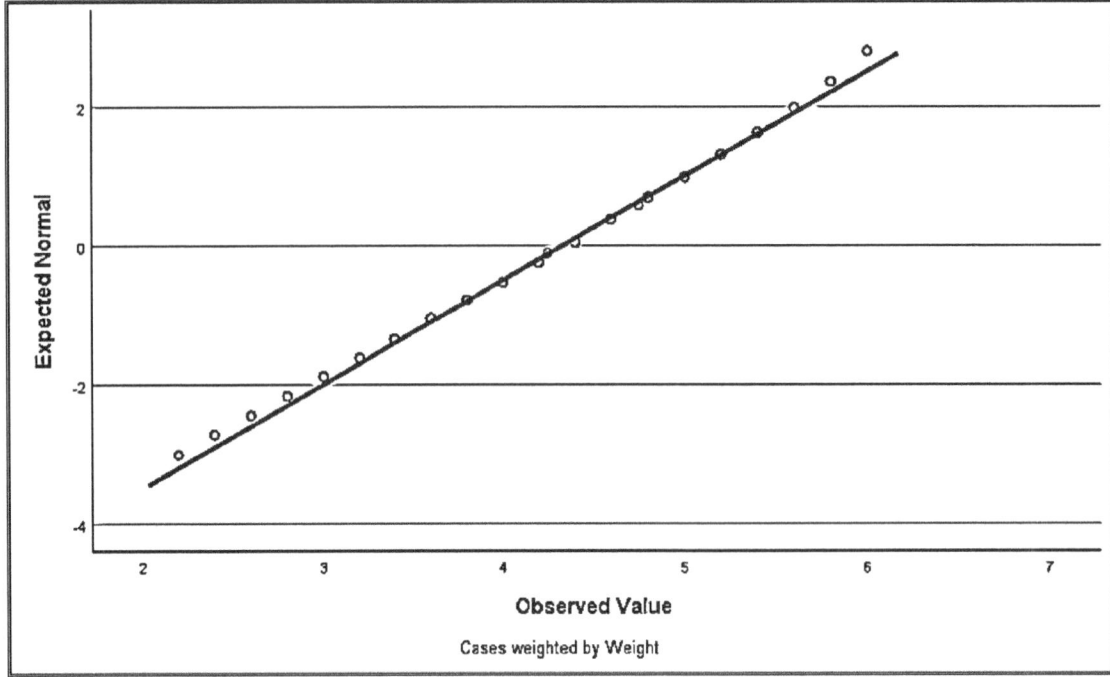

point where the values of one of the variables begins to change more slowly than the other or changes direction entirely. This may be a natural occurrence, but it may also happen as a result of the violation of another assumption—normal distribution.

If the scores for one or both of the variables are not normally distributed, then the intervariable relationship will not be linear, and a scatterplot of the values will not be oval-shaped. In the case of the roller coaster data in Figure 3.38, the Kolmogorov-Smirnov and Shapiro-Wilk tests report that all variables (speed, height, and length) are significantly non-normally distributed; hence, the apparent nonlinearity of the variables in Figure 3.38 may be a result of the violation of the normality assumption; the scatterplots appear to be a mix of linear and curvilinear relationships.

On the other hand, nonlinearity may be due to the physics of the item measured. For example, in the scatterplot of the height and speed of roller coasters in Figure 3.38, the beginning of the scatterplot is linear up to the point where height continues to increase, while speed levels off. Regardless of the source of nonlinearity, violation of the linearity assumption makes data ineligible for techniques that are based on it (e.g., Pearson correlation). Statistical techniques designed to analyze nonlinear relationships are outside the scope of this text.

Scatterplots are the primary tool for assessing linearity between two variables. To reduce the number of scatterplots needed, Tabachnick and Fidell (2018) suggest using skewness indices to screen variables for normality and using scatterplots to examine the relationships of variables likely to depart from linearity due to non-normal distributions. Variables with a natural nonlinear relationship may also benefit from a scatterplot examination before pursuing advanced analyses.

3.4.2.2.3 Homoscedasticity/Homogeneity of Variance

Samples of scores used in statistical analyses are based on the assumption that the populations of scores from which the samples are drawn have the same variance. This is the assumption of homoscedasticity

(Figure 3.47a), and it is equally important for statistical analyses involving ungrouped data and grouped data. A classic example of an analysis of ungrouped data includes Pearson correlation between age and food expenditure. The assumption of homoscedasticity for ungrouped data stipulates that the variance of a continuous variable (e.g., income) should be roughly the same as the variance of another continuous variable (e.g., food expenditure).

The assumption of homoscedasticity for grouped data is the same as homogeneity of variance, which assumes that the variance of the continuous dependent variable should be approximately the same across all levels of the categorical independent variable. Therefore, an investigation of differences in entertainment expenditure (dependent variable) between established age groups (independent variable) may be affected by errors of measurement at some levels of the independent variable; respondents of some age groups may be reluctant to produce accurate reports of expenditure on recreational activities, resulting in heterogenous variance in the dependent variable across age group levels.

Figure 3.47 Homoscedastic and Heteroscedastic Scatterplots of Data

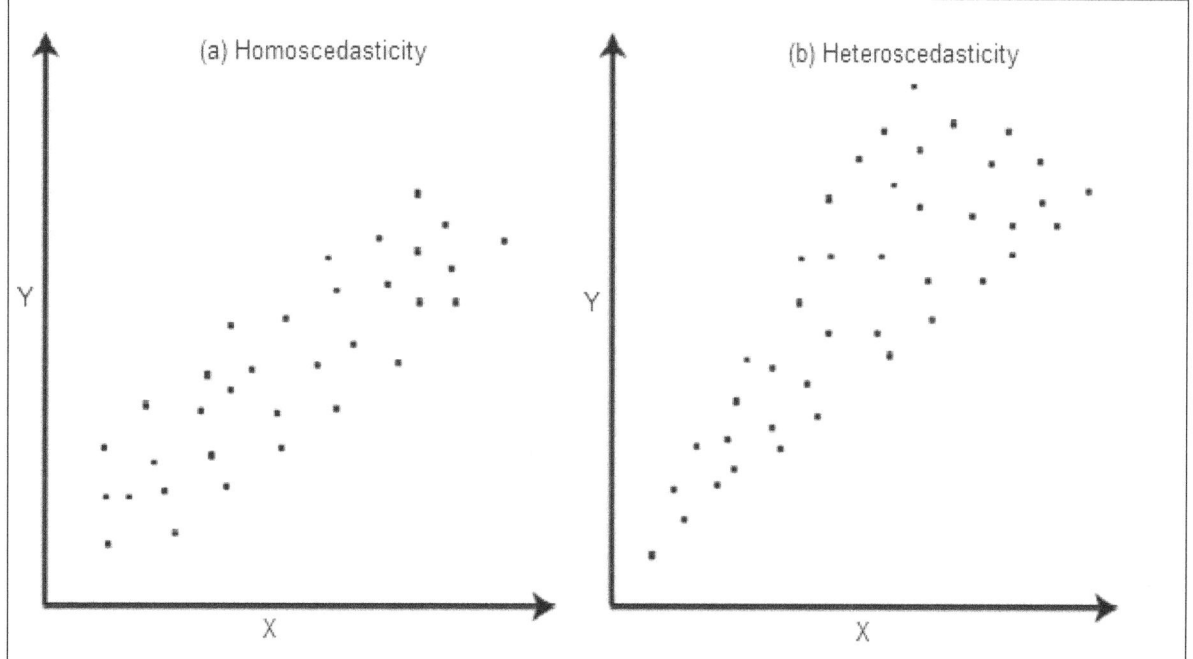

Another frequent source of the violation of homoscedasticity is the violation of the assumption of normality. When variables are normally distributed, the relationship between the variables is homoscedastic. Therefore, in a scatterplot of two normally distributed variables, the distribution of values is about the same width all over. If at least one of the variables has a non-normal distribution due to skewness, then a scatterplot of the variable values will be an uneven, heteroscedastic scatter of the data points as in Figure 3.46b. Therefore, in our example of the relationship between income and food expenditure, people at lower income levels may start out with the same expenditure on food, but as income increases, food expenditures may become more variable, resulting in a skewed distribution of values and a heteroscedastic relationship between income and food expenditure.

Fortunately for applied researchers, heteroscedasticity does not ruin a research study. For example, in a linear regression study of the predictive relationship between hours spent studying and course grade, the linear relationship between both variables will still be evident even with heteroscedastic data. Heteroscedasticity may lead to biased standard errors in correlation or linear regression which reduce the power of the analysis and result in an increased likelihood of a Type I error. In other words, significance tests are weakened as a result of heteroscedastic data.

In analyses involving grouped data (e.g., comparing the test performance of two groups of students), violation of homoscedasticity may also result in an inflated Type I error rate. However, the effect of the violation is said to be reduced with equal or nearly equal cell sizes.

Multiple methods are available to detect violation of the homoscedastic assumption. For ungrouped data, a frequently employed method is the examination of scatterplots of pairs of variables for departures from homoscedasticity as indicated. Likewise, a scatterplot of residuals (errors) against predicted values that is cone-shaped is evidence of heteroscedasticity.

For statistical techniques, like the analysis of variance, used to analyze grouped data, one frequent method of assessing the violation of the assumption of homogeneity of variance is Levene's test, which tests the hypothesis of equal variance across groups (or levels) of the independent variable. Rejection of the null hypothesis is evidence of unequal variance.

3.4.2.2.4. Independence of Errors

The assumption of independence of errors (or observations) means that the scores for each subject in a study are independent of other subjects' scores; they do not depend on each other. The occurrence of one set of responses is not dependent on another set of responses. Violation of the independence of errors assumption may lead to inaccurate interpretations of research results due to inflated Type I or Type II error rates. This assumption violation occurs when:

- Study participants in different treatment groups communicate with each other about the study, resulting in data contamination. Assumption violation may occur as a result of other interactions among subjects: cheating, competition or group learning situations.
- Study participants are included in more than a single treatment group.
- Similar affiliations of study participants exist that influence the performance of intact groups of participants. These affiliations may include students in the same classroom, members of the same family, or co-workers in the same business.
- Research study subjects are from the same geographical locations, resulting in similar performance compared to participants from distant geographical locations.
- Research designs require measuring study participants several times on the same dependent variable. This results in scores for a single subject that are more strongly related than the scores between subjects.

Both graphical and statistical methods are available to test for lack of independence of observations. Graphical methods include the examination of plots of residual scores versus:

- independent (predictor or categorical) variable values,
- ordered numeric values, such as case numbers or ordered participant numbers.

Plotted residuals should be randomly and symmetrically distributed around zero. A nonrandom pattern that exhibits a funnel-shaped distribution suggests lack of independence.

A common statistical method for detecting the violation of the independence assumption in linear regression is the computation of the Durbin-Watson statistic (Durbin & Watson, 1971) used to detect autocorrelation in residuals in regression analyses. Autocorrelation can lead to underestimates of the standard error that may lead to increased Type I error.

The Durbin-Watson statistic ranges from 0 to 4. If the statistic is between 1.5 and 2.5, assume that the residuals are independent (the goal). Other values of the statistic suggest possible violation of the independence assumption.

Preventing dependence in research data lies in proper design and implementation of research. Stevens (2009) recommends the following to avoid violation of the independence assumption:

- If investigators suspect that the design of a study will lead to correlated observations, it is best to test hypotheses at a more stringent level of significance (e.g., $p < 0.01$).
- If multiple groups are involved in each treatment, and there are reasons to suspect that observations may be correlated within the groups but uncorrelated between groups, consider using group means as the unit of analysis. This strategy will reduce the sample size, but the stability of group means is unlikely to create a reduction in statistical power.

3.4.2.2.5 How to Handle Violations of Statistical Assumptions

In addition to strategies already discussed, other common methods of dealing with violations of assumptions include the substitution of more conservative tests of significance or the use of nonparametric tests. For example, imagine that a researcher planned to use one-way analysis of variance to investigate the existence of significant differences among groups, but the computation of Levene's test statistic suggested violation of the assumption of homogeneity of variance. The researcher may choose to use ANOVA at a more stringent alpha level. Alternatively, the researcher may choose to use statistical procedures, like the parametric Welch's ANOVA or the nonparametric Brown-Forsythe test, neither of which requires homogeneity of variance.

If there are violations of multiple assumptions, and sample sizes are relatively small, nonparametric tests, like the Kruskal-Wallis test and Friedman's test, may be viable options. Both have fewer assumptions than their parametric cousins—one-way ANOVA and repeated measures ANOVA, respectively. However, they tend to be less powerful than parametric tests. Consult Chapter 12 of this volume for more about nonparametric tests.

Finally, data transformations are viable options for handling violations of the assumptions of linearity, normality, and homoscedasticity. Data transformation is the application of a mathematical function (e.g., logarithmic or square root functions) to each value of a variable so that the variable better meets statistical assumptions associated with statistical techniques. For example, a common assumption of linear regression is a linear relationship between dependent and independent variables. If a researcher wishes to examine the predictive relationship between hours of study and test scores, but the independent or dependent variable fails to meet the linearity assumption, one option is to transform one or both of the variables to improve linearity between the two variables so that they meet the linearity assumption. The researcher can then conduct the regression analysis using transformed values rather than raw values.

Variables that meet data assumptions improve the precision of statistical analyses. Therefore, typically, data transformations are applied so that variable values that previously violated assumptions come closer to meeting them. Transformation techniques are widely available in statistical applications, including IBM SPSS, through its **Compute** procedure.

In spite of the benefits and availability of data transformation options, they are not universally favored. The source of reservations is that transformed data can be challenging to interpret. For example, a logarithmic transformation of a total raw test score (e.g., 0 – 100) will look very different quantitatively than a raw test score and subsequently will be somewhat difficult for the typical research report consumer to interpret. Thus, it is important to be confident that data transformations are necessary before proceeding with them.

Research methodologists (e.g., Tabachnick & Fidell, 2018; Lomax & Hahs-Vaughn, 2012) provide some general guidelines for researchers considering data transformations:

- Consider the transformation of all data that breach the normality assumption unless interpretation of the transformed values poses an insurmountable challenge.
- If data are heteroscedastic and non-normal, it is best to conduct variance transformation first, as that might cure non-normality.
- When transforming data to meet the linear regression assumption, transform the independent, dependent, or both sets of variables. This is also likely to resolve problems with heteroscedasticity and normality.
- After transforming variable values, rerun the assumption test to see whether transformed variable values meet the assumption.

The most common data transformations, largely based on Tabachnick and Fidell (2018), to correct assumption violations of normality appear in Figure 3.48. Researchers also recommend similar strategies to prevent violations of linearity assumptions (e.g., Berman, 2023).

Figure 3.48 Common Data Transformations Based on Severity of Assumptions Violations

Data Transformation	Description	Common Use
Square root transformation $NEWX_i = SQRT(X_i)$	Find the square root of each variable value	Corrects moderate positive skewness and kurtosis
Base 10 logarithm transformation $NEWX_i = LG10(X_i)$	Find the logarithm of a set of variable values	Corrects substantial positive skewness and kurtosis
Inverse transformation N $NEWX_i = 1/(X_i)$	Divide 1 by each variable value.	Corrects severe positive skewness and kurtosis
Reflect and square root transformation. $NEWX_i = SQRT(K - X_i)$	Subtract each value from the maximum score + 1, then find the square root of each result.	Corrects moderate negative skewness and kurtosis
Reflect and base 10 or natural logarithm transformation. $NEWX_i = LG10(K - X_i)$	Subtract each value from the maximum score + 1, then take the log of each reflected value	Corrects moderate negative skewness and kurtosis
Reflect and inverse transformation. $NEWX_i = 1/(K - X_i)$	Subtract each value from the maximum score + 1, then find the inverse of each reflected value	Corrects severe negative skewness and kurtosis

Note: *i* represents each study participant. Therefore, X_i represents the score or value for each study participant. K = the maximum score + 1.

3.4.2.2.6 SPSS Specifications for Investigating Assumptions

Previous sections demonstrate how to use SPSS to conduct basic data cleansing and screening, as well as the detection of outliers. This section is dedicated to SPSS procedures that test and evaluate the assumption of normality. The section also instructs how to transform data to improve alignment with the assumption of normality.

The data for the SPSS exercise in this section are from the World Values Survey (WVS). The data for this demonstration is a random sample of survey responses ($n = 325$) from respondents in the Wave 6 data collection ($n = 999$) in Trinidad and Tobago in 2010. The variables selected for the SPSS screening exercise are displayed in Figure 3.49.

Figure 3.49 Variables for Screening Exercise

Variable Name	Variable Label	Measurement Scale	Description
M_InGroup_Trust	Mean In-Group Trust	Interval/Ratio	Score based on a subscale of 3 Likert-response items on the extent to which respondents trust family, neighbors, and people they know personally.
M_OutGroup_Trust	Mean Out-Group Trust	Interval/Ratio	Score based on a subscale of 3 Likert-response items on the extent to which respondents trust people outside of their immediate circle (from another nation, another religion, or someone they meet for the first-time)
M_Self_Protective	Mean Self-Protective Values	Interval/Ratio	Mean score for 5 Likert-response items that measure values that serve as anxiety coping mechanism due to uncertainty in the social/physical world (e.g., "Tradition is important to this person").
M_Self_Expansive	Mean Self-Expansive Values	Interval/Ratio	Mean score for 5 Likert-response items that measure anxiety-free motivations and values (e.g., Adventure and taking risks are important to this person).
Sex	Sex	Nominal	Gender of survey respondent

3.4.2.2.6.1 SPSS Specifications to Detect Normality Violations

Section 3.4.2.2.1 presented graphical vehicles for assessing normality, specifically histograms and stem-and-leaf plots. This section illustrates how to use a comprehensive application, **SPSS Explore**, to conduct general screening and cleansing of data, and also to examine whether data meets statistical assumptions.

To assess the assumption of normality for quantitative variables, begin by examining skewness and kurtosis. To do that, open the dataset, **WVS_TT_Screening_Exer.sav**. In the **IBM SPSS Statistics**

Data Editor window, select **Analyze → Descriptive Statistics → Explore** to open the **Explore** dialog window. Move the variables, **M_InGroup_Trust**, **M_OutGroup_Trust**, **M_Self_Protective**, and **M_Self_Expansive**, into the **Dependent List** panel (Figure 3.50).

The **Factor List** panel is reserved for categorical independent variables used to split dependent variable scores into groups (e.g., gender or educational level). Leave that panel blank for now to facilitate the analysis of the ungrouped variables. Below the **Factor List** panel is the **Label Cases by** panel, which permits the identification of individual records in the reports (e.g., boxplots) by assigned IDs rather than the default IBM SPSS case number. Leaving the panel blank assumes the selection of the default as indicated in Figure 3.50. Below the list of variables on the left is the **Display panel**, which presents the options of (descriptive) **Statistics** or **Plots** (box and stem-and-leaf plots). Retain the default—**Both** (both options).

Figure 3.50 IBM SPSS Explore Main Dialog Window

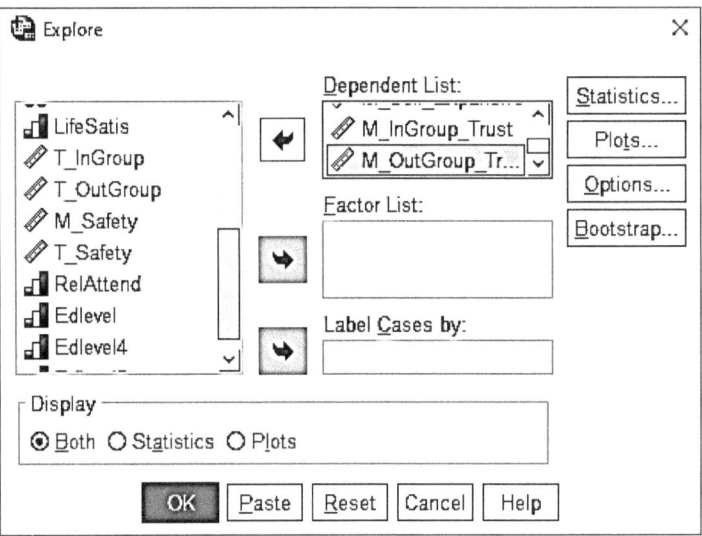

Select the top right **Statistics** pushbutton to open the **Explore Statistics** dialog window (Figure 3.51a). **Descriptives**, the default selection, generates the following measures of central tendency: dispersion, and distribution: mean, 5% trimmed mean, median, variance, standard deviation, minimum, maximum, range, interquartile range (IQR), skewness, kurtosis, and standard errors for the mean, skewness and kurtosis. **Descriptives** also reports the 95% confidence interval for the mean by default, which may be adjusted, as needed, but retained for this exercise. **M-estimators** produces alternatives to the mean and median by applying weights to specific records. Leave the **M-estimators** specification blank for this exercise.

Below **M-estimators**, select **Outliers**, which prints the five highest and five lowest values for each variable along with the case numbers with which they are associated. The final option, **Percentiles**, prints the percentage of records falling within the 5th, 10th, 25th, 50th, 75th, 90th, and 95th percentiles of the data sample and will remain unselected for this exercise. Click **Continue** to return to the main dialog window.

Select the **Plots** button in the **Explore** window, to open the **Explore Plots** dialog window (Figure 3.51b). The options in the **Boxplots** panel on the left presents display options when more than one dependent variable is involved. In the **Boxplots** panel, confirm that the default setting, **Factor levels together**, has been selected. This specification means that SPSS will produce a single boxplot for each quantitative dependent variable. If an independent variable (also called factor or grouping variable) had been specified in the **Explore** main dialog window, the graph for each dependent variable would include multiple boxplots, one for each group (e.g., male and female) for the dependent variable (e.g., **M_Self_Protective**). The next boxplot display option, **Dependents together**, requests boxplots for all dependent variables in the same space. When an independent variable is specified, along with the **Dependents together** option, dependent variable values for each group appear in the same report space.

Figure 3.51 The IBM SPSS Explore Statistics and Plots Dialog Windows

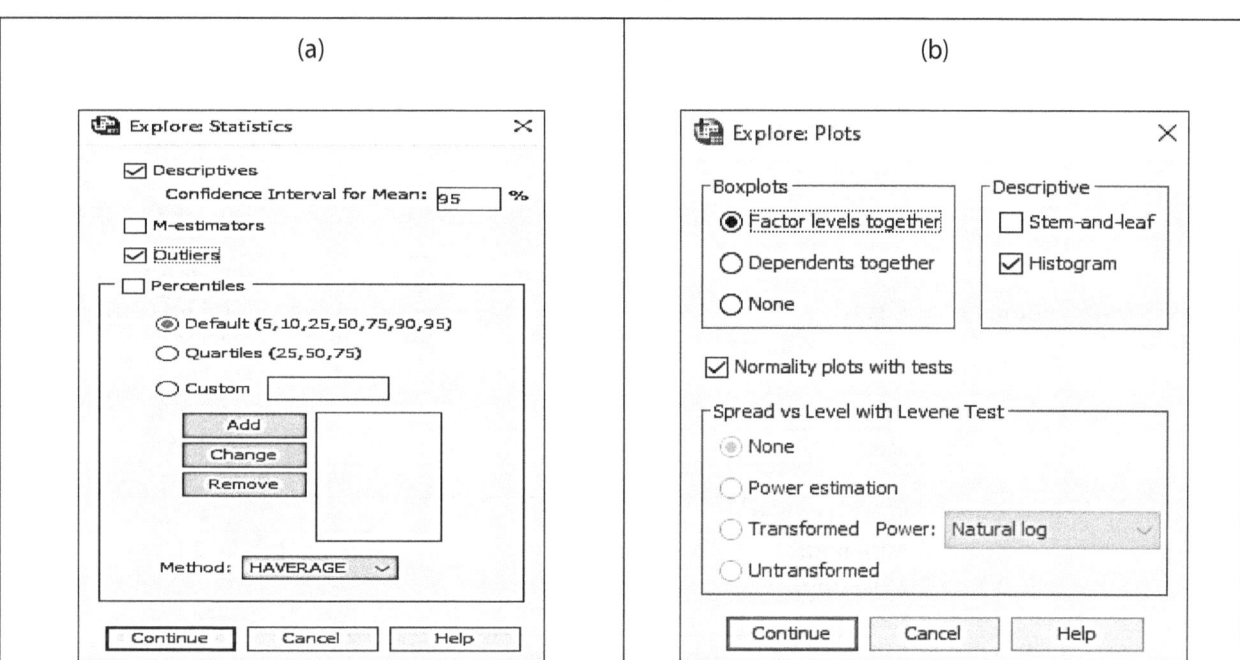

The **Descriptive** panel in the **Explore Plots** window, includes **Stem-and-leaf** and **Histogram** options. Select **Histogram** for display. See Section 3.4.2.2.1 for details regarding both types of plots. Next, select **Normality plots with tests**, which will produce the inferential test of the assumption of normality (the Kolmogorov-Smirnov and Shapiro-Wilk tests) as well as a normality plots specification, which includes a normal Q-Q plot, and a detrended normal Q-Q plot for each dependent variable.

Notice that the **Spread vs Level with Levene tests** panel has been grayed out. If an independent variable had been specified for this exercise, this option would have been available to specify Levene's test for the homogeneity of variance—equal variance across groups. Click **Continue** to return to the main **Explore dialog** window and click **OK** to perform the analysis.

Figure 3.52 displays descriptive statistics for a portion of the four quantitative variables. Statistics include those previously discussed in Section 3.4.1 as well as the following: 95% confidence interval for the mean, 5% trimmed mean, variance, range, and interquartile range.

Of the 325 records in the dataset, 321 were analyzed, with four missing (1.2%) cases for each variable. The descriptive statistics appear reasonable and within expected ranges. For example, the minimum PVQ Self-Protective Values variables score is 1, and the maximum score is 6. The PVQ Growth/Self-Expansive Values minimum is 1.6, and the maximum is 6.0. Trust variables have a minimum score of 1 and a maximum of 4, the same as the analyzed data sample. With no out-of-range values apparent and a very low percentage of missing values, the dataset appears clean.

Figure 3.52 Descriptive Statistics for Quantitative Variables

			Statistic	Std. Error
Mean PVQ Self Protective Values	Mean		4.3517	.04102
	95% Confidence Interval for Mean	Lower Bound	4.2710	
		Upper Bound	4.4324	
	5% Trimmed Mean		4.3754	
	Median		4.4000	
	Variance		.540	
	Std. Deviation		.73498	
	Minimum		2.00	
	Maximum		6.00	
	Range		4.00	
	Interquartile Range		.80	
	Skewness		-.469	.136
	Kurtosis		.403	.271
Mean PVQ Growth/Self Expansive Values	Mean		4.2141	.04142
	95% Confidence Interval for Mean	Lower Bound	4.1326	
		Upper Bound	4.2956	
	5% Trimmed Mean		4.2198	
	Median		4.2000	
	Variance		.551	
	Std. Deviation		.74215	
	Minimum		1.60	
	Maximum		6.00	
	Range		4.40	
	Interquartile Range		.80	
	Skewness		-.171	.136
	Kurtosis		.232	.271
Mean In-Group Trust	Mean		3.0898	.03030
	95% Confidence Interval for Mean	Lower Bound	3.0302	
		Upper Bound	3.1494	
	5% Trimmed Mean		3.1078	
	Median		3.0000	
	Variance		.295	
	Std. Deviation		.54293	
	Minimum		1.00	
	Maximum		4.00	
	Range		3.00	
	Interquartile Range		.67	
	Skewness		-.544	.136
	Kurtosis		.501	.271
Mean Out-Group Trust	Mean		2.2726	.03659
	95% Confidence Interval for Mean	Lower Bound	2.2006	
		Upper Bound	2.3446	
	5% Trimmed Mean		2.2844	
	Median		2.3333	
	Variance		.430	
	Std. Deviation		.65553	
	Minimum		1.00	
	Maximum		4.00	

Skewness and kurtosis values in Figure 3.52 provide the first measures of the shape of the distribution of scores for the variables. The results disclose moderate levels of skewness and kurtosis. Skewness values for our sample data range from -0.171 to 0.544, and -0.494 to 0.501 for kurtosis. These values are all within the relatively liberal +/- 1.00 range proposed by George et al. (2019). In addition, based on the central limit theorem, the relatively large sample size makes larger skewness *z* scores tentatively acceptable.

Figure 3.53 displays a familiar device for assessing normality—frequency histograms. With the addition of a normal curve overlay on each graph, the shape of the distribution of each variable becomes more evident. All variables are slightly negatively skewed, with In-Group Trust (Figure 3.53c) displaying the greatest skew (-0.544), with a slight pile-up of scores on the right.

Figure 3.53 Histograms of the Quantitative Variables

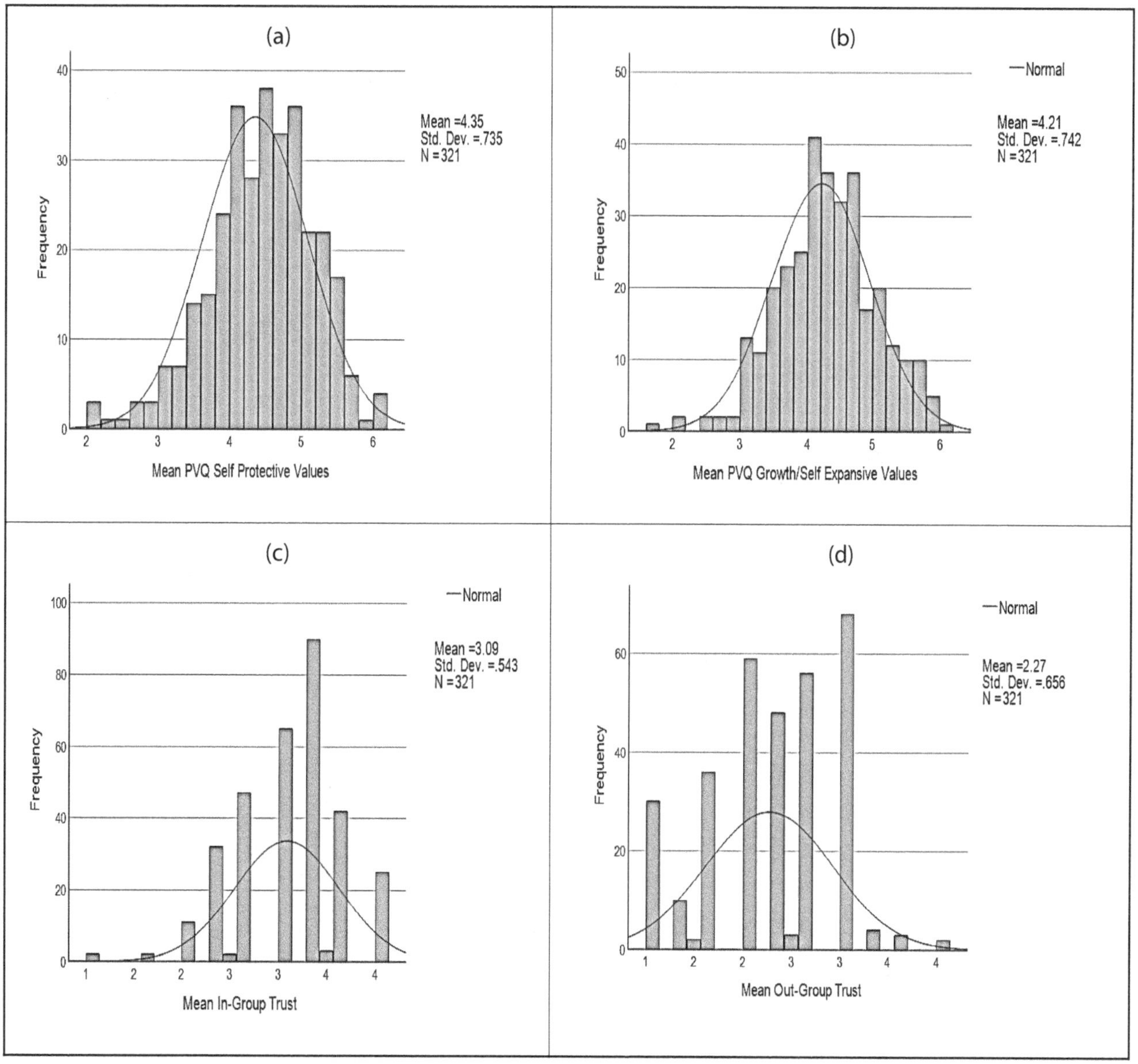

SPSS Explore conducts two hypothesis tests of normality, Kolmogorov-Smirnov (K-S) test and the Shapiro-Wilk (S-W) test, whose results are displayed in Figure 3.54. The K-S and S-W test the null hypothesis that the scores of a quantitative variable are normally distributed. The statistic for each test is displayed in the column, Statistic; the degrees of freedom (which equals the sample size) is in the column, **df**, and the significance level is in the Sig. column. The alpha level for the test is $p < .05$. Typically, the goal of a hypothesis testing is to reject the null hypothesis with significance levels that are less than .05. In this case, the goal is the opposite—to fail to reject the null hypothesis that the scores are normally distributed. Accordingly, the desire is for significance levels that are greater than .05, which means that they are not significantly different from a normal distribution.

In spite of the results of the histograms and skewness indices, the results of the K-S and S-W tests (Figure 3.54) indicate that the distributions significantly deviate from normality ($p < 0.01$). However, both tests are known for their sensitivity to very small deviations from normality, particularly when samples are relatively large.

Figure 3.54 Hypothesis Test of Normality

	Tests of Normality					
	Kolmogorov-Smirnov[a]			Shapiro-Wilk		
	Statistic	df	Sig.	Statistic	df	Sig.
Mean PVQ Self Protective Values	.084	321	<.001	.979	321	<.001
Mean PVQ Growth/Self Expansive Values	.072	321	<.001	.988	321	.010
Mean In-Group Trust	.172	321	<.001	.945	321	<.001
Mean Out-Group Trust	.140	321	<.001	.940	321	<.001

a. Lilliefors Significance Correction

With the **Normality plots with tests** specification, SPSS generates not only the K-S and S-W significance tests, but also two normal Q-Q (or Quantile-Quantile) plots, a standard plot and a detrended plot. The normal Q-Q plot compares the actual quantiles of the data (depicted as dots) against expected quantiles if the data were normally distributed. The expected quantiles are depicted as a solid 45-degreee, nondecreasing diagonal line running from lower to the upper right of the graph. If the data collected are normally distributed, the dots will lie on or close to the diagonal line. Deviation of scores from normality shifts the points away from the diagonal line. If the trend of the Q-Q plot is flatter than the diagonal line, the distribution of the data on the *x*-axis is more variable than the expected data plotted on the *y*-axis. Conversely if the Q-Q-plot is steeper than the diagonal line, distribution of the data is less variable than the normal distribution. If the plot is s-shaped, it is likely that the collected data is skewed.

In the detrended Q-Q plot, the horizontal line, which intersects with the *y*-axis at the zero point, represents a normal distribution. It signifies zero deviation from a normal distribution of values. The dots in the graph are computed deviations from the normal distribution (the horizontal line). If the score distribution of a variable is normal, the scores will be distributed evenly above and below the horizontal line. Normal and detrended Q-Q plots are displayed in Figure 3.55.

Figure 3.55 Normal and Detrended Q-Q Plots for the Trust Variables.

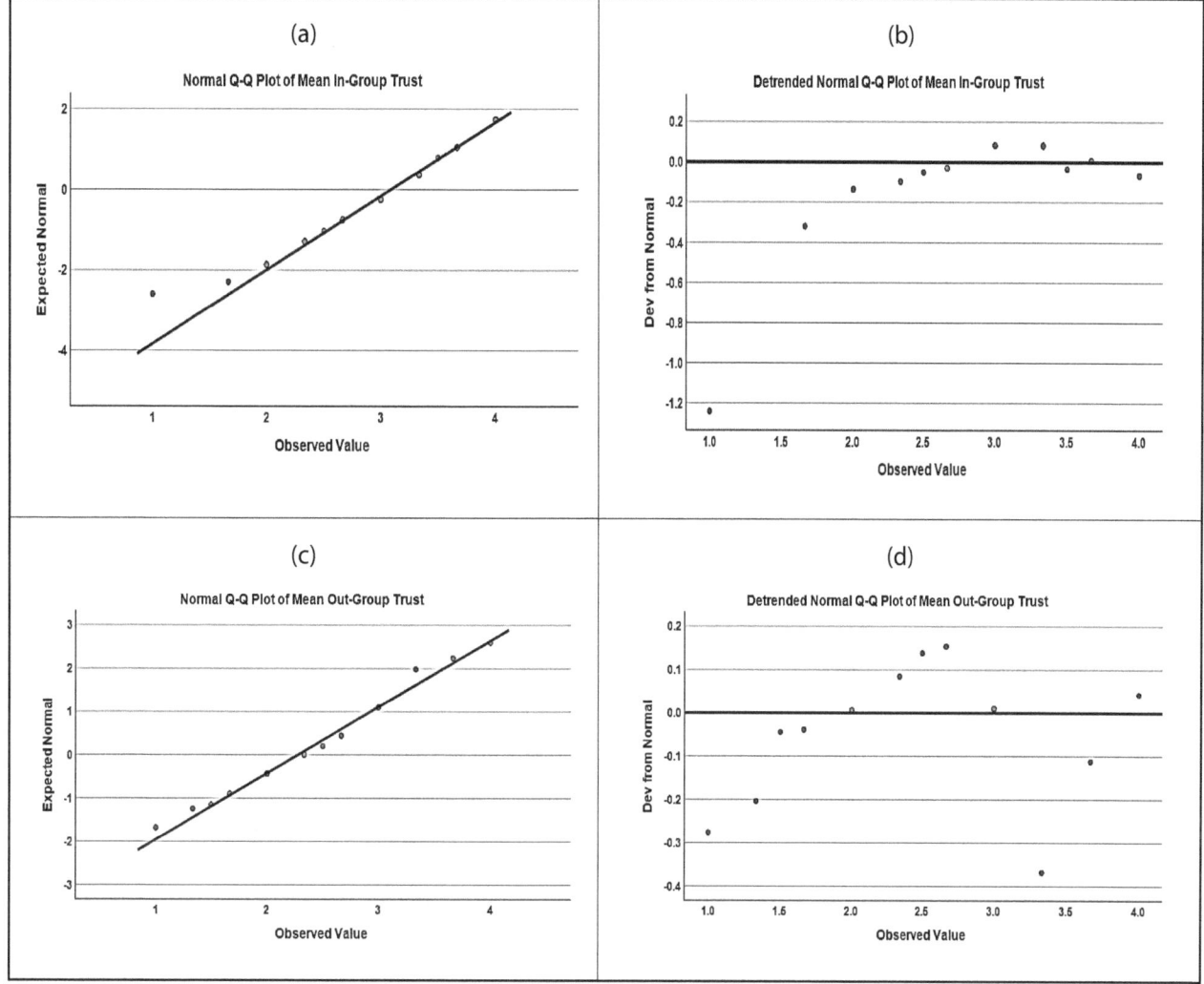

In the displayed normal Q-Q plots, plotted dots for both the In-Group and Out-Group Trust data are located on or close to the diagonal line (Figures 3.55a and 3.55c), with the exception of the lower In-Group Trust scores. This supports the higher skew index of the In-Group Trust data compared to the Out-Group data (-0.544 versus -0.306). The detrended Q-Q plots (Figures 3.55b and 3.55d) exemplify some deviation from the expected normal distribution, particularly in the lowest score region for the In-Group Trust data. Deviations range from -1.2 to 0.1 for In-Group Trust and -0.4 to 0.15 for Out-Group Trust. Differences in the results of the K-S and S-W tests versus the Q-Q plots illustrate the impact that a relatively large sample size can have on significance tests and the importance of supplementing those results with graphical methods of assessing normality.

3.4.2.2.6.2 Transforming Data to Enhance Normality

Although the variables in Figure 3.53 reflect relatively mild skewness and kurtosis, both the K-S and S-W tests of significance found the distributions of the variables to be statistically significantly non-normal (Figure 3.54). The large sample size and the reported sensitivity of the significance tests to large sample sizes likely contributed to the outcome of the significance tests. Nevertheless, for demonstration purposes, this exercise demonstrates how to transform the values of one of the two most highly skewed variables, **M_Self_Protective**, with a negative skew value of -.469 (Figure 3.52).

To complete the data transformation, select **Transform → Compute Variable** from the **SPSS Data Editor** main menu to open the **Compute Variable** dialog window (Figure 3.56).

Figure 3.56 SPSS Statistics Compute Variable Dialog Window

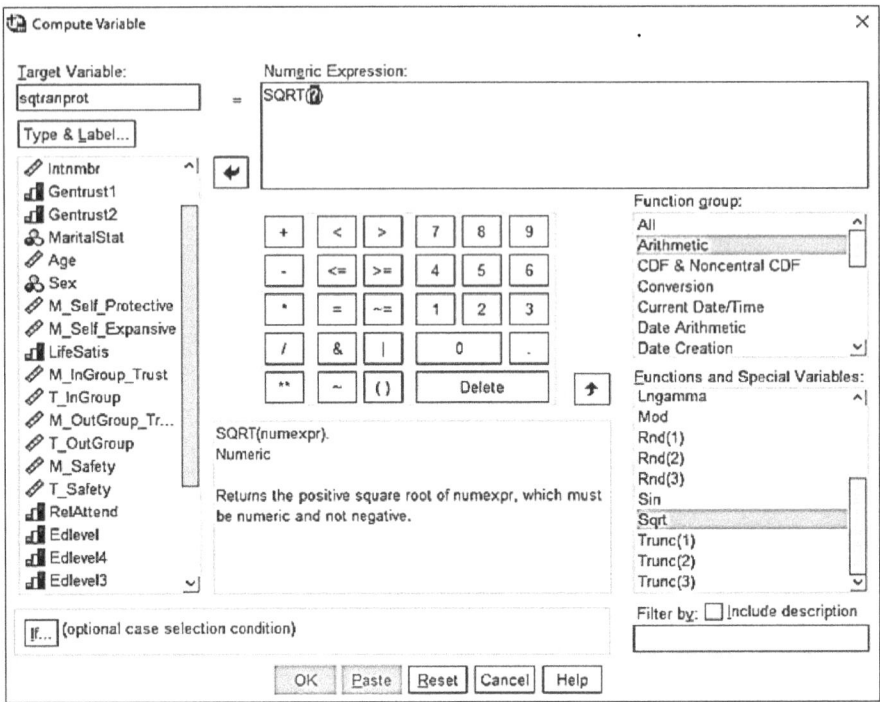

In the panel below **Target Variable**, enter the name, **sqtranprot**, for the new variable that will be created. On the right, below **Function Group**, highlight **Arithmetic**, and it will trigger a list of arithmetic functions in the **Functions and Special Variables** panel. Scroll down the **Functions and Special Variables** list and highlight the **Sqrt** function, the first option recommended for moderately skewed variable values (Figure 3.48). Click on the arrow to the left of the panel to move the **Sqrt** expression into the **Numeric Expression** panel (Figure 3.56). The question mark next to **SQRT** is a placeholder for the remainder of the numeric expression.

According to Figure 3.48, the formula for transforming moderate, negative skewness is:

$NEWX_i$ = SQRT(K - X_i), where

$NEWX_i$ = the target variable, **sqtranprot**,

X_i = the variable to be transformed (**M_Self_Protective**), and

K = 1 + the maximum value recorded for the variable to be transformed. According to Figure 3.52, the maximum value for **M_Self_Protective** (Mean PVQ Self Protective Values) is 6. Therefore, K = 1 + 6 = 7.

The name of the new variable is in the **Target Variable** panel. The remainder of the expression for the **Numeric Expression** panel is **SQRT(7 - M_Self_Protective)**.

Figure 3.57 The Compute Variable Dialog Window with the Completed Compute Statement

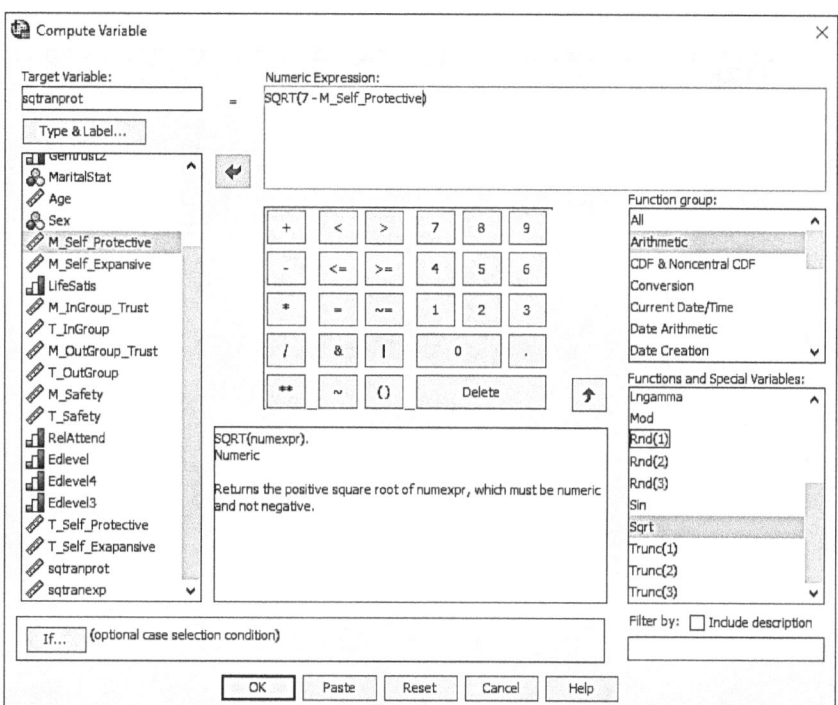

To insert the remainder of the expression in the **Numeric Expression** panel, highlight the question mark and enter the number 7 to replace the question mark, or use the keypad below the panel to enter the numeral. Click on the minus sign on the keypad. Scroll down the list of variables on the left until you arrive at the variable, **M_Self_Protective**. Double-click on that variable to move it to the **Numeric Expression** panel immediately following the minus sign (Figure 3.57). Click **OK,** and the new variable, **sqtranprot**, will be in the last column of the dataset (Figure 3.58).

Figure 3.58 The Newly Transformed Variable, **sqtranprot**, in the Last Column

	Wave	Ccode	Intnmbr	Gentrust1	Gentrust2	MaritalStat	Age	Sex	M_Self_Protective	sqtranprot
1	6	780	1	2	1	5	80	1	4.80	1.48
2	6	780	4	2	8	6	23	0	4.60	1.55
3	6	780	5	2	3	2	41	1	3.60	1.84
4	6	780	9	2	1	1	64	1	4.60	1.55
5	6	780	10	2	1	5	66	0	4.75	1.50
6	6	780	11	2	1	5	61	0	4.00	1.73
7	6	780	12	2	1	1	43	0	5.20	1.34
8	6	780	14	2	7	1	22	1	3.00	2.00
9	6	780	19	2	10	6	18	0	4.40	1.61
10	6	780	21	2	5	6	37	1	4.40	1.61
11	6	780	22	2	2	6	18	0	4.80	1.48
12	6	780	24	2	2	5	67	0	4.40	1.61

To assess the effects of the transformation, from the SPSS Data Editor's main menu, select **Analyze → Descriptive Statistics → Explore** to open the SPSS **Explore** dialog window (Figure 3.59a). From the variable list on the left, highlight and click over the original variable, **M_Self_Protective** and the transformed variable, **sqtranprot**, to the **Dependent List** panel in the top right of the dialog window.

Figure 3.59 SPSS Explore Dialog Windows

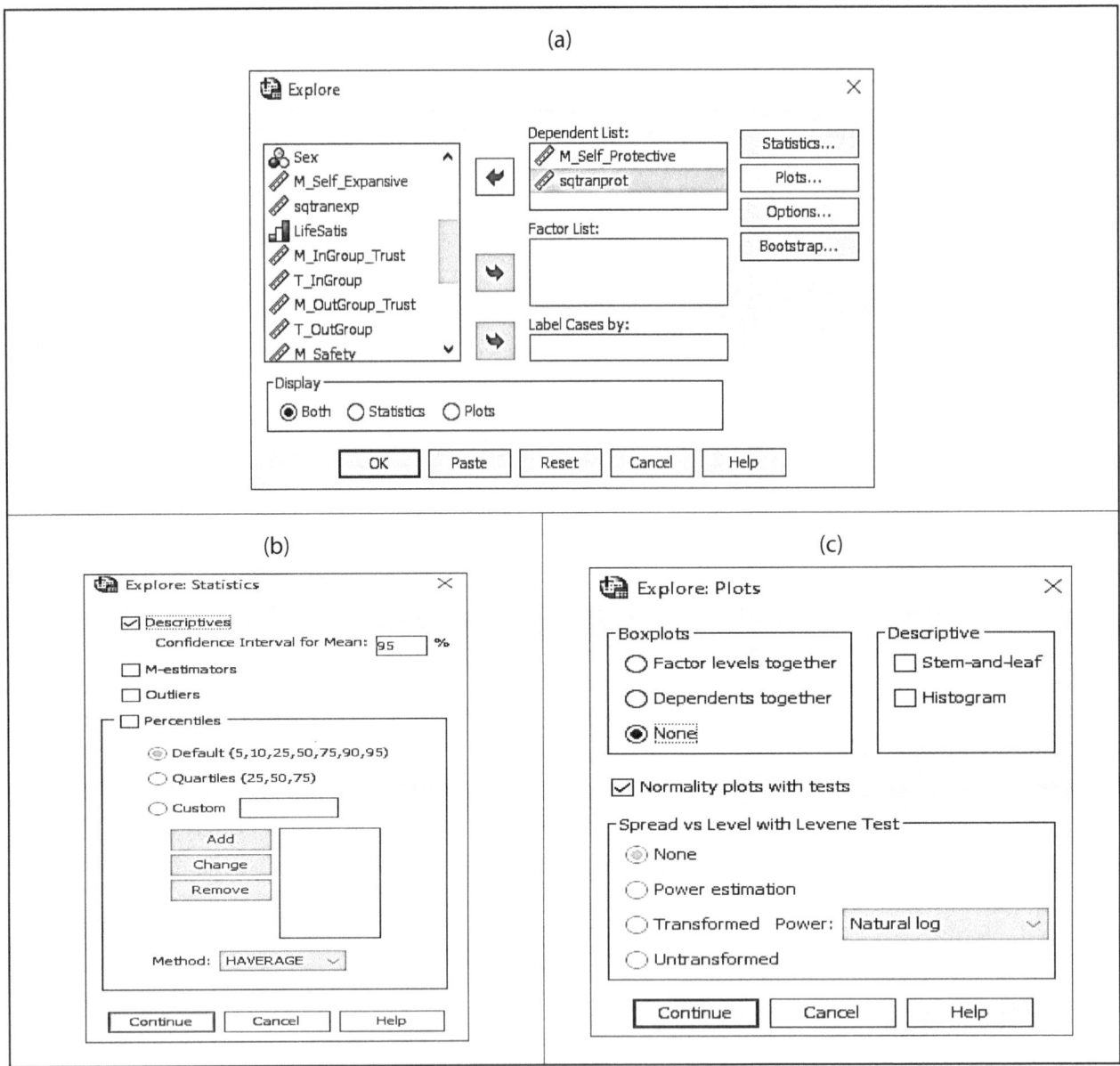

Limit the reports to descriptive statistics as well as the K-S and S-W normality significance tests. To do that, select **Statistics** to open the **Explore Statistics** dialog window, and confirm that **Descriptives** has been selected for the descriptive statistics report (Figure 3.59b). Remove the check mark from the **Outliers** box. Click **Continue** to return to the **Explore** dialog window. Click the **Plots** pushbutton to open that dialog window (Figure 3.59c). There, click on the option **None** under **Boxplots,** and remove the check mark from the **Stem-and-Leaf and Histogram** options. Select **Normality plots with tests**. Click **Continue** to return to the main menu, and click **OK** to run the analysis. Results comparing statistics for the original variable and the transformed variable are presented in Figure 3.60.

Figure 3.60 Descriptives and Normality Test Results for Original and Transformed Variables

(a)

			Statistic	Std. Error
Mean PVQ Self Protective Values	Mean		4.3455	.04076
	95% Confidence Interval for Mean	Lower Bound	4.2653	
		Upper Bound	4.4257	
	5% Trimmed Mean		4.3685	
	Median		4.4000	
	Variance		.540	
	Std. Deviation		.73478	
	Minimum		2.00	
	Maximum		6.00	
	Range		4.00	
	Interquartile Range		.80	
	Skewness		-.454	.135
	Kurtosis		.371	.270
Sqrt Transformed Self-Protective Values	Mean		1.6135	.01255
	95% Confidence Interval for Mean	Lower Bound	1.5888	
		Upper Bound	1.6382	
	5% Trimmed Mean		1.6125	
	Median		1.6125	
	Variance		.051	
	Std. Deviation		.22627	
	Minimum		1.00	
	Maximum		2.24	
	Range		1.24	
	Interquartile Range		.25	
	Skewness		.017	.135
	Kurtosis		.106	.270

(b)

Tests of Normality

	Kolmogorov-Smirnov[a]			Shapiro-Wilk		
	Statistic	df	Sig.	Statistic	df	Sig.
Mean PVQ Self Protective Values	.083	325	<.001	.980	325	<.001
Sqrt Transformed Self-Protective Values	.061	325	.006	.991	325	.036

a. Lilliefors Significance Correction

Comparing the skewness for **Mean PVQ Self-Protective Values** (the original variable) and the **Sqrt Transformed Self-Protective Values**, it is evident that the square root transformation reduced the skewness and kurtosis of the original variable from -0.454 to 0.017 and from 0.371 to 0.106, respectively (Figure 3.60a). However, results of both the K-S and the S-W tests of significance continue to report significant departures from normality based on an alpha level of .05 (Figure 3.60b).

3.4.2.2.7 Sample Report: Cleaning and Screening

Before conducting central analyses, all variables underwent cleansing for data accuracy and consistency as well as screening for missing values, outliers, and violation of statistical assumptions. The study sample comprised a random sample of 325 records from the 2010 World Values Survey conducted in Trinidad and Tobago. Data analysis included four variables (**M_Ingroup Trust**, **M_Outgroup_Trust**, **M_Self_Protective**, and **M_Self_Expansive**).

Of the 325 records, one variable, **M_Outgroup_Trust** recorded four missing values. Although moderate outliers were found on all continuous variables, no extreme outliers were found. Nevertheless, for exploratory purposes, a square root data transformation was performed on **M_Self_Protective**, one of two variables with the highest skewness value (the other being **M_Ingroup_Trust**) to determine whether data transformation might improve data distribution. Skewness improved from -0.454 (K-S and S-W $p < .001$) to 0.017 (K-S $p = .006$ and S-W $p = .036$.

3.5 Key Terms

Boxplot	Homoscedasticity	Outliers
Chart builder	Independence of errors	P-P plots
Codebook	Interquartile range	Q-Q plots
Crosstabs analysis	Kolmogorov-Smirnov test	Quality assurance
Data cleansing	Kurtosis	Quality control
Data dictionary	Levene's test	Scatterplot
Data editor	Linearity	Skewness
Data transformation	Mahalanobis distance	Shapiro-Wilk test
Data screening	Missing data	Stem-and-leaf plot
Data view	Normality	Variable view
Histogram	Normal probability plots	Z scores

4

Bivariate Correlation

Key concepts from previous chapters or courses:
    independent variable: the variable that a researcher manipulates or is studying
    dependent variable: the variable that a researcher measures using its data values
    x **axis**: the horizontal axis on a graph, also called the abscissa
    y **axis**: the vertical axis on a graph, also called the ordinate

> **THIS CHAPTER INTRODUCES**
>
> - how to express the relationship between two variables statistically
> - how to calculate two measures, covariance and the Pearson product moment correlation (Pearson's r), to express the relationship between two variables
> - the circumstances under which Pearson's r is used
> - how to carry out and interpret correlation using SPSS
> - differences between Pearson's r and other correlation coefficients

4.1 Introduction

For many children growing up in the Caribbean, the parish library used to be a haven where they were introduced to ideas and transported to exotic locations and cultures around the world, mostly through the printed word in printed media or (these days) digital media. It is said that Albert Einstein once declared, "The only thing that you absolutely have to know, is the location of the library." And most Caribbean children knew exactly how to get there. A trip to the library was, and may still be, an exciting excursion that begins with walking with friends, riding or being driven to the destination, and once there, enjoying listening to an oral story teller, paging through magazines, or strolling through book stacks to select the maximum number of books from that special, new, and captivating mystery/adventure book series.

    Libraries are often still "an open door to wonder and achievement" (Asimov, 2009) where children develop their love of reading and, as a byproduct, improve reading achievement, commonly acknowledged as a linchpin to achievement in subjects like mathematics. After all, you must read in order to understand word problems. Empirical research (e.g., Grimm, 2008; Nortvedt et al., 2016) tends to support that there is a relationship between achievement in the language arts and mathematics performance; in this case as writing achievement (Communication Task performance) improved on a high school placement test, so did achievement in mathematics. In other words, high performance in language arts is associated with high performance in mathematics (Figure 4.1). Researchers describe this relationship as a correlational relationship.

Figure 4.1 Relationship Between Communication Task (Writing) and Mathematics Performance

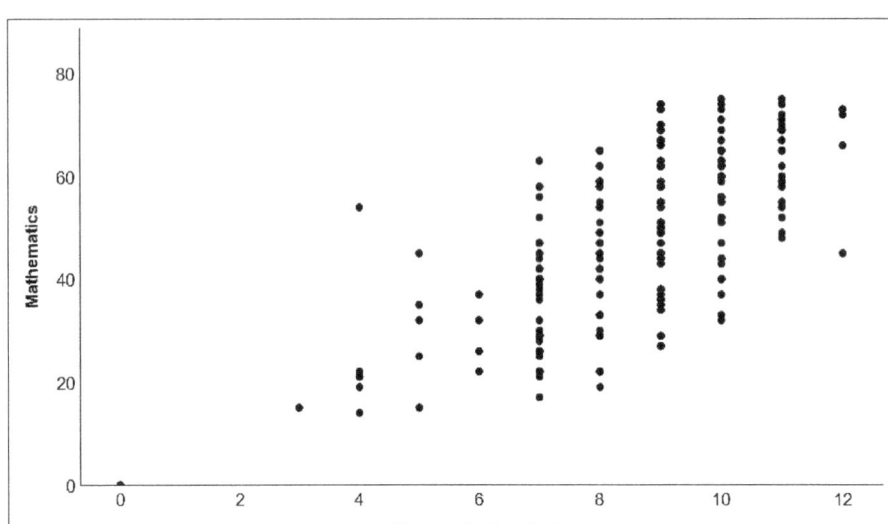

There are other kinds of correlational relationships. Consider the relationship between the weight of passenger cars and their associated fuel efficiency based on data taken from Quinlan (1993). Correlational analysis of the data for all automobiles revealed that as vehicle weight increases, fuel efficiency (miles per gallon) declines. Figure 4.2 displays the relationship between vehicle weight and miles-per-gallon for four-cylinder cars.

Figure 4.2 Relationship Between Miles per Gallon and Passenger Car Weight

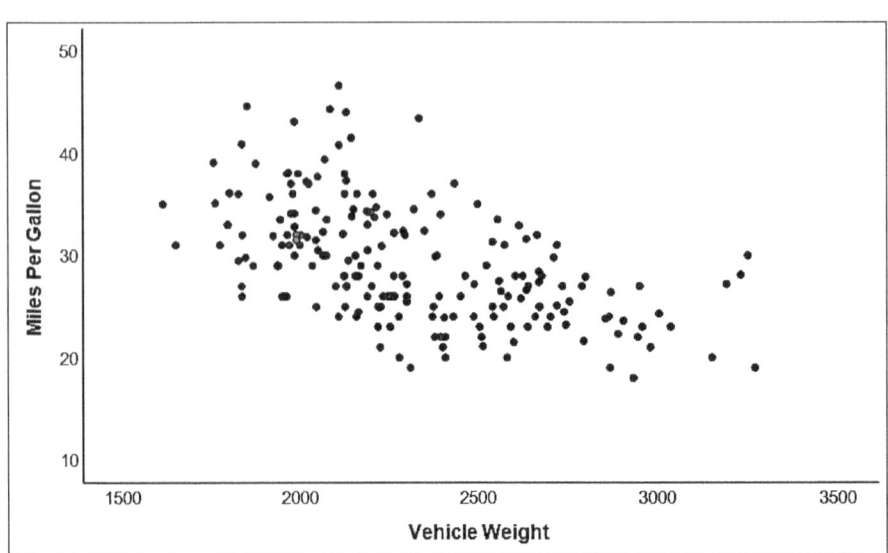

Dua, D. and Graff, C. (2019). UCI Machine Learning Repository. Irvine, CA: University of California, School of Information and Computer Science. https://archive.ics.uci.edu/ml/datasets/Auto_mpg.

Finally, there are times when researchers discover that, in effect, there is no detectable association between the values of one variable and the values of the other. In those situations, the depiction of that relationship will look like the scatterplot of Life Satisfaction and Age in Figure 4.3 from Trinidad and Tobago nationals' responses to the 2010 World Values Survey (Inglehart et al., 2014). This chapter will explore bivariate relationships like the three described and the correlation coefficients that are used to measure and describe them.

Figure 4.3 Relationship Between Life Satisfaction and Age in Trinidad and Tobago

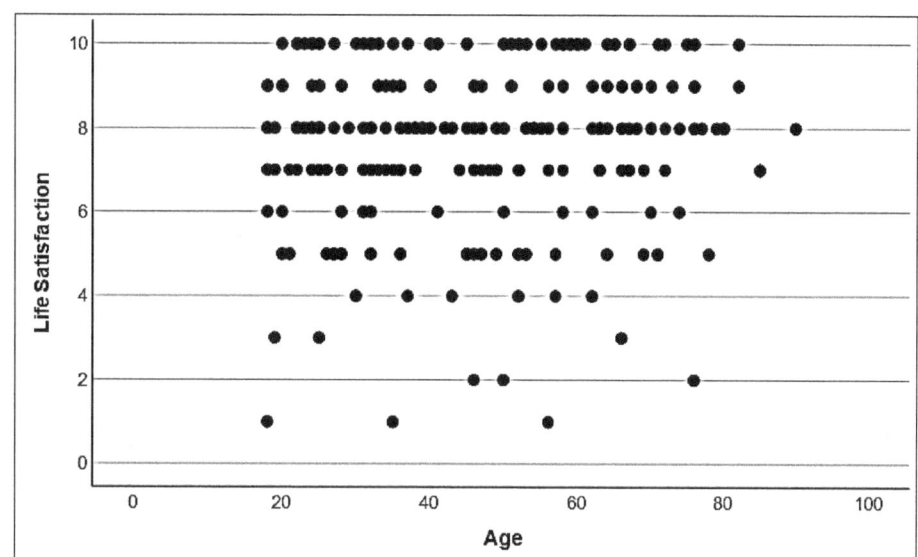

4.2 The Case of Writing and Mathematics Skills

This case includes fictitious written communication and mathematics test scores from 10 12-year-old students. Graham, et al. (2020) and others have consistently found over the years that there is a positive relationship between language arts achievement and mathematics achievement. The objective of this exercise is to determine whether there is a concomitant relationship between mathematics performance and writing proficiency. The data in Table 4.1 will be used to illustrate how to tackle this task. Each row in the table represents the mathematics and written communication scores of a single study participant.

In Box 4.1 is the outline of a research plan that might be used to tackle the investigation related to this case. It includes some of the fundamental concerns likely to be found in the methodology section of a proposal when conducting a correlational study.

4.3 Bivariate Correlation Fundamentals

Bivariate correlation is a statistical method used to describe the relationship between two variables. Therefore, it is a descriptive statistical analysis. The purpose of the analysis is to assess whether the linear relationship between the variables might be described in one of three ways:
- As the values of one variable increase, so do values of the other variable.
- As the values of one variable increase, values of the other decline in magnitude.
- There is no discernible linear pattern of change in the values of the two variables.

Table 4.1 Written Communication and Mathematics Scores of Students

ID	Mathematics	Written Communication
001	56	4
002	46	7
003	20	4
004	27	6
005	65	8
006	53	9
007	66	7
008	47	5
009	77	11
010	29	3

As the name suggests, in bivariate correlation only pairs of variables are involved. Each individual participating in the study has a score or value for each of those two variables. Therefore, for the 10 participants in the case study in Table 4.1, there are two values in Row 1 for participant 001—one value for written communication and one for mathematics. There may be missing data in a dataset; a written communication or mathematics score may be missing for a participant. Ideally, there are no missing values in the dataset, but realistically datasets are rarely complete. Chapter 3 discusses data preparation and analysis options for missing data.

Bivariate correlation refers to relationship between two variables in which the relationship can best be represented by a straight line. These relationships are referred to as linear correlations. For a relationship to be linear, the best-fit line cannot change. If the line that best represents the relationship between variables is not a straight line, the relationship between the two variables is described as curvilinear or nonlinear. This latter relationship cannot be measured using correlational statistics. To illustrate, in Figure 4.4, a line has been added to the original scatterplot in Figure 4.2, that best represents the relationship between MPG and passenger car weight. A careful examination of the diagram divulges that as MPG increases, the weight of passenger cars declines. Those passenger cars weighing the most have the lowest MPG. Conversely, the cars that have the lowest weight have the highest MPG (e.g., the passenger car labeled A).

4.3.1 Scatterplots and Correlation

The plots in Figures 4.1 – Figure 4.4 each depicting the relationship between two variables are called scatterplots (also called scattergrams or scatter diagrams), and they are one technique for identifying whether a linear relationship exists between two variables. You may recall from Chapter 3 that in a scatterplot every study participant is represented by a single point in the plot, and coordinates for each point is each study participant's scores for the variables involved. For example, in Figure 4.4, one car had the highest miles per gallon (65.4 mpg). Car A's weight is 17.5 hundred pounds, resulting in coordinates of 65.4 and 17.5.

Figure 4.4 Relationship Between Miles Per Gallon and Passenger Automobile Weight

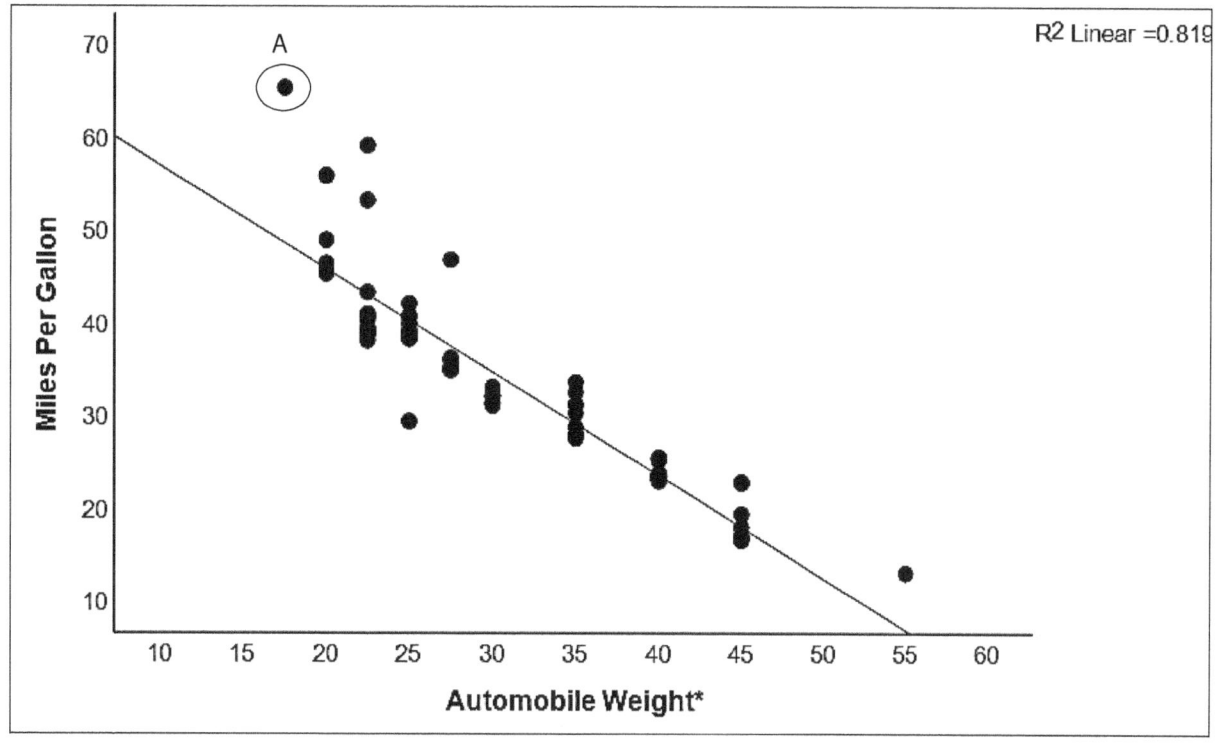

Heavenrich, R.M., Murrell, J.D. and Hellman, K.H. (1991). Light Duty Automotive Technology and Fuel Economy Trends Through 1991. U.S. Environmental Protection Agency (EPA/AA/CTAB/91-02).
*Hundreds of pounds

Scatterplots are useful in depicting whether relationships between variable are positive, negative, or devoid of a relationship by virtue of the direction of the shape of the scatterplot. If the plot rises from left to right, the relationship is positive. If the plot declines from left to right, the variables have a negative relationship.

Scatterplots also provide a graphic indication of whether the inter-variable relationship is linear or nonlinear. Consider the scatterplot in Figure 4.5, which illustrates a curvilinear relationship. Some psychologists (e.g., Keeley et al., 2008; Sarid et al., 2004) suggest that a curvilinear relationship exists between anxiety and test performance; as anxiety increases so does test performance up to a point; then anxiety changes from a positive association with performance to a negative one. Scatterplots revealing a curvilinear relationship is an indication that bivariate correlation is not an appropriate measure, and so it would be unwise to proceed with such an analysis. If correlation is used to represent curvilinear relationship between two variables, it will underestimate the actual relationship between the variables.

4.3.2 Scatterplots and Prediction

In correlation, there is no easily identifiable dependent or outcome variable. In the case of the data presented in Table 4.1, we will investigate whether mathematics achievement is correlated with writing achievement. In this situation, either written communication or mathematics could be the outcome variable if the purpose of the analysis was to predict an outcome.

When the purpose is to predict an outcome, the predictor is typically placed on the *x* axis in a correlational scatterplot, and the outcome or dependent variable is placed on the *y* axis. If, however researchers merely wish to determine whether a concomitant relationship exists between two variables, there is no clear independent or dependent variable; in that situation it is irrelevant which variable is placed on the *x* or *y* axis in a correlational scatterplot.

Figure 4.5 An Example of a Curvilinear Relationship

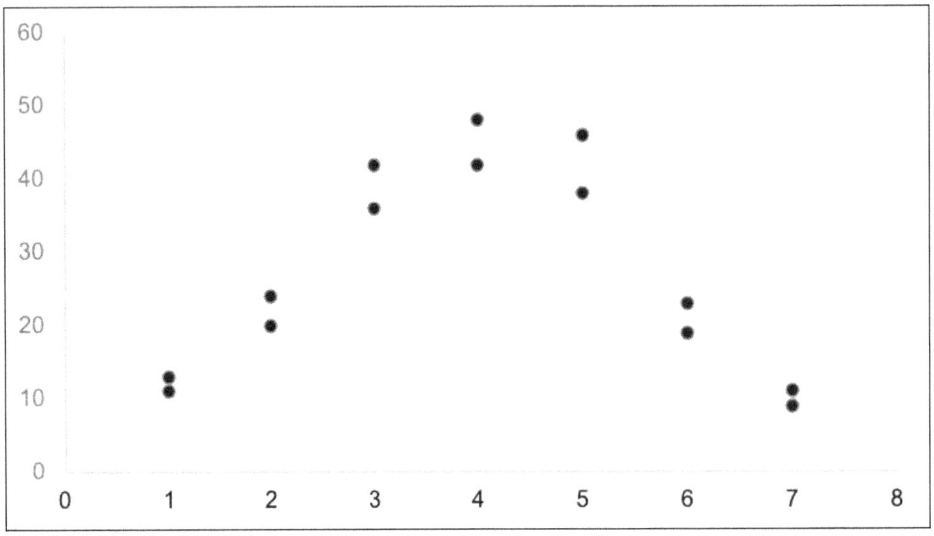

4.4 Planning a Bivariate Correlational Study

Once a researcher settles on the general outlines of the type of analysis that will be conducted in a study, other research related activities have already occurred. As discussed in Chapters 2 and 3, before collecting data or acquiring extant data for a correlational study, the investigator should have already read previous related research in the area of interest and identified appropriate theoretical constructs to guide the general outlines of the study. Without this approach, researchers may secure usable pieces of information, but the likelihood of getting substantially useful information is far less than if the researcher is well-acquainted with previous research and theory surrounding the variables to be correlated.

Assume, however, that the foundational work has been done, and you want to better understand student achievement by exploring potential relationships between achievement in elementary school subjects, like mathematics and writing (written communication). An outline of some of the chief components of the research plan that typically precedes a bivariate correlational study of that sort is presented in Box 4.1.

4.4.1 Eligible Data and Measurement Levels

As a quantitative statistics technique, a prerequisite for conducting a bivariate, correlational analysis is the use of quantitative or numerical data. The type of quantitative data required depends on the correlation coefficient being calculated. As presented in Box 4.1, for the correlational case study, the two variables being analyzed are measured at the interval or ratio level (continuous scores). Exam performances, like mathematics and writing performance used for this case, are considered interval-level data as they are continuous scores without a meaningful zero-point level.

Other types of quantitative data that may be used to conduct bivariate correlation analyses include nominal and ordinal-level data. Those and their associated correlation coefficients will be explored in Section 4.12.

An additional consideration about eligible data in correlational studies is whether the two variables must be the same unit of measure. For example, can you correlate the variable age with the variable blood pressure level which uses millimeters of mercury (mmHg) to measure the pressure in your blood vessels? Fortunately, Pearson's r is not affected by the use of variables with different units of measurement. Therefore, variables with different units of measurement may be correlated.

Box 4.1 Research Plan Outline: The Case of Writing and Mathematics Skills

Research Problem
Graham et al. (2020) and others suggest that for students at all levels of education in the Caribbean, achievement in the content areas, like mathematics, the sciences, and social studies, is related to students' proficiency in the language arts. With students' mathematics performance lagging, a language arts teacher who is a researcher and an avid writer is interested in determining whether writing achievement applies in this situation.

Variables and Measurement Levels
Variable 1: Mathematics Test Score
Variable 2: Written Communication Test Score
Level of Measurement: Interval (continuous scores)

Research Question
Is there a statistically significant linear relationship between mathematics (MS) and written communication (WCS) scores?

Hypotheses: Two-Tailed Test of Significance
Null Hypothesis 1: There is no significant relationship between mathematics and written communication performance.
Null Hypothesis 2: The population correlation coefficient for the relationship between mathematics and written communication performance is not significantly different from 0.
$H_0: \rho = 0$

Alternative Hypothesis 1: There is a significant relationship between mathematics and written communication performance.
Alternative Hypothesis 2: The population correlation coefficient for the relationship between mathematics and written communication performance is significantly different from 0.
$H_a: \rho \neq 0$

Research Design
Nonexperimental, bivariate correlational research design

Data Analysis
Included in preliminary data analysis are descriptive statistics, including means and standard deviations. Mathematics and written communication scores are correlated using Pearson product moment correlation (r). Significance testing using the t statistic and the t distribution determines whether the correlation coefficient is significant at the a priori alpha level of $\alpha = .05$.

4.4.2 Research Question

After isolating and constructing the research problem, using the guidelines in Chapter 2, you should be in a better position to clearly identify the variables you wish to study that appear to be interrelated. It is at this juncture that you can frame a well-formulated research question that is essentially a restatement of the research problem in summarized question form.

Research questions for bivariate correlation may be nondirectional like the question in Box 4.1, in that it inquires whether there is a linear relationship between two variables of interest. Alternatively, the research question may be directional or pose a question about a direct relationship between the two variables; for example,

Is there a positive linear relationship between mathematics test performance and written communication test performance?

Or

Is there a negative linear relationship between mathematics test performance and written communication test performance?

4.4.3 Bivariate Correlation Hypotheses

As is the case for most statistical techniques, hypotheses for bivariate correlation are restated forms of research questions that focus on the expected relationship between the two variables in a single sample of individuals. Hypotheses are educated guesses about the relationship between the variables. They are starting points for the investigation of the nature of the intervariable relationship.

Hypotheses are used in conjunction with the correlational statistical techniques to determine the nature of an intervariable relationship in a sample. Hypotheses and associated statistical tests allow researchers to conduct investigations using a representative sample of a population when they do not have access to an entire target population. They are starting points for significance testing that determine the likelihood that what is observed in a representative sample occurs in the population as well.

Like research questions for correlational studies, hypotheses for these investigations may be nondirectional or directional depending on the objectives of the study. Nondirectional hypotheses for the case study are presented in Box 4.1 as well as the hypotheses in notation form. Here, the symbol, ρ (the Greek letter "rho"), refers to the population correlation coefficient, which is the basis of correlational significance testing. The null indicates that the strength of the correlation coefficient in the population equals zero (no relationship), and the alternative indicates that the correlation does not equal zero, but signifies some relationship in the population. An alternative hypothesis stating that the correlation was either positive ($H_a: \rho > 0$) or negative ($H_a: \rho < 0$), would be called a directional hypothesis and use associated one-tailed rather than two-tailed significance testing of the hypothesis. More about hypothesis testing and significance testing later in this chapter.

4.4.4 Research Design: When to Use Bivariate Correlation

Simple, bivariate correlation is a statistical technique that is frequently used in nonexperimental designs where the variables involved cannot be manipulated by the researcher. While nonexperimental designs are not as powerful as experiments conducted in a lab setting, nonexperimental designs like correlational studies provide important options when experimental designs are not feasible. In some cases, correlational studies may be exploratory precursors to quasi- or laboratory experiments.

The Case of Writing and Mathematics Skills (Box 4.1) is an example of a retrospective (pre-existent data) correlational research design that seeks to explore whether a monotonic (rather than a causal) linear relationship exists between mathematics performance and writing performance. The design does not identify the variables as independent or dependent. Accordingly, a predictive relationship is not being hypothesized, but study outcomes could be the basis for future research that examines the likelihood of a predictive relationship.

4.4.5 Data Analysis

Simple bivariate correlation may be conducted as a parametric or nonparametric test depending on the data collected. If one variable uses ordinal-level or nominal data, the data are best analyzed using a nonparametric test. (More on this in Chapter 12). However, in this case study the researcher collected data measured on an interval scale; therefore, a correlational, parametric test may be used for analysis if the data meet the required assumptions (Section 4.8).

The most frequent parametric method used to examine the strength of association between two variables is the Pearson's product moment correlation coefficient or Pearson's r, and that is what is planned as the primary method of statistical analysis for this case study. The associated coefficient, r, will describe the relationship seen in the sample data. Hypothesis testing will reveal the likelihood that the relationship will occur in the population.

4.5 The Pearson Correlation Coefficient

Although scatterplots facilitate visualization of the association between two variables, scatterplots do not assist with quantification of the nature of the relationship. Scatterplots indicate the direction of the relationship: whether the relationship is positive, negative or probably nonexistent. However, scatterplots cannot provide critical statistical information about the strength of a relationship between two variables. Is the relationship positive or negative? Is it a strong, moderate, or weak relationship? Scatterplots also cannot relate whether the relationship is significantly different from what might be expected by chance in the population. For that level of investigation, researchers need the assistance of correlational statistics.

Researchers can investigate the relationship between variables with several correlational statistics, but as noted in the previous section, perhaps the most common statistic used to measure such relationships is Pearson's product moment correlation coefficient, commonly called Pearson's r when it is applied to a sample of a target population. The index known as Pearson's r was not named after the original developer of the measure. Francis Galton first conceived the index in the 1880s, and it was further developed by Karl Pearson and Auguste Bravais in 1844 (Pearson, 1920; Stigler, 1989).

The Pearson correlation coefficient is generated from the data sets for two variables (e.g., mathematics and written communication scores in Table 4.1) and describes the extent to which the two variables are associated with each other. The Pearson's r index has several characteristics of note.

- It is based on the covariance statistic, which quantifies the extent and direction in which two variables vary together.

- The index quantifies the direction and magnitude of the relationship of the variables.
- The index has a restricted range.
- It is used along with significance testing.
- The use of the index is limited to certain types of data.

4.5.1 Covariance and Pearson's r

Pearson's *r* is based on another statistic known as covariance, whose values reflect the degree to which two variables vary together. For example, if high (above average) scores in language arts pairs with high (above average) scores in mathematics, the covariance statistic will be large and positive (e.g., Figure 4.1). On the other hand, when high (above average) scores on weight are paired with low (below average) scores on exercise, the covariance is negative (e.g., Figure 4.2). When high scores on one variable are paired about equally frequently with high and low scores on the other variable, the covariance will be near zero (e.g., Figure 4.3). Accordingly, like correlation, covariance allows us to identify the direction of the bivariate relationship between variables.

Although the direction of a bivariate relationship is easily interpretable using a scatterplot or covariance, it is more challenging using covariance to interpret the magnitude of relationships. A large covariance which means a strong relationship between two variables in one study may mean a weak relationship in another study simply because different units of measure are used for the same variables. Imagine that a researcher, investigating the covariance of the heights and weights of college freshmen measured first in inches and pounds, and later converted the measures to centimeters and kilograms. In this situation the covariance would change even though all that has changed is the units of measure for height and weight. The larger the values of the data involved (e.g., 152.4 centimeters = 60 inches), the larger the covariance, making it difficult to interpret strengths of relationships.

Another challenge with covariance is that the values are not standardized. As a result, the index has a very wide range, from negative infinity to positive infinity. Consequently, the value of a perfect linear relationship depends on the data. Imagine attempting to interpret the magnitude of a relationship with such a wide potential range. Pearson's *r* fixes this limitation by standardizing the covariance index; that is dividing the covariance by the standard deviation of the two data sets in the study.

4.5.2 Direction and Magnitude of Pearson's r

Figures 4.1 and 4.2 of the linear relationships between two variables represent the positive and negative (respectively) directions of those relationships. Although the general directions of the relationships are illustrated, a perfect positive and negative relationship between two variables is depicted by all points in each scatterplot on a straight line as presented in Figures 4.6a and 4.6b. In these situations, every change in x is accompanied by a corresponding, proportional change in y. A scatterplot in which the points do not have any discernible direction or are circular or nearly circular as in Figure 4.3, illustrates zero or near-zero relationship between two variables.

The Pearson's *r* index measures the quantitative value of the magnitude of the relationship between two variables, and it does so in standardized form in which the range is -1.0 to +1.0. The coefficient for the magnitude of the perfect positive correlation between x and y in Figure 4.6a is +1.0 and the coefficient for the perfect negative correlation between x and y illustrated in Figure 4.6b is -1.0.

Figure 4.6 Examples of Perfect Linear Relationship Between Two Variables

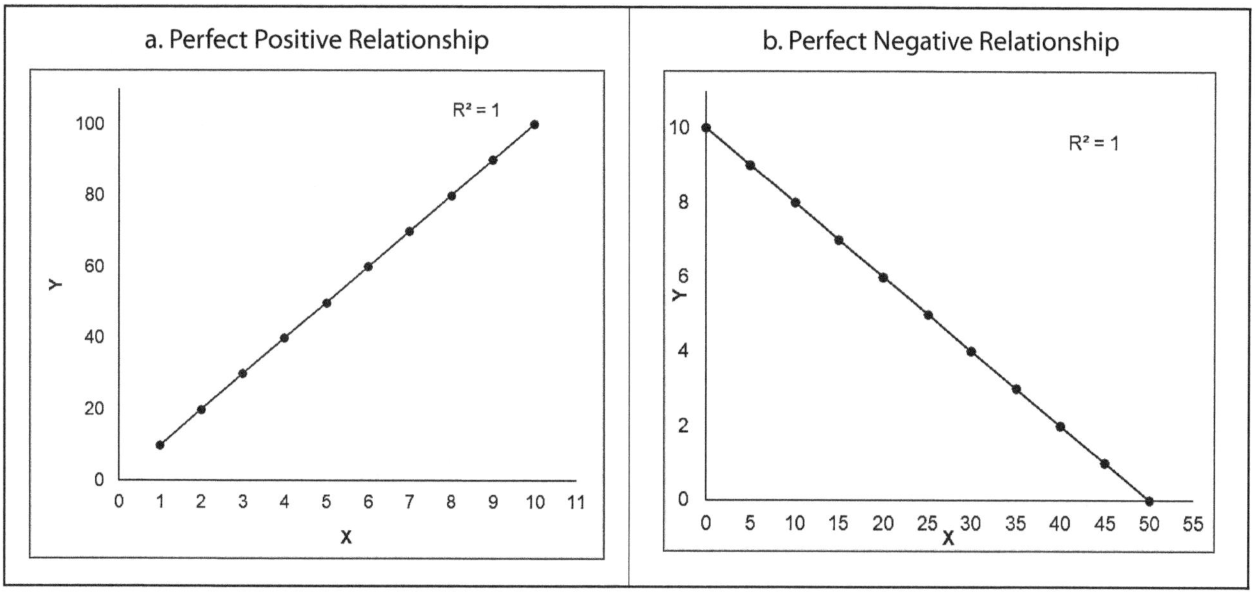

a. Perfect Positive Relationship

b. Perfect Negative Relationship

The value of the correlation coefficient is based on the slope of the pattern of points in a scatterplot and the extent to which the points are spread out or are close to the line of best fit. If the slope of the mass of points is in a lower left to upper right pattern, the correlation coefficient is positive. If the mass of points is in an upper left to lower right direction, the correlation is negative. In addition, the narrower the mass of points, the stronger the association between the two variables; the closer they are to the line of best fit; and the larger the correlation coefficient (closer to 1 or -1), regardless of the direction of the pattern. By contrast, the greater the spread of data points around the line of best fit, the weaker is the association between the two variables; the farther the data points are from the line of best fit, and the closer the Pearson coefficient index is to zero. For example, consider the scatterplot in Figure 4.4, where the data points are more spread out; the corresponding Pearson's r is .9, while the data points in Figure 4.6 line up in perfect correlation.

Aside from the direction of the line, Pearson's r does not tell us anything about the value of the slope of the line of best fit (called the regression line in linear regression). Therefore, a Pearson correlation coefficient of +1 does not mean that for every unit increase in one variable there is a unit increase in the other. However, it does mean that there is no variation between the data points and the line of best fit as illustrated in Figures 4.6.

Pearson's r, like other correlation coefficients, is only interpretable when each participant in a study has a value from each of the two variables to be correlated. Therefore, a correlation coefficient may not be used to investigate the correlation between mathematics scores for Group 1 and reading scores for Group 2. Similarly, correlation coefficients may not be used to examine the correlation between satisfaction with life for Group 1 and happiness for Group 2. Data may only be taken from a single group of study participants and each participant should have a score for each variable.

4.5.3 Interpreting the Pearson's r Index

An important part of interpreting Pearson's r is understanding the scale for that index. The range of possible values for Pearson's r is on an ordinal scale. Therefore, it represents a ranking of the values of the coefficient, and it does not have the characteristics of an interval or ratio scale. Accordingly, unlike values on interval or ratio scales of measurement, it is incorrect to say that an increase from $r = .30$ to $r = .50$ is the same as an increase from $r = .50$ to $r = .70$. Furthermore, it is incorrect to say that an index of $r = .60$ is twice as large as $r = .30$.

Deciding how to interpret Pearson's r must be done within the context of other relevant considerations. For example, is the correlation significant? A Pearson's r that indicates strength of association between two variables in a study sample may not be reflective of what is occurring in the target population from which the sample was taken, and therefore, the correlation might be only due to chance coincidence (random sampling error). To avoid that situation, a correlation analysis must be accompanied by a significance test to determine the extent to which the sample correlation is reflective of what is likely in the population. More on this in Section 4.5.5.

Another concern is the size of the study sample. The larger the sample size the greater the likelihood that sample correlations will reflect the population correlation value. The smaller the sample size, the greater the likelihood of obtaining a spuriously-large correlation coefficient. This is especially true with study samples with homogeneous data values. Proper study design as well as reporting other indices along with Pearson's r (e.g., significance testing results and shared variance results, r^2), allows researchers to present a balanced picture of the results of their study. Guidelines for interpreting the Pearson's r is presented in Table 4.2.

Table 4.2 Correlation Coefficient Interpretation Guidelines

Strength of Association	Coefficient r Direction	
	Positive	Negative
Very high correlation	.90 to 1.00	-.90 to -1.00
High correlation	.70 to .90	-.70 to -.90
Moderate correlation	.50 to .70	-.50 to -.70
Low correlation	.30 to .50	-.30 to -.50
Little to no correlation	.00 to .30	.00 to -.30

A third concern is the tendency to interpret the relationship between two variables as having a cause-and-effect relationship. As noted in Section 4.3.2, often there is no clearly identifiable dependent or independent variable. Furthermore, bivariate correlation is designed to examine linear monotonic associations between two variables rather than a cause-and-effect relationship or even a predictive relationship. More about correlation and causality in Section 4.7 and about bivariate prediction in Chapter 5.

4.5.4 Interpreting Pearson's r as Shared Variability (r^2)

Pearson's r describes the direction and strength of the association between two variables. For example, in Figure 4.4, the Pearson's r index describes the relationship between auto weight and miles per gallon as -0.905, strong, negative relationship between the two variables. While it is fair to conclude that as auto weight increases, miles per gallon decreases, the changes in miles per gallon might be related to other factors as well, such as poor maintenance, or speeding. Though Pearson's r is an index of the relationship between weight and miles per gallon, it tells only a part of the story. Knowing what proportion of the variance in car weight is associated with the variance in miles per gallon, provides evidence of the practical significance of Pearson's *r*. The square of Pearson's *r* provides that additional information.

The square of the Pearson correlation coefficient *r* is known as the coefficient of determination, r^2. This index indicates the proportion of variance in one variable that is shared by the variance in the other variable in bivariate correlation. It provides a measure of the amount of variation that can be explained by the model (Here, the correlation is the model) or how close the mass of datapoints is to the line of best fit; the closer the points are to the line of best fit, the larger the coefficient of determination. So, for example, a Pearson correlation coefficient of 0.905 would result in a coefficient of determination of 0.819, (i.e., r^2 = 0.905 x 0.905 = 0.819), indicating that data values are close to the line of best fit.

The coefficient of determination is an index that is measured on the ordinal scale. Therefore, it has limitations in interpretation that are similar to those of Pearson's *r*, discussed in Section 4.5.3. It is often expressed as a percentage (e.g., 82% instead of .819) to describe the proportion of variance shared by the variables. However, in practice, researchers do not write r^2 = 82%, or any other percentage. They report the coefficient of determination as a proportion (e.g., r^2 =.819).

R squared or the coefficient of determination, always falls between 0 and 1 or 0% and 100%:
- 0% indicates that the variables do not share any variability. Variable A does not explain any of the variance in variable B.
- 100% indicates that variable A shares (explains) all of the variability in variable B. However, it is highly unlikely that an independent variable will account for all of the variance in a dependent variable, meaning that all the data points fall on the line of best fit.

The more variance that is accounted for by the correlation model the closer the data points will hang together.

4.5.5 Testing the Statistical Significance of Pearson's r

As noted earlier, most research studies are based on representative samples of the population. The statistical techniques used in those situations produce results based on a sample rather than a population. This happens in correlation studies as well. Both Pearson's *r* and the coefficient of determination (r^2) summarize the direction and strength of a linear relationship *in samples only*. However, the outcomes of studies based on different samples do not always agree. Different correlation studies might realize divergent correlation coefficients and coefficients of determination values, resulting in different conclusions. To avoid this problem, the best practice is to *draw conclusions about populations*, not just samples. To accomplish this task, researchers conduct significance tests or calculate confidence intervals. This section explains how to conduct a statistical significance test for the population correlation coefficient ρ (the Greek letter "rho"). The significance test reveals the probability that what is observed in a study sample will occur in the population.

4.5.5.1 Before Significance Testing

For illustration purposes, consider the case study described in Section 4.2 and Box 4.1.
- Data has been collected from 10 study participants with scores for two variables: mathematics and writing performance.

132 Bivariate Correlation

- Assume the sample correlation coefficient (*r* or Pearson's *r*) is known. (This will be computed in Section 4.6.
- The population correlation coefficient (ρ or rho) between mathematics and writing is unknown.
- The objective is to make an inference about the value of ρ based on *r*.

4.5.5.2 The Hypotheses

Hypothesis testing is a core part of what is known as statistical significance testing. The objective of statistical significance testing is to make inferences about a population based on the representative sample correlation coefficient (*r*) and the sample size (*N*).

Every statistical significance test begins with a null hypothesis and an alternative hypothesis like the ones in Box 4.1. The hypothesis test is based on the following inferences about the population correlation coefficient rho (ρ):

- It is **not** significantly different from zero (H_0: ρ = 0).

OR

- It is significantly different from zero (H_a: ρ ≠ 0).

If, using predetermined guidelines, the significance test shows that ρ is close enough to zero, it means that the null is true and there is insufficient statistical evidence that there is a significant linear relationship between mathematics and writing in the target population. Hence there is no significant difference between the population correlation coefficient, ρ, and zero. Any evidence in the sample to the contrary occurred as a result of chance.

If, however, ρ proves to be different enough from zero, the null hypothesis is not supported. The statistical evidence supports the claim that there is a significant correlation between mathematics and writing in the target population.

In effect, the objective is to prove that the null hypothesis is false (i.e., ρ ≠ 0). However, it is important to be careful in interpreting the statistical significance of a correlation. A correlation coefficient that is statistically significant does not mean that there is a strong association between the variables. It is the result of testing the null hypothesis that there is no relationship between two variables. Rejecting the null and accepting the alternative hypothesis means there is evidence that a significant relationship exists. However, significance testing does not provide any information about the strength of the relationship of the variables or its importance.

4.5.5.3 The Statistical Test: Student's T Test

A *t* test, called Student's *t* test, is an inferential statistic that allows researchers to use sample data (specifically the sample correlation coefficient, *r*) to test the null hypothesis about the target population and make a generalization about that entire population. In the matter of the case study, it is the vehicle to uncover whether the sample correlation between mathematics and writing is likely to occur in the entire population.

The following formula may be used to calculate the value of the *t* test statistic, a standardized value called the *t*-value. The larger the *t* value, the more likely it is that the correlation will occur in the population.

$$t = \frac{r(\sqrt{N-2})}{\sqrt{1-r^2}}$$

where:
N is the sample size,
r is the sample correlation coefficient

4.5.5.4 The Probability Value or P Value

Every t value has a matching p value. A p value is the probability that the null hypothesis is true. In our case, it represents the probability that the correlation between x and y in the sample data occurred by chance.

A p value of .05 means that there is only 5% chance that results from your sample occurred due to chance. A p value of .01 means that there is only 1% chance. The lower the p value, the stronger the evidence that we should reject the null hypothesis that there is no significant correlation between the two variables (e.g., mathematics and writing). However, in most research studies, the threshold for considering test results statistically significant is a p value lower than $\alpha = .05$.

Typically, during the planning phases of a study, the researcher sets a standard for evaluating study results. This predetermined standard is called the significance level or alpha (α). Perhaps the most frequently used alpha levels are $\alpha = .05$ and $\alpha = .01$. Once the researcher secures the p value for the statistic, the next step is to compare it against the alpha. If the p value is smaller than the alpha level, the researcher can reject the null. If the p value exceeds the alpha, then the null hypothesis stands.

Finding the p value requires two pieces of information, the t- value using the formula in 4.5.5.3 and the number of degrees of freedom (df) for the study, which is $df=N - 2$ (where N is the sample size). With the t value, degrees of freedom and preselected alpha level of .05 in hand we can calculate the p value by:
- Using statistics software like SPSS
- Using hand calculations and the critical value tables in Appendix A

4.5.5.5 Reject or Fail to Reject—Software

As Box 4.1 indicates, the case-study plan calls for an alpha of $\alpha = .05$. Therefore, a p value that is smaller than the alpha ($\alpha = .05$), leads to a **rejection** of the null hypothesis in favor of the alternative hypothesis and which means that the correlation between mathematics and writing is statistically significant; there is a strong likelihood at the stated alpha level that there is a monotonic, linear relationship between mathematics and writing in the target population.

A p value that is larger than the alpha ($\alpha = .05$), means a **failure to reject** the null hypothesis, meaning there is insufficient statistical evidence that the correlation between mathematics and writing is statistically significant; rather, there is a strong likelihood at the stated alpha level there is not a significant linear correlation between mathematics and writing in the target population.

4.5.5.6 Calculation to Reject or Fail to Reject—Method 1

$$t = \frac{r(\sqrt{N-2})}{\sqrt{1-r^2}}$$

SPSS and other statistical software use the above equation to test the significance of a correlation coefficient. Calculate the t-value and compare that with the critical value in the statistics table (Appendix A) at the appropriate degrees of freedom and the desired alpha level.

134 Bivariate Correlation

If the *t* value exceeds the critical value, then you may reject the null hypothesis that the correlation between the two variables is not significantly different from zero. If the calculated *t* value does not exceed the critical value, then you cannot reject the null hypothesis that there is no linear relationship between the two variables.

4.5.5.7 Calculation to Reject or Fail to Reject—Method 2

A faster shorthand way to test the statistical significance of a correlation coefficient is to calculate the relationship between the sample size and the correlation coefficient:

$$|r| \geq \frac{2}{\sqrt{N}}$$

where $|r|$ is the absolute value of a correlation coefficient and N is the sample size.

The absolute value of a Pearson correlation coefficient ($|r|$) that is equal to or greater than the calculated critical value demonstrates that there is a statistically significant linear relationship at approximately the .05 level of significance between the two variables. As the above formula shows, there is an inverse relationship between the sample size and the required correlation coefficient for significance of a linear relationship. The larger the sample size the smaller the required correlation coefficient for statistical significance. With a sample size of 10, the required correlation for significance is .6325 at an alpha level of .05; with a sample size of 50, the required correlation for significance is .273.

For an even faster method of calculation, use the statistical table of critical values for correlation coefficients in Appendix A to look up the critical value for the calculated correlated coefficient. For example, assume a correlation coefficient of .723 for a sample size of 11 and therefore a degrees-of-freedom of 9 ($N-2$). Imagine also a selected alpha level of .05 for a two-tailed test. Locate the critical value for 9 degrees of freedom at the pre-set alpha level, which is .602. The correlation coefficient of .723 exceeds the critical value of .602. Therefore, the coefficient ($r = .723$) represents a significant linear relationship between two variables.

4.6 Calculations for the Pearson Product Moment Correlation

4.6.1 Computing Pearson's Correlation Coefficient

With a foundational understanding of how to interpret Pearson's *r* and the population correlation coefficient, this exercise uses the data in Table 4.1 to illustrate how to calculate both statistical indexes.

Though some practitioners defer to statistics software or Excel to calculate the Pearson *r* coefficient, it is relatively straightforward to compute the coefficient index manually if there are only a few data points as there are in Table 4.1. There are multiple ways of calculating the Pearson correlation coefficient. However, as Howell (2014) notes, the raw score formula is useful when calculating Pearson's *r* by hand. The Pearson's *r* formula is as follows:

$$r = \frac{N\Sigma XY - \Sigma X \Sigma Y}{\sqrt{[N\Sigma X^2 - (\Sigma X)^2][N\Sigma Y^2 - (\Sigma Y)^2]}}$$

where:

r = Pearson correlation coefficient
N = the sample size (10)
Σ = the sum of
X = mathematics score
Y = written communication score

The data table is presented again in Table 4.3. The variable, mathematics, is represented by the symbol X in the formula and in Table 4.3, while the variable, writing communication, is represented by the symbol Y in the formula and in Table 4.3.

Using the above formula, the procedures for computing the Pearson correlation coefficient are presented in four steps.

1. Calculate the values for X^2, Y^2, and XY.
2. Calculate values for the numerator.
3. Calculate values for the denominator.
4. Using the results from Step 2 and 3, calculate Pearson's r.

Table 4.3 Variables and Values for Computing the Pearson Correlation Coefficient

Row Number	Mathematics (X)	Written Communication (Y)
1	56	4
2	46	7
3	20	4
4	27	6
5	65	8
6	53	9
7	66	7
8	47	5
9	77	11
10	29	3

The values for Step 1 appear in Table 4.4. Calculation of the values for the numerator (Step 2) and denominator (Step 3) are displayed in Figure 4.7. Use the information in Figure 4.7 to enter the values in the equation.

Table 4.4 Values for Calculating the Pearson Correlation Coefficient

(1) Row	(2) X	(3) X²	(4) Y	(5) Y²	(6) XY
1	56	3136	4	16	224
2	46	2116	7	49	322
3	20	400	4	16	80
4	27	729	6	36	162
5	65	4225	8	64	520
6	53	2809	9	81	477
7	66	4356	7	49	462
8	47	2209	5	25	235
9	77	5929	11	121	847
10	29	841	3	9	87
11	ΣX = 486	ΣX² = 26750	ΣY = 64	ΣY² = 466	ΣXY = 3416

Figure 4.7 Values Needed for Steps 2 and 3 of Pearson *r* Formula

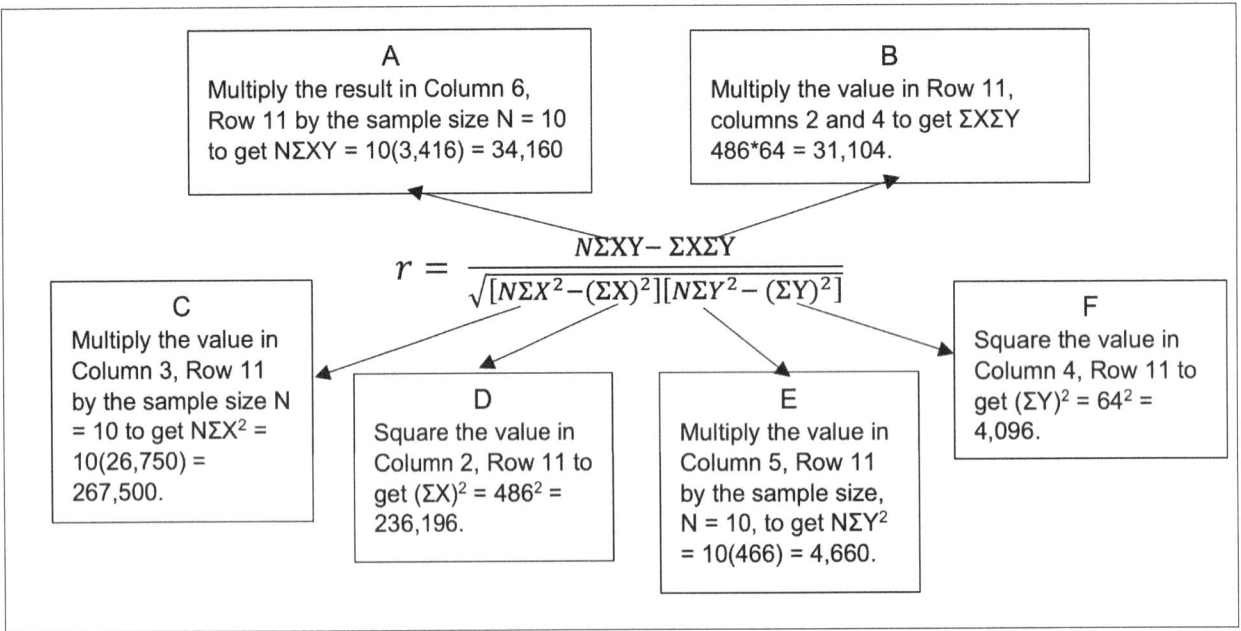

Step 4: Insert the values for the Pearson's *r* formula in Figure 4.7 into the formula.

$$r = \frac{34{,}160 - 31{,}104}{\sqrt{(267{,}500 - 236{,}196)(4{,}660 - 4{,}096)}}$$

$$r = \frac{3{,}056}{\sqrt{31{,}304 * 564}}$$

$$r = \frac{3{,}056}{\sqrt{17{,}655{,}456}}$$

$$r = \frac{3{,}056}{4{,}201.840}$$

$$r = .727$$

4.6.2 Shared Variability (r²) of Mathematics and Written Communication

Section 4.5.4, explored shared variability or the coefficient of determination as the index of the proportion of variance shared by the two variables being analyzed. It is one way of divining the practical significance of the relationship between the variables in bivariate correlation. The Pearson's r of mathematics and written communication performance at .727 means that, based on the guidelines in Table 4.2, the variables are moderately highly correlated; just over the range for moderate correlation. To determine what proportion of the variance they share, find the square of the Pearson's r. Therefore, $r^2 = .727^2 = .529$. Accordingly, the coefficient of determination for written communication and mathematics performance is .529. Written communication accounts for 52.9% or more than half of the variance in mathematics performance.

4.6.3 Finding the Significance of the Correlation Coefficient

According to Section 4.5.5, both Pearson's r and the coefficient of determination (r^2) summarize the direction and strength of a linear relationship in samples only. However, the objective is to draw conclusions about the target population of the study. To accomplish this task, it will be necessary to test the statistical significance of the sample correlation coefficient.

Student's t test is the inferential significance test designed to determine the probability of the sample correlation between mathematics and written communication occurring in the target population. The value of the t-test statistic (t-value) may be calculated, using following formula:

$$t = \frac{r * \sqrt{N-2}}{\sqrt{1-r^2}}$$

where:
N is the sample size,
r is the sample correlation coefficient

Substituting in the values for N and r, replace N with the sample size, 10, and r with Pearson's $r = .727$, which results in:

$$t = \frac{.727 * \sqrt{10-2}}{\sqrt{1-.727^2}}$$

$$t = \frac{.727 * \sqrt{8}}{\sqrt{1-.529}}$$

$$t = \frac{.727 * 2.828}{\sqrt{.471}}$$

$$t = \frac{2.056}{.686}$$

$$t = 2.997$$

Using the *t*-value of 2.997, we may compare it against the critical value found in the statistical tables in Appendix A. The degrees of freedom are $df = N - 2$ where N is the size of the sample. The critical value for 8 degrees of freedom at an alpha level of .05 (two-tailed test) is 2.306. The *t* value exceeds the critical value. Therefore, we may reject the null hypothesis in favor of the alternative hypothesis and conclude that the correlation between mathematics and written communication is statistically significant.

4.6.4 Calculating the Confidence Interval for the Correlation Coefficient

Chapter 2 described a confidence interval as a range of values which is likely to include the true population value. In this case, the range of values are correlation coefficients calculated from a study sample (r) within which the true correlation coefficient (ρ) for the target population lies.

Correlation coefficients present a special challenge in calculating confidence intervals, which are based on the distribution of standard scores (z) that are normally distributed. However, the sampling distribution of Pearson's r is not normally distributed. Rather, it is negatively skewed because Pearson's r cannot take on values greater than 1.0, and therefore, the distribution cannot spread as far in the positive direction as it can in the negative direction (Figure 4.8).

Figure 4.8 Example of the Sampling Distribution of Pearson's *r*

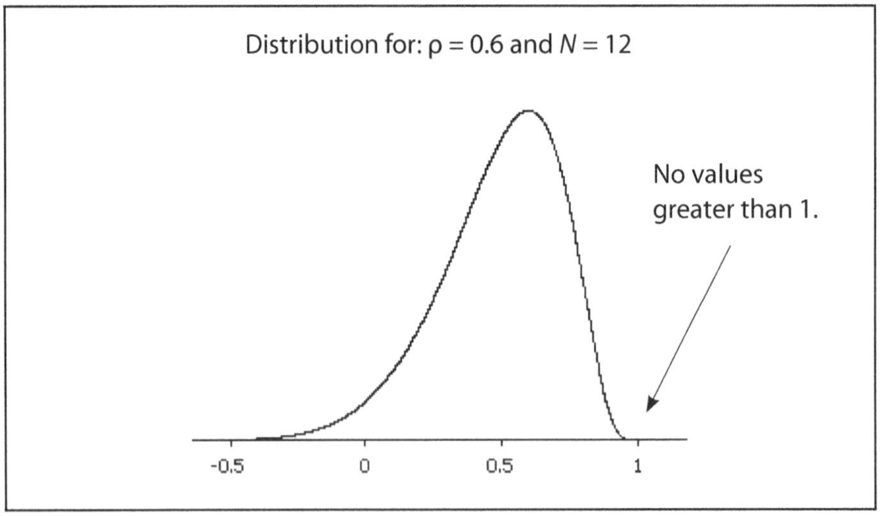

Online Statistics Education: A Multimedia Course of Study (http://onlinestatbook.com/).
Project Leader: David M. Lane, Rice University.

As a result of the negative skew of the sampling distribution of r, a work-around is necessary to compute the confidence interval for the sample correlation coefficient, .727 (Section 4.6.1)
1. Use Table A.3 in Appendix A to convert r to its standardized score, in this case Fisher's z_r which is normally distributed.
2. Hand-calculate the standard error for Fisher's z_r, which is:

$$SE_{z_r} = \frac{1}{\sqrt{N-3}}$$

where N is the sample size.

3. Calculate the confidence interval for the lower boundary (CI_L) and the upper boundary (CI_U) level using the following equations:

   $CI_L = z_r - (1.96 * SE_{z_r})$
   $CI_U = z_r + (1.96 * SE_{z_r})$
   1.96 represents the z score for a 95% ($\alpha = .05$) level.

4. Convert the boundaries of the confidence intervals back to correlation coefficients for easier interpretation.

Steps 1 and 2: Calculate Fisher's z_r and its standard error.

- For the hand-calculated example, the Pearson correlation coefficient (r) was 0.727. Looking up that correlation in Appendix A, Table 3 shows that the Fisher's z transformed value is $z_r = 0.929$.
- Using the equation in Step 2, calculate the standard error for Fisher's z_r:

$$SE_{z_r} = \frac{1}{\sqrt{10-3}}$$

$$SE_{z_r} = \frac{1}{\sqrt{7}}$$

$$SE_{z_r} = 0.378$$

Step 3: Calculate the confidence interval for Fisher's z_r.

- Applying the confidence interval formulas in Item 3, results in the following calculations.
  $CI_L = 0.929 - (1.96 \times 0.378) = 0.188$
  $CI_U = 0.929 + (1.96 \times 0.378) = 1.670$

Step 4: Convert the confidence intervals back to correlation coefficients

- Interpreting the confidence intervals correlation coefficients when they are standard score values can be challenging. To ease interpretation, convert the endpoints of the confidence intervals in standard score form back to correlation coefficients by using Table A.3 in Appendix A.
- Based on Table A.3, the approximate value of Pearson's r associated with a z_r of 0.188 is .18 (exact calculated value = .186). So, the lower bound of the confidence interval of the Pearson's r is 0.19.
- In Table A.3, the closest value of Pearson's r associated with the z_r of 1.670 is .94 (exact calculated value = .932). Therefore, the upper bound of the confidence interval of Pearson's r is .93.

According to the tabled values, the population correlation (ρ) between mathematics and written communication is likely to be between .19 and .93, given a Pearson's r of .727 with a sample size of 10. The 95% confidence interval is: $.19 \leq \rho \leq .93$. Yes, this is a very wide range for a confidence interval. That is due to the tiny sample size used in this demonstration. The closer a sample size is to the population size, the more precise the estimations and the smaller the confidence interval.

4.7 Correlation and Causation

The finding of a high moderate correlation between mathematics and written communication ($r = .727$) provides strong evidence of a significant positive association between the two school subjects in the target population. This outcome could be a tantalizing temptation to regard the relationship as causative—performance in writing **caused** the performance in mathematics—and that might well be an accurate inference. However, while such an inference is enticing, the relationship is limited to a statistical association. Alternative explanations of correlational results include:

- Rather than writing achievement causing mathematics achievement, the reverse may be true—mathematics achievement could influence writing especially if the focus is on problem solving with written explanations.
- Both variables might be affected by a third confounding variable excluded from this analysis. Reading achievement might be the influencer of Mathematics **and** writing achievement.
- Writing achievement could be a necessary cause, but only in the presence of a third or more unmeasured variables. Therefore, writing achievement may lead to mathematics achievement, but only if other conditions are present, e.g., reading ability or a supportive classroom climate.
- Bivariate correlations may have no logical relationship, but are both affected by changes in associated variable values over time. Any two variables that change over time could be correlated, positively or negatively; e.g., rise in crime rate and home ownership or the rise in online learning and the decline in church attendance.

Correlations between variables show that there is a pattern in the data: that the variables tend to move together. However, correlations alone are not evidence the data are moving together because one variable causes the other. To determine causation, certain conditions must be present:

- The cause (independent variable) must occur before or at the same time as the response (outcome variable).
- Other possible explanatory variables must be ruled out before a causative relationship can be established.
- The study must be based on a thorough consideration of the logical and theoretical basis for a causative interpretation.

If a researcher cannot explain the causation indicated by correlation studies through rigorous logical and theoretical means, such results are best further explored by robust experimental studies. With the use of random selection and assignment that are characteristic of experimental studies, experimental controls can be incorporated to exclude the threat of confounding or extraneous variables that might produce biased results.

4.8 Assumptions of Bivariate Correlation

Typically, researchers using bivariate correlation to investigate the relationship of two or more variables are unable to do so using data from an entire target population. Ideally, they use a representative sample of the target population to conduct the investigation so that they can use the sample value of the correlation coefficient (e.g., Pearson's r) to make inferences about the value of the population correlation coefficient (ρ) with the help of significance testing. In order to draw accurate inferences, researchers must establish that the study sample is representative of the population of interest. Testing whether the study sample meets certain assumptions about the target population provides evidence that analysis results based on the sample may be used to generalize back to the population.

For the Pearson correlation coefficient test, there are six fundamental assumptions. The first three may be assessed by the researcher as part of the research design, data collection and data preparation phases of the study. The last three may be addressed using statistical techniques.

- The two variables being correlated should be measured at the interval or ratio level.
- Each study participant (or case) must have a pair of values, one for each of the variables.
- Study participants should be independent of each other.
- There should be a linear relationship between the two variables.
- There should not be any significant outliers in the data.
- Each variable should be approximately normally distributed.

4.8.1 Variable Measurement Level

In order to use Pearson's r in bivariate correlation, variables must be measured at the interval or ratio level. Examples of such variables include test scores, measures of intelligence, time (measured in hours and/or minutes), age, weight and height.

If one or both of the variables in the study are measured at an ordinal level or nominal level, Pearson's r cannot be used to analyze such data. However, there are a multitude of correlation coefficients designed to handle such data. More on those correlation coefficients in Section 4.12 and Chapter 12.

4.8.2 Variable Pairs

In bivariate correlation, each study participant should have a value for each of the variables being correlated. Therefore, if the variables to be correlated are height and weight, each study participant must have a value for each of those variables to be retained in the analysis. If the variable candidates for correlation are miles per gallon and horsepower, or miles per gallon and vehicle weight, each case must have a value for each of those variables.

4.8.3 Independence of Cases

This assumption is also commonly known as the independence of observations, where the observations are the values (e.g., scores) used in the correlational analysis. Accordingly, the pairs of scores from one study participant should be independent of the score pairs from other individuals in the investigation. Data points of two or more study participants that are connected in some way could skew your results. If data points from two or more study participants are related, the result is data that are poor candidates for Pearson's r correlation.

Scores from students in the same study group or the same family could be the source of scores that are more similar than those of other study participants. These situations represent a violation of the assumption of independence of cases.

4.8.4 Linearity

A basic assumption underlying the use of Pearson's r is that there is a linear relationship between the two variables being correlated (Section 4.3.1). This assumption does not mean that the points representing each pair of values must fall on a straight line. Rather, the trend of the points in a scatterplot should be linear. A curvilinear, or nonlinear, relationship between two variables cannot be accurately estimated using Pearson's r. If the Pearson r coefficient is used for a nonlinear relationship, the result will be an underestimated correlation coefficient. For example, imagine that a researcher examines the correlation between two variables and finds that the $r^2 = .0$, suggesting that the two variables do not share any variance. In addition, consider that an associated computation finds that $r = 0$, signifying that there is no linear relationship between the two variables. All of this is possible, while a perfect curved (or "curvilinear" relationship) exists between the variables.

Due to the impact of linearity or nonlinearity on Pearson's r, it is important to verify prior to correlational analysis that the variables to be analyzed have a linear relationship. Scatterplots are effective methods of assessing graphically whether the relationship between the two variables is truly linear or nonlinear.

4.8.5 Absence of Outliers

There should not be any significant outlier values in the variables to be correlated. Outliers are data points that fall outside of the normal pattern of values. Pearson correlation coefficient is sensitive to the presence of outliers, and therefore, outliers have a substantial effect on the line of best fit by pulling it too far in one direction or another. This is likely to result in inaccurate correlation coefficients, driving the coefficients higher or lower than they might have been without the offending outliers.

SPSS may be used to detect outliers using P-P plots as well as scatterplots discussed in Chapter 3. Once identified, there are multiple methods for dealing with outliers, including removing them from analysis, transforming the data, or capping the data at a maximum value to bring outliers into line. In the absence of other solutions, it may be necessary to use a different correlation coefficient to measure variable associations—one that is less sensitive to outliers. More on this in Section 4.12 and Chapter 12.

4.8.6 Normality

The values of variables to be correlated should be **approximately normally distributed**. When one or both variables have non-normal distributions, the validity of Pearson's r becomes questionable. Larger sample sizes are robust to violations of the assumption of normality, but smaller samples may result in significance test results and confidence intervals that are questionable.

In order to assess the statistical significance of a Pearson r correlation, data values need to exhibit bivariate normality, but this assumption is difficult to assess, so a simpler method is more commonly used. This method involves determining the normality of each variable separately—univariate normality.

To test for normality in SPSS graphically, use histograms, normal P-P plots or Q-Q plots to graphically depict score distribution. Another option is to apply the Kolmogorov-Smirnov or Shapiro-Wilk test of normality. If the data are significantly non-normal, consider using another method to compute the correlation coefficient, some of which are discussed in Section 4.12.

4.9 SPSS Correlation Specifications

To recap, bivariate correlation analysis procedure is used to investigate whether there is a linear relationship between two quantitative variables (e.g., mathematics and written communication). The variables used in this procedure need not be independent or dependent, nor is the purpose of the analysis necessarily to establish a causal relationship unless that can be rigorously substantiated by the research literature, but rather the purpose is to determine whether there is a concomitant association between the two variables.

The bivariate correlation procedure to be used in this demonstration exercise is Pearson product moment correlation or Pearson's *r*. To complete the exercise, use the dataset in Table 4.1 which includes the variables, written communication task and mathematics.

The written communication task is an extended writing task based on a single prompt. The maximum mark possible on this task is 12, and the minimum is 2. The mathematics test includes a total of 80 questions. Box 4.1 includes the research questions and hypotheses associated with this case.

4.9.1 SPSS Specifications for Testing Correlation Assumptions

Testing the assumptions for statistical techniques provides researchers the opportunity to evaluate characteristics of the data that might prevent them from drawing accurate inferences about the results. Before examining the bivariate correlation of variables in the case study, for demonstrative purposes, a preliminary exercise will evaluate the extent to which the data meet two Pearson's *r* correlation assumptions that were discussed in Chapter 3—normality and linearity.

4.9.1.1 Assessing the Assumption of Data Linearity

As Section 4.8.4 states, scatterplots are a common and effective method of assessing the extent to which the variables to be correlated have a linear relationship. To specify instructions to assess linearity for the two case-study variables (mathematics and written communication) in SPSS, perform one of the following:

1. Navigate to the location of the dataset, **Math_Comm_Scores.sav** (taken from Table 4.1), and double-click the filename to bring the file into the IBM SPSS Statistics Data Editor.
2. Open SPSS, then select **File → Open → Data** to reveal the **Open Data dialog** window. Navigate to the location of the dataset, and double-click on the dataset filename to open it in the **IBM SPSS Statistics Data Editor** window. Click on **Variable View** (Figure 4.9).

Figure 4.9 IBM SPSS Statistics Data Editor Window

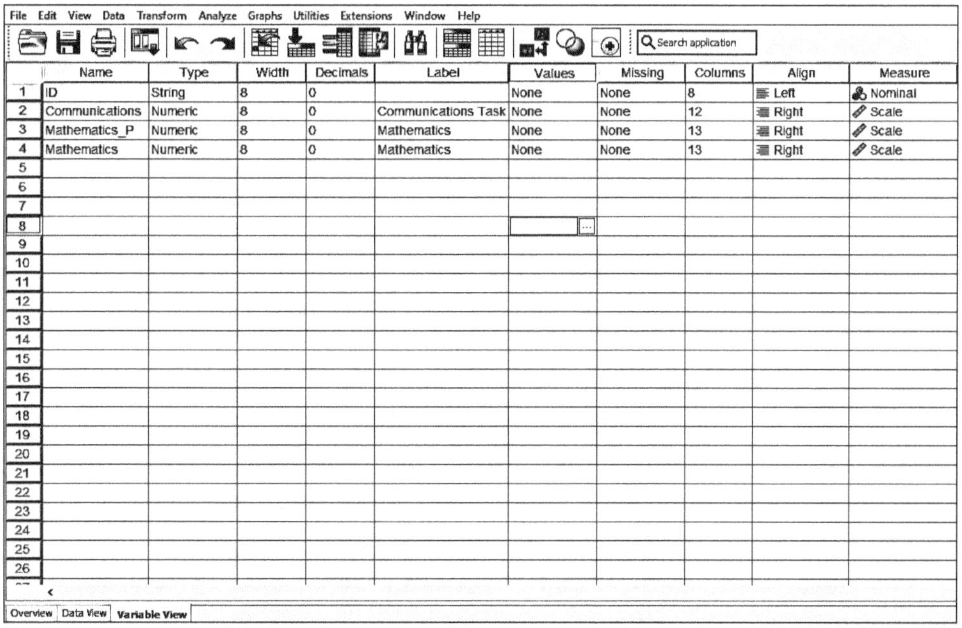
File Edit View Data Transform Analyze Graphs Utilities Extensions Window Help
Search application
Name Type Width Decimals Label Values Missing Columns Align Measure
1 ID String 8 0 None None 8 Left Nominal
2 Communications Numeric 8 0 Communications Task None None 12 Right Scale
3 Mathematics_P Numeric 8 0 Mathematics None None 13 Right Scale
4 Mathematics Numeric 8 0 Mathematics None None 13 Right Scale
Overview Data View Variable View

Before specifying the procedure, confirm that both variables to be analyzed (**Mathematics** and **Communication**) are scale-level variables (SPSS term for interval/ratio) by finding each variable in the **Name** column in the **Variable View** window.

Move the mouse along the row to the column, **Measure**, and confirm that **Scale** appears in the appropriate cell. Optionally, select the **Data View** tab, and check to ensure that a picture of a ruler, the SPSS icon for **Scale**, shows up in the Communication and Mathematics columns.

SPSS provides two choices to complete a scatterplot of the two variables: (a) **Chart Builder** and (b) a legacy scatterplot chart. To use **Chart Builder**, select **Graphs** → **Chart Builder**, the first option shown in Figure 4.10.

Figure 4.10 Selecting the Scatterplot Option

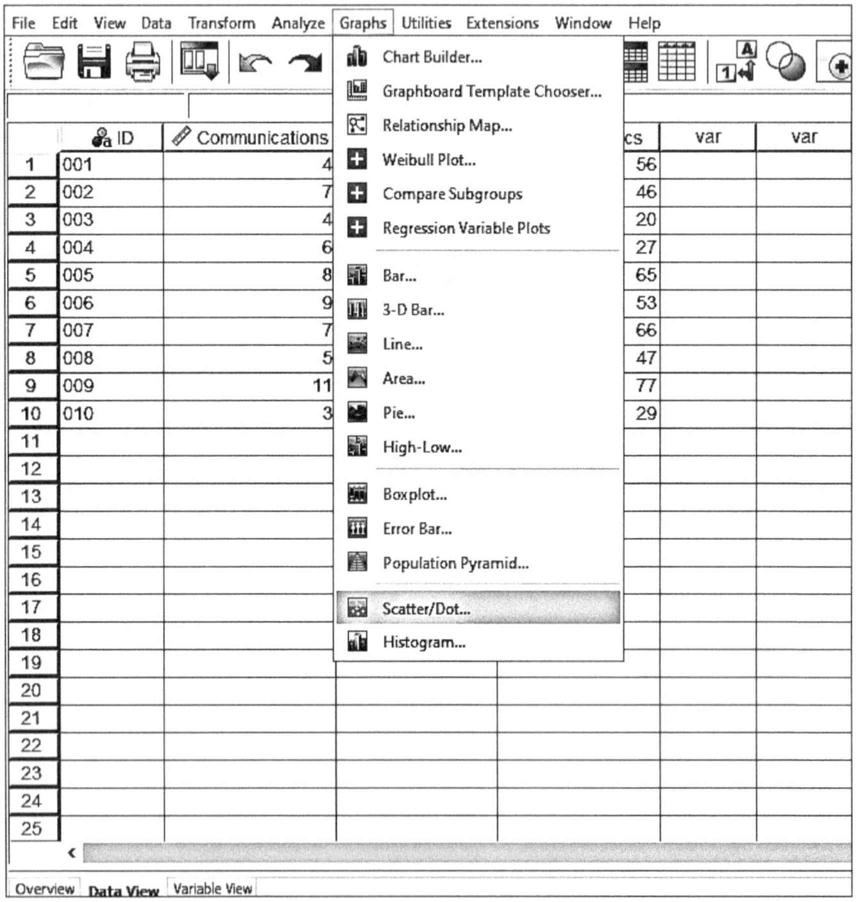

To use the legacy chart, navigate to **Scatter/Dot** on the pull-down menu and double-click to open the **Scatter/Dot** window (Figure **4.11**).

In the **Scatter/Dot** dialog window there are five different types of scatterplots to choose from:
- **Simple Scatter** produces a scatterplot of values of one continuous variable against another.
- **Matrix Scatter** produces a scatterplot grid of relationships between each pair of variables.
- **Simple Dot** (called density plot, dot chart or strip plot) is similar to a histogram and used with small data sets where the values belong to different categories. Each "bar" of dots equals the number of items in that category.

Figure 4.11 IBM SPSS Scatter/Dot Window

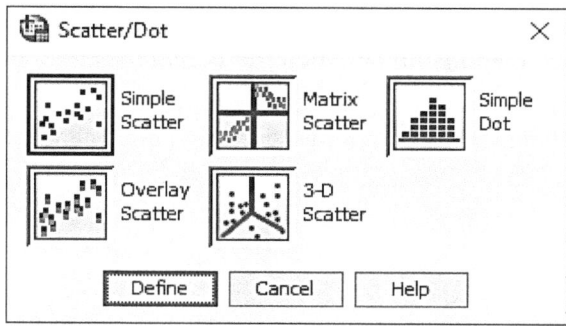

- **Overlay Scatter** charts two plots one on top of the other, each distinguished by color.
- **3-D Scatter** graphs values of three continuous variables in a three-dimensional space.

For this exercise, select **Simple Scatter**. Then click on **Define**, to open the **Simple Scatterplot** dialog window (Figure 4.12a).

Figure 4.12 IBM SPSS Simple Scatterplot and Options Windows

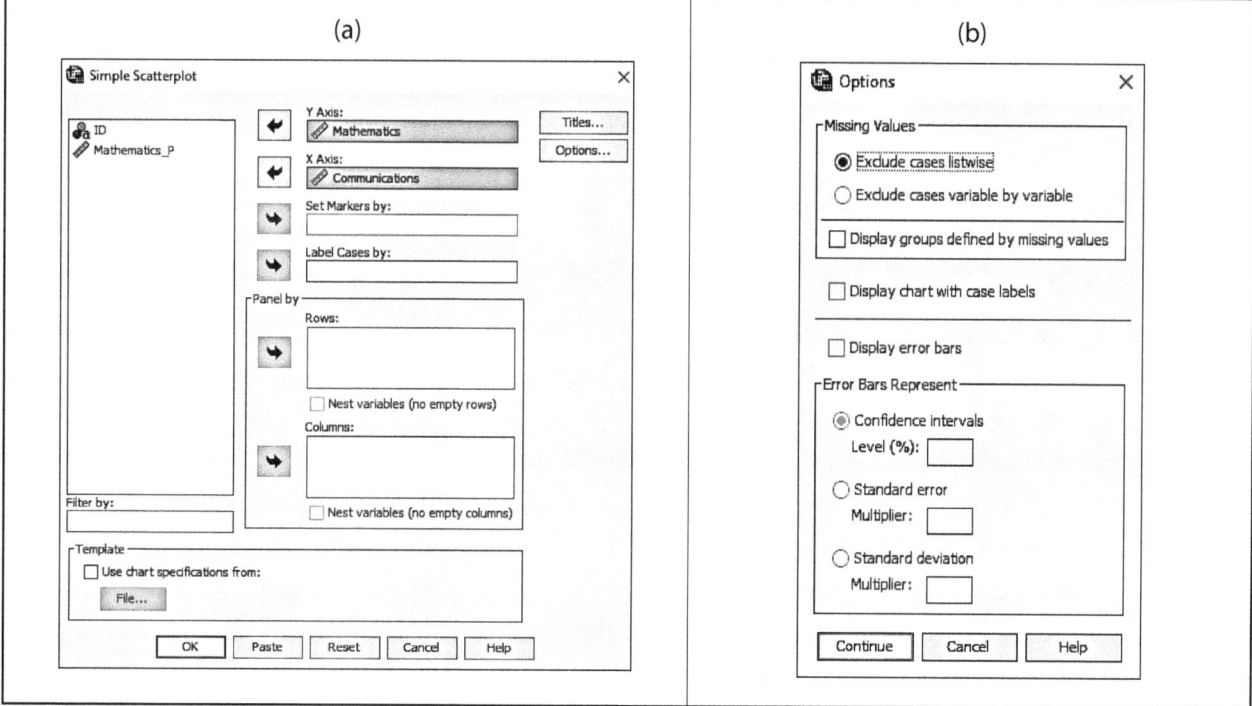

In the **Simple Scatterplot** dialog window, highlight **Mathematics** in the left panel, and click the arrow next to the panel, *Y-Axis,* to place the variable in that panel. Then highlight **Communication**, and click it to the panel, *X-Axis* (Figure 4.12a). No need to select any of the remaining settings. Select **Options** in the upper right corner of the **Simple Scatterplot** dialog window to open the **Options** dialog window (Figure 4.12b). Notice that the default setting for **Missing Values** is **Exclude cases listwise**, where a case with a missing value for either of the variables is excluded from the graph. Two values are needed for each dot, so retain the default. Click **Continue** to return to the **Simple Scatterplot** window. Click **OK** to run the analysis.

4.9.1.2 Scatterplot Results

Figure 4.13 displays the scatterplot. Even with a limited number of values, the plot displays a trend that is clearly a linear positive relationship between mathematics and communication.

Figure 4.13 Scatterplot of Mathematics and Communication Task Performance

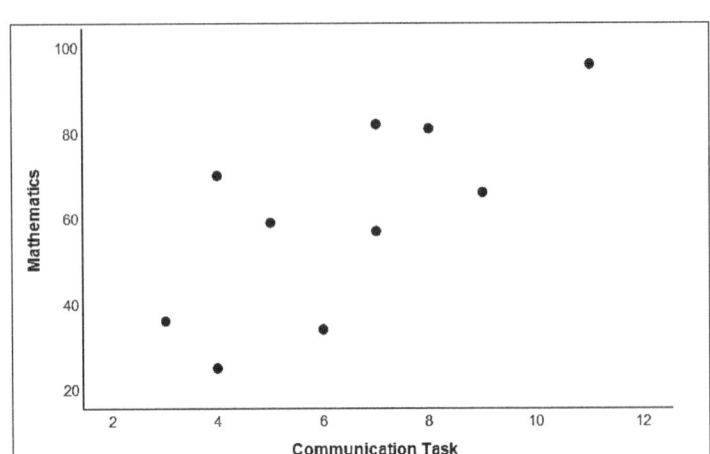

4.9.1.3 Assessing the Assumption of Data Normality

Recall that the following are methods for assessing the normal distribution of bivariate correlation variable values:

- Normal probability plots (P-P plots) or quantile plots (Q-Q plots)
- Kolmogorov-Smirnov Test of normality, and
- Shapiro-Wilk test of normality

All of these methods may be completed using the **SPSS Explore** procedure described in Chapter 3. To analyze the dataset using **SPSS Explore**, select **Analyze → Descriptive Statistics → Explore** to open the **Explore** window.

In the **Explore** dialog box (Figure 4.14), select **Mathematics** and **Communication Task** in the left panel, and click them over to the **Dependent** List panel on the right. Continuous variables are always placed in the **Dependents** List panel.

If the objective was to subset the analyses by one or more categorical variables, the categorical variables would be placed in the **Factor List** box. Since there are no categorical variables, leave that panel blank. Similarly, the **Label Cases by** panel was designed to identify outliers by specifying participant IDs to include on SPSS reports. However, no outliers will be identified in this exercise, therefore, the **Label Cases by** panel remains blank. Under **Display** at the bottom of the dialog window, the specification, **Both**, is the default setting, indicating that SPSS will generate both plots and statistics. Selecting **Both** is also the only means of getting the **Kolmogorov-Smirnov** and **Shapiro-Wilk** statistical tests of normality. Therefore, retain the default setting.

Selecting the **Statistics** pushbutton opens the **Explore Statistics** dialog window (Figure 4.15a). Here, **Descriptives** (default) produces basic descriptive statistics, including measures of skewness and kurtosis. Click **Continue** to return to the **Explore** menu.

Figure 4.14 SPSS Explore Dialog Window

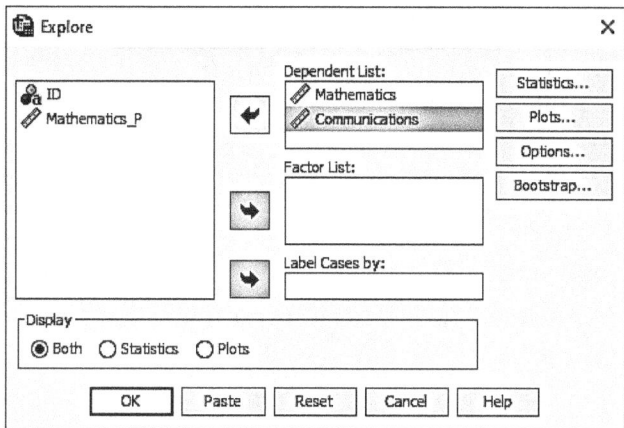

Figure 4.15 Explore Dialog Windows: Statistics, Plots and Options

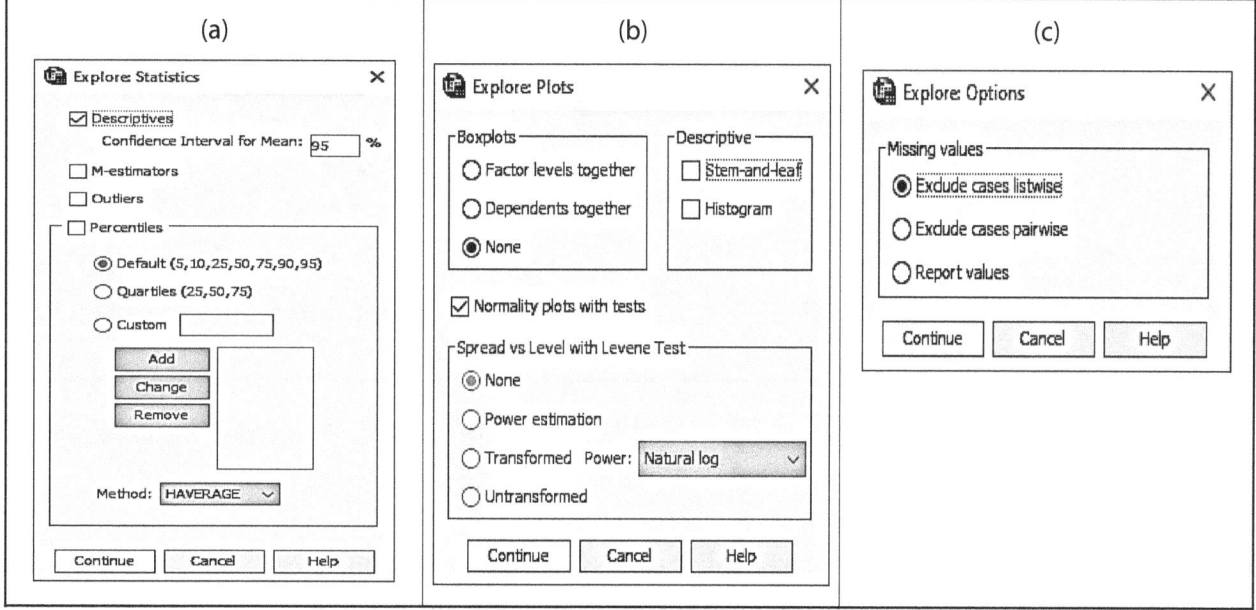

Next, select the **Plots** pushbutton to open the **Explore Plots** dialog window (figure 4.15b). Boxplots are particularly useful in identifying outliers in a dataset. Besides providing a graphic depiction of the frequency and shape of the distribution of scores in a dataset, stem-and-leaf plots and histograms are also useful for identifying outliers. However, of primary interest is the **Normality plots with tests** setting below **Boxplots** and **Descriptive**, which will provide the Kolmogorov-Smirnov and Shapiro-Wilk statistical tests of normality. Therefore, select **None** below **Boxplots** and unselect **Stem-and-Leaf** below **Descriptive**. Finally, select **Normality plots with tests** which will provide Q-Q Plots and results of the significance tests of normality. Click **Continue** to return to the main **SPSS Explore** menu.

To complete specifications for the normality significance tests, select the **Options** pushbutton in the main **Explore** window (Figure 4.15c). Retain the default setting, **Exclude cases listwise**. Click **Continue** to return to the main menu, and click **OK** to run the analysis.

4.9.1.4 Normality Test Results

In addition to descriptive statistics, specifications for this exercise produced three measures of normality; Q-Q Plots, which are graphic representations of the proximity of data values to a normal distribution as well as Kolmogorov-Smirnov, and Shapiro-Wilk tests of significance. The Q-Q plot for communication task is displayed in Figure 4.16.

If each point in the plot, which represents each communication task score, is exactly where it should be, the distribution would be perfectly normal, and every point would fall on the line. The farther a point is from the line (normal distribution), the farther it is from a normal distribution. The Q-Q plot for communication and mathematics appear relatively normally distributed.

Figure 4.16 Normal Q-Q Plot of Communication and Mathematics Scores

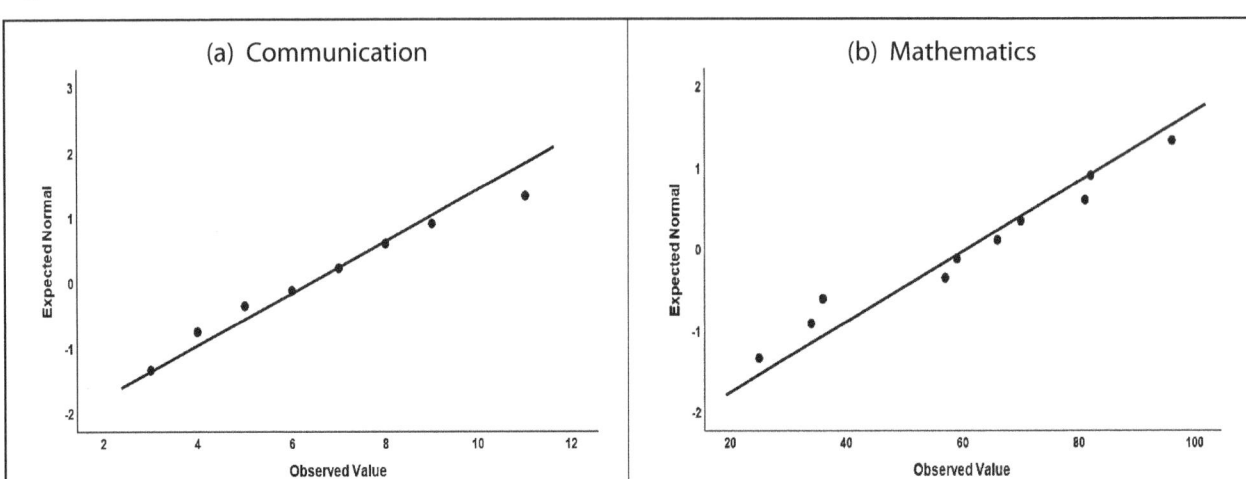

Figure 4.17 displays the Kolmogorov-Smirnov test of significance and the Shapiro-Wilk test of significance. The hypotheses for both tests are as follows:

Null hypothesis: The sample of scores is from a normal distribution.

Alternative hypothesis: The sample of scores is from a non-normal distribution.

A p value (significance level) less than $\alpha = .05$ (for this exercise) necessitates rejecting the null which means there is insufficient evidence of normal distribution of the data. A p value greater than $\alpha = .05$, results in failing to reject the null, meaning that there is sufficient evidence to conclude normal distribution of the data.

Figure 4.17 Results of Significance Tests of Normality

	Tests of Normality					
	Kolmogorov-Smirnov[a]			Shapiro-Wilk		
	Statistic	df	Sig.	Statistic	df	Sig.
Communication Task	.131	10	.200*	.965	10	.842
Mathematics	.153	10	.200*	.956	10	.744

*. This is a lower bound of the true significance.
a. Lilliefors Significance Correction

The *p* values (**Sig.** columns) for both significance tests for mathematics and communication exceed the alpha level of .05. Therefore, there is not enough evidence to conclude data distribution non-normality. Bear in mind that these two significance tests are especially sensitive to minor departures from normality in large datasets. Accordingly, with the small size of the dataset, test results indicating that the data are normally distributed are convincing evidence that the dataset did not violate the assumption of normality for bivariate correlation. In addition, the tests confirm graphs in Figure 4.16 that illustrate the scores are normally distributed.

4.9.2 SPSS Specifications for the Omnibus Pearson's *r* Correlation Analysis

With tests of assumptions completed and interpreted, the next phase is to conduct the omnibus test for the Pearson's *r* correlation. To analyze the dataset, select **Analyze → Correlate → Bivariate** (Figure 4.18) to open the **Bivariate Correlations** dialog window.

Figure 4.18 Selecting Bivariate Correlation Option

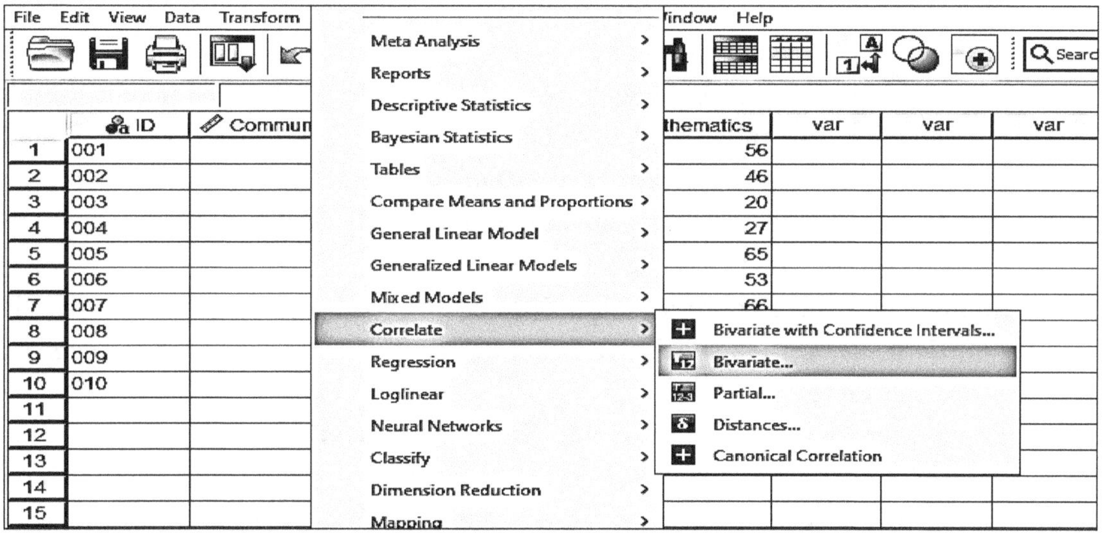

In the **Bivariate Correlations** dialog window (Figure 4.19a), move both variables (**Mathematics** and **Communication**) to the **Variables** panel on the right. There is room to select more than two (the minimum) variables for analysis. However, even if more were selected, the procedure would correlate all possible pairs of variables and produce a coefficient for each pair.

Below the heading **Correlation Coefficients** are the three coefficients this procedure can generate: **Pearson**, **Kendall's tau-b** and **Spearman**. The last two are non-parametric correlation coefficients that measure the relationship between ordinal level variables and will be discussed in Section 4.12 and Chapter 12. The default (**Pearson**) is designed to measure the relationship between continuous variables (interval and ratio). Since both mathematics and communication task are continuous variables, retain the default correlation coefficient, which is Pearson.

Figure 4.19 Bivariate Correlations Dialog Window

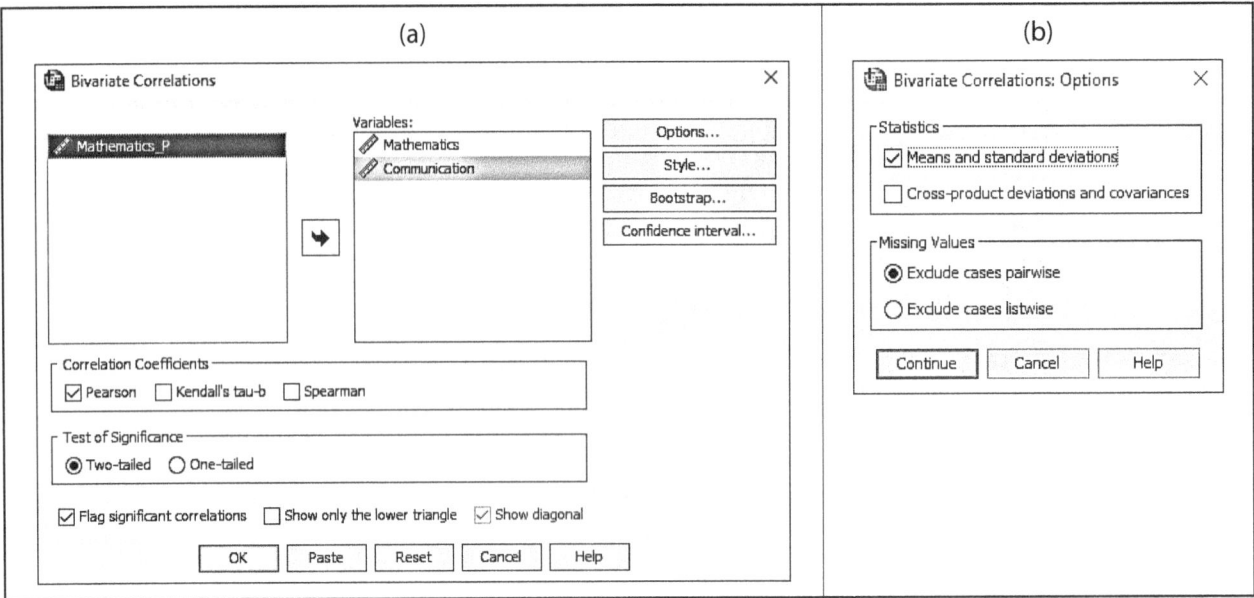

Imagine that there is no basis on which to hypothesize the direction of the association between the mathematics and communication task. In such instances, hypotheses must be nondirectional (Box 4.1). Therefore, ensure that **Two-tailed** (default test of significance) is selected. In addition, retain the check mark (default) next to **Flag significant correlations** which produces an asterisk or two next to statistically significant correlation coefficients in the output table. By default, SPSS generates statistical significance at α = .05 (*) and α = .01 (**) levels.

Clicking **Options** in the upper right corner of the **Bivariate Correlations** window opens the **Bivariate Correlations: Options** dialog window (Figure 4.19b). Here, specify which statistics to generate and how to address missing values.

The default option for **Statistics** is **Means and Standard Deviations,** which produces the mean and standard deviation of each variable as well as the number of missing values. The **Cross Product Deviations and Covariances** option generates the numerator of the Pearson correlation coefficient formula and the covariance, which is an unstandardized measure of Pearson's *r* (Section 4.5.1). In most cases, the **Cross Product Deviations and Covariances** option is not needed and will not be needed for this analysis. Select **Means and Standard Deviations**.

Below **Missing Values**, choosing the pairwise or listwise option does not affect SPSS computations with only two variables involved, but choosing one or the other can make a big difference if more than two variables are analyzed. By default, SPSS applies **pairwise deletion of missing values** so that each bivariate correlation uses all cases with valid values for both variables.

Listwise deletion excludes from analysis all cases that have one or more missing values on any of the variables involved. Accordingly, listwise deletion can substantially reduce sample size when analyzing a correlation matrix with more than two variables involved. To minimize impact on sample size for correlation matrices, the default noted in Figure 4.19b is best. With only two variables involved in this analysis, it does not matter which option is used. Nevertheless, retain the default option. Click **Continue** to return to the main window, and click **OK** to run the procedure.

4.9.3 Results of the Pearson r Procedure

Figure 4.20 displays results from the correlational analysis. The first table, **Descriptive Statistics**, presents the means and standard deviations of the variables as well as the number of cases in the analysis. As in the hand-calculation of the correlation coefficient in Section 4.6 and in Table 4.1, this small demonstration dataset contains 10 cases (N column). The mean and standard deviations of the scores for communication and mathematics are 6.4 (SD = 2.50) and 48.6 (SD = 18.65).

Figure 4.20 Results of the Pearson *r* Correlation Analysis

Descriptive Statistics			
	Mean	Std. Deviation	N
Communication Task	6.40	2.503	10
Mathematics	48.60	18.650	10

Correlations			
		Communication Task	Mathematics
Communication Task	Pearson Correlation	1	.727*
	Sig. (2-tailed)		.017
	N	10	10
Mathematics	Pearson Correlation	.727*	1
	Sig. (2-tailed)	.017	
	N	10	10

* Correlation is significant at the 0.05 level (2-tailed).

The second table in Figure 4.20, **Correlations**, presents a square correlation matrix, so called because it presents redundant correlation results. Accordingly, draw a diagonal line from the upper left to the lower right (as shown), it will be evident that the correlation coefficients and the significance levels above and below the diagonal are virtually identical. A quick way of finding the diagonal of a correlation is to locate 1 in the table where a variable is correlated with itself.

When one variable is correlated with another variable, three pieces of information are typically provided by SPSS:
- the Pearson's *r* value,
- results of the significance test to determine whether the population correlation (ρ), based on the sample coefficient, is significantly different from 0, and
- confirmation of the sample size.

152 Bivariate Correlation

The sample size of 10 cases seen in the descriptive statistics table is confirmed in the correlation matrix (third row). The Pearson r value, which is .727 in the table, means that as scores of the communication task rise, so do mathematics scores. Given the magnitude and direction of the correlation coefficient, performance on the communication task would be characterized as having a high moderate positive correlation with performance on the mathematics test. SPSS's generated sample correlation coefficient is identical to the hand-calculated coefficient of .727 in Section 4.6.1.

The significance test evaluates the probability of getting the reported sample correlation coefficient in the target population if the null hypothesis is true. The significance test was conducted based on an alpha of .05 (Box 4.1), and based on that alpha level, the correlation was found to be statistically significantly different from 0 at $p = .017$ (Figure 4.20). As in the worked example in Section 4.6, use the correlation coefficient to compute the strength of this significant relationship between the variables by squaring the correlation coefficient ($.727^2$) which is .529 as in Section 4.6.2.

4.9.4 Sample Report: The Case of Writing and Mathematics Skills

The research literature suggests that for students at all levels of education (from elementary to graduate institutions), achievement in the content areas, like mathematics, the sciences and social studies, is related to students' proficiency in the language arts (e.g., Grimm 2008; Nortvedt et al., 2016). With primary-level students' mathematics performance lagging in a small island nation, the purpose of this study was to determine whether writing was linearly related to mathematics achievement among Grade 6 students.

After confirming that the data met the statistical assumptions for bivariate correlation, a Pearson product moment correlation was performed to determine whether there was a significant linear relationship between written communication skills and mathematics performance. A Pearson's r index of .727 was obtained, indicating a high moderate positive correlation existed between writing and mathematics. Accordingly, as writing skills performance increased so did mathematics performance. Furthermore, the correlation between mathematics and written communication was found to be statistically significant, $r(10) = .727$, $p = .017$, CI(95) $= .18 \leq \rho \leq .94$, $r^2 = .529$. Accordingly, mathematics and written communication share 52.9% of the available variance.

4.10 Factors that Influence Correlation

Section 4.8 discussed six factors that can impact the accuracy of the Pearson correlation coefficient as well as the associated data assumptions that researchers must take into consideration before conducting the Pearson r procedure. Prominent among the factors discussed were some that might affect the magnitude or the interpretability of Pearson's r, including violation of the assumptions of linearity, absence of outliers, and normal distribution. This section will focus on three additional factors that can influence the size of Pearson's r.

- sample size
- restriction of range
- correlating aggregated data

4.10.1 Sample Size

Sample size is an important consideration when conducting any quantitative research study, including correlational studies. Although including the entire target population is the best option for any study, it is rarely the most practical solution. Given the practical restrictions of using an entire population, the next best option is a representative sample. Of course, the larger the sample size is, the greater the likelihood that it will be representative of the target population, leading to more stable and reliable results.

Researchers are less likely to experience large differences in correlation coefficients across studies when they use large representative samples. However, all things being equal, the number of study participants does not influence the magnitude or stability of correlation coefficients. Rather, the degree to which the smaller sample size is representative is a more important factor, and smaller samples are less likely to be sufficiently representative to prevent spurious results.

Consider the correlation of ratings of happiness and satisfaction with life from the Trinidad & Tobago edition of the World Values survey (Inglehart et al., 2014). The entire dataset of 883 respondents yielded a correlation coefficient of .506 ($p < .01$) and a 99% confidence interval of $.439 \leq \rho \leq .568$ as Figure 4.21 illustrates. However, as sample sizes decline, correlation coefficients result in ever-widening confidence intervals.

Figure 4.21 Correlation between Happiness and Satisfaction with Life in Trinidad & Tobago

		Correlation	
		Happiness	Satisfaction with Life
Happiness	Pearson Correlation	1	.506**
	Sig. (2-tailed)		.000
	N	884	883
Satisfaction with Life	Pearson Correlation	.506**	1
	Sig. (2-tailed)	.000	
	N	883	883

**. Correlation is significant at the 0.01 level (2-tailed).

As Figure 4.22 illustrates, where the original sample generated a moderate positive correlation between the variables, attempting to secure a similar correlation coefficient with a sample size of 20 from the original group could result in coefficients that range from no correlation between the variables ($r = -.067$, CI = .99) at the lower end of the confidence interval, to a very high positive correlation ($r = .828$, CI = .99) at the upper end of the confidence interval. It is important, therefore, to be wary of small sample sizes in correlational studies. Large sample sizes provide a more accurate reflection of the target population correlation coefficient.

Although large sample sizes are preferred in research studies, that is not always possible. Accordingly, a general rule of thumb is a sample size of no less than 50 participants for a correlational study (Green, 1991; Harris, 1985). Larger samples are recommended when score distributions are skewed or when effect sizes are expected to be small (Tabachnick & Fidell, 2018).

Figure 4.22 Confidence Intervals for Varying Sample Sizes for a Coefficient of .506

Sample Size (n)	99% Confidence Interval (CI)
883	$0.439 \leq \rho \leq 0.568$
440	$0.409 \leq \rho \leq 0.592$
220	$0.365 \leq \rho \leq 0.624$
110	$0.299 \leq \rho \leq 0.668$
50	$0.180 \leq \rho \leq 0.732$
20	$-0.067 \leq \rho \leq 0.828$
10	$-0.394 \leq \rho \leq 0.919$

Sometimes correlation analyses indicate that two variables are statistically significantly correlated with a $p < .01$, such as the correlation between the importance of family and happiness on the Trinidad and Tobago World Values Survey, yet the r^2 value (shared variance) is noticeably small, $r^2 = .01$). The explanation for this peculiar occurrence is that statistical significance does not imply practical significance.

In general, the larger the dataset (e.g., $n = 883$ in the above example), the easier it is to reject the null hypothesis that the slope is not significantly different from 0, even when in reality shared variance is not different from 0. Therefore, as in the above example, it is possible to get a significant p-value (.004) when $r(883) = .096$, a quantity that is not meaningfully different from 0.

Large sample sizes can become a snare. Therefore, for meaningful interpretation, it is essential to look beyond significance testing to effect sizes and to the magnitude of correlation coefficients.

4.10.2 Restriction of Range

Restricting or constraining the normal range of the values of either variable in a bivariate correlation typically results in the alteration of the correlation between the two variables. Depending on the nature of the data, a restriction in range could either result in the reduction or an increase in the magnitude of the correlation coefficient. More often than not, the restriction results in a drop in the magnitude of the correlation coefficient. The restriction in range will only increase Pearson's r when an initial curvilinear trend in the data is eliminated.

Imagine that the scatterplots depicted in Figure 4.23a, and 4.23b represent the relationship between anxiety and test score. Incorrectly using Pearson's r to estimate a curvilinear relationship between anxiety and test performance results in $r = 0$ (Figure 4.23a). If, however, the ranges of the values for both variables were inadvertently restricted as displayed in Figure 4.23b, the resultant correlation between the variables would become a high-positive correlation of $r = .940$.

As discussed earlier, restricting the range of variable values may result in lower correlation coefficients than would normally occur. Consider the scatterplots in Figure 4.24.

The data depicted are from a study commissioned by the U.S. Department of Transportation to study the legibility and visibility of highway signs. The researchers determined the maximum distance (in feet) at which each of 30 study participant drivers could read a newly-designed sign. Figure 4.24a portrays a scatterplot, using data from the original study and shows a strong, negative correlation ($r(30) = -0.801$) between the age of drivers and the legibility of street signs. Figure 4.24b illustrates that if a

Figure 4.23 Restriction of Range that Increases Pearson Correlation Coefficient

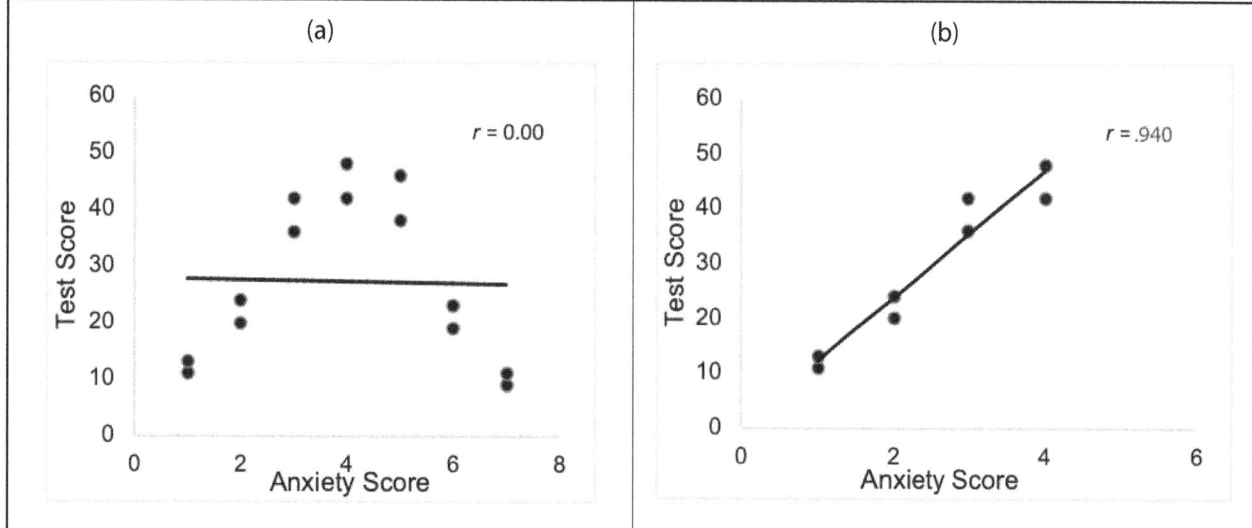

Figure 4.24 Restriction of Range that Reduces Pearson Correlation Coefficient

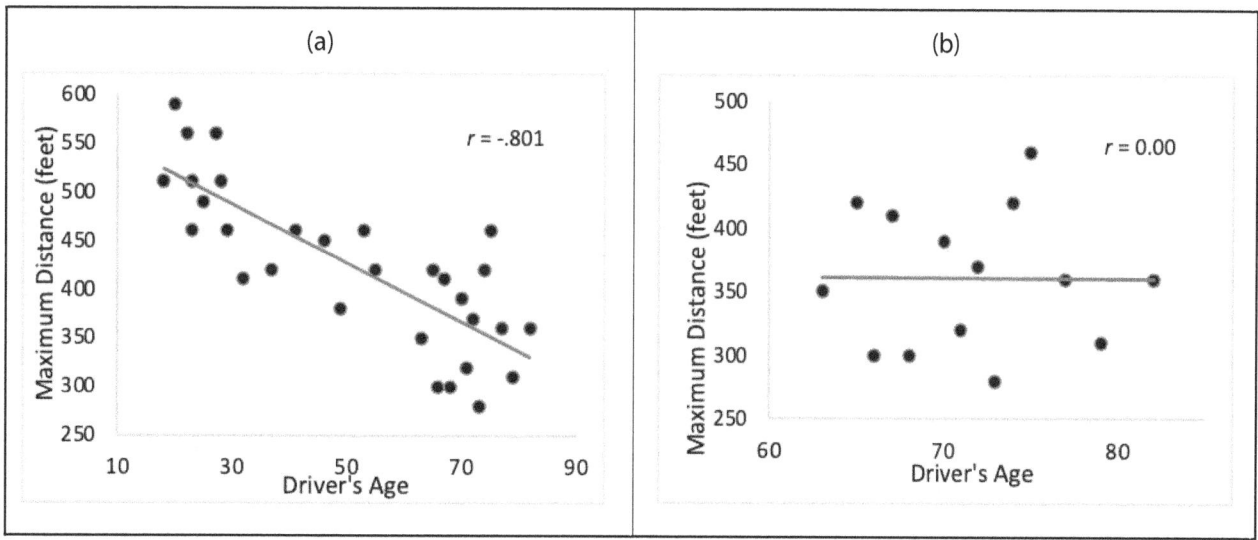

Source: Utts, J. and Heckard, R. F. (2002) Mind on Statistics. Duxbury. Original source: Data collected by Last Resource, Inc, Bellfonte, PA.

less representative sample with a restricted age range had been used in the study, researchers might have ended up with no correlation ($r(30) = .00$) between the two variables.

Figures 4.23 and 4.24 illustrate the importance of taking into account the effect of range restriction when a coefficient is based on a limited study sample. The study outcome may not support the study hypothesis all because of a sample with a restriction in range. While it is sometimes necessary to restrict the range of values for one or more variables (e.g., no need for drivers younger than age 16 in the street sign study), it is critical that study samples are representative of the target population.

4.10.3 Correlating Aggregated Data: Ecological Correlation

Variable values (often scores) from individual cases in a study is perhaps the most widely used source of data in correlational studies. The results are used to make inferences about the individuals in the associated target population. For example, the correlation of biology and mathematics scores on related achievement tests from a random sample of students across Barbados may be used to draw conclusions about the relationship between performances on those subjects for students at the individual level.

Sometimes researchers wish to make inferences at a higher unit of analysis than the individual level, like high schools in Barbados. For example, researchers might correlate values of the same variables (mathematics and biology) but from a random sample of high schools across the nation of Barbados, and using school-level average scores for each subject. In this case, a correlation between aggregated data like school group means would be described as ecological correlation. These results could subsequently be used to make inferences about a different target population—high schools across Barbados.

In individual-level correlations, the unit of analysis is the individual, so data are collected from individuals and inferences are made about individuals. Similarly, in ecological correlations, the unit of analysis is the group; data come from group means, and inferences are made at the group level.

Difficulties occur when researchers use group-level data to make individual-level inferences. For example, a 2018 World Bank Development report (World Bank, 2018) concluded that throughout the world more schooling is systematically associated with higher wages. However, if researchers assume that because there is a correlation between average income and years of schooling at the country level, there is a similar relationship at the individual level, this would be regarded as an ecological fallacy; that is, an error in reasoning which happens when researchers make inferences about individuals based on aggregated group data.

The term, ecological fallacy, is associated with Robinson (1950) one of the earliest scholars to identify a flaw in using group-level data to make inferences about individuals. In his article, Robinson reported that the correlation of variables based on group-level data is generally much higher than the correlations of the same variables, but based on individual-level data. In the case of schooling and earnings, results may be dependent on the context, values, and professions of study participants.

There are advantages and limitations of ecological studies:

Advantages
- Aggregate data is often readily available at low cost and can be quickly analyzed.
- Because of ready availability, they are useful for explorational studies to develop hypotheses for future research.
- Large numbers of people across wide ranges of populations and sites can be included in analyses.
- Large numbers of factors can be included in the study.
- Some research interests can best be studied at the aggregate population rather than the individual level, such as relationships of smoking bans and health outcomes, or seatbelt use and vehicular deaths.

Drawbacks
- Associations of factors in groups of people may not accurately reflect associations in individuals within those groups.
- For one variable to "cause" another, both must occur in the same individual. However, ecological studies do not have data on individuals, so causality becomes unclear. For example, it is not clear that people with more schooling are those with higher wages.

- There is no way of adjusting for other factors that influence the association between variables. For example, in the study of the relationship between earnings and schooling, illegal sources of income may affect the correlational relationship.
- Curvilinear or other nonlinear relationships between variables may dampen the relationship between variables, causing misleading results.

4.11 Intercorrelation Matrices

A research investigation can be limited to the relationship between two variables. However, it is more often the case that more than two variables are included in a correlational study.

Imagine that the objective is to investigate the relationships between five variables included in a dataset of department store employees:
- years of education
- current salary
- beginning salary,
- months of previous experience
- age

To conduct this correlational analysis of multiple variables, using SPSS requires following the same procedures outlined in Section 4.9.2. However, instead of limiting the number of variables entered to two variables as shown in Figure 4.19, all five of the variables listed above may be specified. All remaining procedures remain the same as outlined in Section 4.9.2. The outcome is displayed in the intercorrelation matrix, **Correlations**, in Figure 4.25.

Some differences between the two correlational analyses should be immediately evident in Figure 4.25. Rather than correlation coefficients and significance testing for two variables, the result is now a square matrix for five variables, with correlation coefficients and significance testing results for all possible pairs of variables. As with a single bivariate correlation analysis, the correlation matrix provides the Pearson correlation coefficient for each variable pair as well as the statistical significance of the correlation and the sample size. The asterisks indicate whether the variables are significant at $p < .01$ (**) or $p < .05$ (*). In some settings, probabilities less than .001 (***) may be flagged as well.

Regardless of the number of variables included in an intercorrelation matrix, bivariate correlation is restricted to examining variable relationships only in a pairwise manner. Accordingly, the correlation between years of education and current salary ($r = .695$, $p < .01$) is not connected with the correlation between years of education and age ($r = -.233$, $p < .01$). These two sets of correlations are in the same matrix for efficiency and convenience, but they are unrelated otherwise. The scatterplot matrix in Figure 4.26 presents a graphic depiction of the row and column variables for each bivariate correlation in Figure 4.25.

The scatterplot of the values for the variables, current salary and years of education are located in the upper left corner of Figure 4.26. The direction and slope of the scatterplot confirms the positive, moderately high correlation between the variables, and the line-of best-fit provides a clearer picture of the direction and slope of the relationship.

Figure 4.25 Matrix of Pairwise Correlations of Department Store Employee Data

Correlations		Years of Education	Current Salary	Beginning Salary	Months of Previous Experience	Age
Years of Education	Pearson Correlation	1	.695**	.649**	-.142	-.233**
	Sig. (2-tailed)		.000	.000	.102	.007
	N	134	134	134	134	133
Current Salary	Pearson Correlation	.695**	1	.903**	-.016	-.088
	Sig. (2-tailed)	.000		.000	.851	.315
	N	134	134	134	134	133
Beginning Salary	Pearson Correlation	.649**	.903**	1	.104	.030
	Sig. (2-tailed)	.000	.000		.232	.736
	N	134	134	134	134	133
Months of Previous Experience	Pearson Correlation	-.142	-.016	.104	1	.774**
	Sig. (2-tailed)	.102	.851	.232		.000
	N	134	134	134	134	133
Age	Pearson Correlation	-.233**	-.088	.030	.774**	1
	Sig. (2-tailed)	.007	.315	.736	.000	
	N	133	133	133	133	133

**. Correlation is significant at the 0.01 level (2-tailed).

As with results for a single pair of variables, the series of 1s along the diagonal of the intercorrelation matrix in Figure 4.25 represent the expected perfect correlation between a variable and itself. In the scatterplot matrix, depiction of the perfect correlation between each variable and itself is excluded leaving empty cells along the diagonal. As with the correlation matrix, the scatterplots on either side of the diagonal are mirror images of each other.

4.12 Other Correlation Coefficients

Pearson's r is one of the most widely used correlation coefficients and is designed to measure relationships between two variables. However, Pearson's r has some characteristic limitations that prohibit its use in analyzing some types of data.

As discussed in Section 4.8, data values must meet specific assumptions in order to obtain accurate Pearson's r coefficients. One important assumption underlying Pearson's r is that both variables must be measured on an interval-level or ratio-level scale. Thus, variables such as height, weight, temperature, test scores, or age are suitable. However, some data values are not measured on interval or ratio level scales. They are not continuous variables.

Figure 4.26 Scatterplot Matrix of Pairwise Correlations of Department Store Employee Data

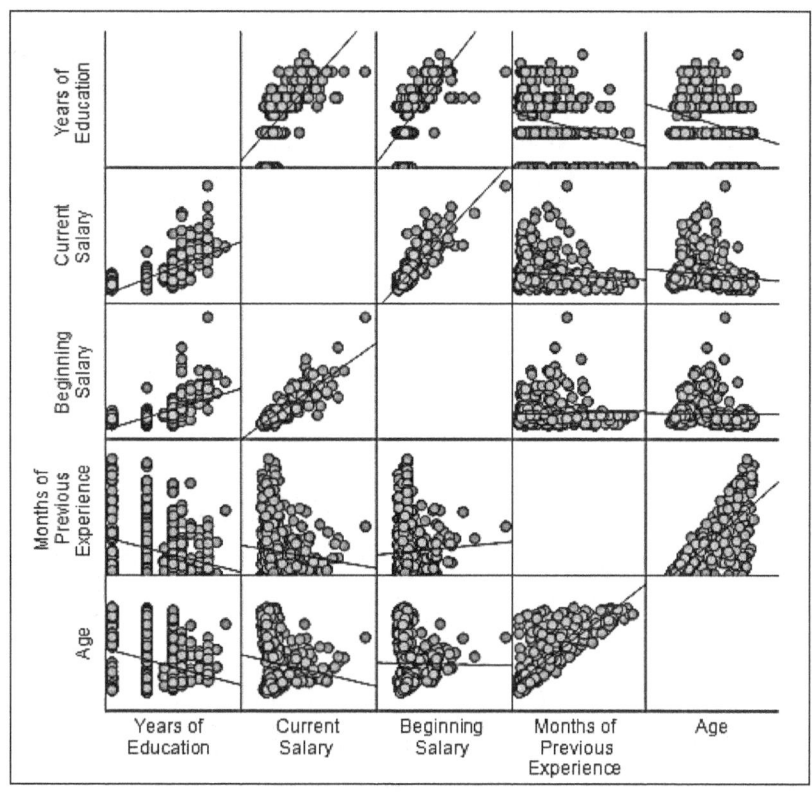

In some bivariate correlations, at least one of the variables is measured on a nominal scale that is dichotomous (or binary), such as marital status (unmarried = 0, married = 1), gender (male = 0, female = 1), or answer correctness (incorrect = 0, correct = 1). For example, in the World Values Survey (Inglehart et al., 2014), male and female survey respondents in Trinidad and Tobago were asked whether most people can be trusted. In this case, both variables are dichotomous. Male and female would be coded 0 and 1, respectively, while trust and mistrust might be coded 1 and 0, respectively. A related research question might be: Is gender related to ability to trust others?

The binary values 1 and 0 are not always used to indicate the presence or absence of a characteristic. For example, 1 and 3, or 2 and 4 are equally suitable. Regardless of the values used, SPSS converts the values to 0 and 1 for processing.

In some cases, the values for one or both of the correlated variables may be ranked, so they are measured on an ordinal level scale. For example, imagine that a researcher wishes to investigate the relationship between educational level and salary. Here, while salary is a continuous variable, educational level is measured on an ordinal-level scale of rankings (e.g., no education = 0, primary = 1, secondary = 2, college = 3, graduate/professional school = 4).

Although there are exceptions to the rule, when one or both variables are dichotomous or ranked, an alternative correlation coefficient is recommended to index an accurate relationship between the two variables. To accommodate situations with noncontinuous variables, several correlation coefficients have been developed to measure the relationship between variables that are measured on a dichotomous or ordinal scale. Two of the most widely used correlation coefficients are Spearman's Rho (also known as Spearman's rank correlation, Spearman's correlation or Spearman's ρ) and Kendall's Tau (also known as Kendall's rank-order correlation).

Both Spearman's Rho and Kendall's Tau are nonparametric versions of the Pearson correlation and are indicated by the Greek letters, ρ (rho) and τ, (tau). Both are nonparametric statistics because data need not meet some of the assumptions (e.g., linearity and normality) required in parametric statistics like Pearson's *r*. To generate these correlation coefficients in SPSS, use the same procedures followed for Pearson correlation, except select **Kendall's tau-b** or **Spearman** in the **Bivariate Correlations** dialog box (Figure 4.19) and unselect **Pearson**. More about nonparametric correlation coefficients in Chapter 12.

Some of the most commonly-used correlation coefficients are displayed in Figure 4.27 along with the type of data used in such bivariate correlation analyses. Empty cells in the table identify situations where specific correlations coefficients are not available for these variable value combinations (e.g., one ranked and one dichotomous variable). Correlations of all of the data shown may be generated using Pearson's *r*, but with associated interpretative cautions.

Figure 4.27 Correlation Coefficients for Variables X and Y measurement levels.

Variable Y		Variable X		
		Interval/Ratio	Nominal/Dichotomous	Ordinal
	Interval/Ratio	Pearson *r*	Point Biserial	
	Nominal/Dichotomous	Point Biserial	Phi	
	Ordinal			Spearman Rho Kendall's Tau

Source: Adapted from Howell, D.C. (2014). Fundamental statistics for the behavioral sciences. Wadsworth CENGAGE Learning.

4.13 Key Terms

bivariate correlation
coefficient of determination
continuous data (interval & ratio levels)
correlation coefficient
covariance
dichotomous (binary) variables
intercorrelation matrix

Kendall's tau
magnitude of relationship
measurement levels
direction of relationship (negative, positive)
Pearson product-moment correlation coefficient (*r*)
phi coefficient
point-biserial correlation

population correlation coefficient rho (*p*)
range restriction
ranked data (ordinal level)
scatterplot (scatter diagram, scattergram)
Spearman rho correlation coefficient

5
Simple Linear Regression

Key concepts from previous chapters or courses:
- **independent variable**: the variable that a researcher manipulates or is studying; also, a variable that may be used to predict values of another variable.
- **dependent variable**: the variable that a researcher is measuring or a variable whose values are predicted by the independent variable.
- **scatterplot**: the pattern of points resulting from plotting the values of two variables on a graph. Each dot on the graph represents the intersection of the values of two variables from the same source.
- **correlation coefficient**: a number representing the extent to which two variables are related.
- **regression line**: the line that best represents the pattern of points in a scatterplot and summarizes the relationship between the two variables.
- **standardization**: converting raw scores to z scores, which have a mean of 0 and a standard deviation of 1.

THIS CHAPTER INTRODUCES

- the differences between correlation and regression
- how to craft and test hypotheses for simple linear regression studies
- how to express the predictive relationship between two variables
- the circumstances under which simple linear regression is used
- how to fit a regression line
- how to calculate the equation for a regression line
- how to use SPSS to conduct simple linear regression

5.1 Introduction

Hurricanes are a big deal in the Caribbean. Every year storm watchers and forecasters predict how many hurricanes the Caribbean is likely to experience and the intensity of those storms. Six to eight were predicted for the Atlantic hurricane season in 2004. However, no matter how many were forecast it took only one to make a hurricane season memorable. In the Cayman Islands that was Hurricane Ivan on September 11, 2004. It was the costliest (US$2.8 billion in damages) and the most intense storm (wind gusts up to 217 mph, with waves 20-30 feet high). Hurricane Ivan, a Category 5 hurricane, devastated multiple countries in the Caribbean region, including the Cayman Islands, where 95% of the buildings were damaged and 90% of the crops in the islands were destroyed.

Forecasters had predicted the landfall of Ivan in the Cayman Islands, and with the help of the National Hurricane Committee that issued an alert 48 hours before landfall, Caymanians prepared, but not enough to forestall disaster in 2004. However, the hurricane prediction and its outcome helped Cayman and other island countries to identify strengths and weaknesses of their disaster preparedness systems and to upgrade hurricane-ready policies to better withstand future hurricanes. Hurricane-related systems and policies included improved hurricane predictions, building codes, and disaster management.

Chapter 4 discussed bivariate correlation, a statistical analysis in which the relationship between a pair of variables is examined. Each pair of variables can be analyzed separately; for example, the correlation between mathematics achievement and reading achievement or miles per gallon and vehicle weight. This correlational relationship can be depicted in a scatterplot of the two variables. In addition, superimposing what is called a "line of best fit" on that scatterplot summarizes the linear relationship between the two variables.

Although correlation provides useful information about how pairs of variables covary, as well as the nature and strength of their relationship, this statistical technique is limited. It is restricted to revealing how much and in what direction one variable tends to change when the other one does.

Linear regression is an extension of correlation that also examines the relationship between variables. However, it is designed to go beyond correlation to build a statistical model that predicts the values of one variable based on the values of one or more other variables.

People make predictions or are the recipients of predictions on a daily basis. Medical researchers use regression to understand how they can use a patient's height to predict a healthy weight. Business leaders use regression to forecast future sales. Educators may use student test performance to predict future performance or areas for deficit remediation. Weather forecasters make predictions about what the weather will be like each day, and the accuracy of those predictions can mean life or death. In the Caribbean, weather predictions are of particular importance during the hurricane season.

Linear regression is designed to compute predictions with the least error possible. In this instance accurate predictions of hurricanes literally save lives.

5.2 The Case of the Height and Weight of College Students

Over the years, studies have consistently shown a significant predictive relationship between height and weight (e.g., Gutin, 2017), hence its broad use in computing body mass index (BMI), an important factor in efforts to maintain a healthy weight. This case study focuses on self-reported heights and weights of 33 college students in the Caribbean. The purpose of the study is to determine whether height is a significant predictor of weight among college students in the Caribbean. An outline of the study may be found in Box 5.1. Data for the study may be found in the data file, **Carib_Height_Weight.sav,** which excludes the duplicate cases removed in Chapter 3. The data are also presented in Table 5.1. Height is measured in centimeters, and weight is measured in kilograms.

Table 5.1 Height and Weight of Caribbean College Students

ID	Height	Weight	ID	Height	Weight	ID	Height	Weight
1001	185	90	2009	183	82	3900	145	47
1002	185	109	2010	195	79	3910	150	55
1010	191	79	3810	167	62	3920	139	40
2001	175	87	3820	164	58	3930	148	50
2002	185	111	3830	155	55	3940	158	60
2003	180	90	3840	157	82	3950	180	82
2004	173	68	3850	190	90	3960	166	68
2005	175	84	3860	142	43	3970	158	60
2006	170	65	3870	160	60	3980	188	95
2007	168	58	3880	170	70	3990	177	80
2008	185	79	3890	192	90	4000	163	65

5.3 Simple Linear Regression Fundamentals

Simple linear regression is so named because it is the most basic of the linear regression procedures. It is used to study the relationships between two variables—a predictor and a criterion variable. The predictor is also known as the independent variable. The criterion variable is also called the dependent or outcome variable. The predictor variable is denoted by the letter x. The criterion variable is denoted by the letter y.

Because simple linear regression is limited to the investigation of two variables, it is also termed bivariate regression. It is called linear regression because it is used to investigate solely linear relationships between two variables.

Although simple linear regression is similar to correlation (examining the direction and strength of the relationships between two variables), it is primarily used for prediction studies; predicting something that will happen in the future based on values of a single predictor. Statistics from the regression procedure form the basis for equations that are then used to predict future occurrences. For example, growth charts were developed by the United States Centers for Disease Control (Kuczmarski et al., 2000) and others to predict physical growth in children based on quantile regression methods.

An important component of correlation is the use of a scatterplot to graphically depict the relationship between two variables. The points or dots on the scatterplot represent the intersection of each pair of variable values. The direction of the scatterplot can be summarized by a straight line that is called the "line of best fit." However, the "line of best fit" has a relatively minor role in correlation. It quantitatively summarizes the overall trend of the relationship depicted by the scatterplot, including the magnitude of the correlation and the amount of variance both variables share. In simple linear regression, the line of best fit takes on added importance.

5.3.1 Line of Best Fit

A key component of linear regression is to compute a "line of best fit," also known as a regression line. It is used to build an equation that predicts values of a dependent variable based on the values of the

164 Simple Linear Regression

independent variable. To a great extent, a vital purpose of regression is to calculate the line of best fit for a set of data, which is then used with new data in the future to make predictions.

Imagine that a researcher wished to establish a regression equation to predict students' mathematics performance on an achievement test based on their performance in language arts. Furthermore, envision that the researcher already had mathematics and language arts scores from students who previously took the tests. In this case, the researcher had mathematics and language arts scores from a total of 207 students.

To calculate the the line of best fit, the researcher plotted students' scores; each point on the scatterplot represented the intersection of students' scores on the language arts and mathematics tests. The researcher then used this data to compute a line of best fit. The line is the graphic representation of a regression equation that may be used to predict mathematics performance from language arts performance in the future.

As Figure 5.1 illustrates, the variable values for language arts (LA Raw Score) are on the x-axis. The values represent the independent or predictor variable in the study scenario and is typically located on the x-axis of a scatterplot. The variable values for mathematics (Math Raw Score), are on the y-axis, as is customary for the dependent variable in regression studies. The maximum raw score possible on both tests is 80, and the minimum is 0. The points on the graph represent the intersection of students' scores in mathematics and in language arts.

The line of best fit has specific characteristics that facilitate its use as a prediction tool. In the mathematics and language arts example, it is sloped upward. This means that as language arts performance increases, so does performance on the mathematics test. Accordingly, the best fit line depicts the direction of the relationship between the independent and dependent variables. In this case, it indicates that there is a positive relationship between both variables. As noted in Chapter 4, the direction of the line could also indicate a negative relationship between two variables or no relationship at all.

The slope or steepness of the line signifies the ratio of change in the dependent variable (mathematics scores) with one unit of change in the predictor (language arts scores). The slope estimates the magnitude of the relationship between the variables: the more moderate the incline of the slope, the smaller the magnitude of the slope, and the weaker the relationship between the variables. Similarly, the sharper the incline of the slope is, the larger the magnitude of the slope, and the stronger the relationship between the variables. A correlation of .8 indicates a much stronger relationship between two variables than a correlation of 0.1, and therefore, the associated slope for the stronger correlation will be markedly steeper than that for the poorer correlation.

Another characteristic of the line of best fit is that in linear regression, it is always a straight line. The points that the line attempts to represent do not change direction. Accordingly, the line does not reflect a curvilinear relationship between the variables.

Even though the goal of regression is to calculate the line of best fit, the line is not a perfect fit. It is an estimate of the direction and magnitude of the predictive relationship. No prediction is perfect. As Figure 5.1 illustrates, there are points on the line, but also scattered on either side of the line. While most scores are clustered together close to the line, there are some instances in which students' scores are farther away from the line than others. In one outlying case (circled), a student's language arts score is 52, and the same student's mathematics score is 77. Therefore, although the line of best fit is the best

Figure 5.1 Scatterplot of Language Arts and Mathematics Performance

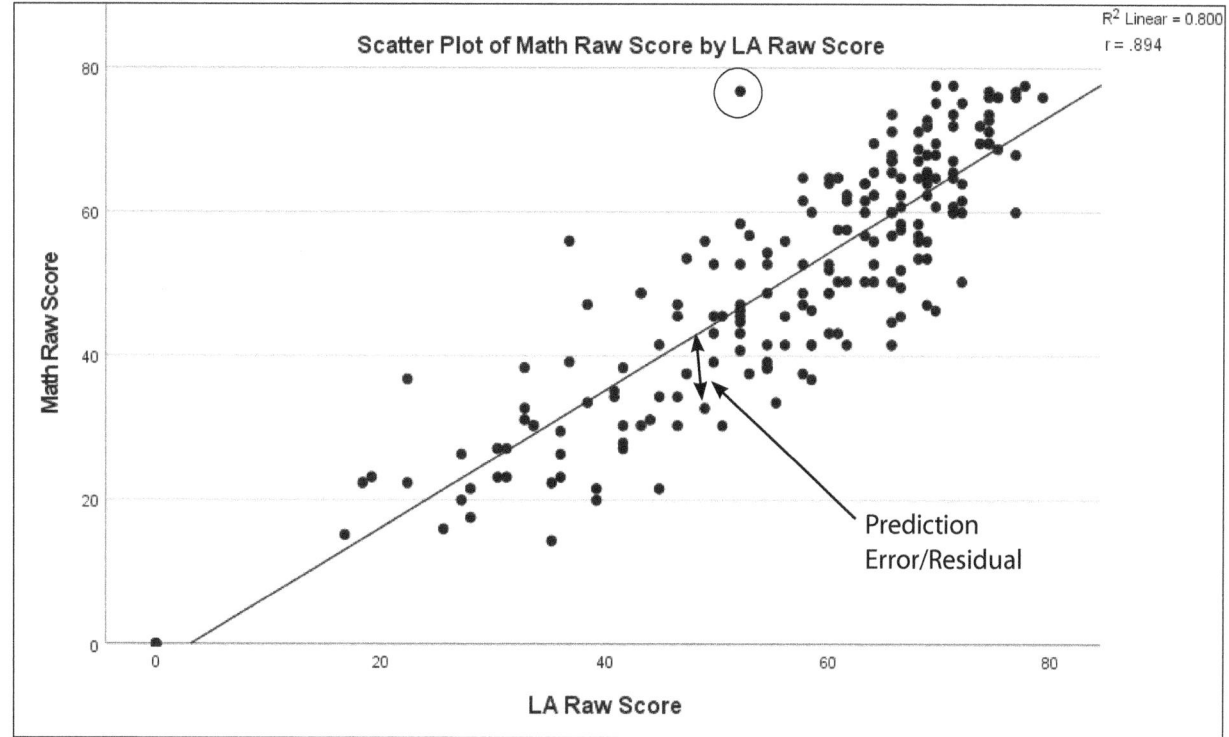

line that illustrates the predictive relationship between the variables, there is error in the prediction of the dependent variable because some points do not sit directly on the line. The spaces between the data points and the regression line (illustrated by the arrows in Figure 5.1) are indicative of prediction error, also known as the residual.

The regression line is computed so that it reduces to a minimum the differences between a regression line and the scores shown in the scatterplot. In so doing, it minimizes the differences between actual math scores and the predicted math scores that are represented by the regression line. In effect, the line is designed to keep prediction error at a minimum; in this case, the regression line minimizes the error in the prediction of mathematics scores. Using weather as an example, a storm may arrive a little earlier or later at a geographical location than originally forecast, or it may travel in a slightly different direction, but in most instances, weather forecasts are quite accurate.

5.3.2 The Strength of the Predictive Relationship (Goodness of Fit)

As in correlation, it is possible to measure the strength of the relationship between two variables in simple linear regression. The square of the correlation (e.g., Pearson r), also known as r^2, r-squared or the coefficient of determination, expresses how much of the variance of the dependent variable (DV) the independent variable (IV) explains or shares as depicted in Figure 5.2. Pearson's r is typically expressed as a proportion. Accordingly, as noted in the upper right corner of the scatterplot in Figure 5.1, the r^2 for mathematics and language arts is .80. Translated to a percentage, r^2 means
- Language arts and mathematics share 80.0% of each other's variance.
- Language arts accounts for or explains 80.0% of the variance in mathematics.

Figure 5.2 Shared Variance Between the Independent and Dependent Variables

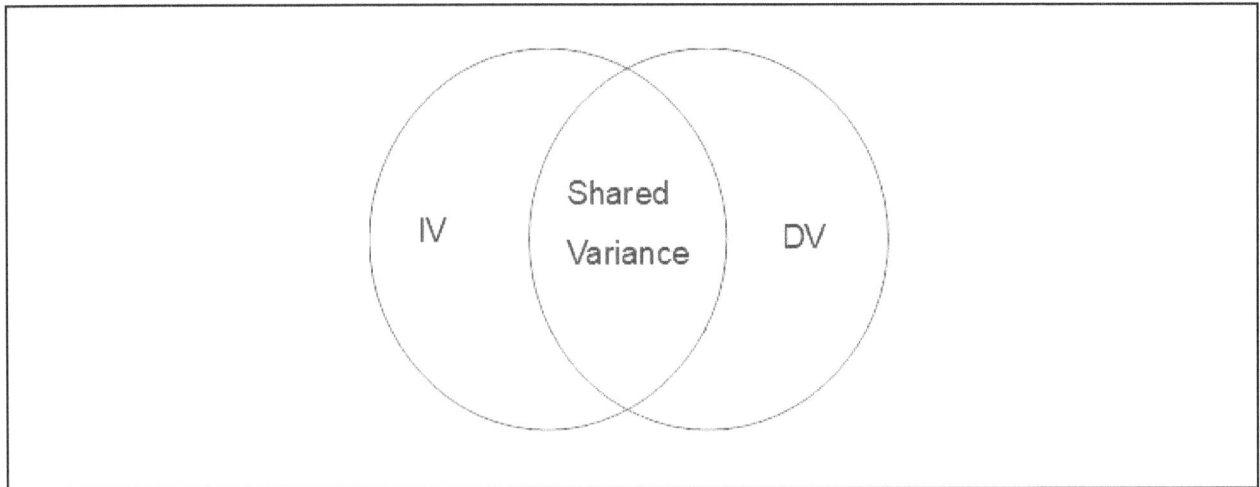

Higher r^2 values mean smaller differences between the actual data and fitted or predicted values. Lower r^2 values represent larger differences between actual data values and predicted values. Therefore, r^2 is a measure of how well the data fit the regression model.

In the test score example in Figure 5.1, language arts scores explain much of the variance in mathematics scores. Nevertheless, there is some variance that is unexplained, which is described as the residual (Figure 5.1) or the coefficient of nondetermination. The residual is expressed by subtracting the explained variance from all of the variance (which is equal to 1) or $1 - r^2$. Accordingly, $1 - r^2 = 1 - 0.8 = 0.2$. Converting 0.2 to a percentage, results in 20.0%, which is the amount of variance in mathematics that remains unexplained after shared variance with language arts is accounted for.

There is one drawback to the use of r^2, and that is it is often somewhat inflated due to the influence of two factors:
- the presence of error variance, and
- the number of predictors in the model relevant to the sample size.

To correct for r^2 inflation, an adjusted r^2 is computed that is based on the sample size and the number of predictors in the model. Adjusted r^2 is regarded by researchers as an indicator of the population r^2. Adjusted r^2 increases only if an additional predictor improves the model more than would be expected by chance. It declines when a predictor improves the model by an amount that is less than expected by chance.

5.3.3 The Linear Regression Model

The primary purpose of simple linear regression is to establish a model that is used to predict the values of one variable (dependent variable) based on the values of another variable (independent variable). In this simple model, a single predictor (independent variable) predicts a single outcome, and the model is linear, which means that a straight line summarizes the predictive relationship between the two variables.

The general model for simple linear regression is:

Outcome = predictor + error

Notice that the outcome is the dependent variable, which is predicted by the independent variable, and some error is always associated with a prediction. The general model breaks down into two versions—unstandardized and standardized. They are typically presented as prediction equations, with $\hat{Y}$ representing the predicted outcome and excluding the error component (ε). The error equals the actual value minus the predicted value or Y - $\hat{Y}$

5.3.3.1 Unstandardized

Researchers use unstandardized regression models when variable values are in their original form rather than standardized or z scores. The associated regression model for research samples includes two additions to the general form of the equation: a and b, known as regression coefficients or regression weights.

$\hat{Y} = a + bX$

where

$\hat{Y}$ = the predicted value of the dependent variable—pronounced "Y-hat"

a = the constant (intercept), which is the predicted value of the dependent variable when the independent variable (X) = 0.

b = the slope of the regression line—the amount of change in y associated with a one-unit change in X.

Notice that $\hat{Y}$ rather than Y represents the dependent variable in the equation. Y represents actual scores obtained (e.g., mathematics scores). These values (typically on the y-axis of a scatterplot) are used to assemble the regression equation that will be used later to make predictions. $\hat{Y}$ represents estimated, predicted values that come as close as possible to observed values of mathematics.

In the case of the test scores, X is the value of the independent variable, language arts, which is typically on the x-axis of a scatterplot. In order to predict estimated values of mathematics for future students, the task of the researcher is to use the regression procedure to produce the values of the constant (a), and the slope of the regression line (b).

The constant (a) and the regression coefficient (b) can be calculated using the equations reproduced in Figure 5.3, in which,

$\hat{Y}$ = the values of the dependent variable,

X = the values of the independent variable, and

N = the sample size

Once the values for the constant and regression coefficient are calculated, they may be substituted into the regression equation to find the predicted value, $\hat{Y}$.

Although researchers can derive the constant (a) and regression coefficient (b) using the equations in Figure 5.3, the values can be generated using widely available statistical software. Once a researcher has conducted the regression procedure and secured the constant and the regression coefficient, those values may be inserted into the equation. The researcher can subsequently plug in actual scores for the predictor variable, in this case language arts scores, in order to secure the associated predicted value for mathematics.

So far, the discourse has been limited to data that are unstandardized; that is, they are in the original units of measure in which they were collected. The regression coefficient (b) associated with unstandardized scores are called unstandardized coefficients or raw score coefficients.

The language arts and mathematics test scores are examples of unstandardized data. They are different tests developed using different testing objectives and having different content. They have different standards as well. Consequently, their scores have different meanings. A score of 60 for mathematics

Figure 5.3 Equations for the Regression Constant, a and the Regression Coefficient, b

$$a = \frac{(\sum Y)(\sum X^2) - (\sum X)(\sum XY)}{N(\sum X^2) - (\sum X)^2}$$

$$b = \frac{N(\sum XY) - (\sum X)(\sum Y)}{N(\sum X^2) - (\sum X)^2}$$

does not have the same meaning as a score of 60 for language arts, so performance comparisons across the tests are verboten.

There are benefits and disadvantages of using unstandardized data. One benefit is that they are in the original unit of measure, and therefore, have substantive intrinsic meaning. Accordingly, they may be easier to understand and readily interpretable. However, when dependent and independent variables are in different units of measure, it can be difficult to keep track of the differences in interpretation. One method of dealing with interpretation challenges is to use standardized data.

5.3.3.2 Linear Regression Equation: Standardized

Because the test scores are unstandardized, the mean and standard deviation for mathematics and language arts are different. For the scores included in the scatterplot in Figure 5.1, the mean mathematics score is 49.97, with a standard deviation of 18.32. The mean language arts score is 55.37, with a standard deviation of 17.13.

Imagine that scores in Figure 5.1 were standardized. If that were the case, both mathematics and language arts scores would have the same mean and standard deviation—a mean of 0 and a standard deviation of 1. Except for the change in the mean and standard deviation, the pattern of the data would be unaffected. However, the advantage of standardized data is the acquisition of a useful benefit—the feasibility of comparing performance across subjects. For example, it would be acceptable to describe a student's performance as better in language arts than in mathematics because the student's performance was one standard deviation above the mean in language arts and one standard deviation below the mean in mathematic.

A regression coefficient that is computed from standardized data is called a standardized regression coefficient, β, or beta. As with the unstandardized model, the standardized regression coefficient specifies the slope of the regression line that best represents the predictive relationship between two variables. The line is constructed using standardized data for the two variables. The equation for the standardized regression model is:

$$\hat{Y}_z = \beta X_z$$

where
X_z = the standardized (z) score that is the predictor or independent variable.
β = the standardized regression coefficient for the standardized independent variable X. It is also known as a beta weight or beta coefficient.

There is no constant in standardized regression equation because the regression line always passes through the point of origin in the scatterplot. The point of origin is the location where both the independent variable, X, and the dependent variable, Y, are equal to zero. Therefore, in standardized regression equations, the constant, a, is always equal to 0. Accordingly, there is no need to include it in the equation.

5.3.4 Hypothesis Testing in Bivariate Regression: How Good Is Your Model?

Hypothesis testing involves investigating whether there is sufficient evidence that a claim (null hypothesis) is true. If not, the claim is rejected. Two major tests are conducted in simple linear regression:
- a test of the significance of the model
- a test of the significance of the predictor

5.3.4.1 Testing the Significance of the Model

In testing the significance of a simple linear regression model, researchers seek to determine whether the entire regression model is significant; that is, whether the improved regression model is significantly better than the baseline regression model. As the equations below illustrate, there is no independent variable in the baseline model. The model comprises solely the constant and the residual. Alternatively, the null hypothesis model may be symbolized by an equation in which the regression coefficient (b) equals 0. In this case, the predicted value is equal to the value of the constant (the intercept) as denoted in the null hypotheses below.

Null Hypothesis: The baseline model with no independent variables fits the data as well as your improved model.

H_o: $\hat{Y} = a + e$

Or

H_o: $b = 0$

The improved model for simple linear regression is the model in which a single independent variable, X, which has a value greater than 0 is added to the regression model along with the associated regression coefficient, b, with a value greater than 0.

Alternative Hypothesis: The improved model fits the data significantly better than the intercept-only model.

H_a: $\hat{Y} = a + bX + e$

Or

H_a: $b \neq 0$

Researchers use the *F* test of significance to test the overall significance of linear regression models based on the null hypothesis stated above. If the probability level is less than the a priori alpha level (e.g., $p \leq .05$), that suggests that there is less than a five percent probability that the slope of the regression line equals zero. Another way of expressing this notion is that there is less than a five percent probability that the value of the unimproved model is not significantly different from that of the improved model. In either case, the researcher may reject the null and conclude that there is sufficient evidence to support the alternative hypothesis.

The *F* test may be computed by hand, but most frequently with statistical software like SPSS. Chapter 7 discusses this statistical technique in detail. However, Section 5.7 provides a preview of ANOVA and how to interpret one aspect of the results of that procedure (the ANOVA table).

170 Simple Linear Regression

5.3.4.2 Testing the Significance of the Predictor

In the equation for simple linear regression, the predictor or the independent variable (X) is linked with a weight, known as a regression coefficient (b) that represents the slope of the regression line. If the independent variable is a poor predictor, as the value of the independent variable changes the dependent variable realizes little to zero change in value. Little or no change in the dependent variable results in a flat or close to flat regression line: the flatter the regression line the poorer the prediction, the smaller the regression coefficient, and the greater the likelihood that the regression coefficient (b) for the independent variable (X) will equal zero or a value that is not significantly different from zero.

If the independent variable significantly predicts the dependent variable, the incline (or decline) of the regression line will increase as the values of the independent variable increases. Similarly, the value of the regression coefficient will increase, enhancing the probability of being significantly different from zero.

The t statistic tests whether the regression coefficient is significantly different from zero. In simple linear regression where there is a single independent variable, if the regression coefficient for the independent variable is nonsignificant, the entire model is statistically nonsignificant—no better than the baseline model. The significance of the value of the regression coefficient determines whether the independent variable contributes significantly to the dependent variable as well as to the entire model.

Imagine that a student received a score of zero for language arts. In that scenario, the associated predicted value for mathematics would be (error term excluded):

$$\hat{Y} = a + b0$$
$$\hat{Y} = a$$

Since the product of the regression coefficient (b) and the value for language arts (0) = 0, the value of the dependent value, $\hat{Y} = a$ (the constant or the value of $\hat{Y}$ when $X = 0$).

In simple linear regression, the hypotheses for the significance test of the regression coefficient looks similar to the hypotheses for the entire regression model.

Null hypothesis: The regression coefficient is not significantly different from 0.
H_o: $b = 0$
Alternative hypothesis: The regression coefficient is significantly different from 0.
H_a: $b \neq 0$
The equation for hand-calculating the t statistic for the b coefficient is as follows:

$$t = \frac{b(s_x)\sqrt{N-1}}{s_y\sqrt{(1-r^2)\frac{N-1}{N-2}}}$$

where
b = regression coefficient
s_x = the standard deviation for the independent variable
s_y = standard deviation of the dependent variable
r^2 = the correlation coefficient squared or the amount of variance in the dependent variable that the independent variable accounts for.
b = regression coefficient
N = sample size

Once the calculation of the t statistic is complete, the next step is to compare the result with the critical value (Section 2.3.6), using the degrees of freedom for simple linear regression:
$$df = N - p - 1$$
where
N = the sample size, and
p = the number of independent variables.

With one independent variable in simple linear regression, the formula reduces to:
$$df = N - 1 - 1 = N - 2$$

If the *t* statistic calculated is larger than the critical value, then reject the null hypothesis; the regression coefficient is not equal to 0. Consequently, there is sufficient evidence that the independent variable is a significant predictor of the dependent variable.

5.3.5 Solving the Regression Equation

Once regression coefficients have been computed and the regression model evaluated, researchers may produce the regression equation and test its accuracy using existing or new data.

5.3.5.1 Using Unstandardized Data

Imagine that a researcher conducts a bivariate linear regression on language arts and mathematics scores. Language arts performance is intended as a predictor for future mathematics performance. Evaluation of the data discloses that language arts is a significant predictor of mathematics. In addition, analysis of the data reveals that language arts predicts 80% of the variability in mathematics. Using the regression procedure, the researcher derives a constant of -2.98, which is the theoretical value of mathematics (the dependent variable, Y) when language arts (the independent variable, X) is equal to zero. Finally, the researcher derives a value of 0.96 for the regression or b coefficient associated with language arts. This means that for every 1-point change in the score for language arts, there is a predicted 0.96 change in the score for mathematics.

With the value for the constant and the regression coefficient available, all the information needed to predict students' scores is available. Imagine that the researcher wishes to cross-validate the regression equation by predicting the mathematics score of a student who receives a language arts score of 52.

$$\begin{aligned}\text{Predicted mathematics score} &= -2.98 + 0.96(52) \\ &= -2.98 + 49.92 \\ &= 46.94\end{aligned}$$

In this instance, the predicted score is different from that of the outlier student who received a score of 52 on the language arts test. The predicted mathematics score is 46.94. The actual math score of the student receiving a score of 52 in language arts was 77. The difference between predicted mathematics score and actual score (30.06) is due to prediction error or the difference between actual obtained scores (Y) and predicted scores ($\hat{Y}$). Prediction error tends to be greater with outlier score values.

For nonoutlier data, however, regression models are useful tools to make predictions about the future. With that in mind, imagine that the following year, a student from the same population of students as the year before receives a raw score of 60 in language arts (X) on a preparation test that is statistically

172 Simple Linear Regression

equivalent to the actual test. Below is the calculation for the predicted mathematics score ($\hat{Y}$) for that student.

$$\hat{Y} = a + bX$$
$$\hat{Y} = -2.98 + 0.96(60)$$
$$\hat{Y} = -2.98 + 57.6$$
$$\hat{Y} = 54.62$$

Verify this result by locating the score (60) for language arts on the x axis of the scatterplot in Figure 5.1 and identifying the associated mathematics score on the y axis, by charting a straight line to the regression line and from the regression line, making a 90° turn to the y axis and to the predicted mathematics score.

5.3.5.2 Using Standardized Data

Continuing with the example of the prediction of mathematics performance based on language arts results, assume that scores are standard scores rather than raw test scores. Under these circumstances, researchers use standardized rather than unstandardized regression weights. Therefore, if the standardized regression weight for language arts is .91. The equation would be

$$\hat{Y}_z = .91X_z$$

Inserting a language arts score of half standard deviation ($s = -0.5$) below the mean into the equation, results in a predicted mathematics standard score of

$$\hat{Y} = .91(-0.5)$$
$$\hat{Y} = -0.41$$

Recall that half of a standard deviation below the mean is equal to -0.5. Therefore, a predicted mathematics standard score of -0.41 equates to less than half of a standard deviation below the mean mathematics score.

Converting unstandardized data to standardized or z scores in order to compute standardized regression coefficients can be tedious. For example, in the previous example, the mean language arts score is 55.37 with a standard deviation of 17.131. The mean language arts standard score is 0 with a standard deviation equal to 1. Therefore, it is fortuitous that standardized data are not necessary to determine what would happen with z scores. Standardized regression coefficients are standard output in most statistical software packages. This is particularly handy when more than one predictor is involved. More about that in Chapter 6.

It is important to note that when researchers use standardized data in bivariate regression, the correlation coefficient is the same as the standardized regression coefficient. Therefore, if the correlation of mathematics and language arts is .89, the standardized coefficient is also .89.

5.4 Planning a Simple Linear Regression Study

Chapter 4 pointed out that bivariate correlation does not imply a causative relationship between the variables. In addition, correlation does not indicate that there is always a predictive relationship between the variables. The two procedures are distinctively different. An outline of some of the components of the research plan for a bivariate regression study is presented in Box 5.1.

Box 5.1 Research Plan Outline: The Case of the Height and Weight of College Students

Research Problem
Over the years, studies have consistently shown a significant predictive relationship between height and weight (e.g., Gutin, 2017), hence its broad use in computing body mass index (BMI) an important factor in efforts to maintain a healthy weight. This case study focuses on self-reported heights and weights of 33 college students in the Caribbean. The purpose of the study is to determine whether height is a significant predictor of weight among college students in the Caribbean.

Variables and Measurement Levels
Independent Variable: Height
Dependent Variable: Weight
Level of Measurement: Ratio (Continuous Scores)

Research Question
1. Is there a significant linear predictive relationship between height and weight?
2. How much of the variance in the dependent variables does the independent variable account for?
3. What is the regression equation for the model?

Hypotheses
Null Hypothesis 1: The baseline model without the independent variable fits the data as well as the improved model. (Height is not a significant predictor of weight.)
$H_o: \hat{Y} = \alpha + e$ **or** $H_o: b = 0$

Alternative Hypothesis 1: The improved model fits the data significantly better than the intercept-only model. (Height is a significant predictor of weight.)
$H_a: \hat{Y} = \alpha + bX + e$ **or** $H_a: b \neq 0$

Null Hypothesis 2: The regression coefficient for height is not significantly different from 0.
$H_o: b = 0$

Alternative Hypothesis 2: The regression coefficient for height is significantly different from 0.
$H_a: b \neq 0$

Research Design
Nonexperimental, predictive correlational research design

Data Analysis
Generating descriptive statistics and correlation of the variables are typically the first steps in regression analysis, followed by tests of the assumptions associated with simple linear regression. In simple linear regression, only the standard regression is possible. That method includes a test of significance of the entire model (*F*-test for Hypothesis 1) as well as the test of the significance of the independent variable as a predictor (*t*-test for Hypothesis 2). The Pearson's *r* statistic measures the direction and magnitude of the relationship between height and weight. Results include R^2 and adjusted R^2 which provide the amount of variance in weight that height accounts for. An important part of the analysis is to generate the constant, and regression coefficient needed for the regression equation.

5.4.1 Variables and Measurement Levels

Like correlation, simple linear regression assumes that both variables involved in the regression analysis are quantitative. However, while Pearson product moment correlation requires both variables to be at the interval or ratio level, in simple linear regression, that requirement is restricted to the dependent variable. There is more flexibility for the measurement level of the independent variable, which is mostly interval or ratio level, but may be categorical or ordinal as well.

As Chapter 4 explained, some types of correlation methods (excluding Pearson) may be suitable for categorical or ordinal level variables. However, in simple linear regression, categorical variables must be transformed into dummy variables (0, 1 variables) before proceeding with the analysis (e.g., the categorical variable, sex, must be transformed from, say male = 1 and female = 2, to male = 0 and female = 1). In the case of ordinal independent variables, there is ongoing debate (e.g., Pasta, 2009; Williams, 2020) regarding how these variables should be treated. However, in general, including them without further modification appears to be the most common practice.

In the Case of the Height and Weight of College Students, our independent variable, height, is a ratio-level variable. The dependent variable, weight, is also a ratio-level variable.

5.4.2 Simple Linear Regression Research Questions and Hypotheses

Arguably social scientists primarily conduct statistical analyses in order to find the answers to specific research questions. Some of those questions in a prediction study are likely to include:

1. Is the improved regression model (with the independent variable included) significantly different from the baseline model (without the independent variable)?
2. Does the independent variable provide better-than-chance prediction of the dependent variable?
   a. What is the value of the slope of the line of best fit (or the regression coefficient, b, used to predict the dependent variable?
   b. Is the regression coefficient, b, significantly different from 0?
3. What is the value of the constant (the value of the dependent variable when the independent variable equals 0)?
4. How much of the variance in the dependent variable does the independent variable account for?
5. What is the equation for the regression model predicting the dependent variable from the independent variable?

In the Case of the Height and Weight of College Students, researchers wish to predict the weights of college students based on their heights. To develop research questions for the case, start with the generic questions and modify them to reflect the variables in the case.

1. Is the improved regression model with the variable, height included, significantly different from the baseline model without the variable, height?
2. Does the variable, height, provide better-than-chance prediction of the variable weight?
   a. What is the value of the slope of the line of best fit (or the regression coefficient, b, used to predict weight from height?
   b. Is the regression coefficient for height, significantly different from 0?
3. What is the value of the variable, weight, when the variable, height, equals 0 (the constant)?
4. How much of the variance in the variable, weight, does the variable, height, account for?
5. What is the regression equation for the model, predicting weight from height?

It should be evident that Questions 1, 4, and 5 reflect the research questions in Box 5.1. Questions 2 and 3 (particularly 2) reflect the hypotheses in Box 5.1, which contribute to answering Question 1: Is the improved regression model (with variable, height, included) significantly different from the baseline model (without the variable, height)? This information is necessary to determine whether the improved model is better than the base model. Questions 2a and 3 contribute to answering Research Question 3 in Box 5.1. Question 4 reflects Question 2 in Box 5.1 and provides the overall fit of the model.

In general, simple linear regression hypotheses are nondirectional; for example, for the hypothesis related to the b coefficient, the null hypothesis claims that the coefficient is not significantly different from zero ($H_o: b = 0$), while the alternative hypothesis claims that it is significantly different from zero ($H_a: b \neq 0$). Directional hypotheses might have claimed that the coefficient is either greater than or less than zero or another value ($H_o: b < 0$, $H_a: b > 0$). Non-directional hypotheses require two-tailed, rather than one-tailed significance testing.

5.4.3 Research Design

Simple, linear regression is a statistical method that is typically used in nonexperimental correlation research designs that are used for predicting an outcome variable based on one or more independent variables. The research design is similar to that used for bivariate correlation discussed in Chapter 4, with the exception that, in this case, there is a single independent variable used to estimate (predict) the values of a single dependent variable.

5.4.4 Data Analysis

Typically, the central data analyses conducted for simple linear regression include correlation, which Chapter 4 discussed as well as analysis of variance (ANOVA), which Chapter 7 will examine in greater detail, and the t test, which should be a familiar technique from previous statistics courses. There are three steps to data analysis for simple linear regression after screening the data and testing that the data meet the assumptions for regression.

1. Conduct a correlation of the two variables, including a scatterplot, to identify the direction and magnitude of the relationship. In addition, use the correlation coefficient (often Pearson r), to find the amount of variance in weight that height accounts for or the practical significance of the relationship, by squaring the coefficient to get R^2.

2. After calculating the correlation coefficient, one can estimate the slope of the regression line (b) and determine whether it is significantly different from zero. However, because there is only a single predictor in simple linear regression, the most straightforward approach to testing the significance of the slope is to base your conclusion on the significance of the correlation between the independent and dependent variables. A significant correlation means that the slope of the regression line is significant as well.

    There are two options for assessing the significance of the correlation coefficient. The first is described in Step 3a, below. The second option is using SPSS to conduct an ANOVA, which is designed to assess whether the correlation between the independent and dependent variables is statistically significant. See Section 5.7.9.

176   Simple Linear Regression

3. Estimate the two regression coefficients:
   a. Estimate the b coefficient (slope of the regression line) by conducting a t-test to determine whether the regression coefficient is significantly different from zero in the target population.
   b. Estimate the constant or a coefficient (the value of the dependent variable when the independent variable equals zero) by using the t test to test the null that $H_o: a = 0$. For both statistics a and b, the goal is to secure values that will allow rejection of the null.

If the results of the ANOVA show the independent variable is a significant predictor, the values of the b and a coefficients may be used to construct the regression equation.

5.5 Calculations for Simple Linear Regression

The purpose of this hand-calculated example is to calculate the a and b coefficients needed for the regression equation to predict weight from height. Height and weight values in Table 5.1 for the Case of the Height and Weight of College Students will serve as the data in the demonstration of how to manually calculate the a and b coefficients as well as how to conduct significance testing for those coefficients. A demonstration of the hand-calculation of the F test to test the significance of the prediction equation may be found in Chapter 7.

Table 5.2 Values for Calculating Regression Coefficient and Constant

(1) ID	(2) X	(3) X²	(4) Y	(5) Y²	(6) XY	(1) ID	(2) X	(3) X²	(4) Y	(5) Y²	(6) XY
1001	185	34225	90	8100	16650	3840	157	24649	82	6724	12874
1002	185	34225	109	11881	20165	3850	190	36100	90	8100	17100
1010	191	36481	79	6241	15089	3860	142	20164	43	1849	6106
2001	175	30625	87	7569	15225	3870	160	25600	60	3600	9600
2002	185	34225	111	12321	20535	3880	170	28900	70	4900	11900
2003	180	32400	90	8100	16200	3890	192	36864	90	8100	17280
2004	173	29929	68	4624	11764	3900	145	21025	47	2209	6815
2005	175	30625	84	7056	14700	3910	150	22500	55	3025	8250
2006	170	28900	65	4225	11050	3920	139	19321	40	1600	5560
2007	168	28224	58	3364	9744	3930	148	21904	50	2500	7400
2008	185	34225	79	6241	14615	3950	180	32400	82	6724	14760
2009	183	33489	82	6724	15006	3960	166	27556	68	4624	11288
2010	195	38025	79	6241	15405	3970	158	24964	60	3600	9480
3940	158	24964	60	3600	9480	3980	188	35344	95	9025	17860
3810	167	27889	62	3844	10354	3990	177	31329	80	6400	14160
3820	164	26896	58	3364	9512	4000	163	26569	65	4225	10595
3830	155	24025	55	3025	8525						

The first and second steps in simple linear regression data analysis is to determine whether a significant, linear relationship exists between the independent and dependent variables. Chapter 4 illustrated how to accomplish that by calculating the Pearson product moment correlation coefficient and assessing its significance. Therefore, there is no need to repeat that component of the analysis. In addition, for this computation, assume that the data meet the assumptions required for simple linear regression.

Data for a total of 33 study participants will be included in the computations. Values required to calculate the regression coefficients are presented in Table 5.2, where X represents height and Y signifies weight. The summary descriptive statistics for height and weight in Table 5.2 may be found Table 5.3. The correlation coefficient of height and weight is $r(33) = .856$ ($p < .001$). The amount of variance in weight that height accounts for is $R^2 = .733$ or 73% of the variance in weight.

The final step in conducting simple linear regression analysis is to calculate the values for the prediction equation. Therefore, to find the value of the b coefficient by using the calculated expressions in Table 5.3 and Figure 5.4.

Table 5.3 Descriptive Statistics for Table 5.1

	Height (X) (cm)	Weight (Y) (kg)	Height*Weight (XY)	Height² (X²)
Sum (Sum Squared)	5,619 (31,573,161)	2,393 (5,726,449)	415,047	964,561
Mean	170.273	72.515		
St. Dev.	15.611	17.850		

Figure 5.4 Values to Compute the b Coefficient

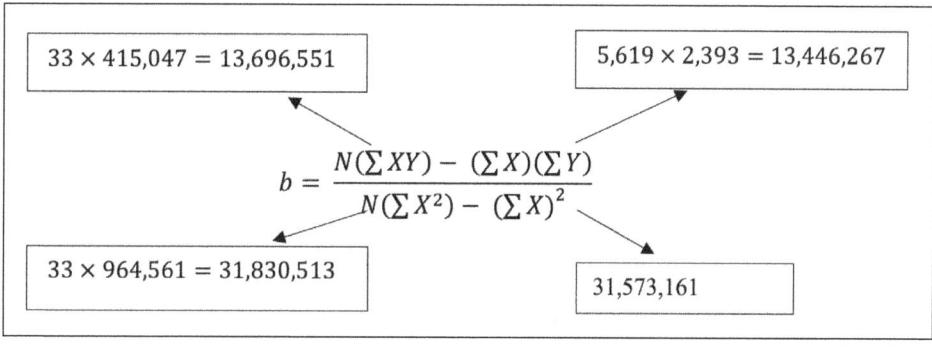

$$b = \frac{13{,}696{,}551 - 13{,}446{,}267}{31{,}830{,}513 - 31{,}573{,}161}$$

$$b = \frac{250{,}284}{257{,}352}$$

$$b = .973$$

The regression coefficient (b) has a calculated value of .973, but is the coefficient (the slope of the regression line) equal to zero as the null hypothesis states in Box 5.1? To test the significance of the slope of the regression line, begin by substituting the values for the expressions in the *t*-statistic formula in Section 5.3.4.2.

$$t = \frac{b(s_x)\sqrt{N-1}}{s_y\sqrt{(1-r^2)\frac{N-1}{N-2}}}$$

where

b = regression coefficient
s_x = the standard deviation of height
s_y = standard deviation of weight
r^2 = the correlation coefficient squared or the variance in weight that height accounts for.

Figure 5.5 Values to Compute the *t* Statistic

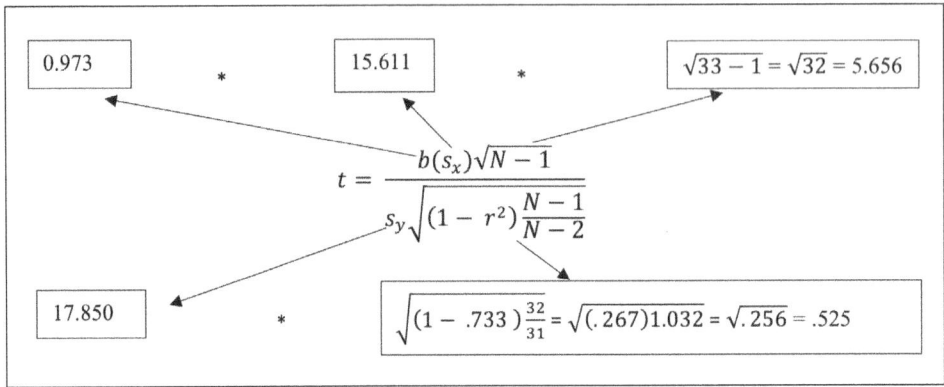

$$t = \frac{0.973 * 15.611 * 5.656}{17.850 * .525}$$

$$t = \frac{85.912}{9.371}$$

$$t = 9.168$$

Using the *t* distribution table in Appendix A, follow the row for 31 degrees of freedom (n − 2), and the column for the alpha level of .05 for a two-tailed test to find the critical value of 2.040. Since the calculated *t* value is 9.168, it exceeds the critical value of 2.040, meaning that the *b* coefficient for the slope of the regression line is significantly different from zero. Since the *b* coefficient for height is significantly different from zero, height is a significant predictor of weight.

Because height is the only predictor of weight, the regression model, with height included, is significantly better than the model with no predictor. Nevertheless, for demonstration purposes, calculate coefficient *a*, which is the constant or the value of weight when height equals 0. The calculation of coefficient *a* is illustrated in Figure 5.6, using the values in Table 5.3.

Figure 5.6 Values to Compute Coefficient *a* (Constant)

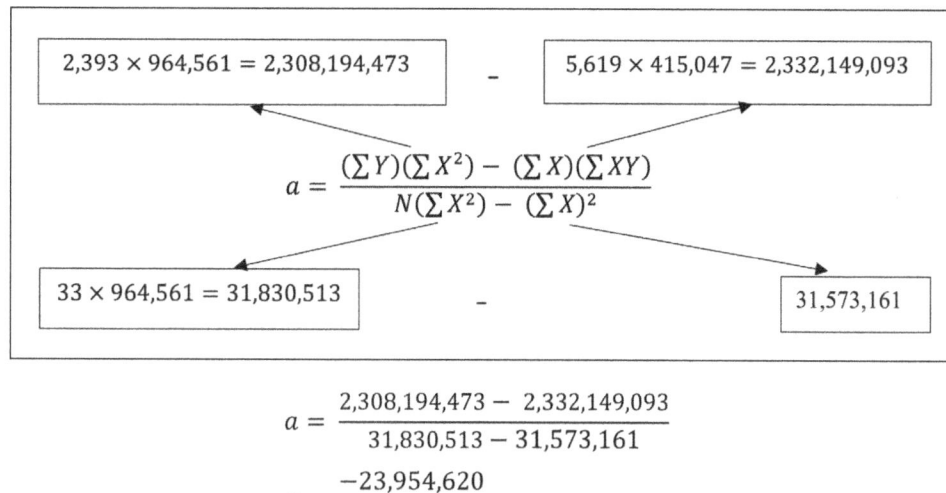

$$a = \frac{2{,}308{,}194{,}473 - 2{,}332{,}149{,}093}{31{,}830{,}513 - 31{,}573{,}161}$$

$$a = \frac{-23{,}954{,}620}{257{,}352}$$

$$a = -93.081$$

The calculated constant (*a* coefficient) equals -93.081. Therefore, in theory, when height equals 0, weight equals -93.081. Now, apply the calculated coefficients to the regression equation for predicting weight from height:

$$\hat{Y} = a + bX$$
$$\text{Predicted weight} = -93.081 + 0.973(\text{height})$$

5.6 Assumptions of Linear Regression

As with all statistical techniques in this text, linear regression makes some assumptions about the data being analyzed, some of which were discussed in Chapter 3. If the assumptions are violated, the results of the tests may be questionable. Therefore, it is necessary to evaluate whether the data meet associated assumptions before central regression procedures are conducted. If the data do not meet the assumptions for the given test, alternative statistical methods should be used to conduct the analysis, some of which are discussed in Chapter 12.

Like other statistical techniques, in linear regression, some assumptions are met by the research design chosen by the investigator. Others must be assessed using statistical methods. The assumptions that follow apply to both simple linear (bivariate) and multiple linear regression. Chapter 6 describes additional assumptions for multiple regression.

5.6.1 Characteristics of Variables

In bivariate regression, the dependent variable is always a continuous variable (i.e., measured at the interval or ratio level). Examples of continuous variables include test scores, height, weight, salary, and temperature. In addition, dependent variables may include summarized values of data at the ordinal level of measurement (e.g., subscale or scale scores of Likert response data).

The independent variable in bivariate regression may be either continuous or categorical. Although categorical variables can be used as independent variables in simple linear regression analysis (and multiple regression), if they have more than two unordered categories (e.g., varying models of cars or eye color), it may be difficult to accurately interpret the relationship between the variables.

To analyze categorical variables, use the technique called 'dummy' coding. If an independent variable is dichotomous, assign 0 to one category and 1 to the other. If an independent variable has more than two categories, create a series of dummy coded variables, one for each variable category. This technique allows direct comparison of each dummy-coded variable with the others. It also facilitates separate assessments of the influence of each category of the independent variable on the dependent variable.

5.6.2 Sample Size

In linear regression the sample size rule of thumb is at least 20 cases per independent variable. However, with only one independent variable, limiting a study to 20 cases may result in lowered statistical power and effect size. In addition, a small sample size may not provide the data necessary to assess whether the data are normally distributed (another assumption).

To increase the statistical power of bivariate regression, aim for a sample size of 60 – 105, which is likely to provide reliable results for medium to large effect sizes (e.g., Green, 1991;). The best course of action is to calculate the required sample size depending on the expected power and effect, using one of many internet tools available to complete the computation.

5.6.3 Independence of Observations

When collecting data, each study participant is counted as one observation. The statistical assumption of independence of observations in regression requires the collection of data from each study subject only once. When this occurs, typically the errors associated with one observation are not correlated with errors of any other observation and the data meet the assumption of independence. However, subjects who are dependent in some way (e.g., having the same instructor) may exhibit a lack of independence in responses, resulting in correlated prediction error. This phenomenon in regression is known as autocorrelation. The existence of autocorrelation in a dataset may result in invalid significance tests.

The Durbin-Watson test was designed to test the assumption of independence of observations by detecting autocorrelation—whether errors of adjacent data points are correlated. As a rough rule of thumb, if Durbin–Watson is less than 1.0 or greater than 3, there may be cause for concern. A demonstration of two methods of evaluating this assumption is provided in Section 5.7.1.3.

5.6.4 Range Restriction

A restricted range of scores is a range of values that has been condensed or shortened. For example, the entire range of GPA scores is 0.0 to 4.0. A restricted range could be 3.0 to 4.0, or 2.0 to 3.0. The result of the restriction of the range of scores is the alteration of the correlation between the independent and dependent variable. The most likely effect is a reduction in the correlation, but the correlation can also be artificially increased. Either way, the result is a reduction in predictive accuracy.

5.6.5 Linearity

In simple linear regression relationships, any given change in an independent variable will always produce a corresponding change in the dependent variable, resulting in a straight line. Accordingly, for linearity to exist, as values of the independent variable increase, values of the dependent variable also increase. Or as independent variable values increase, dependent variable values concomitantly decrease. Linearity is required for prediction precision. Any other pattern results in curvilinearity, which cannot be measured using linear regression (Chapter 3). Not only is the relationship between the independent and dependent variable values expected to be linear, but also the relationship between the values of the dependent variable and its residuals (difference between actual and predicted values). More on this in Section 5.7.1.1.

Check to ensure a linear relationship exists between the independent and dependent variable by producing scatterplots of the data and examining the plot for indications of curvilinearity. If the data are not linearly related, consider using SPSS to transform the data to improve linearity. Alternatively, analyze the data using nonlinear regression.

5.6.6 Homoscedasticity

Chapter 3 presents a detailed discussion of homoscedasticity and heteroscedasticity (Section 3.4.2.2.3). As described there, another term for homoscedasticity is equal variance of the residuals along a range of values. In regression, this means that variances of prediction error along the regression line mostly remain similar all across the line. Variance does not increase and/or decrease in places along the regression line if the distribution of the residuals is constant. When variances are inconsistent along the regression line, the data are said to be heteroscedastic.

Violating the homoscedasticity assumption can lead to poor estimates of the slope of the regression line (b coefficient) and less confidence in regression significance tests. If your residuals are normally distributed and homoscedastic, the data are likely to be linear.

To test for homoscedasticity, use SPSS Regression to create a scatterplot of your predicted values and your residuals. If there is significant departure from homoscedasticity, a scatterplot of your data will have a cone- or fan-shape. An example of how to test for homoscedasticity is provided in Section 5.7.1.2.

5.6.7 Normally Distributed Residuals

In linear regression the values of the independent variable need not be normally distributed (unless sample sizes are very small, in which case, significance test results may not be valid). However, it is assumed that theoretical residuals are normally distributed. Since theoretical residuals cannot be measured, observed residuals are substituted. Therefore, it is assumed that the differences between predicted values and actual values (observed residuals) are very close to zero, and that differences much greater than zero happen only occasionally. In other words, the residuals in the regression model are randomly and normally distributed with a mean of zero.

Some of the same methods used to test the assumption of normality of variable values may be used to test for the normality of residuals. Among those methods are the P-P plot the Shapiro-Wilk test, and the Kolmogorov-Smirnov test, all of which were discussed in Chapter 3 and Chapter 4. For the P-P plot, evidence of adherence to the assumption consists of data points that are closely clustered around the diagonal line, which represents the expected or predicted values. For the Shapiro-Wilk test and the Kolmogorov-Smirnov test, a nonsignificant outcome ($p > .05$) is evidence of nonsignificant differences from a normal distribution of the residuals.

5.6.8 Sources of Bias: Outliers

An outlier is a score on the dependent variable that is very different from those of other study participants. Accordingly, in a scatterplot of data for simple linear regression, the score of the outlier participant will show up as comparatively distant from the regression line. Moreover, there is likely to be a larger difference (residual) between the participant's actual score and the predicted score (the regression line) than other data points (Figure 5.7).

Figure 5.7 Scatterplot with an Outlier Score

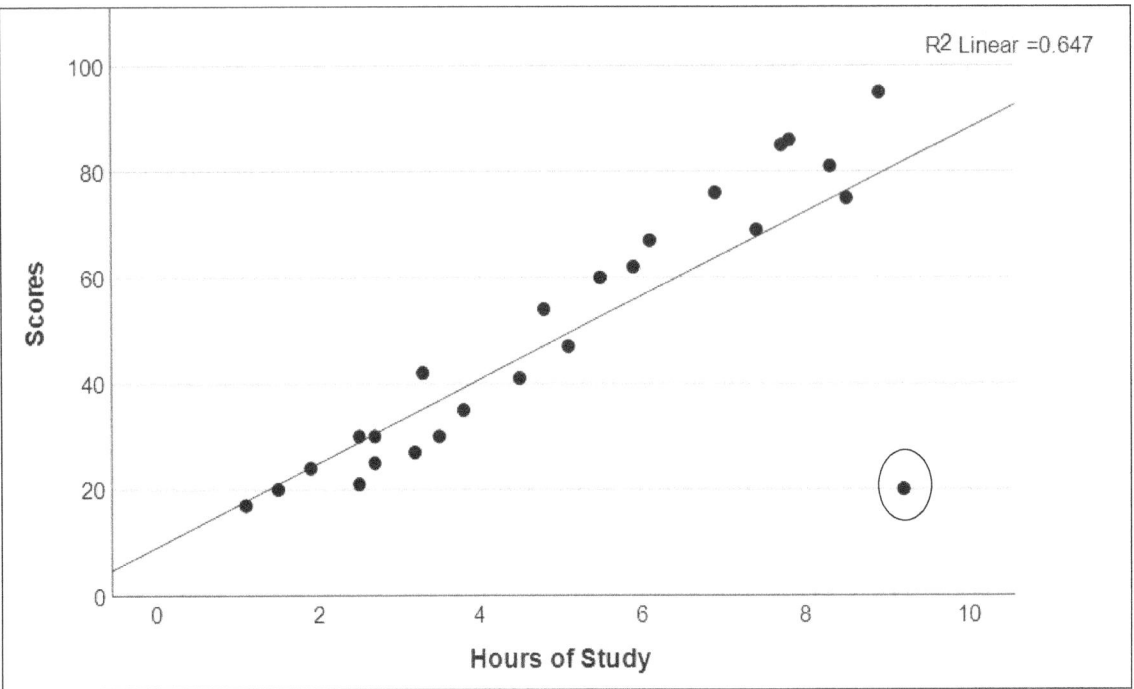

The problem with outlier scores in a regression analysis is that they often highly influence the estimates of regression coefficients (b weights); for example, outliers may result in the reduction of regression coefficients due to the flattening of the regression line. The intercept may increase for the same reason. Figure 5.7 and 5.8 illustrate what this looks like in a scatterplot of hours of study and test scores of college students. Figure 5.7 illustrates the regression line of the plot when an outlier is present. Figures 5.8 shows the change in the regression line when the outlier is absent (black line) compared to the regression line (dashed line) when the outlier is present. In Figure 5.7, the intercept is close, but does not equal 0. In Figure 5.8, the intercept (for the solid black line) would be theoretically less than 0 when extended to the y axis. The regression line in Figure 5.7 is also flatter than the line in Figure 5.8 as the outlier pulls the regression line toward itself. Outliers affect the accuracy of the predicted value of the dependent variable and change the predictive capacity of the independent variable as indicated by the difference in the R^2 value for each scatterplot.

Because outliers have dependent variable values that are very different from other study participants, a regression model is less likely to predict outlier values accurately. Furthermore, the poorer the fit of the model is to the data, the larger the residuals for all data points in the scatterplot. The presence of outlier values can affect the accuracy of all predictions due to the poor fit of the regression model.

Figure 5.8 Scatterplot Without an Outlier Score

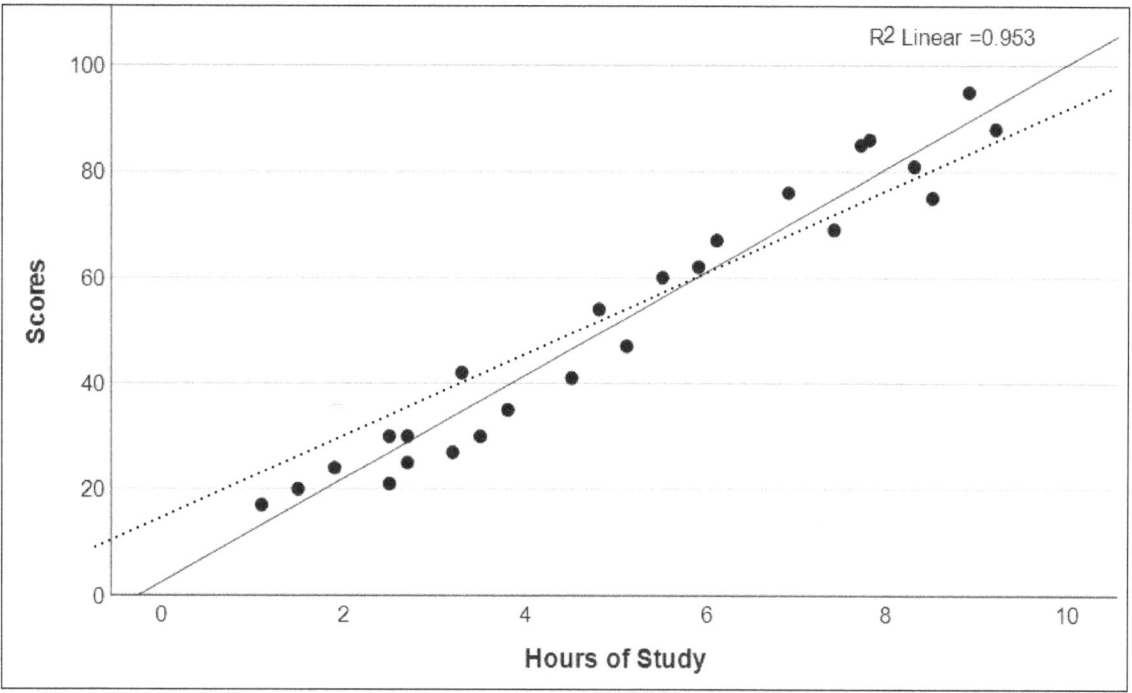

The discourse thus far, has focused on outliers and the associated residuals (distance from the regression line) that are larger than those of other scores. These are unstandardized residuals because they are measured in the same units as the dependent variable. However, units of measure change across models. As researchers evaluate different models, it becomes challenging to use unstandardized measures to identify large residuals. To avoid the use of different guidelines for different models, common practice is to convert unstandardized residuals to standardized residuals which are z scores that have a mean of zero and a standard deviation of 1. With standardized residuals, the same guidelines can be used across models to detect unacceptably large residuals. The following general guidelines may be used to identify outlier dependent variable values with residuals of concern in a regression model:
- standardized residuals with an absolute value greater than three standard deviations because they fall within a tiny percent— about one percent— of the population;
- more than one percent of a sample with standardized residuals greater than 2.5 standard deviations;
- more than five percent of a sample with standardized values greater than two standard deviations.

5.6.9 Sources of Bias: Other Influential Cases

In addition to the detrimental effect of outliers on regression outcomes, regression models may be biased by a few unusual cases in a data sample. The presence of these influential cases could affect the stability of a prediction model across a sample of cases. Diagnostic statistics that may be engaged to identify influential cases include Mahalanobis distance (Chapter 3), leverage, Cook's distance, and DF beta. The following section discusses leverage, and Cook's distance.

Leverage is a measure of how far an observation deviates from the mean of that variable. An observation with an extreme value on an independent variable is called a point with high leverage. Data points with high leverage can have an unusually large effect on the estimate of a regression line or model fit. However, it does not necessarily have great influence on regression coefficients that reflect the contribution of independent variables. The average leverage is defined as (K + 1)/N in which K is the number of predictors and N is the number of study participants. The average leverage for the Case of the Height and Weight of College Students is (K + 1)/N = (1 + 1)/33 = 2/33 = 0.06. A value of 1 indicates that a case has complete influence over predicted values.

Guidelines for estimating influence using leverage:
- All leverage values should be close to the average value of (K + 1)/N.
- Hoaglin and Welsch (1978) recommend investigating cases with values greater than twice the average or 2((K + 1)/N), which is 0.12 for the Case of Height and Weight of College Students.
- Stevens (2002) recommends using three times the average or 3((K + 1)/N).

Cook's distance (D) is a measure of the overall influence of a case on the model, where the values for the independent and dependent variables for a case are sufficiently extreme to influence a regression model. It measures how much influence an independent variable value has on the dependent variable. It also measures how much the dependent variable will change if a specific record is dropped from the data set. The lowest value that Cook's D can assume is zero, and as the Cook's D of a record increases, the influence of the record increases. Guidelines for estimating the influence of individual cases using Cook's distance:

- All values should be less than 1. Examine cases with values greater than 1 (Cook and Weisberg, 1982).
- Examine extreme Cook's distances relative to other values by plotting Cook's distances against participant IDs.

5.7 SPSS Specifications for Simple Linear Regression

This section uses the heights and weights of college students in the Caribbean to demonstrate how to employ SPSS to conduct a bivariate linear regression. Height and weight data for this demonstration appear in Table 5.1. Assumptions for simple linear regression recommend a minimum sample size of 60. Therefore, this analysis is for demonstration purposes only.

5.7.1 SPSS Specifications to Test Linear Regression Data Assumptions

Section 5.6 described factors that can unduly impair the outcome of regression results and the associated regression model for the height and weight of college students in the Caribbean. However, there are methods for identifying and correcting data that do not quite meet the assumptions of linear regression. This section provides instructions in how to use SPSS to examine some statistical assumptions and how to conduct a bivariate linear regression analysis.

5.7.1.1 Linearity

The linearity assumption is perhaps most frequently tested using scatterplots. Two commonly used methods of evaluating data linearity are to examine a scatterplot of the independent and dependent variable and to examine scatterplots of the dependent variable and its residuals. The first method should result in a scatterplot that shows a linear relationship between the variables. A nonlinear scatterplot suggests a nonlinear relationship. The desired outcome of the second method is a horizontal spread of the data points. A nonhorizontal spread cloud is diagnostic of nonlinearity

5.7.1.1.1 Scatterplot Specifications

To check for linearity between the independent (**Heightcm**) and dependent (**Weightkg**) variables, graph a scatterplot of both variables and examine the plot for cases that are farther away from the regression line than other cases; that is, they record greater error than the other cases. To accomplish that open the data set, **Carib_Height_Weight.sav**.

In SPSS, users may specify a scatterplot by choosing **Chart Builder** or **Legacy Dialogs** from the **Graphs** drop-down menu. This exercise will use **Chart Builder**. Before opening the Chart Builder window, check to ensure that the variables, **Heightcm** and **Weightkg**, are identified as **Scale** level variables by clicking on the **Variable View** tab at the bottom of the main screen and checking the **Measure** column. SPSS uses the term "Scale" for interval and ratio-level variables. Confirm that **Scale** appears in the rows for **Heightcm** and **Weightkg**. After confirming the measurement level of the variables, select **Chart Builder** from the **Graphs** drop-down menu and the initial **Chart Builder** window appears. To define the chart in the main **Chart Builder** window, click **OK** to certify the measurement level of the variables has been set.

In the bottom half of the **Chart Builder** menu, click on the **Gallery** tab, then click on the **Scatter/Dot** option under **Choose From.** This will unlock a selection of **Scatter/Dot** options. Click on the first **Scatter/Dot** option to bring it to the preview panel as shown (Figure 5.9).

Figure 5.9 Specifications for Chart Builder Preview Window

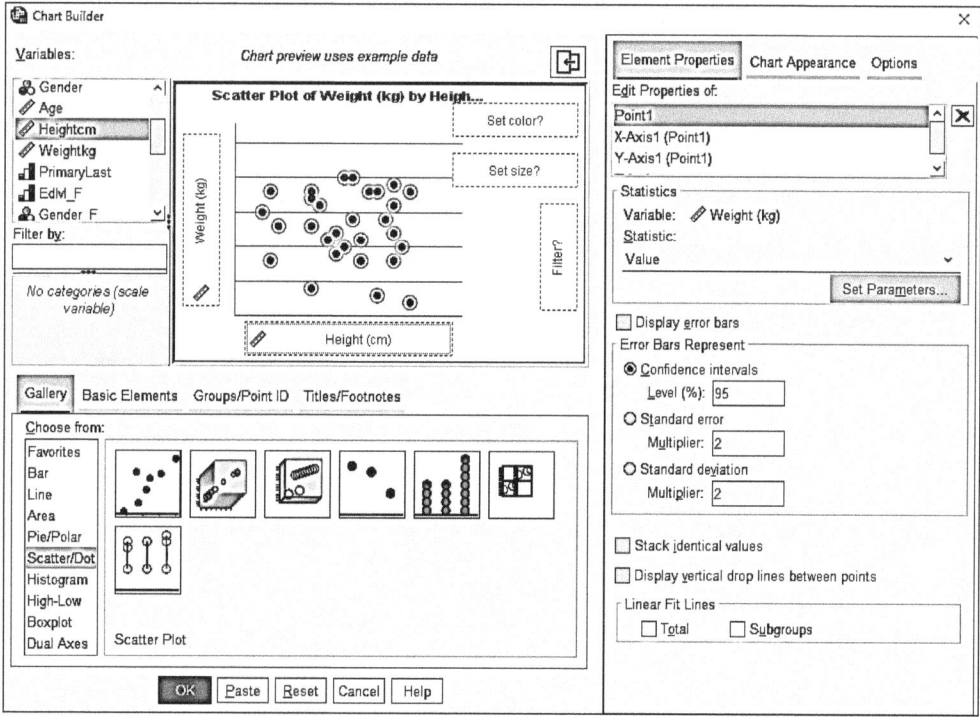

To complete the scatterplot specifications, select and drag the dependent variable label, **Weight (kg)**, from the **Variables** panel on the left to **Y-Axis** on the chart. Then select and drag the independent variable label, **Height (cm)**, from the **Variables** panel to **X- Axis**. Additional specifications that appear on the right side of the **Chart Builder** window are not needed at this time. Click **OK** to run the Scatterplot procedure.

186 Simple Linear Regression

5.7.1.1.2 Scatterplot Results

The scatterplot of **Height (cm)** and **Weight (kg)**, presented in Figure 5.10, illustrates a linear relationship between height and weight as indicated by the direction of the data points on the scatterplot. The direction of the scatterplot is a visual confirmation that a concomitant, positive relationship exists between height and weight in that as height increases, so does weight.

Figure 5.10 Scatterplot of Height and Weight

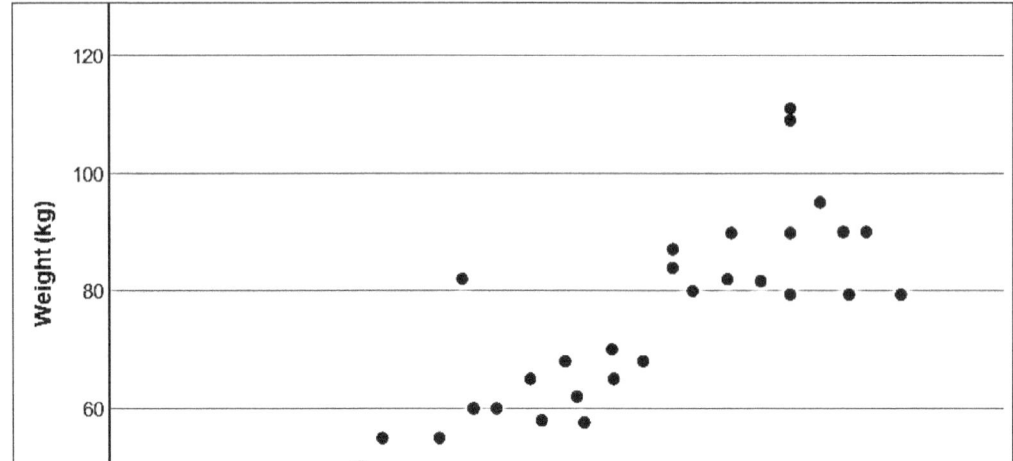

To confirm linearity, double-click on the output in the **Output** window to open the **Chart Editor**. Click on **Elements** at the top of the **Chart Editor** window, and select **Fit Line at Total** to request the regression line for the scatterplot (Figure 5.11). Close the **Chart Builder** window.

Figure 5.11 Scatterplot of Height and Weight with Regression Line

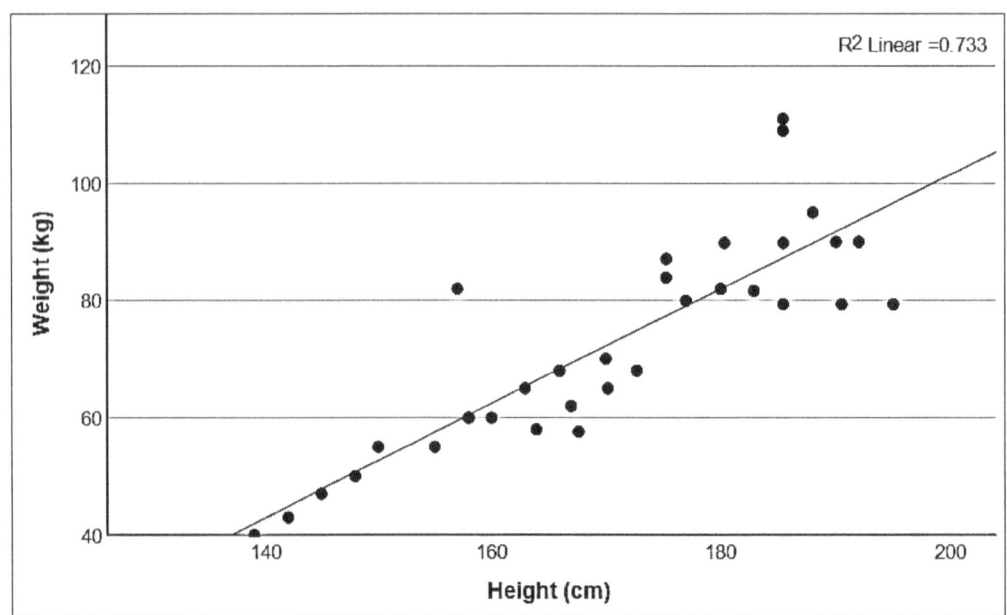

Despite the limited sample size, the regression line confirms the linearity of the scatterplot (Figure 5.11). In addition, the specifications produced the regression equation for predicting weight from height as well as R^2 (.733 or 73.3%) that expresses the proportion of variance in weight that height explains. Because the lowest weight in the data set is 40 kg and the shortest individual is 140 cm, the intercept (when the independent variable, height = 0) is not visible in the chart, since no one is 0 cm tall.

5.7.1.1.3 Using Standardized Residuals to Evaluate Linearity

Not only is the relationship between the independent and the dependent variables expected to be linear, the relationship between the dependent variable and its residuals (differences between actual and predicted values) should be linear. Constructing a bivariate plot of predicted values and its residuals discloses whether a linear relationship exists.

To produce the bivariate plot of the dependent variable values and their residuals, select **Analyze → Regression → Linear**. In the subsequent **Linear Regression** window, highlight and click over **Weightkg** to the **Dependent** panel (Figure 5.12a). Click over **Heightcm** to the **Independent(s)** panel below **Independent(s)**. Select the **Plots** pushbutton on the right to open the **Linear Regression: Plots** window (Figure 5.12b). Highlight **ZRESID** (standardized residuals for the dependent variable, weight, computed by SPSS), and click it over to the panel below **Y**. Next, highlight **ZPRED** (standardized predicted values for weight also computed by SPSS), and click it over to the panel below **X**. Select the **Normal probability plot** option below **Standardized Residuals Plots** as well. Section 5.7.5 will discuss results for the normal probability plot. Click **Continue** and **OK** in the Linear Regression window to run the procedure.

Figure 5.12 SPSS Linear Regression Window

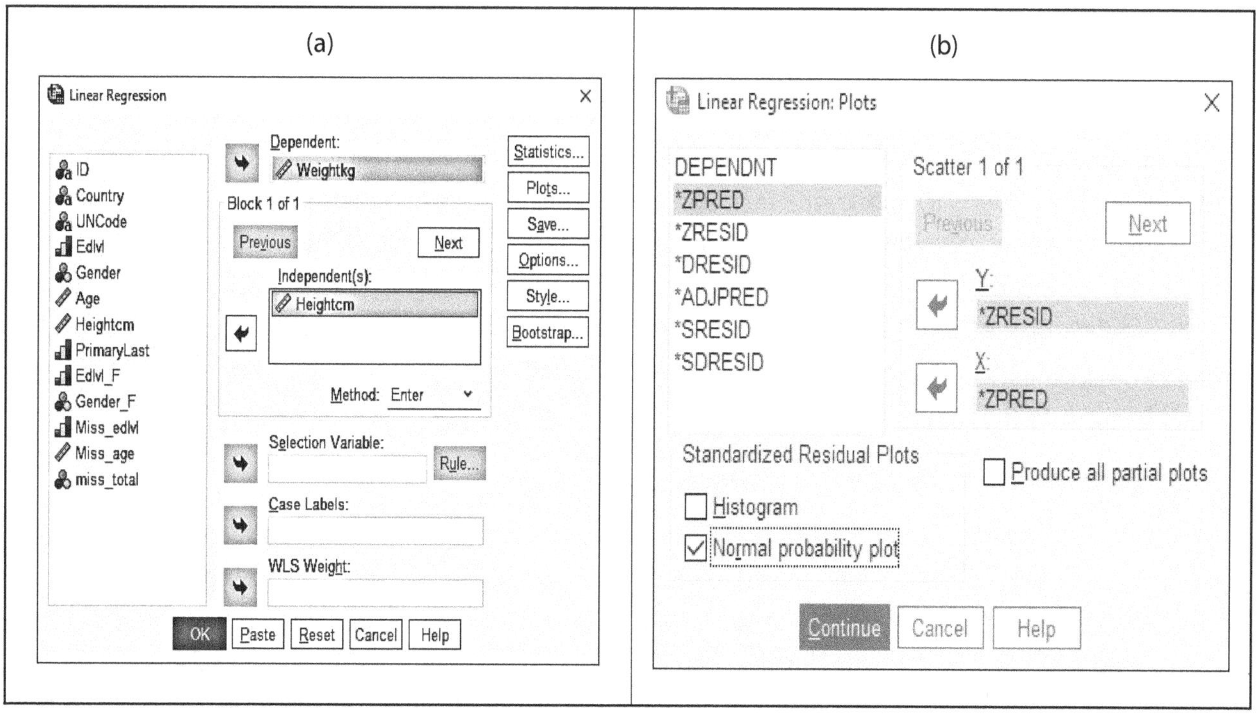

The result of the scatterplot specifications (Figure 5.13) for wight is challenging to interpret due to the small sample size. The objective is to determine whether the data points are evenly distributed about the mean standardized residual score, zero. Careful review of the figure shows that the plotted points are somewhat inconsistently distributed around zero on the *y* axis. Perhaps the two most effective corrective actions for the nonlinearity disclosed is to enhance the sample size or identify and delete outliers. However, while deleting outliers may improve the fit of the data used to build the regression model, it may lead to poor predictions for data in the future.

Figure 5.13 Scatterplot of Standardized Residuals and Predicted Values for Weight

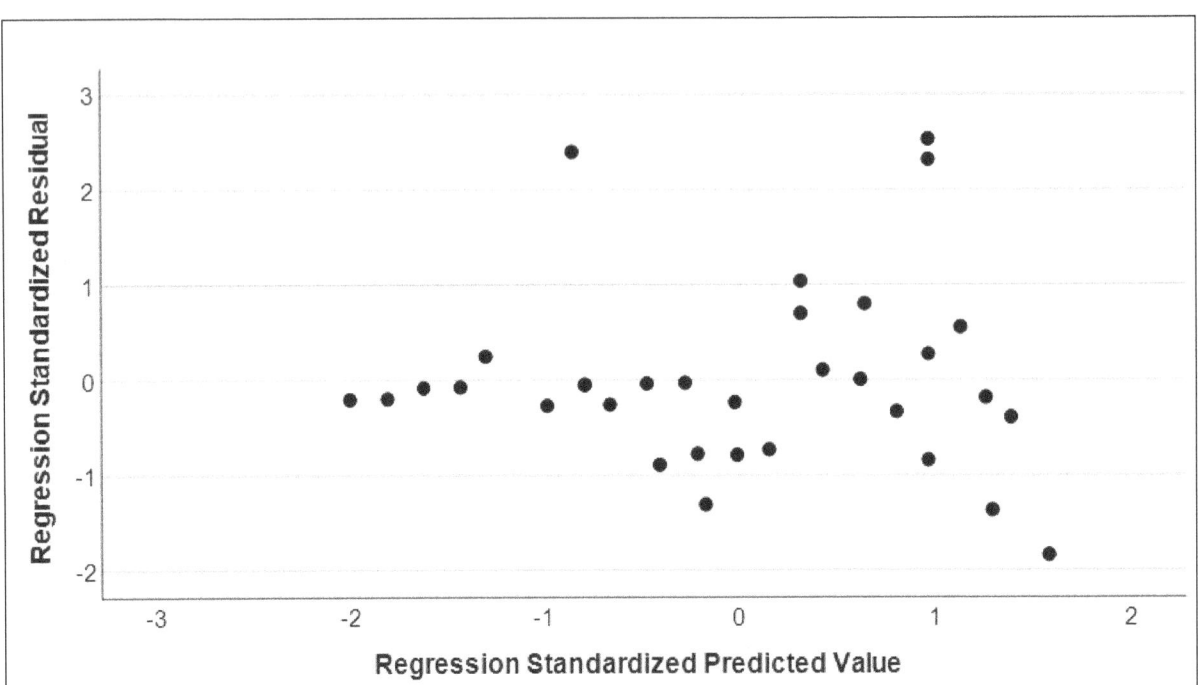

5.7.1.2 Homoscedasticity

If a regression equation fits the data well, there should be no pattern to the residuals plotted against the predicted values. If the variance of the residuals is not constant, the residual variance is in violation of the assumption of homoscedasticity. To investigate this assumption using SPSS, review the scatterplot pattern in Figure 5.13. The pattern of the plot of residuals against the predicted (dependent variable) should be relatively constant about zero on the *y*-axis all along the various levels of the predicted value. Based on visual inspection, the pattern of data points is not totally consistent, particularly due to the single predictor in the regression model and the small sample size.

5.7.1.3 Independence of Observations (Errors)

This assumption states that the errors associated with one observation should not correlate with the errors of other observations. Violating this assumption can invalidate significance tests. One method of investigating the independence of observations (also known as independence of errors) is to plot standardized residuals against standardized prediction values as presented in Figure 5.13.

The method used to assess the assumption of homoscedasticity is also used to assess the independence of observations. Visually examine the plot in Figure 5.13 to determine the extent to which the plot is rectangular in shape and between the values of -3 and 3 on the *x* and *y* axes. An inspection of the shape of the plot should reveal that all data points are within the specified parameters. However, the shape is not entirely rectangular.

Computing the **Durbin-Watson statistic** (*D*) is another method of testing the independence of observations assumption. The Durbin-Watson statistic is based on the order of the observations (rows) in a dataset. It tests whether adjacent residuals are correlated. The null hypothesis is that correlation between adjacent residuals equals zero (uncorrelated).

To produce the Durbin Watson statistic in SPSS, select **Analyze** → **Regression** → **Linear**. In the **Linear Regression** window (Figure 5.14a), highlight, then click and drag **Weightkg** to the **Dependent** panel. Click and drag **Heightcm** to the **Independent(s)** panel. Click on the **Statistics** pushbutton to open the **Linear Regression Statistics** window (Figure 5.14b), and select **Durbin Watson** in the **Residuals** panel. Click **Continue** and click **OK**.

Figure 5.14 SPSS Linear Regression Statistics Window: Durbin-Watson Statistic

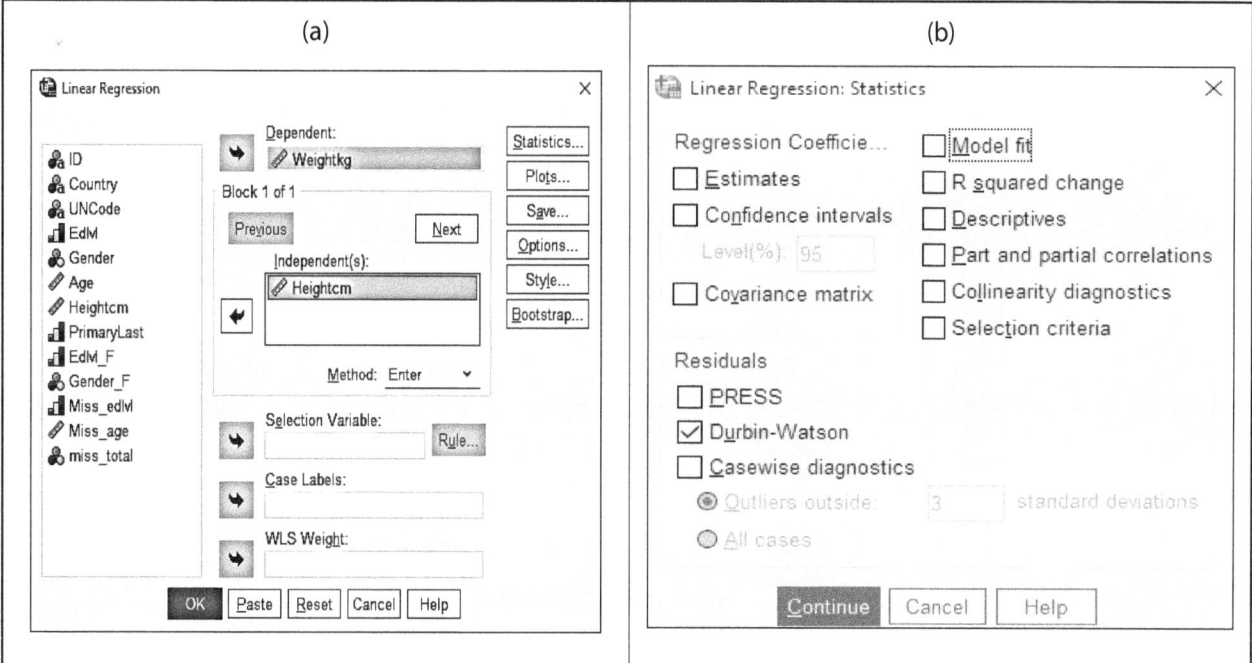

The Durbin-Watson statistic appears in the Model Summary table in Figure 5.15. For now, focus only on the location and value of the Durbin-Watson statistic in the far-right column of the Model Summary table which indicates that the Durbin-Watson test statistic is 2.162, which is relatively normal.

Figure 5.15 The Durbin-Watson Statistic

Model Summary[b]					
Model	R	R Square	Adjusted R Square	Std. Error of the Estimate	Durbin-Watson
1	.856[a]	.733	.724	9.382	2.162

a. Predictors: (Constant), Height (cm)
b. Dependent Variable: Weight (kg)

It is important to note that the Durbin-Watson statistic changes based on the order of the observations in the dataset. The residuals scatterplot is consistent even when the order of observations changes. Accordingly, scatterplots may be a more consistent method of investigating independence of observations for regressions studies. For more information, see Durbin & Watson's paper on autocorrelation (1951).

5.7.1.4 Normally Distributed Residuals

To investigate normal distribution of dependent variable residuals (Section 5.6.7), use results from specifications for the **Normal Probability Plot** of regression standardized residuals produced from Section 5.7.1.1.3. Normal probability plots are available from the SPSS **Linear Regression** procedure in the **Linear Regression Plots** window.

Figure 5.16 Normal Probability Plot of Regression Standardized Residuals

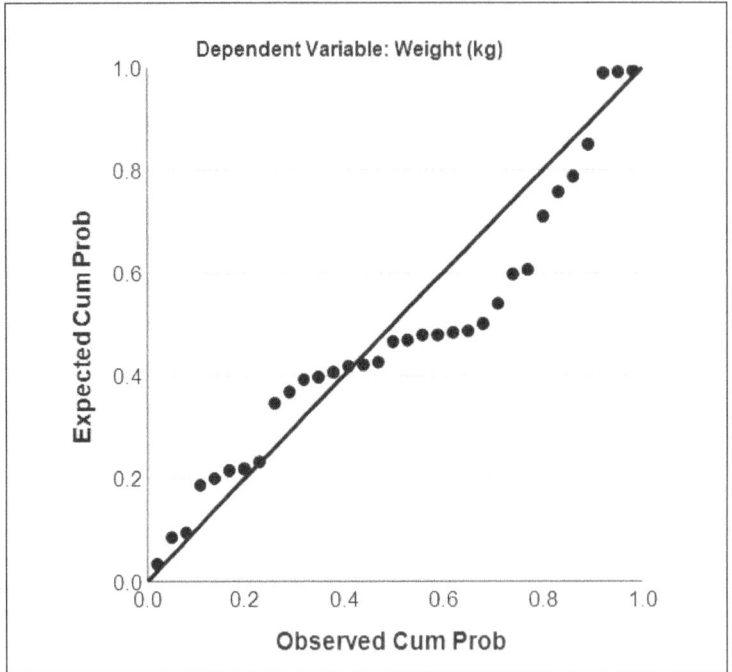

Figure 5.16 presents the results of the normal probability plot specifications. The P-P plot compares the observed cumulative distribution function (CDF) of the standardized residuals to the expected CDF of the standardized residuals. The data points should be closely clustered around the diagonal line. However, notably there are some outliers in Figure **5.16**, suggesting departures from normality.

Recall that a second option for testing the assumption of normal distribution of residuals is to conduct the Shapiro-Wilk test of normality. To do that, select **Analyze → Regression → Linear** to open the **Linear Regression** dialog window. Move **Weightkg** to the **Dependent** panel and **Heightcm** to the **Independent(s)** panel (Figure 5.14a).

Select the **Statistics** pushbutton to open the **Linear Regression: Statistics** window. It is not yet time to test the regression model, so uncheck the box next to **Estimates** and **Model Fit** (Figure 5.17a) for this procedure. Click **Continue** to return to the main dialog window. Then select the **Save** pushbutton to open the **Linear Regression: Save** window (Figure 5.17b).

Figure 5.17 Linear Regression Statistics and Save Windows

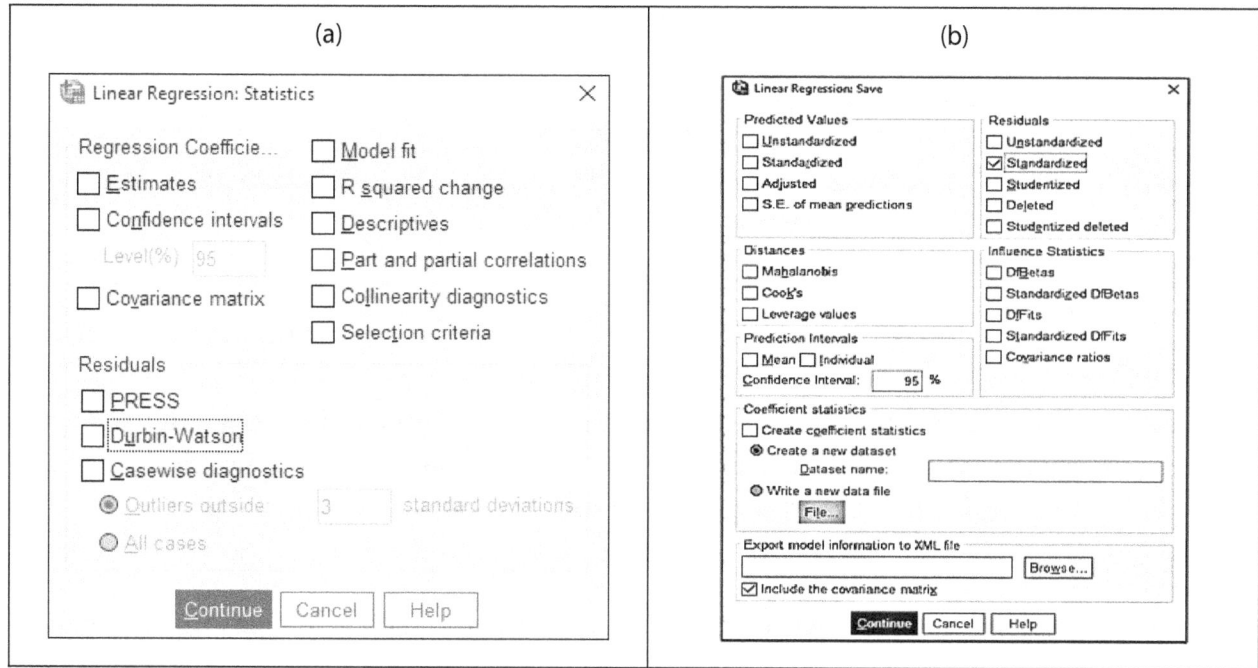

The previous exercise generated a plot of standardized residuals of the dependent variable (Figure 5.13). For this analysis, save the standardized residuals to the dataset so that they might be used to conduct the Shapiro-Wilk test of normality. To do that, check the **Standardized** box below **Residuals** (Figure 5.17b). Click **Continue** and **OK** to run the procedure. Click the **Data View** tab in **Data Editor** to confirm the addition of standardized residuals with the variable name, **ZRE_1**.

To test whether the standardized residuals are normally distributed, select **Analyze → Descriptive Statistics → Explore** to open the **Explore** dialog window. Move the variable **ZRE_1** to the **Dependent** panel (Figure 5.18a). Click the **Plots** pushbutton to open the **Explore: Plots** window (Figure 5.18b). To restrict output to the significance test, uncheck all boxes with the exception of **Normality plots with tests**. Click **Continue,** and **OK.**

Figure 5.18 Explore Dialog Windows

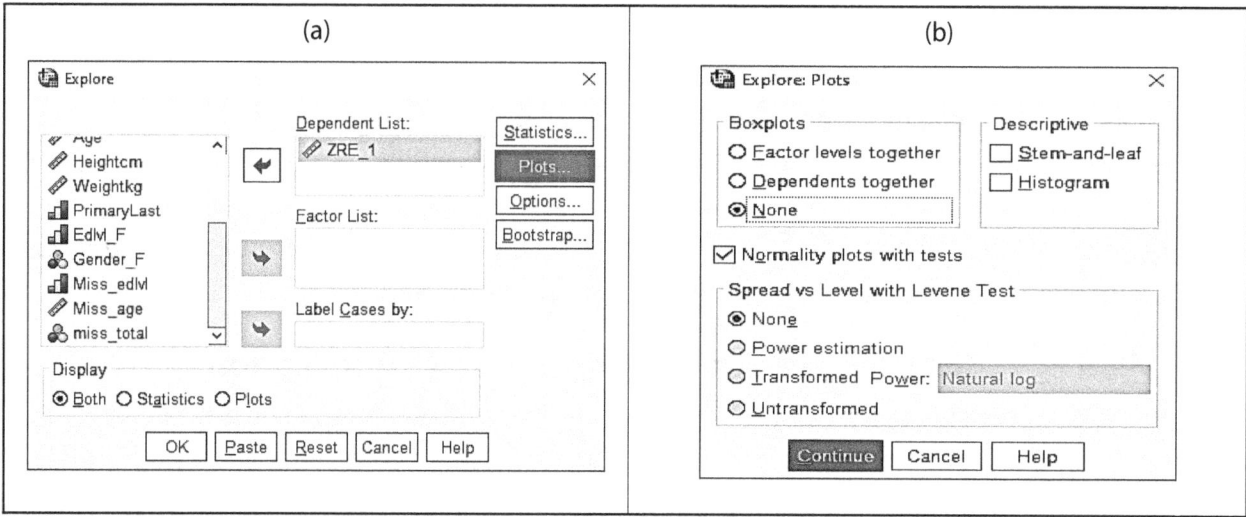

Analysis results (Figure 5.19) uncover non-normal distribution of the standardized residuals for Weight ($p = 0.002$ and $p < 0.003$), which supports the finding of the P-P plot in Figure 5.16.

Figure 5.19 Results of Test of Normality of Residuals

	Tests of Normality					
	Kolmogorov-Smirnov[a]			Shapiro-Wilk		
	Statistic	df	Sig.	Statistic	df	Sig.
Standardized Residual	.196	33	.002	.890	33	.003

a. Lilliefors Significance Correction

Researchers may consider transforming the independent, dependent, or both types of variables to produce a more normal distribution. Given the small sample size, increasing the sample size is likely to result in a more normal distribution, without losing an efficacious regression model.

5.7.1.5 Sources of Bias: Outliers

As Section 5.6.8 noted, outliers are cases with values on the dependent variable that are extreme compared with other values in the dataset. An important assumption of linear regression is the absence of extreme outliers in the data due to the impact of the outliers on the accuracy of regression models. Evaluating scatterplots of height and weight and of standardized residuals against standardized predicted values of the dependent variable are two methods of detecting extreme outliers.

A scatterplot of the height and weight dataset (Figure 5.10) reveals three potential extreme outliers—ID=1002 (height: 185 cm, weight: 109 kg), ID=2002 (height: 185 cm, weight: 111kg), and ID=3840 (height: 157 cm, weight: 82 kg) (table 5.1). However, it is important to note that values reported in the scatterplot in Figure 5.10 are unstandardized values (Section 5.6.8); they are not standard scores. Therefore, it is difficult to use these scatterplots to assign universal guidelines to detect a large residual, except to visually examine the scatterplot to identify cases that appear to stand out. However, using the guidelines described in Section 5.6.8, an examination of Figure 5.13 indicates that ID values 1002, 2002 and 3840 all have standardized residuals for weight greater than two but fewer than three standard deviations from the mean. All of these cases are outside of the range of Guideline 1 (Section 5.6.8). In addition, the three cases fall within the range of Guideline 3 in that these weight values that are greater than two standard deviations from the mean represent 10% of the small sample of 33, five percent more than the guideline stipulates. This indicates that the three cases are outliers and are likely to be contributors to a poor fit of the data to the regression model.

Under normal circumstances, the three outlier cases could be candidates for one of two interventions: (a) data transformation to secure a more normal distribution of values before the central analysis, or (b) removal of the outlier cases from the study. Nevertheless, for demonstration purposes, this explorative exercise will retain all cases until it is time to evaluate the model using regression diagnostics.

5.7.2 SPSS Specifications for Conducting Bivariate Regression

With assumptions testing completed, the next phase of data analysis is to conduct the central linear regression procedure. The primary objective of this analysis is to predict the weights (**Weightkg**) of 33 Caribbean college students based on their heights (**Heightcm**).

To specify the regression procedure, select **Analyze** → **Regression** → **Linear** to open the **Linear Regression** main dialog window (Figure 5.20). From the variables list in the left panel, select **Weightkg**, and move it to the **Dependent** panel. Select the independent variable **Heightcm**, and move it to the panel below **Independent(s)**.

Confirm that the **Method** drop-down menu below the Independent(s) panel says **Enter**; that is the default setting for linear regression procedures that request a standard regression analysis. A standard regression analysis is one in which all the independent variables are entered into the analysis at the same time. In this instance, with a single independent variable, no other method is possible. Chapter 6 will introduce other methods of entry.

Figure 5.20 Linear Regression Main Dialog Window

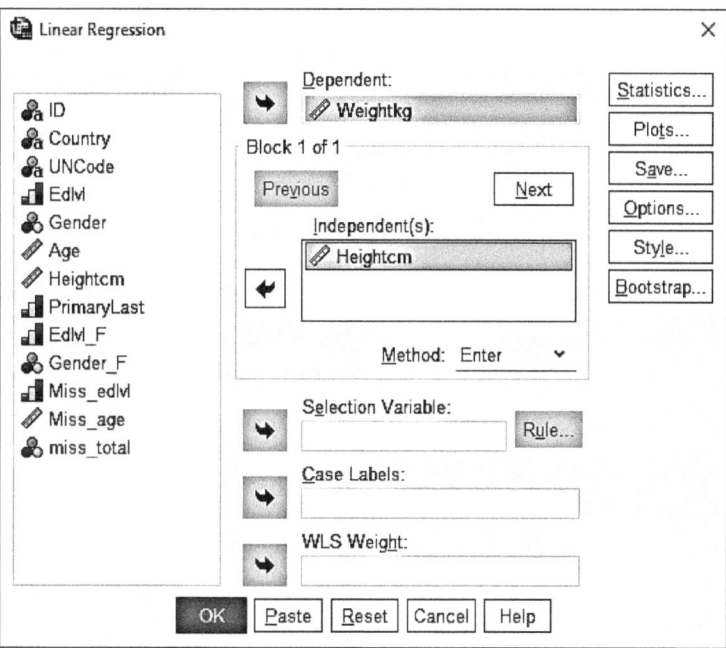

Select the **Statistics** pushbutton to open the **Linear Regression: Statistics** dialog window (Figure 5.21a). Check to confirm that the default selections have been made; **Estimates** in the **Regression Coefficients** panel should be prechecked. This default selection instructs SPSS to print the value of the regression coefficient estimates, the standard error for the estimates, as well as the significance level of the estimates and the correlation estimates.

Check **Model fit**, which provides **R-Square** (R^2). **R-square** is a measure of the strength of the relationship between the independent and dependent variables and is also known as the coefficient of determination. This command also provides **adjusted R-square**, a statistic which takes into account the sample size and number of predictors in estimating the strength of the predictive relationship. Finally, **Model Fit** calculates the standard error and generates an ANOVA table that provides results of the significance test of the regression model.

Select **Descriptives**, which provides the means and standard deviations of the independent and dependent variables in the model as well as the correlations table. Click **Continue** to return to the main dialog window.

194 Simple Linear Regression

Figure 5.21 Linear Regression: Statistics and Options Windows

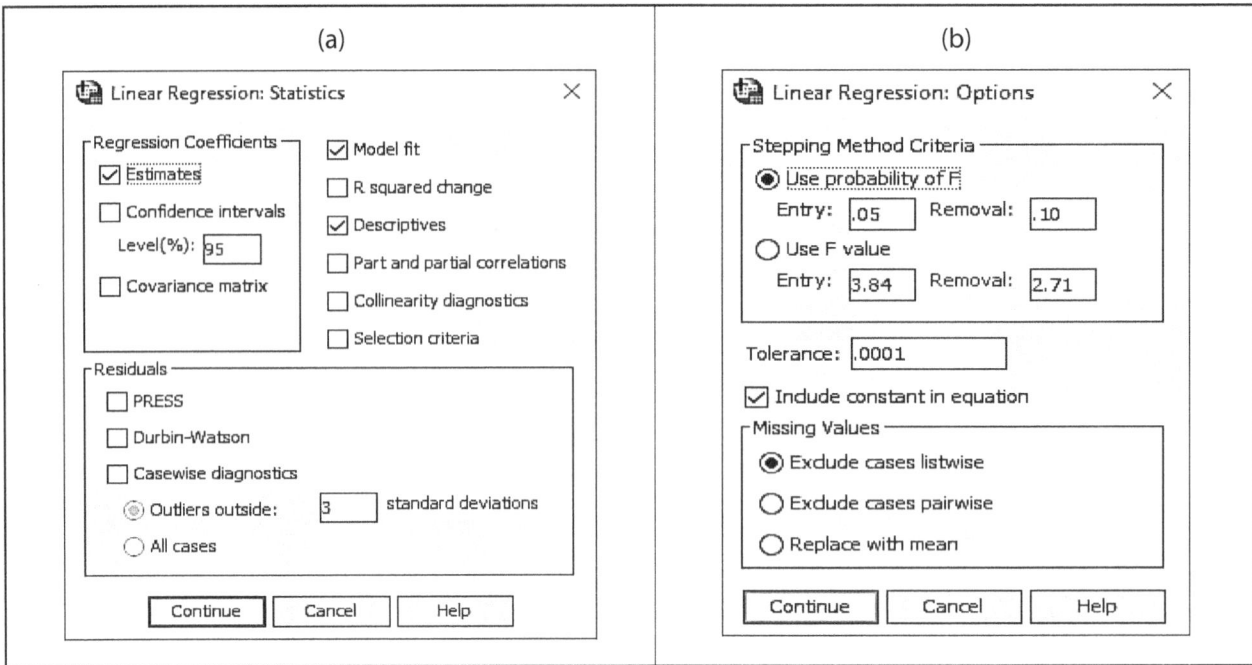

Select the **Options** pushbutton to open the **Regression: Options** window shown in Figure 5.21b. The default **Stepping Method Criteria** presented in the upper panel is not relevant to bivariate regression, but it will not affect the results. Confirm that the following defaults have been selected: **Include constant in equation,** and **Exclude cases listwise**. The former is needed for unstandardized regression models, which includes the value for the intercept, and the latter instructs SPSS to drop cases from the analysis that have a missing value for at least one of the specified variables. Only cases which have a complete set of data are included in the regression procedure. Accordingly, each case must have data for the independent and dependent variables in order to be retained for analysis. Retaining cases with missing data can result in biased results. Click **Continue** and click **OK** to run the analysis.

5.7.3 Simple Linear Regression Results

It is always important to view and interpret any case processing summary information first. Case processing provides confirmation of the number of valid cases included in the analyses. In the case of regression, it is also helpful to view the **Variables Entered/Removed** table that tracks which and how many variables entered the analysis. The practice of reviewing the **Variables Entered/Removed** table is an effective method of checking for data entry and analysis specification errors in the procedure.

Figure 5.22 displays the **Variables Entered/Removed** table which reports that only one variable, height, entered the bivariate regression model. Height is the only independent variable. If the model included more than one independent variable, all variables would appear in this table.

Descriptive statistics and Correlation Results presented in Figure 5.23 describe the study sample and the relationship between the variables involved. The upper table (5.23a) reports the means, standard deviations and total cases included in the analysis of the variables, height and weight. Note that the sample size (N) equals the total number of cases analyzed. Since no cases had missing data, a total of 33 cases were included in the analysis.

Figure 5.22 Variables Entered/Removed

Variables Entered/Removed[a]			
Model	Variables Entered	Variables Removed	Method
1	Height (cm)[b]	.	Enter

a. Dependent Variable: Weight (kg)
b. All requested variables entered.

Figure 5.23 Descriptive Statistics and Correlation Results

(a)

Descriptive Statistics			
	Mean	Std. Deviation	N
Weight (kg)	72.27	17.857	33
Height (cm)	170.32	15.650	33

(b)

Correlations			
		Weight (kg)	Height (cm)
Pearson Correlation	Weight (kg)	1.000	.856
	Height (cm)	.856	1.000
Sig. (1-tailed)	Weight (kg)		<.001
	Height (cm)	.000	.
N	Weight (kg)	33	33
	Height (cm)	33	33

Figure 5.23b presents **Correlations**, a square correlation matrix for height and weight. The correlation matrix provides three pieces of information. The top entry is the Pearson's r value, which is $r = .856$. This indicates a positive and strong relationship between height and weight. The second set of results is the probability of obtaining the given correlation if the null hypothesis ($r = 0$) is true and the sample size is as reported. The null hypothesis correlation is evaluated against an alpha of $\alpha = .05$ in a one-tailed significance test. Correlation results report statistical significance of the correlation of height and weight ($p < .001$). Accordingly, reject the null that the correlation equals 0. The third row of results confirms a sample size of 33.

Figure 5.24a, **Model Summary**, provides an overview of the regression results. The table presents the multiple correlation coefficient, **R**, which represents the correlation between a set of independent variables and a dependent variable. However, in bivariate regression with a single independent variable, the multiple correlation coefficient is the same as the bivariate correlation coefficient reported in Figure 5.24b ($r = .856$).

Figure 5.24 Statistical Significance of the Bivariate Regression Model

(a)

		Model Summary[b]		
Model	R	R Square	Adjusted R Square	Std. Error of the Estimate
1	.856[a]	.733	.724	9.382

a. Predictors: (Constant), Height (cm)
b. Dependent Variable: Weight (kg)

(b)

		ANOVA[a]				
Model		Sum of Squares	df	Mean Square	F	Sig.
1	Regression	7475.032	1	7475.032	84.916	<.001[b]
	Residual	2728.897	31	88.029		
	Total	10203.929	32			

a. Dependent Variable: Weight (kg)
b. Predictors: (Constant), Height (cm)

Of greater interest are results for **R-square** and **Adjusted R-Square** in Figure 5.24a which indicate the percent of variance in the dependent variable (weight) that the independent variable (height) explains; these statistics express how well the existing data fit the regression model. The results report that height explains about 73% ($R^2 = .733$), or almost three-quarters, of the variance in weight. Adjusted R-Square is a more precise estimate of the variance explained by increasing as additional independent variables improve the regression model. As Adjusted R-Square (Adj. $R^2 = .724$) illustrates, there is little loss in the strength of R-square (.009%) despite a small sample of 33 and a single predictor in the model.

Figure 5.24b, labeled **ANOVA**, displays results of the test of significance for the regression model. The objective of ANOVA is to test the null hypothesis that the independent variable does not add any significant predictive value to the regression model beyond the base model. With only one independent variable, height, in the model, ANOVA assesses whether height predicts weight significantly better than might be expected by chance. In effect, it is a test of whether the correlation coefficient for the two variables is significantly greater than 0.

ANOVA results report a total of 32 degrees of freedom ($N - 1$, or $33 - 1$). With only one predictor in the model, the regression effect (**Regression** row) has one degree of freedom. The residual degrees of freedom equal 31 (N - 2 or 33 – 2). As the significance (**Sig.**) column shows, in simple linear regression, significance testing results are the same as that reported for the correlation between the two variables ($p < .001$). Therefore, the prediction model that includes the independent variable, height, represents a significant improvement over the base model; that is, there is a greater than 99% probability that height significantly improved the model beyond what might be expected by chance. The results lead to rejecting the null that the correlation coefficient is equal to zero. The remaining statistics in Figure 5.24b (**Sum of Squares**, **df**, **Mean Square**, and **F**) are used to calculate the significance of the model manually and will be discussed in detail in Chapter 7.

Figure 5.25, **Coefficients**, presents more details of the regression results. **Constant** (*y*-intercept) reports a theoretical value of –93.820, which means that the regression line crosses the *y*-axis below zero. The **Sig.** column reports that the constant makes a significant ($p < .001$) contribution to the model. In reality, height is never less than zero. Accordingly, although this negative intercept remains in the equation, it has no intrinsic meaning and is of no practical value.

Figure 5.25 Regression Coefficients Table

		Coefficients[a]				
		Unstandardized Coefficients		Standardized Coefficients		
Model		B	Std. Error	Beta	t	Sig.
1	(Constant)	-93.820	18.124		-5.176	<.001
	Height (cm)	.977	.106	.856	9.215	<.001

a. Dependent Variable: Weight (kg)

The analysis has established the overall significance of the regression model, but to what extent does the independent variable contribute to the prediction of the dependent variable? The statistics in the columns **B** (.977) and **Beta** (.856) provide the unstandardized and standardized regression weights for the independent variable, height. The *t* statistic (**t** column) tests whether the regression coefficient for the independent variable is statistically significant. The Significance (**Sig.**) column reports that the independent variable (height) makes a significant contribution to the prediction model ($p < .001$). This is not surprising since both height and weight are significantly correlated. The outcome of the significance test supports the conclusion that height is a significant, direct, positive predictor of weight.

The raw-score coefficient (*b*) expresses the change in the dependent variable for every unit change in the independent variable. Accordingly, for every unit increase in height, there is a predicted 0.977 increase in weight. Based on regression results, the bivariate linear regression equation is as follows:

$$\text{Predicted Weight} = -93.820 + 0.977(\text{Height})$$

Note that the beta coefficient (**Beta**), the standardized regression coefficient, is the same as the Pearson *r* correlation. For a standardized regression equation, substitute the beta weight (.856) for the raw score coefficient, and exclude the constant.

How do manually-computed values compare with SPSS generated values? The results are similar. Hand-calculated values were -93.081 and .973 for the constant and *b* coefficient, respectively, making the constant slightly larger than the SPSS value (a difference of .739) and the b-coefficient slightly smaller than the SPSS statistic (a difference of .004).

5.7.4 SPSS Specifications to Identify Sources of Bias

After creating a regression model, it is important to evaluate the model with a goal of enhancing its efficacy. Regression diagnostics may be used to assess the model's alignment to assumptions and to investigate whether there are cases with large undue influence (bias) on the model. This section will use SPSS to investigate sources of bias. Methods include casewise diagnostics, leverage, and Cook's distance to statistically identify extreme outliers and influential cases. These cases can result in biased estimates of regression coefficients and affect prediction accuracy.

5.7.4.1 Casewise Diagnostics

Scatterplots of the unstandardized values for the independent and dependent variables as well as plots of standardized residuals and predicted values of the dependent variable are two important methods of identifying extreme outliers (Section 5.6.8). Using the **Casewise diagnostics** option in **SPSS regression** is another method of generating cases of extreme outliers. This specification produces a table of the actual and predicted value of the dependent variable, the residual (the difference between the actual and predicted value), as well as the standardized residual for all cases in the dataset. However, the default is to generate statistics only for those cases with standardized residuals greater than the absolute value of 3, or 3 standard deviations from a mean of zero.

To identify outliers using **Casewise diagnostics**, select **Analyze → Regression → linear** to open the **Linear Regression** dialog window. There, move the independent (**Heightcm**) and dependent (**Weightkg**) variables to the appropriate panels (Figure 5.20). Also, move **ID** to the **Case Labels** panel to identify outlier cases. Select the **Statistics** pushbutton to open the **Linear Regression: Statistics** window (Figure 5.26a), and choose **Casewise diagnostics** below **Residuals.**

Instead of retaining the default diagnostic criterion of **3 standard deviations**, insert **2** in the panel next to **Outliers outside** to print cases with standardized residuals of 2 standard deviations or more. **Regression coefficients** and **model fit** statistics are not needed for this exercise. Click **Continue** to return to the main **Linear Regression** window.

For purposes of efficiency, include specifications for **Leverage values** and **Cook's** (for Cook's Distance) in this analysis as well. To do that, select the **Save** pushbutton to open the **Linear Regression: Save** dialog window (Figure 5.26b). For this exercise, select **Cook's** and **Leverage values** below **Distances**. Click **Continue** to return to the main window, and click **OK** to run the analyses.

Results for **Casewise Diagnostics** is displayed in Figure 5.27. As expected, **Casewise Diagnostics** identified the same three cases identified in Section 5.7.1.5 with standardized residuals greater than 2—Cases 1002, 2002 and 3840.

5.7.4.2 Influential Cases: Leverage

As a result of the specifications for **Cook's distance** and **leverage values**, SPSS added two new columns to the dataset for the new statistics with the variable names **COO_1** for Cook's Distance and **LEV_1** for centered leverage values. Figure 5.28 displays summary statistics, including the minimum, maximum, mean, and standard deviation, for leverage and Cook's Distance.

Figure 5.29 presents influential cases based on leverage results from two sets of guidelines to identify potential influential, high leverage cases—one from Hoaglin and Welsch (1978), and the other from Stevens (2002).

Hoaglin and Welsch: $2((K + 1)/N) = 2((1 + 1)/33) = 2*0.06 = 0.121$
Stevens: $3((K + 1)/N) = 3((1 + 1)/33) = 3*0.06 = 0.18$.
where:
K = number of independent variables
N = sample size

Figure 5.26 SPSS Statistics and Save Windows

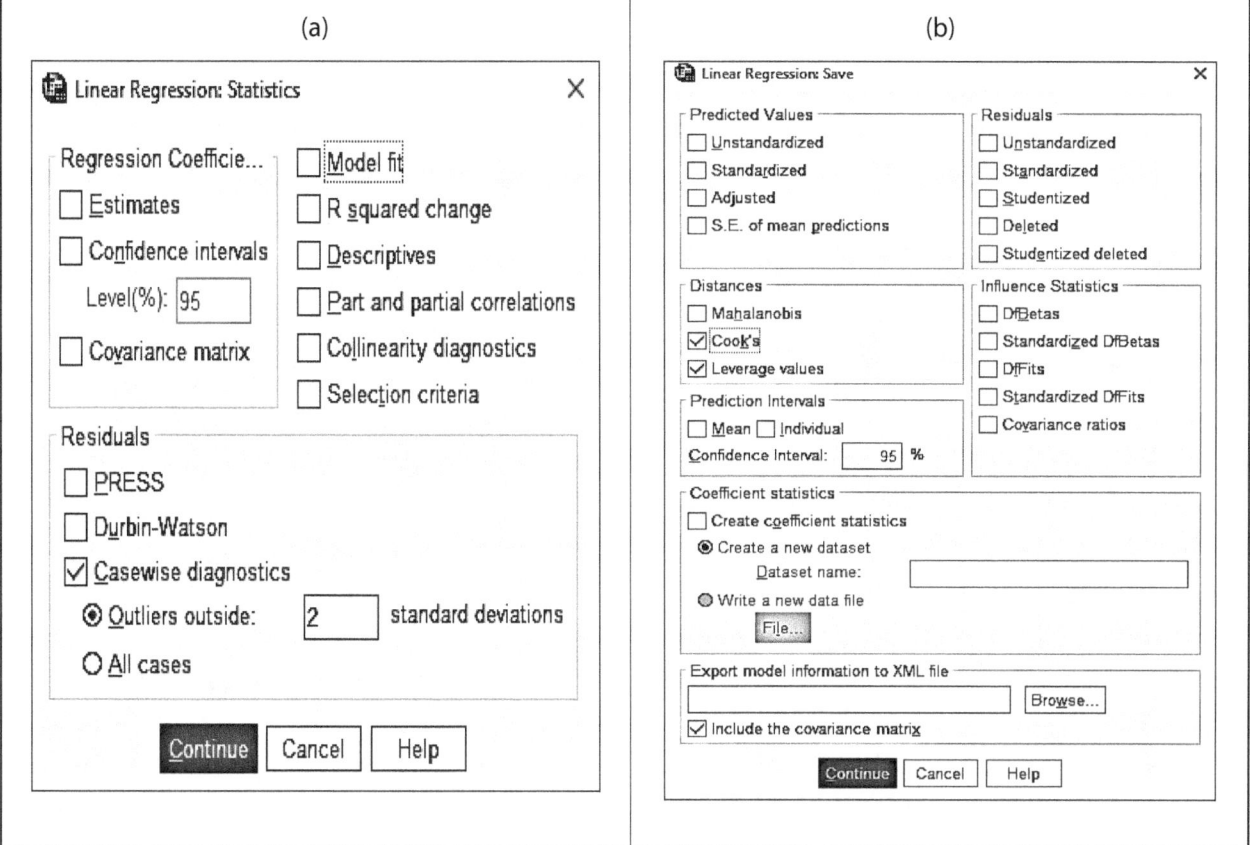

Figure 5.27 Casewise Diagnostics Results

Casewise Diagnostics[a]					
Case Number	Identification Code	Std. Residual	Weight (kg)	Predicted Value	Residual
5	2002	2.530	111	87.27	23.734
2	1002	2.317	109	87.27	21.734
17	3840	2.397	82	59.51	22.490

a. Dependent Variable: Weight (kg)

Figure 5.28 Residual Statistics for Leverage and Cook's Distance[a]

Statistic	Minimum	Maximum	Mean	Std. Deviation
Centered Leverage Value	0.00	0.125	0.030	0.032
Cook's Distance	0.00	0.229	0.032	0.065

a. Dependent Variable: Weight(kg)

Figure 5.29 Influential Case Based on Guidelines from Hoaglin and Welsch and Stevens

Influential Case Guidelines	Case ID	Values
Exceeds Twice Average = 0.121	3920	0.125
Exceeds Three Times Average = .18	-	-

Results of the influential data points analysis disclosed that the average leverage for the dataset is 0.03 (Figure 5.28), well below the two criteria for identifying influential cases. Figure 5.29 reports that one case (ID = 3920) exceeded Hoaglin and Welsch's standard for influential cases with a leverage value of 0.125. Case 3920 has the lowest value for the independent variable, height (139 cm). However, in spite of the unusually low value for height, weight (40 kg) follows the predicted regression line; it is the initial point on the line. Therefore, although Case 3920 may strengthen the regression relationship, it is unlikely to have any influence on the regression. None of the cases in Figure 5.27 had leverage values that exceeded the guidelines.

5.7.4.3 SPSS Specifications for Identifying Influential Cases: Cook's Distance

Using the Cook's Distance guidelines (Section 5.6.9) to evaluate the newly-computed values in the dataset (variable, **COO_1**) reveal that the maximum Cook's Distance is 0.229 (Figure 5.28). Only one case in the dataset has a Cook's Distance of .29 (ID 2010) and it is well below the cut-off guideline of 1.00 (Cook and Weisberg, 1982).

The second Cook's Distance guideline suggests that Cook's distances may be plotted against participants' IDs to identify influential cases relevant to all other cases. To do that, select **SPSS Graphs** → **Chart Builder** (Figure 5.30). Below **Gallery** select **Scatter/Dot** and drag the scatterplot icon (the first option) to the **Chart Preview** panel. Drag **COO_1 (Cook's Distance)** to **Y-Axis** and **ID** to **X-Axis**. Click **OK**.

The scatterplot is presented in Figure 5.31. It illustrates that the two cases with the most extreme Cook's Distances values are ID 2010 with a Cook's distance value of 0.229 and ID 2002 with a Cook's Distance of 0.215. These cases appear to be the most influential cases. However, excluding these cases results in a loss of data for what is already a small sample size. Another option is to increase the sample size to better meet the assumptions of linear regression.

Figure 5.30 SPSS Chart Builder

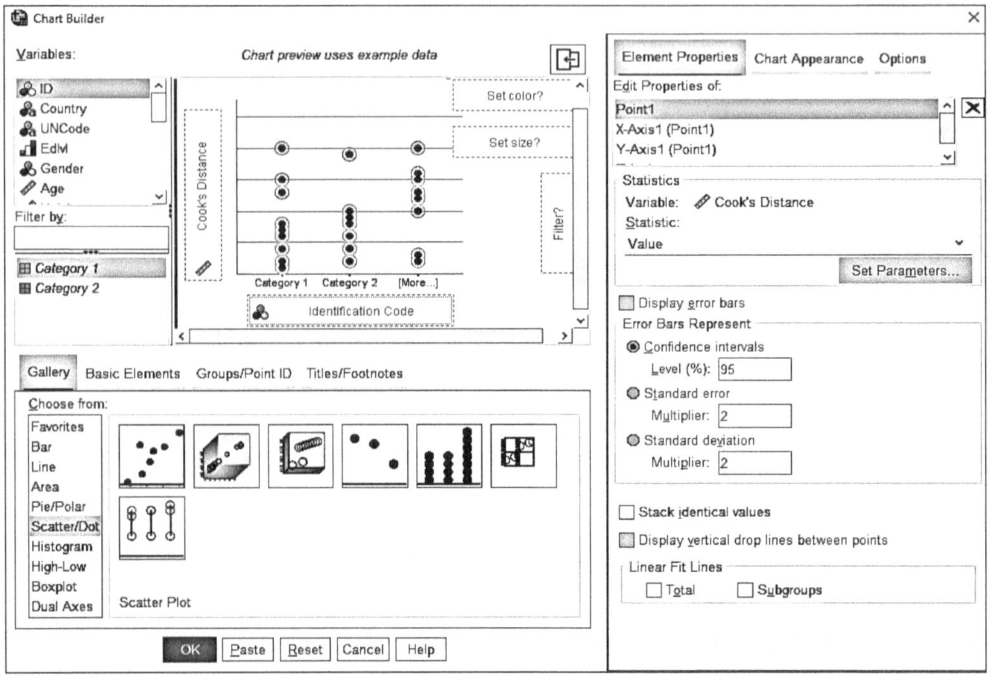

Figure 5.31 Scatterplot of Cook's Distance and Study Participants' Identification Codes

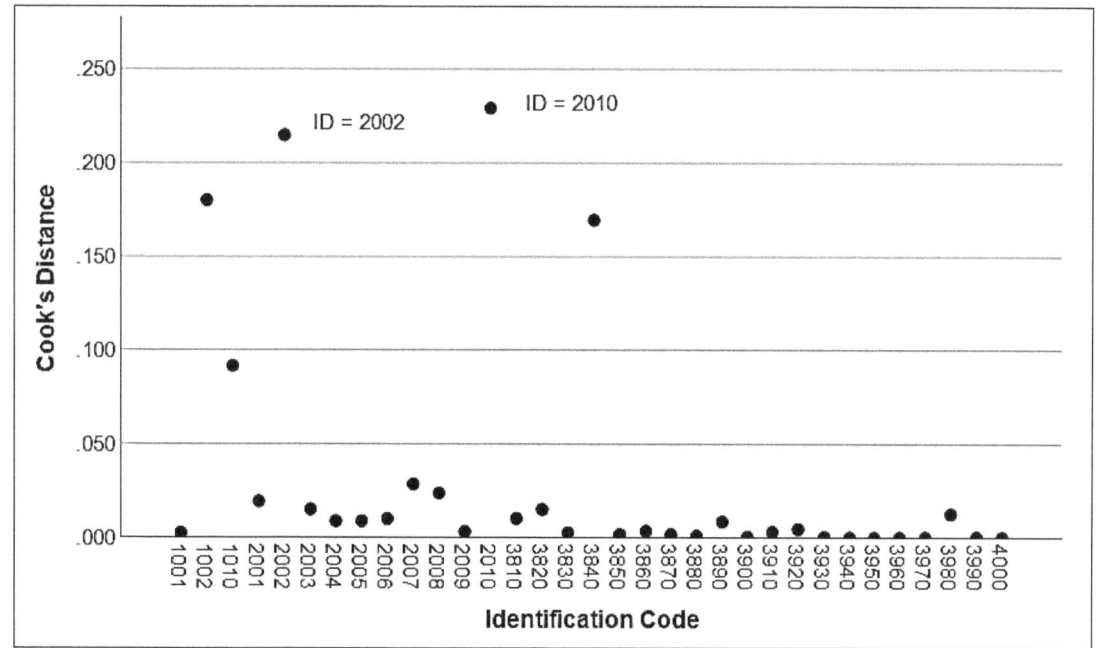

5.7.5 Sample Report: The Case of the Height and Weight of College Students

A total of 33 Caribbean college students reported their heights and weights. The values were used to investigate whether students' height significantly predicted their weight (that is, whether the model derived was significantly different from zero). In addition, the study sought to determine the amount of variance in weight that height account for.

A standard bivariate regression was conducted on a total of 33 participants after screening the data and evaluating the data for violations of associated assumptions. A statistically significant model was obtained from the analysis $F(1, 31) = 84.916$, $p < .001$, $R^2 = .733$. The raw regression coefficient was .977 ($SE = .106$), with a constant of -93.820 ($SE = 18.124$). The standardized regression coefficient was .856. Height explained to almost three-quarters (.733) of the variance in weight.

5.8 Key Terms

Adjusted R-square (R^2)
Autocorrelation
Cook's Distance (D)
Durbin-Watson test
F test
Goodness of fit
Heteroscedasticity

Homoscedasticity
Leverage
Mean square
Multiple regression
Predictor
Predicted value

Residual
Simple linear regression
Standardized residuals
Sum of squares
t statistic
Unstandardized

6
Multiple Regression

Key concepts from previous chapters or courses:
  regression line: the line that best summarizes the relationship between two variables.
  slope: the rate at which a regression line rises (positive relationship) or falls (negative relationship).
  regression constant: value of the dependent variable when the independent variables equal zero.
  unstandardized regression coefficient: the value of the slope of a regression line in the original metric that has not been standardized (transformed to a z score).
  standardized regression coefficient: the value of the slope of the regression line when the values of the independent variables have been converted to standard scores or z scores.
  regression equation: an algebraic equation expressing the predictive relationship between a single dependent variable and one or more independent variables.
  ***t* statistic**: In regression, this number is calculated in a t test to determine whether the independent variable is a significant predictor of the dependent variable.
  ***F* statistic**: this number derives from an F test used to determine whether independent variables in regression represent a significant improvement over a model without any independent variables.
  ***r*-square (r^2)**: the coefficient of multiple determination, a statistic that represents the amount of variance in the dependent variable explained by one or more independent variables.
  adjusted *r*-square: In regression, this statistic is a corrected r^2 that better represents how much of the independent variable(s) explains the dependent variable.

THIS CHAPTER INTRODUCES:

- the differences between simple linear and multiple linear regression
- how to craft and test hypotheses for multiple regression studies
- how to express the predictive relationship between two or more independent variables and a single dependent variable
- the circumstances under which multiple regression is used
- how to evaluate the unique contribution of each independent variable to the prediction of the dependent variable.
- how to use SPSS to conduct multiple regression

6.1 Introduction

The goal of linear regression is to predict values of a dependent variable (outcome variable) based on one or more independent variables (predictors). A dataset of the heights and weights of male and female college students was used to investigate whether a bivariate predictive relationship existed between the independent variable, height, and the dependent variable, weight.

While height turned out to be a significant predictor of weight, accounting for two-thirds of the variance in weight, there may be room for a regression model that has greater precision. One option is to add another predictor to the model to enhance its predictive power. By adding more predictors, we enter the realm of multiple regression, which is an expansion of simple linear regression (Figure 6.1).

Figure 6.1 Simple and Multiple Linear Regression Models

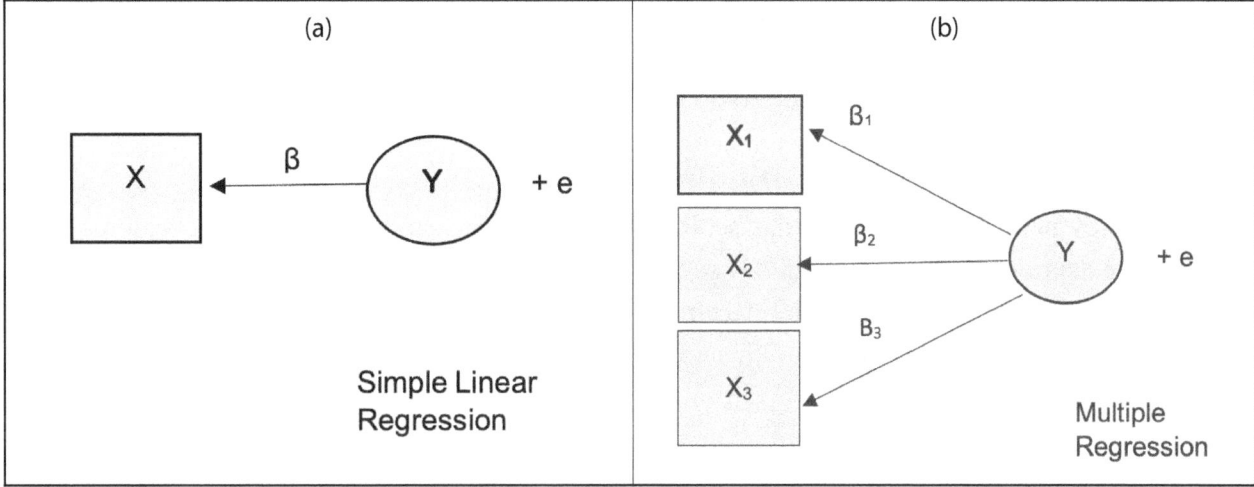

Perhaps the most distinguishing feature of multiple regression is the inclusion of more than one independent variable. In multiple regression, a set of independent variables is used to predict values of a dependent variable. The weighted linear combination of variables is also known as a variate. In spite of the differences between simple linear and multiple regression, the procedures share the same central purposes: to build a prediction equation that may be used to predict the outcome variable (Y).

6.2 Cases for Multiple Regression SPSS Demonstrations

6.2.1 The Case of the Weight of U.S. Adolescents

The Case of the Weight of U.S. Adolescents is similar to the case in Chapter 5—the Case of the Height and Weight of College Students. In this case, the data come from the National Health and Nutrition Examination Survey (NHANES), a major program of the National Center for Health Statistics (NCHS), which is part of the Centers for Disease Control and Prevention (CDC) in the United States. NHANES was developed to assess the health and nutritional status of adults and children in the United States. For more information about NHANES datasets and the NHANES program, go to https://www.cdc.gov/nchs/index.htm.

The original NHANES dataset includes 9,3702 study participants and 72 health and nutrition variables. For this example, a total of 1,142 individual participants in Grades 6 – 12 were extracted from the original dataset for this exercise.

The variables for the case include

- weight (BMXWT_L10)
- height (BMXHT)
- gender (SEX)
- age (RIDAGEYR)"

The SPSS demonstration will explore the extent to which a combination of independent variables, age, gender, and height, are significantly predictive of body weight in U.S. adolescents. Accordingly, a total of three independent variables are predictors in this case, and a single dependent variable is the criterion. The dataset for this case is NHANES.sav available via Appendix B. An example of a general outline of the research approach for the study may be found in Box 6.1.

6.2.2 The Case of Employee Salaries

This case uses an SPSS hypothetical dataset that includes salaries and other employee-specific information. The datafile contains a total of 474 subjects in three job categories: managerial, custodial and clerical. The case will be limited to clerical employees, a total of 363 cases. The dependent variable of interest for this case is current salary. Other variables for the case are as follows:

- employee code (ID),
- gender (gen)
- years of education (educ)
- current salary (salary)
- beginning salary (salbegin)
- months since hire (jobtime)
- age (age3)

The data for the Case of Employee Salaries may be found in Employee_Fin.sav. A general outline of a potential research plan for this case may be found in Box 6.2.

6.3 Fundamentals of Multiple Regression

Chapter 5 presented simple linear regression as the most basic of the linear regression procedures. Multiple regression is an extension of simple linear regression. It is one of the most common regression techniques used by social scientists to explore and evaluate predictive relationships between more than one independent variable and one dependent variable. The purpose of multiple regression is to determine the extent to which independent variables explain the variance in scores of the dependent variable.

Box 6.1 Research Plan Outline: The Case of the Weight of U.S. Adolescents

Research Problem
A common means of assessing health in adolescents is to estimate their body mass index (BMI). The equation for calculating BMI incorporates the variables height and weight. However, according to the CDC, BMI is age- and gender-specific; hence the development of age and gender-related growth charts. Therefore, the purpose of this study is to determine whether the combination of height, age, and gender are significant predictors of weight in adolescents.

Variables and Measurement Levels
Independent variables: height, age, gender
Dependent variable: weight

Level of Measurement:
Weight: ratio
Height and age: ratio
Gender: nominal

Research Questions
1. Does the weighted combination of height, age, and gender significantly predict weight in U.S. adolescents?
2. How much of the variance in weight does the combination of height, age, and gender explain?
3. Does one or more of the independent variables uniquely and significantly contribute to the prediction of weight?
4. What is the regression equation for the model?

Hypotheses (Research Question 1)
Null hypothesis: The baseline model that excludes all independent variables fits the data as well as the improved model. (The coefficients, which represent the slopes of the regression lines for height, age, and gender equal 0.)
H_0: $b_1 = b_2 = \ldots = b_{n-1} = 0$, where n equals the total number of independent variables in the model.

Alternative hypothesis: The improved model fits the data significantly better than the intercept-only model. (At least one of the coefficients, which represent the slope of the regression line, for height, age, or gender does not equal 0.)
H_a: $b_i \neq 0$, where i equals any independent variable.

Hypotheses (Research Question 3)
Null hypothesis: The correlation of the dependent variable and the corresponding independent variable is not significantly different from zero when the effects of the other independent variables are taken in account.
H_0: $b_j = 0$, where j equals any variable.

Alternative hypothesis: The correlation of the dependent variable and the corresponding independent variable is significantly different from zero when the effects of the other independent variables are taken in account.
H_a: $b_j \neq 0$.

Research Design
Nonexperimental, predictive correlational research design. See Figure 6.1b

Data Analysis
Generating descriptive statistics and correlation of the variables are typically the first steps in regression analysis, followed by tests of the assumptions associated with multiple linear regression data analysis. The central analyses of the standard method of multiple regression includes a test of significance of the entire model (F test for Hypothesis 1) as well as the test of the significance of the regression weight associated with each independent variable (t test for Research Question 2). Other statistics include: the multiple r statistic which measures the direction and magnitude of the relationship between the combined set of independent variables and weight; R^2 and adjusted R^2 which represent the amount of variance in weight that the independent variables explain; the constant and regression coefficients for the independent variables used to build the regression equation.

206 Multiple Regression

Box 6.2 Research Plan Outline: The Case of Employee Salaries

Research Problem
According to the United Nations an enduring income gap exists between men and women across the globe. Other variables found to be related to income include, years of education and age. The purpose of this investigation is to examine whether a predictive relationship exists between age, gender, education level, months on the job, beginning salary and current salary.

Variables and Measurement Levels
Independent variables: Age (Age); Gender (Gen); Years of Education (Educ); Beginning Salary (Salbegin), Months on the Job (Jobtime).
Dependent variable: Current Salary (Salary).

Level of Measurement:
Dependent variable: Ratio
Independent variable: Nominal, ordinal and ratio

Standard Multiple Regression Research Questions
1. Does the combination of blocks of variables that include gender; age; education level, beginning salary; months on the job predict current salary?
2. How much of the variance in current salary does the combination of independent variables explain?
3. Does the addition of a block of independent variables significantly change the amount of variance in current salary explained?
4. Does adding a block of independent variables uniquely and significantly add to predicting current salary?

Hypotheses (Research Question 1)
Null hypothesis: The baseline model excluding all independent variables fits the data as well as the improved model. (The coefficients for the blocks of variables gender, age, education level, beginning salary, months on the job are not significantly different from zero.)
H_0: $b_1 = b_2 = \ldots = b_{n-1} = 0$, where n equals the total number of independent variables in the model.

Alternative hypothesis: The improved model fits the data significantly better than the intercept-only model. (At least one independent variable coefficient within a block of variables is significantly different from zero.)
H_a: $b_i \neq 0$, where i equals any independent variable.

Hypotheses (Research Question 4)
Null hypothesis: The entry of a block of independent variables **does not** add significantly to the prediction of current salary over and above those previously entered.
H_0: $b_j = 0$, where j equals any variable.

Alternative hypothesis: The entry of a block of independent variables significantly adds to the prediction of current salary over and above those previously entered. (For each model, at least one regression coefficient statistically, significantly explains unique variance in current salary after controlling for previously entered blocks of independent variables.)
H_a: $b_j \neq 0$.

Research Design
Nonexperimental, predictive correlational research design. See Figure 6.30

Data Analysis
Descriptive statistics of the variables is typically the first step in regression analysis, followed by tests of the assumptions associated with multiple linear regression (Section 6.3). The hierarchical method of multiple regression includes a test of significance of each block of variables entered into the model (f test) as well as a test of the significance of the regression weight associated with each of the independent variables (t test). Other generated statistics include: multiple r which measures the direction and magnitude of the relationship between the combined set of independent variables and the dependent variable; R^2 and adjusted R^2 for each step in the model-building process.

6.3.1 Simple and Multiple Linear Regression Similarities

Multiple linear regression is a quantitative, statistical procedure that has all of the statistical features of the simple linear regression technique. It is most frequently used in correlational, nonexperimental research designs which explore complex relationships among naturally occurring variables. Naturally occurring variables are those that have not been directly manipulated by the researcher. For example, height, weight, and age are all naturally occurring variables. Hours of rest, music volume, instructional methods are manipulated variables.

Like simple linear regression, multiple regression is not used to investigate causal relationships. However, it may serve as preliminary research that point to the need for causal investigations. For instance, a researcher may hypothesize a predictive relationship between grade point average (GPA) and a weighted combination of hours of study, hours of watching television, and gender. The results of this initial multiple regression study may isolate all or some of the variables as significant predictors of GPA, which may be used in subsequent randomized controlled studies to examine causal relationships between GPA and one or more of the predictors.

Of course, the selection of variables for any study is not a random process based on hunches and opinions. Rather, their identification is informed by relevant research literature, a vital component of investigators' research design for simple and multiple regression. Both types of regression become viable statistical analysis options in a study's research design when the research literature and, subsequently, the research problem imply prediction.

Like bivariate regression, multiple linear regression is limited to predicting a single dependent variable. When a research design prescribes the prediction of more than one dependent variable, alternative statistical techniques must be used.

The primary purpose of a multiple regression study is to develop an accurate predictive model (regression equation) that may be used in the future to make new predictions. A training dataset for all variables (independent and dependent) is used to develop the model.

The accuracy of the new model and its components are evaluated using significance testing (t test and F test). In addition, investigators use the coefficient of determination (R^2) statistic to evaluate the magnitude of the predictive relationship.

Once investigators are satisfied with the accuracy and fit of the newly developed model, it may be implemented in the future, using variable values from new candidates to predict the outcome. For example, Khamis and Roche (1994) developed a model for predicting the future adult height of children in the United States, using children's age, gender, weight, and height, along with the heights of both parents, as predictors.

6.3.2 Properties of Multiple Regression

While simple linear regression is used to build a model (regression equation) to predict a dependent variable from a single independent variable, in multiple regression the number of independent variables is extended to two or more variables. A researcher may use the variables, gender, school attendance, high school location, as well as performance on the Caribbean Secondary Education Certificate (CSEC) examinations to predict grade point average in the first year of college. A secondary example is our case study, for which we will use the variables height, gender, and age to predict the dependent variable weight.

6.3.2.1 Dependent Variable

In multiple regression, there is a single dependent variable, also known as the outcome, criterion or predicted variable that is designated as Y in regression equations. The predicted value of a dependent variable may not be the same as the true or observed value. For example, as a result of the inevitability of error in making predictions, a study participant's actual weight is unlikely to be the same as the weight predicted by a regression equation. As the regression equation improves, however, prediction error declines and the accuracy of the predicted value increases.

6.3.2.2 Independent Variables

Deciding which independent variables to include in a study is among the most important decisions in regression research design. A fundamental cornerstone for successful research is relying on the research literature to identify relevant variables for a study. In this context, the likely goal of a literature review may be to identify predictors that are highly correlated with the dependent variable in regression studies.

In addition, to being strongly correlated with the dependent variable, ideally, independent variables should be uncorrelated with each other. In some instances, predictors are correlated to some degree, but if they are highly correlated, the regression procedure can be compromised due to multicollinearity or singularity (Section 6.52).

As a part of selecting variables supported by the literature, to optimally develop a regression model, researchers should use the fewest uncorrelated independent variables possible. Too many predictors for the sample size could lead to an overfitted model, which means that the model represents the noise in the data rather than the intervariable relationships and is unlikely to work well in implementation. Judiciously selecting independent variables, each of which predicts a substantial and unique segment of the variance in the dependent variable, enhances the likelihood of a powerful model (Tabachnick & Fidell, 2018).

6.3.2.3 Measurement Scales of Variables

The measurement level of the independent variable in simple linear regression may be nominal, ordinal, interval, or ratio-level. Chapter 3 discussed the importance of dummy coding nominal variables before analysis to ensure valid interpretations of results (See Chapter 3 and Chapter 5.). In multiple regression, independent variables in the model may include a mix of measurement levels. For example, in our case study, weight, height, and age are all ratio-level variables. However, gender is a nominal-level variable.

While researchers are free to use independent variables of different measurement levels in multiple regression, like correlation and simple linear regression, the dependent variable must be measured at the interval or higher levels. Therefore, variables with values of equally spaced intervals (e.g., Celsius or Fahrenheit used to measure temperature, grade point average [GPA], or intelligence quotient [IQ]) may be incorporated in a regression model. Ordinal level variables, including those measured on a Likert scale (e.g., frequency of school attendance), may be considered along with a note of caution since there remains disagreement among researchers about whether Likert scale intervals are equally spaced.

6.3.2.4 Independent Variables as a Composite Variable

Multiple regression equations embody regression models. Ideally, independent variables in the regression equation bring unique qualities to the prediction model that add to prediction accuracy. Together, the weighted linear combination of the variables signifies an underlying dimension (composite variable) that serves as a predictor of the dependent variable. In the Case of the Weight of U.S. Adolescents, that dimension might be conceived of as human physical characteristics represented by the weighted combination of gender, age, and height.

Prior to a regression study, researchers might have scoured the literature and studied related scientific theories to determine which available indicators of physical characteristics might do the best job to predict weight. Other physical characteristics that might have been considered include mother's and father's weight. Regardless of the variables tapped for inclusion in a study, the composite variable is used to calculate the best predicted value given the circumstances.

In some cases, the model is determined not only in the combination of independent variables that are associated with the outcome variable, but also the sequence in which the variables are introduced into the model to predict the outcome. In these cases, regression procedures might be managed by the computer software based on prespecified criteria (e.g., stepwise regression). Alternatively, researchers may take a more active role in the statistical analysis process than in other situations. More about this in Section 6.3.3, hierarchical multiple regression.

6.3.2.5 Reliability and Validity of Independent Variables

As a part of the design and preparation for a regression study, it is crucial to ensure that variables to be included in the study have the technical qualities that will support study conclusions. Therefore, the greater the reliability and validity of independent variables, the more trustworthy the regression model is. For some variables, that is easier to do than others. For example, measuring height, age, and gender can be relatively uncomplicated. Certainly height, in many instances, can be measured directly and with a consistent outcome. Age, defined as chronological age, can be taken from birth records or based on birth date. Gender, defined as the sex of an individual, or male and female, can be secured from a questionnaire. Other variables, such as perception of weight, cannot be measured directly and are not directly observable.

Measuring more complicated variables often requires a lengthier process; for example, measuring the perception of weight may include the acquisition and use of a psychological measure to indirectly measure the variable. Acquiring the measure may involve developing a data collection instrument in a way that adheres to established data collection guidelines or selecting one that already exists. Regardless, these measures must exhibit specific technical qualities that illustrate they are valid and reliable. More on reliability and validity in Chapter 2.

6.3.2.6 The Predictive Power of Independent Variables

In evaluating the overall predictive model, a critical question to answer is: Do the combined independent variables predict the outcome variable better than might be expected by chance alone? If the model is not statistically significant, the predictors do not have sufficient explanatory or predictive power, and researchers must return to the research design to determine whether modifications will be useful or

6.3.2.7 The Size of the Predictive Relationship

In addition to considering the statistical significance of the predictive relationship between independent variables and the dependent variable, the magnitude of the predictive relationship is equally important. If the predictive relationship is significant, but the linear combination of the predictors accounts for or explains only a small portion of the variance in the dependent variable (e.g., 10%), then the utility of the regression model may be questionable. In the initial stages of a research study, or in an exploratory investigation, independent variables that combine to account for 10% of the variance in the dependent variable might be acceptable. Cohen's (1988) guidelines for interpreting the percentage of variance in dependent variables that independent variables explain are briefly provided below:

Percentage of Variance Explained	Effect Sizes
10	Small
25	Medium
40 or more	Large

6.3.2.8 Unique Contributions of Independent Variables

Although the overall model is evaluated based on the combined power of the set of independent variables, it is also important to examine the unique contribution of each independent variable. A researcher that selected four independent variables in a regression analysis (e.g., age, gender, height, and nationality) based on the research literature may discover that some of those variables do not make a significant contribution to the prediction of the dependent variable. Multiple regression affords researchers the opportunity to identify those variables that make a significant, unique contribution to a regression model and those that do not do so. In most cases, nonperforming variables may be re-evaluated or dropped from further analyses.

6.3.2.9 Produce a Regression Prediction Equation

The primary purpose of multiple regression is to produce an additive, linear prediction equation that incorporates the best linear combination of independent variables that optimally predicts the dependent variable. Chapter 5 presented two general forms of the simple linear regression prediction equation—the unstandardized (raw score) equation and the standardized equation. The regression models and the equations are presented below.

	Regression Model	Prediction Equation
Unstandardized (raw score)	$Y = a + bX + e$	$\hat{Y} = a + bX$
Standardized	$Y_z = \beta X_z + e$	$\hat{Y}_z = \beta X_z$

In both models there is a single independent variable represented by X or X_z and the residual represented by e. In the unstandardized equation, the constant (a) or y-intercept represents the value of the dependent variable when the independent variable equals zero. As discussed in Chapter 5, the constant is missing from the standardized equation because the constant always equals zero.

In multiple regression, there are also two forms of the equation. However, each equation has multiple independent variables ($X_1, X_2, ..., X_n$, where n represents the number of the final variable):

Unstandardized (Raw Score) Equation: $\hat{Y} = a + b_1X_1 + b_2X_2 + \cdots + b_nX_n$

Standardized Equation: $\hat{Y}_z = \beta_1 X_{z1} + \beta_2 X_{z2} + \cdots + \beta_n X_{zn}$

6.3.2.9.1 Unstandardized Multiple Regression Equations

The unstandardized equation includes each independent variable ($X_1, X_2, ..., X_n$) that significantly and uniquely accounts for variance in the dependent variable. In addition, each independent variable is assigned a regression coefficient ($b_1, b_2, ..., b_n$) based on the extent of its unique contribution to the prediction of the dependent variable. Each regression coefficient represents how many units the predicted value of the dependent variable will change for a one-unit change in the independent variable (X).

To use the unstandardized equation to predict the dependent variable for a target participant in the study, each regression coefficient is multiplied by the score associated with its independent variable (X) value for the participant. The products of each coefficient and its associated independent variable score are then totaled, together with the value of the constant (a) to produce the predicted value of the dependent variable.

As a brief review of the solution of a simple linear regression equation, imagine a researcher uses simple linear regression to investigate the predictive relationship between mathematics and science performance, using the following equation:

$$science_{pred} = a + b \text{ (mathematics)}$$

The regression (b) coefficient for mathematics is 0.67. Participant A's raw mathematics score is 60. The product of 0.67 and a mathematics score of 60 equals:

$$science_{pred} = a + 0.67(60)$$
$$science_{pred} = a + 40.2$$

The constant (a) is the value (on the y-axis) for the dependent variable, science, when the independent variable, mathematics (X), equals 0. To find the constant, follow the regression line (predicted values) in Figure 6.2 to the point on the y-axis at which mathematics performance equals 0, a value of 7.63. Therefore, based on the regression line in Figure 6.2, the predicted value for science for Participant A (Science$_{pred}$) equals:

$$science_{pred} = 7.63 + 40.2$$
$$science_{pred} = 47.65$$

In multiple regression when more than one independent variable is involved, the model becomes more realistic, but finding the constant becomes more challenging. The value of the intercept is the mean value of the dependent variable value when the independent variable values (predictors) equal 0. However, unlike simple linear regression, when more than one independent variable is involved, it is impossible to graphically illustrate the location of the constant using a 2-dimensional figure. For example, imagine that we are investigating whether mathematics and communication performance significantly predict science performance:

Figure 6.2 The Value of the Intercept When the Independent Variable Equals Zero

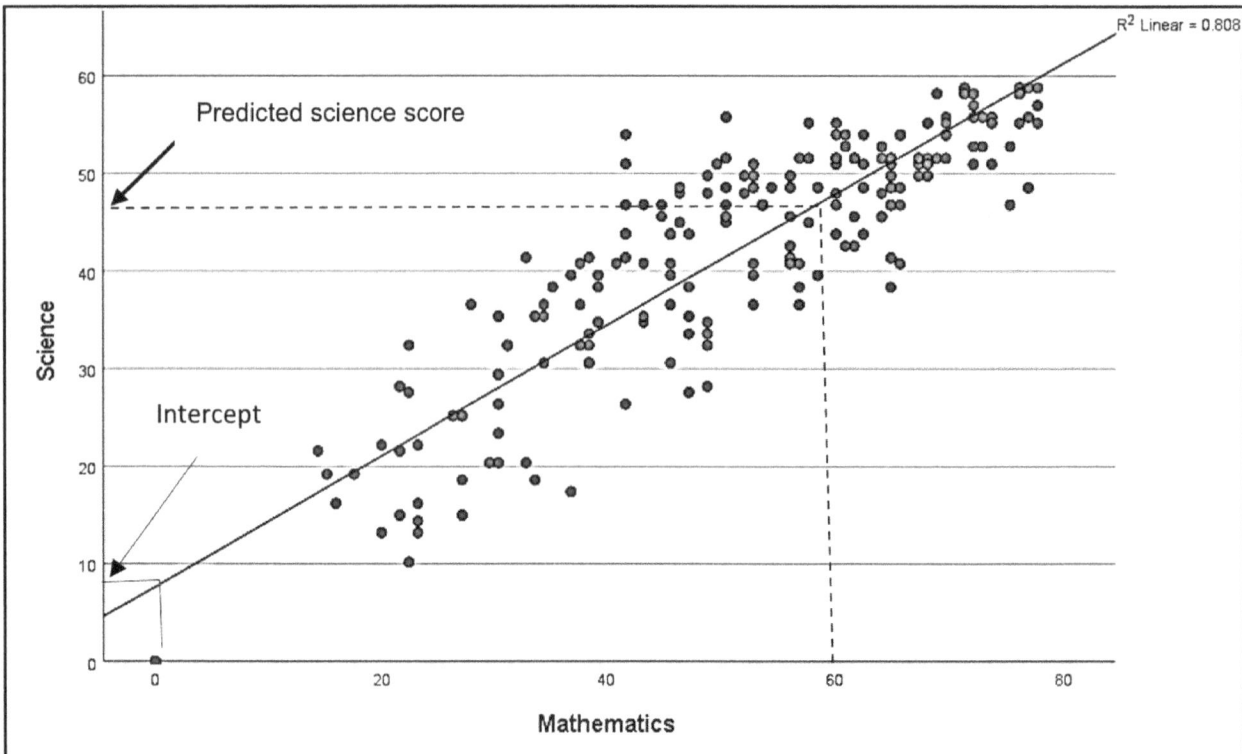

$$science_{pred} = a + b_1(mathematics) + b_2(communication)$$

With two independent variables in the regression model (mathematics and communication performance), a 3-dimentional figure (Figure 6.3) could depict the relationship among the variables. However, as the number of independent variables increases, graphical representations become more complex, and researchers become more reliant on statistical software to calculate the constant in regression equations.

6.3.2.9.2 Standardized Multiple Regression Equations

As the name suggests, unstandardized multiple regression equations are calculated in raw score units. But the independent variables may use different metrics, making interpretations and comparisons difficult. Consider a study of the prediction of body weight using height, age, and gender, where height is measured in centimeters; age is measured in years; and gender is a nominal variable. In this situation, the magnitude of the variable values cannot be viably compared to quantify their relative strength as predictors. Unlike unstandardized regression equations, in standardized equations the values of independent variables are in the same metric. Therefore, it is possible to compare independent variables to identify their relative strength and importance as predictors in the model.

The standardized regression equation is similar to the unstandardized version. Like its unstandardized cousin, each independent variable is assigned a standardized regression coefficient also known as a beta coefficient or beta weight. However, there are some differences between the two types of equations. Perhaps the most obvious is that everything in the standardized equation is specified in standardized (z) score form. As a result, the means for independent and dependent variables is 0, and the standardized deviation is 1. In standardized regression equations, this characteristic is symbolized

Figure 6.3 Scatterplot of the Relationship Among Three Secondary School Placement Tests

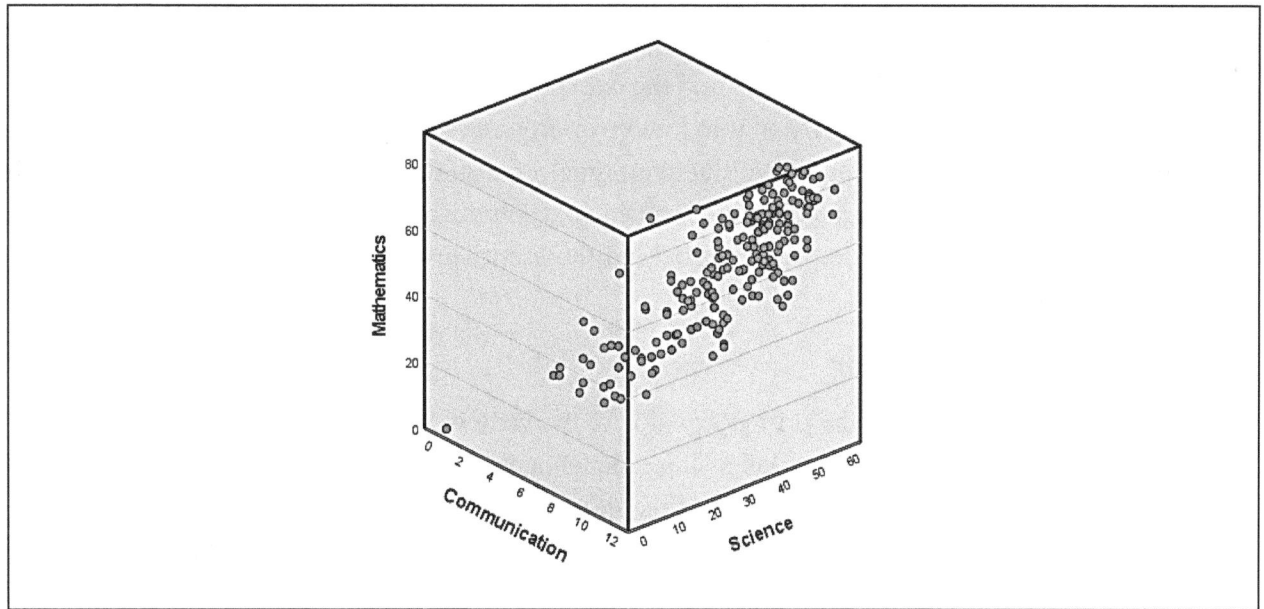

by Y_{zpred} (or $\hat{Y}_z$), which is the predicted z score of the dependent variable. The beta coefficient (β), is the standardized weight for each predictor. Finally, X is the standardized value for each independent variable in the equation. Therefore, the standardized multiple regression equation is:

$$\hat{Y}_z = \beta_1 X_{z1} + \beta_2 X_{z2} + \cdots + \beta_n X_{zn}$$

Here, all the values are standardized or z scores. For example, in building a standardized regression equation for the prediction of science performance, using the generic example above, all variable values in the standardized equation are standardized scores:

$$\text{science}_{z(pred)} = \beta_1 (\text{mathematics}_z) + \beta_2 (\text{communication}_z)$$

As in simple linear regression, the absence of a y-intercept is a distinctive characteristic of a standardized multiple regression equation. With the y-intercept located at the origin of the y- and x-axis in standardized regression equation, the y-intercept equals zero. Because it adds no value to the equation, it is excluded from the equation.

Recall that beta coefficient values are standard scores. Therefore, if a beta coefficient is positive, for every one-unit increase in the independent variable, the dependent variable increases by the value of the beta coefficient. If the beta coefficient is negative, for every one-unit increase in the independent variable, the dependent variable decreases by the beta coefficient value.

6.3.3 Multiple Regression Methods

There are several procedural methods for conducting regression. Chapter 5 explains that with only a single independent variable in simple linear regression, only one method is feasible—standard regression. However, in multiple regression, three of the most common regression approaches are: standard, statistical, and hierarchical regression. Like simple linear regression, the goal of each method is to examine the extent to which the composite independent variable predicts the dependent variable. The process begins with 100% of the variance in the dependent variable unexplained (unpredicted). As scores on independent variables enter into the procedure, the computer application evaluates the correlational pattern between each independent variable and the dependent variable. As the unique predictive relationship improves, the error in prediction (random error around the regression line) shrinks.

Although all regression methods have the same goal of explaining unexplained variance, each answers a slightly different research question, and each accomplishes the goal in a different way. We will examine the differences by comparing the behind-the-scenes processes used by statistical software (e.g., SPSS) for each approach.

6.3.3.1 Standard Regression

In standard multiple regression, the primary goal is to evaluate the predictive relationship between a weighted set of independent variables and a single dependent variable. The distinctive feature of this approach is that the relationship between all of the independent variables and the dependent variable is evaluated at the same time. To accomplish that, all of the independent variables are entered into the regression procedure simultaneously. This approach is most appropriate when there is no logical or theoretical basis for prioritizing the entry of one variable over another.

Imagine that in the Case of Predicting Weight in U.S. Adolescents, social science researchers wish to examine the following research question:

> Does the combination of height, age, and gender significantly predict body weight in U.S. adolescents?

As the research question indicates, the emphasis is on the entire set of independent variables (height, age, and gender) and its predictive relationship to the dependent variable (body weight). In addition, the research question signals that there is no intention to prioritize variable entry into the regression model.

In evaluating the predictive power of the independent variables, the software computes several familiar statistics that the researcher pre-specifies. Some are generated to determine whether the overall model (with all the independent variables entered) predicts a significant amount of the variance in the dependent variable (F-test). Others tell us how much of the dependent variable's variance the entire set of independent variables account for (coefficient of determination or R^2), and others estimate the significant unique contribution of each individual independent variable (t test of the b and beta coefficients). Finally, the a (y-intercept or constant) coefficient provides the value of the dependent variable when the independent variables equal zero.

Statistics that estimate the overall model take precedence over those that evaluate the performance of individual independent variables. If the overall model fails to significantly predict the dependent variable, the functioning of individual predictors becomes irrelevant. If the model is significant, the coefficients are used to build the regression equation that will be used with new data to make predictions later on.

6.3.3.2 Statistical Multiple Regression

Perhaps the most striking difference between statistical and standard multiple regression is that of the change in purpose reflected in the research question. In the Case of the Weight of U.S. Adolescents, the stepwise regression central research question is:

What is the best weighted combination of independent variables (height, age, and gender) that significantly predicts body weight?

The research question for this regression method signals that decisions about the order of variable entry are mostly taken out of the hands of the researcher and turned over to the computer, or to be more precise, to the mathematical algorithms in the statistical software, which make decisions about which independent variables will enter into the procedure and when. Furthermore, including the phrase "best weighted combination" in the research question suggests that some of the independent variables may be excluded from the model.

Once a researcher specifies the variables to be analyzed, mathematical algorithms regulate the sequence of entry of independent variables at each stage of the analysis, by identifying the independent variable that will contribute most to a significant regression model. Largely, this means that depending on the type of statistical regression specified, at each step, independent variables may be entered into the model or removed from the model until a combination of variables remain that best predict the dependent variable. The default entry standard for independent variables (typically, alpha $\leq .05$) is the probability of the variable contributing to a significant regression model. The variable with the lowest probability of contributing to a significant model is the first to be removed.

There are three statistical methods available in statistical regression—forward selection, backward deletion, and stepwise. In **forward selection**, as in all statistical regression methods, the regression procedure starts out with none of the variance of the dependent variable accounted for. The first independent variable to be added to the procedure is based on the probability that it has the most highly significant correlational relationship with the dependent variable. Therefore, the first independent variable selected for entry explains greater unique variance in the dependent variable (called a semi-partial correlation or sr^2) than any of the other independent variables. In forward selection, once an independent variable is entered into the equation, it remains there.

The next variable entered is the one with the best likelihood of making a significantly unique contribution to the dependent variable, after accounting for the first variable entered into the procedure. The remaining variables enter one by one if and only if each makes a significant unique contribution to the regression model after accounting for the variables already there. What this means is that if a variable does not meet the requirements for entry, it will not be entered into the model. The procedure stops when no remaining variables meet entry requirements.

The **backward elimination** method starts out in the same way as the standard method, with all the independent variables in the equation regardless of the worth of each one. However, where the standard approach stops, with all the variables in the model, the backward deletion method continues by sequentially evaluating each independent variable. A variable is deleted if removing it is least likely to significantly decrease the variance explained. If a variable fails to meet the removal standard, it is retained along with a new modified model. In this way, the most expendable variables are removed, leaving behind a model with a subset of predictors that best account for the variance in the dependent variable.

The **stepwise method** is a blend of forward selection and backward deletion. In this approach the equation starts out empty, and independent variables are added one at a time if they meet statistical criteria for entry, but they may also be deleted at any step where they no longer contribute significantly to the regression model. In effect, the multiple regression procedure is conducted repeatedly, entering variables that contribute greatest unique variance to the model or removing variables that contribute least. With the entry or removal of a variable, a new model is born, and variable evaluation continues until algorithms produce the strongest solution (the best combination of independent variables).

Regardless of the statistical regression method used, at the conclusion of the analysis, the best predictors are presented in the conclusive results. In addition, a list of the predictors that did not make the cut is preserved.

6.3.3.3 Hierarchical or Sequential Multiple Regression

Hierarchical or sequential multiple regression is a combination of both standard regression and stepwise regression. However, the associated central research question looks like a research question for standard regression:

> Does the combination of age, gender, years of education, months on the job, and beginning salary significantly predict current salary?

Hierarchical regression examines the relationship between independent variables and a dependent variable by entering independent variables in blocks. The blocks may contain one or several variables. The variable sets enter the model in an a priori sequence that the researcher stipulates. In this situation, the researcher is the final arbiter of the entry sequence of independent variables, and the researcher bases entry decisions on logical and theoretical grounds established in the research literature. If there is more than a single variable within a block of variables, the variables enter simultaneously as in standard regression. In addition, once they are placed in the model, they remain in the model. The purpose of sequentially entering the sets of variables is to evaluate the predictive relationship after controlling for blocks previously entered in the procedure.

Imagine that a researcher wishes to investigate the predictive relationship between reading achievement and a set of six independent variables—hours of leisure-time reading, school type (e.g., primary, preparatory and all-age), gender, nursery school attendance, visual acuity, and prior achievement. On the basis of existing research literature, the researcher stipulates that using hierarchical regression, the independent variables should be entered into the procedure in three blocks of two independent variables each. As Figure 6.4 illustrates, the researcher decides that Gender and School Type should enter the procedure first, followed by the second block of Prior Achievement and Visual Acuity, and finishing up with the third block of Nursery School Attendance and Hours of Leisure Time Reading.

In hierarchical regression, when variables are entered in sets, the procedure evaluates the predictive relationship between the first block of variables (Gender and School Type) and the dependent variable. When the second block enters (Prior Achievement and Visual Acuity), the procedure once again evaluates the predictive relationship after accounting for the first block of items. With entry of the third block of variables, the procedure again evaluates the relationship with Reading Achievement after accounting for the prior two blocks of variables.

In hierarchical multiple regression, a common practice is to enter independent variables in pre-specified blocks (models) as follows:

- The first model typically includes demographic information such as age, gender, ethnicity, school type and education.

Figure 6.4 Hierarchical Multiple Regression Entry Sequence

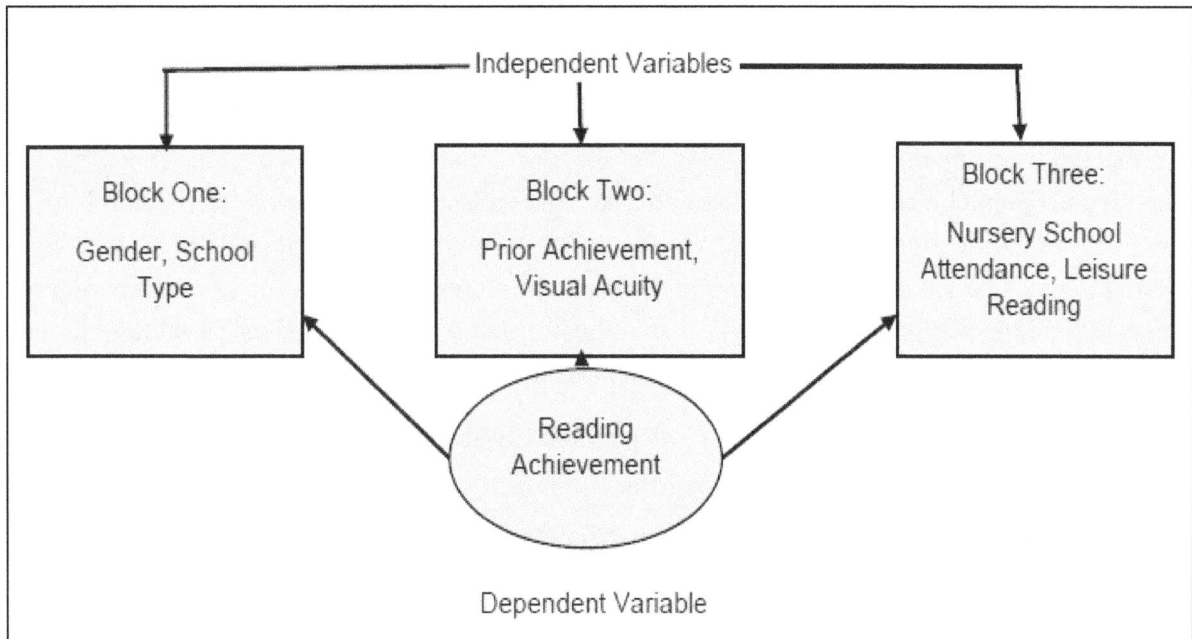

- The second model includes known important variables in the research of interest. In this step, previous research is replicated by adding variables from those research studies.
- In the third step, researchers add the variables in which they have a particular interest and which they believe will make a significant contribution to the model.

As the progression in Figure 6.20 illustrates, hierarchical modeling starts with the "smallest model" with the least important variables and then proceeds by adding selected sets of variables until the "full model" is reached. The least important variables are independent variables whose contributions are known from previous research. The variables a researcher includes will make a difference in the solution, or will add new knowledge, and these variables are added toward the end of the data analysis when the new variables' unique contribution can be evaluated.

While statistical and hierarchical regression appear similar in some respects, there are clear differences. Using statistical regression could result in the same sequence as hierarchical regression if single variables were always added in each step. However, typically blocks of variables, rather than single variables are included in each model. In addition, independent variables are not removed in hierarchical regression as might occur in statistical regression.

6.4 Planning a Multiple Regression Inquiry

An outline of some of the chief components of the research plan that typically precedes a multiple regression study is presented in Box 6.1 and 6.2. Much of the research plan outline for a multiple regression study mirrors those for simple linear regression (Section 5.4). Distinguishing features of multiple regression are directly related to the statistical options available to conduct multiple regression. The method used (standard, statistical/stepwise, or hierarchical) depends on the research questions and hypotheses used.

218 Multiple Regression

6.4.1 Research Questions and Hypotheses

As Box 6.1 demonstrates, the research question and hypotheses for standard multiple regression considers the independent variables as a weighted combination or set of predictors. In addition, with more than one independent variable, a question of the importance of each variable to the prediction of weight becomes essential.

Contrary to simple linear regression, the multiple regression statistical regression method is a viable alternative for making predictions, though controversial in some quarters. In the Case of the Weight of U.S. Adolescents, statistical regression's central research interest is the **best** linear combination of the independent variables that predicts weight—an indicator that some independent variables may not be in the final equation. Below are the hypotheses related to the research question.

> What is the best linear combination of independent variables that significantly predicts body weight?

Null Hypothesis 1: There is no significant predictive relationship between the best weighted combination of independent variables and the dependent variable than would be expected by chance.

Alternative Hypothesis 1: There is a significant relationship between at least one of the independent variables and the dependent variable.

Suppose a researcher chose to use hierarchical regression to predict body weight from height, age, and gender rather than the statistical regression method. Under those circumstances the researcher might enter variables in two blocks: (1) age and gender (demographic variables), (2) height. Figure 6.5 illustrates how the nature of research questions might impact the regression method selected.

Figure 6.5 Differences in Research Questions Across Multiple Regression Methods

Standard	Statistical	Hierarchical
Does the weighted combination of gender, age, and height significantly predict weight?	Of all the independent variables (height, age, and gender), what is the best linear combination of variables to predict weight?	Does adding the variable, height, significantly add to the prediction of weight after controlling for the previously variables, age and gender?
	Does adding or removing one or more independent variables significantly improve the prediction of weight?	Does adding a block of independent variables significantly add to the amount of variance in the dependent variable that the independent variables explain?
Does one or more variables in the set of independent variables uniquely and significantly contribute to the prediction of weight	Which independent variable contributed most to the prediction of weight?	Does one or more block of variables in the set of independent variables uniquely and significantly contribute to the prediction of weight?

6.4.2 Data Analysis

The data analyses conducted in standard multiple regression and accompanying results are similar to those described for simple linear regression in Chapter 5. Nevertheless, as the number of independent variables increases, the amount of information generated also rises. For example, standard multiple regression for a model that includes a single independent variable, generates a single unstandardized regression coefficient and single standardized coefficient for that variable. If a model includes four independent variables, standard multiple regression will generate four times the number of coefficients produced for a simple linear regression model. Statistical regression may generate more than one model (with four independent variables, it may produce as many as four models) as it identifies the best linear combination of variables to predict the dependent variable.

6.5 Data Assumptions of Multiple Regression

6.5.1 Assumptions Review

Chapter 5 discussed some characteristics of research data that may impact the accuracy of linear regression analyses and the ability to make generalizations from a sample to a target population. To enhance accuracy, researchers make some assumptions about the data and the sample selected for the investigation. Here are some that have already been discussed.

- The sample size must be sufficient for the expected power and effect size.
- Dependent variables must be continuous variables at the interval/ratio level. Independent variables include continuous data—ordinal data that have been summarized at the subscale or scale level, interval or ratio level data, or dummy-coded categorical variables.
- Data comprise independent observations; the errors associated with one observation should not correlate with the errors of any other observation.
- There are no extreme outliers in the data due to the influential nature of the outliers on regression models. With influential outlier cases removed from analysis, a very different regression model may ensue.
- Scores do not have a restricted, condensed, or shortened range.
- There is a linear relationship between the independent and dependent variable.
- Data must be homoscedastic so that the variances of the residuals (error) along the regression line remain similar as you move along the line.
- Regression model residuals (difference between predicted and actual values) should be randomly and normally distributed with a mean of zero. This is necessary for the *b* coefficients to be valid.

In addition to the assumptions described, multiple regression also assumes that the data will not be collinear or multicollinear.

6.5.2 Collinearity and Multicollinearity

If a regression model has more than one independent variable, one additional assumption should be evaluated—collinearity or multicollinearity. Multicollinearity is defined as a strong linear correlation between two or more independent variables in a regression study. Two independent variables with negative or positive intercorrelations at or above .90 (e.g., Tabachnick & Fidell, 2018) are said to be collinear; that is, they are measuring the same characteristic. This may occur if two instruments measuring mathematics achievement (e.g., General Certificate in Education [GCE] and Caribbean Advanced Proficiency Examinations [CAPE]) were included in the same regression model. If more than two independent variables are correlated at or above .9 (e.g., by including CSEC, GCE and CAPE scores as predictors), they are said to be multicollinear.

Collinearity or multicollinearity contributes to a less than optimum regression model:

- It can limit the amount of overall variance accounted for in the dependent variable. For highly correlated variables that predict the dependent variable, once the first is entered, it uses up much of the available variance in the dependent variable. Therefore, little unique variance is left for remaining variables that are highly correlated with the first. This results in limiting the overall variance accounted for in the model. By contrast, if uncorrelated variables are included in the model, the likelihood increases that they will account for greater unique and overall variance in the dependent variable.
- The unique contributions of highly correlated independent variables are difficult to extricate if all of them are left in the model. If two or more predictors are highly correlated, and they account for similar variance in the dependent variable, it is unlikely that all affected variables will receive the same weight even though they account for similar variance. Accordingly, regression coefficients may be misleading in evaluating unique contribution of predictors of the model. According to Meyers et al. (2016), distorted regression weights might exceed the expected range of +/-1.
- Regression coefficients (weights) may become untrustworthy. Meyers et al (2016) report that standard errors of regression weights for independent variables may be inflated, making the independent variables more unstable across study samples. If predictors are more variable, they are less likely to represent the population.
- Multicollinearity can also fatally affect the operation of the multiple regression program. If multicollinearity is sufficiently high, internal mathematical operations used to calculate regression models can be disrupted sufficiently to bring the program to a halt.
- Singularity among independent variables can also affect the functioning of multiple regression. When variables are exactly correlated, they measure the same psychological construct. For example, singularity is likely to occur if a regression model includes the overall score for a math test as well as subcomponent scores, like the arithmetic and geometry scores of the same test.

To reduce the likelihood of multicollinearity or singularity, it is important to carefully review the research literature to identify variables that are likely to be significantly related to the underlying construct of the dependent variable, but not intercorrelated with other independent variables. It is best to isolate potentially problematic variables in the preparation and design phase of a study rather than to modify the design later. Meyers et al. (2016) recommend that two variables "correlated in the mid- .7s or higher should probably not be used together in a regression … analysis" (p. 364). One of the independent variables should be removed from the design for replacements that are less intercorrelated. In this way, the model is likely to account for more variance in the dependent variable.

Added methods of dealing with multicollinearity and singularity include summing or averaging or combining collinear variables to form a single, composite variable. For example, instead of using three subscales of an emotional intelligence quotient instrument, use the total score or an average of the subscale scores for a single emotional intelligence predictor.

6.6 SPSS Procedures for Standard Multiple Regression

This exercise explores how to use SPSS to conduct and interpret standard multiple regression. For information on equations and procedures for hand-computing regression coefficients and other regression statistics for multiple regression, consult Tabachnick and Fidell (2018), and Pituch and Stevens (2016) for detailed overviews. Data from the Case of the Weight of U.S. Adolescents, **NHANES.sav**, will be used in this exercise.

6.6.1 SPSS Specifications for Standard Multiple Regression

This exercise demonstrates how to use SPSS to predict the body weight of U.S. adolescents based on the weighted linear combination of height, gender, and age (in years). To begin the specifications to conduct the analysis, open IBM SPSS either by double clicking on the data file **NHANES.sav** or by opening IBM SPSS, clicking on **File → Open → Data**, and navigating to the location of the data file.

Figure 6.6 Main Dialog Window for IBM SPSS

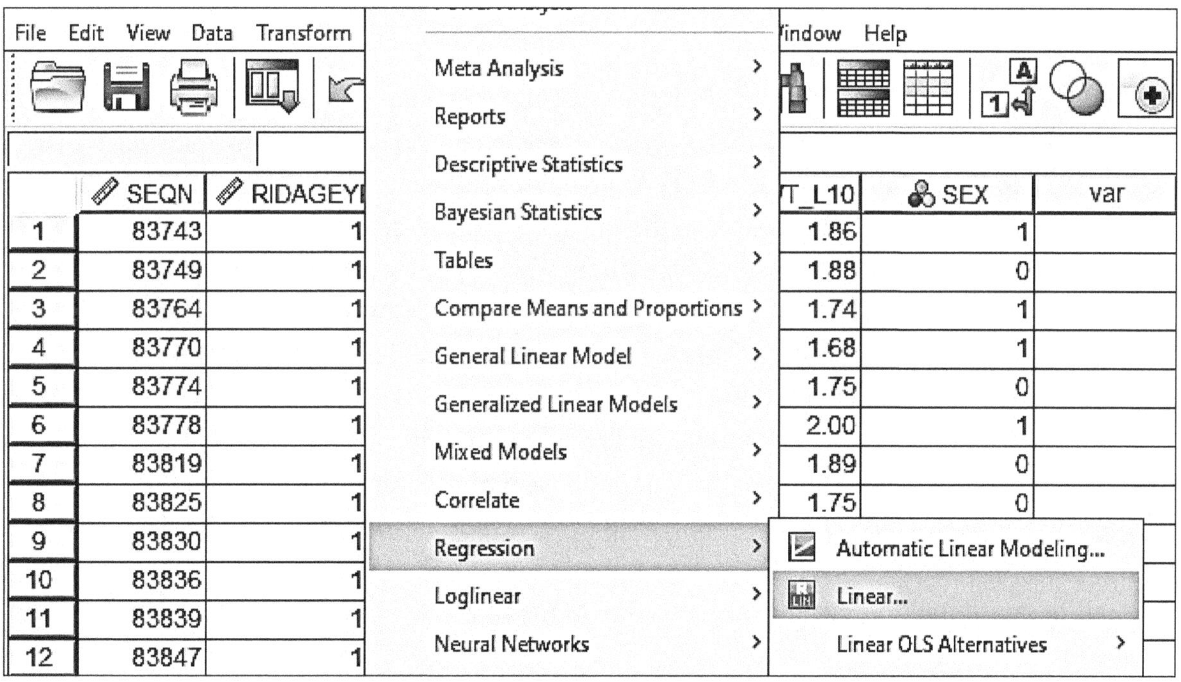

Once the data file is open, select **Analyze → Regression → Linear** (Figure 6.6) to open the **Linear Regression** main dialog window (Figure 6.7). From the variables list on the left, move **BMXWT_L10** (weight) to the **Dependent** panel. Move **Sex** (Gender), **BMXHT** (height [cm]), and **RIDAGEYR** (age in years) to the **Independent(s)** panel. Leave the **Methods** drop-down menu (below the **Independents** Panel) at its default setting of **Enter**, which is the specification for a standard multiple regression analysis.

Next, to specify the statistics needed, select the **Statistics** button in the upper right corner of the dialog window to open the **Linear Regression: Statistics** dialog window shown in Figure 6.8a. By default, **Estimates**, in the **Regression Coefficients** panel is already checked. This instructs IBM SPSS to compute and print the values of the regression coefficients for each independent variable, along with other related statistics that will appear in the **Coefficients** table and will be discussed in Section

Figure 6.7 Linear Regression Main Dialog Window

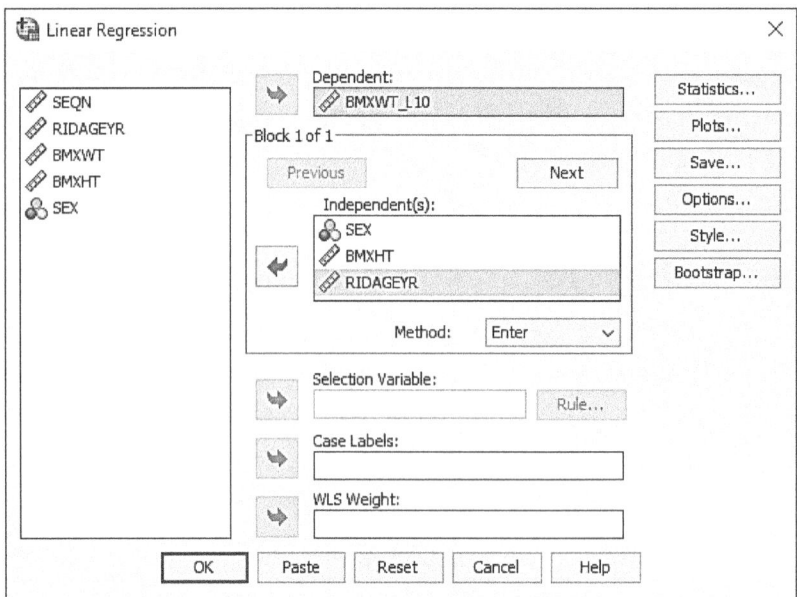

6.6.5. **Model fit** is also selected by default, requesting the computation and printing of a selection of statistics related to how well the standard regression model fits the data. **Model fit** statistics will appear in the **Model Summary** table and the **ANOVA** table, all of which will be discussed in Section 6.6.3 and 6.6.4. Selecting **Descriptives** generates the means, standard deviations, and the correlations for all the variables in the study. These statistics will appear in the **Descriptive Statistics** table and the **Correlations** table, respectively. Finally, click **Part and Partial Correlations** to produce the unique variance accounted for by each independent variable in the full model. Click **Continue** to return to the **Linear Regression** main dialog window.

To complete specifications for standard multiple regression, select the **Options** button in the **Linear Regression** main dialog window to reveal the **Linear Regression: Options** dialog window (Figure 6.8b). Here, three default selections appear, but only two are relevant for this analysis. The default selection, **Use probability of F**, is required with the selection of a step method such as statistical regression. The default will not go into effect for this analysis due to the override effect of selecting **Enter** (standard regression) as the regression method of choice. The default selection, **Include Constant in Equation**, instructs IBM SPSS to calculate the value for the *y*-intercept (constant) for the regression equation and print it in the **Coefficients** table. In addition, the default, **Exclude Cases Listwise**, tells IBM SPSS to exclude cases that do not have valid values for all of the variables in the analysis; a case with a missing value on even one of the variables will be excluded from analysis. In this way, the regression model will be based on the same set of cases and only complete cases. Click **Continue** to return to the linear regression main dialog window, and click **OK** to run the analysis.

6.6.2 Standard Multiple Regression Procedure Results: Descriptives

As discussed in Chapter 5, descriptive statistics provide an opportunity to check the results for data entry and data analysis specification errors as well as the reasonableness of initial results; this is an opportunity to assess whether the procedure was completed successfully. The results, recorded in Figure

Figure 6.8 The Linear Regression: Statistics Dialog Window

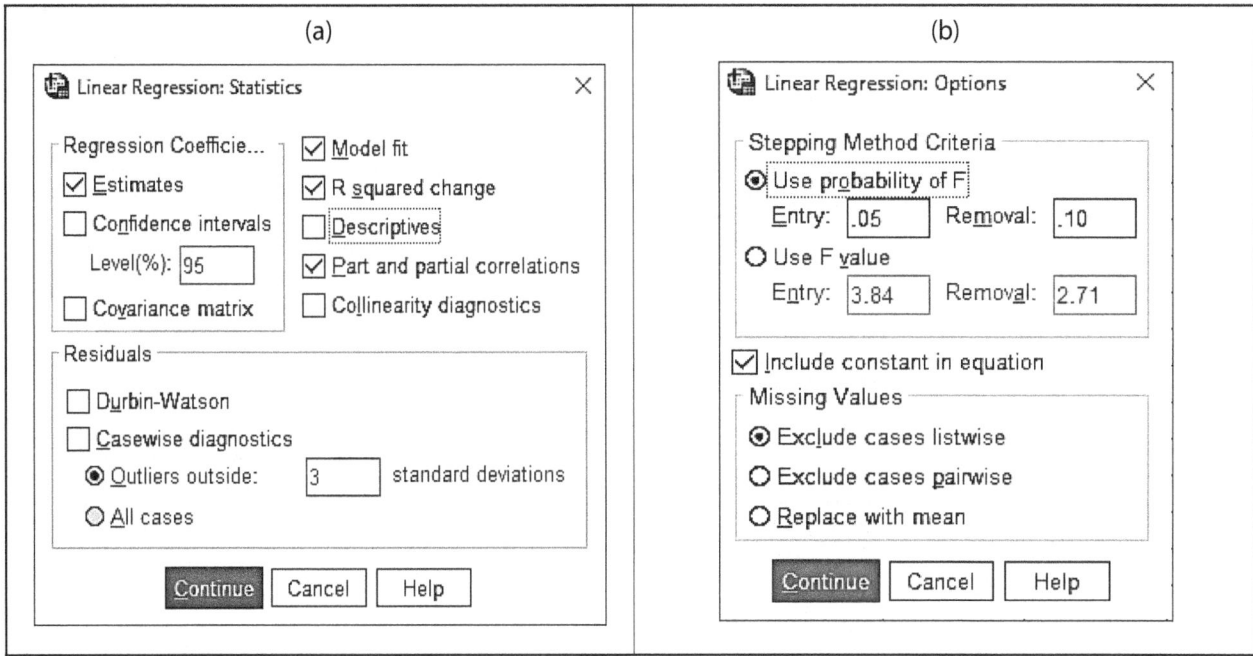

6.9 report that of the original 590 cases in the data file, about 10% was deleted due to the **Exclude case listwise** selection in the **Linear Regression: Options** dialog window; a total of 559 cases remained for the regression analysis. The original unit of measure for the dependent variable, **BMXWT**, was kilograms, with a mean of 66.259 kg and a standard deviation of 19.578 kg. However, due to significant skewness in the body weight data, a natural logarithmic transformation was applied to **BMXWT,** reducing skewness to within the normal range. The mean of the transformed variable is 1.803 kg. According to the results, participants are aged 15 on average, and they are about 165 centimeters tall. In addition, with males coded 1 in the data file, slightly more than 50% of the students are male.

Figure 6.9 Descriptive Statistics: Means and Standard Deviations

Descriptive Statistics			
	Mean	Std. Deviation	N
Weight	1.8033	.11554	559
Gender	.55	.498	559
Height (cm)	164.6420	9.35645	559
Age in years	15.21	2.104	559

Included in IBM SPSS descriptive statistics are the correlations of the variables in the regression analysis reported in Figure 6.10. Results are reported in three major rows. The first reports the magnitude of the Pearson correlations of pairs of variables. Correlations of independent and dependent variables range from .16 (gender and weight) to .56 (height and weight), the highest correlation with the dependent variable, weight. Independent variable intercorrelations range from almost zero correlation between gender and age (.006) to a positive moderate correlation between gender and height (.460).

Hence, no evidence of collinearity is present. The second major row which reports the significance of the correlation of each variable pair shows that all independent variables are significantly correlated with the dependent variable ($p < .001$). In fact, all variables, with the exception of age and gender, are significantly correlated. The third major row reports the sample size of each variable pair.

Figure 6.10 Correlations: The Case of Predicting Weight (N=559)

	Correlations				
		Weight	Gender	Height (cm)	Age in years
Pearson Correlation	Weight	1.000	.156	.562	.333
	Gender	.156	1.000	.460	.006
	Height (cm)	.562	.460	1.000	.377
	Age in years	.333	.006	.377	1.000
Sig. (1-tailed)	Weight	.	<.001	<.001	<.001
	Gender	.000	.	.000	.445
	Height (cm)	.000	.000	.	.000
	Age in years	.000	.445	.000	.
N	Weight	559	559	559	559
	Gender	559	559	559	559
	Height (cm)	559	559	559	559
	Age in years	559	559	559	559

6.6.3 Results: Estimation of the Overall Model

The primary method of evaluating the significance of the overall prediction model is one-way analysis of variance (ANOVA) with the set of predictors defined as the combined independent variable. In the Case of the Weight of U.S. Adolescents, age in years, gender, and height represent the combined independent variable. The objective of ANOVA is to test the following hypotheses:

Null hypothesis: The set of coefficients, which represents the slopes of the regression lines for height, age and gender equal 0 (is not significantly different from 0).
H_0: $b_1 = b_2 = b_3 = 0$
Alternative hypothesis: The set of coefficients, which represent the slopes of the regression lines for height, age, or gender is significantly different from 0.
H_a: $b_i \neq 0$, where i equals any independent variable

ANOVA uses the F statistic (F ratio) to test the statistical significance of the overall model. Typically, an alpha level of $\alpha = .05$ is used to evaluate the F statistic. If the F statistic is significant at a probability level less than or equal to the preset alpha level of .05, then there is a 95% or greater chance that the null hypothesis is false. A false null hypothesis leads to the conclusion that the regression model is significantly better than might be expected by chance.

Figure 6.11 reports the ANOVA results of the Case of the Weight of U.S. Adolescents. The columns, the **Sum of Squares**, **df** (degrees of freedom) and **Mean Square** are used to calculate the F statistic by hand (See Section 5.5 for more on this). The calculated F statistic is in Column, **F**, which may be compared with the critical value for the F distribution (Appendix A) to determine whether the model is significant. Alternatively, the value <.001 in the **Sig.** column means that the F statistic exceeds the critical value, so the correlation between the set of predictors and the dependent variable is significantly different from 0 at a level less than the a priori alpha level of .05, $F(3, 555) = 96.031$, $p < .001$.

Figure 6.11 Test of Significance of Regression Model: The Case of Predicting Weight

Model		Sum of Squares	df	Mean Square	F	Sig.
1	Regression	2.546	3	.849	96.031	<.001[b]
	Residual	4.904	555	.009		
	Total	7.450	558			

ANOVA[a]

a. Dependent Variable: Weight
b. Predictors: (Constant), Age in years, Gender, Height (cm)

6.6.4 Results: Variance Accounted for in Weight

A second set of hypotheses for the Case of Predicting Weight focuses on the extent to which the independent variables account for variance in the dependent variable as follows:

Null hypothesis: The set of independent variables—gender, age, and height—does not account for a significant amount of the variance in the dependent variable, weight ($R^2 = 0$).
Alternative hypothesis: The amount of the variance in the dependent variable that the set of independent variables accounts for is significantly greater than 0 ($R^2 > 0$).

Before describing the statistic used to test this null hypothesis, a brief discussion of a related statistic—Pearson correlation (denoted as Pearson's r) and multiple correlation which is symbolized by an uppercase italic R will provide context for the results. Recall from Chapter 4 that Pearson's r quantifies the magnitude of the linear relationship between two variables in bivariate correlation. When the linear association between one variable and a set of variables is being considered, the correlation coefficient that is used is multiple correlation, symbolized by uppercase R. Accordingly multiple correlation is a measure of the degree of association between a dependent variable and a set of independent variables.

Similarly, r^2 represents the amount of shared variance between two variables in bivariate correlation. When more than two variables are involved, the index of the strength of that relationship (the amount of shared variance) is symbolized by uppercase R^2, also known as the squared multiple correlation or the coefficient of multiple determination.

To test the null hypothesis for the strength of the relationship between the set of independent variables and the dependent variable in the Case of Predicting Weight, it is necessary to compute the shared variance or R^2. Therefore, R^2 can also be considered an estimate of the amount of variance in the dependent variable that the set of independent variables explains.

Figure 6.12 presents the model summary for the Case of the Weight of U.S. Adolescents. The linear association between the weighted combination of gender, age, height, and weight is $R = .585$, a moderate correlation between the set of independent variables and weight. The set of independent variables explains $R^2 = .342$ of the variance in weight. Rounding to two decimal places, that converts to approximately 34% of the variance.

Figure 6.12 Standard Multiple Regression Model Summary: The Case of Predicting Weight

		Model Summary		
Model	R	R Square	Adjusted R Square	Std. Error of the Estimate
1	.585a	.342	.338	.09400

a. Predictors: (Constant), Age in years, Gender, Height (cm)

The model summary also reports another statistic, **Adjusted R Square** (*Adj. R^2* = .338). Variance explained is often somewhat inflated due to the number of independent variables relative to the size of study samples and the presence of error in an estimated model. Small sample sizes exacerbate squared multiple correlation inflation, reducing its effectiveness as an estimate of the target population. This phenomenon is known as R^2 shrinkage. Some mathematical formulae (SPSS algorithms among them) are designed to estimate the amount of R^2 shrinkage to produce better population estimates of R^2 (e.g., Olkin & Pratt, 1958). Others are based on estimates from cross-validation studies (e.g., Browne, 1975). Regardless of the formula used, the outcomes are very similar with adjusted R^2 providing a less biased estimate of variance explained in dependent variables.

The best way to avoid severe R^2 shrinkage is to increase sample size. Meyers et al. (2016) recommend minimum regression study sample sizes of 200 cases. Furthermore, they note that researchers should use at least 20, but preferably 30 cases per predictor. As the sample size to predictor ratio increases, R^2 shrinkage decreases.

With the total sample size for the Case of Predicting Weight at 559, and the number of predictors at 3, the R^2 shrinkage for this analysis was very small:

$$R^2 \text{ shrinkage} = R^2 - \text{Adjusted } R^2 = .342 - .338 = .004 = 0.4\%.$$

6.6.5 Results: Independent Variables' Contribution to the Model

If a model is significant and the independent variables account for some variance in the dependent variable, the next step is to find each independent variable's unique contribution to the regression solution. For this evaluation the hypotheses for each independent variable are as follows:

Null hypothesis: The correlation of the dependent variable and the corresponding independent variable is not significantly different from zero when the effects of the other independent variables are taken in account.

$H_0: b_j = 0$

Alternative hypothesis: The correlation of the dependent variable and the corresponding independent variable is significantly different from zero when the effects of the other independent variables are taken in account.

$H_a: b_j \neq 0$

Section 6.6.4 explained that multiple regression measures the total amount of variance explained in a model. It also measures the relative contribution of each independent variable to the variance explained in the model. Each variable is assessed as though it entered the model after all the other independent variables and can only account for the variance left over in the dependent variable. If an independent variable accounts for a significant amount of the residual variance, it is considered a significant contributor to the model. The contribution of each independent variable is represented by a regression coefficient which denotes its singular influence on the dependent variable. SPSS conducts a *t* test of each regression coefficient to evaluate whether the associated independent variable accounts for a significant amount of the residual variance; that is, whether the partial correlation between each independent variable and criterion variable is significantly different from zero.

Instead of computing the *t* test for each independent variable by hand, statistical applications like IBM SPSS may be used instead. In the process of conducting each *t* test, IBM SPSS generates a *t* statistic as well as the probability that the value of the *t* statistic would occur if the regression coefficient equaled zero.

For example, in Figure 6.13, the unstandardized regression coefficient (*b*) for the variable, age, is 0.007. The *t* statistic for the Age coefficient is 3.196 noted in the **t** column. The **Sig.** column reports that based on the *t* statistic, the coefficient for the variable, age, is significantly different from zero at $p < .001$. Therefore, the *t* statistic (3.196) for the age coefficient (0.007) has less than a probability of .001 of being equal to zero. Consequently, we may reject the null hypothesis; age is significantly different from zero and makes a unique significant contribution to the prediction of weight while controlling for the remaining independent variables. Similarly, as Figure 6.13 reports, gender and height also contribute significantly and uniquely to the prediction of weight.

Values in the **Unstandardized (*b*) Coefficients** column in Figure 6.13 carry important information, but they have a critical disadvantage—they cannot be used to compare the influence of independent variables. Unstandardized coefficients are tied to the units of measure in which they are calculated. Within their own metric the coefficients are interpretable. An increase in 1 unit in age is associated with an increase of .007 in weight, statistically controlling for the effects of the other variables. A similar interpretation may be made for height—an increase of 1 unit in Height equals a gain of .007 in the variable, weight. However, the metrics for each variable is different, making it impossible to compare the contribution of the variables to the model.

Interpreting the regression coefficient for the dichotomous variable, gender, differs from continuous variables. For dichotomous variables there are only two possible values; in this study, female= 0 and male =1. Accordingly, inserting the value for female and the associated coefficient for gender in the expression $b_1$gender results in the expression -0.024*0 = 0, indicating that the gender coefficient has no effect on weight of females. However, for males there is a 0.024-point decrease in weight (-.024*1).

Standardized (beta) coefficients for the Case of the Weight of U.S. Adolescents also appear in Figure 6.13 in the column **Standardized Coefficients**. These standardized values are all *z* scores, and therefore,

Figure 6.13 Standard Regression Model: Coefficients Table

		Coefficients[a]						
		Unstandardized Coefficients		Standardized Coefficients			Correlations		
Model		B	Std. Error	Beta	t	Sig.	Zero-order	Partial	Part
1	(Constant)	.567	.077		7.375	<.001			
	Gender	-.024	.009	-.104	-2.636	.009	.156	-.111	-.091
	Height (cm)	.007	.001	.565	13.204	<.001	.562	.489	.455
	Age (Years)	.007	.002	.121	3.196	.001	.333	.134	.110

a. Dependent Variable: Weight

share the same metric. This characteristic allows direct comparisons of the standardized regression coefficients to assess the relative strength of each independent variable. For example, visual inspection of the beta coefficients presented in Figure 6.13 discloses that height with a beta coefficient of .565 is the most influential of the independent variables in predicting weight. By computing the ratio of the coefficients for height and age, where the former variable is the numerator and the latter the denominator (.565/.121=4.67), the relative importance of the variables can be quantified (Pedhazur, 1982, 1997). The ratio of height and age shows that height is close to 5 times more effective as an independent variable than age. Furthermore, using the absolute value of the beta coefficient for gender (.104), we find that the ratio of age/gender is .121/.104 = 1.16, so that Age contributes somewhat more to the prediction of weight than gender.

Standardized coefficients have limited applicability as well. According to Pedhazur (1997) they are affected by the variability of associated independent variables. Meyers et al. (2013) assert that standardized coefficients are sensitive to the strength of intercorrelations of the independent variables. Independent variables that are very highly correlated can lead to coefficients that exceed +/- 1, in which case the values of these aberrant coefficients are ignored by researchers, and other indicators of coefficient magnitude are used. Due to these limitations, standardized coefficients should be interpreted cautiously and should be used in concert with other indicators of independent variables' importance to predictions.

Zero-order correlations presented in Figure 6.13 as **Zero order** (under **Correlations**) is an alternative term for Pearson product-moment correlation discussed earlier in this section. It is the bivariate correlation between each independent variable and the dependent variable. Therefore, gender has a low, positive correlation of .156 with weight, indicating that males (coded 1) weigh more than females (coded 0). If gender had a negative correlation with weight, that would suggest that males, on average, weighed less than females. As expected, age and height have positive correlations with weight of .333 and .562, respectively.

Partial correlations, under **Partial** in the **Correlations** column of Figure 6.13, are the correlations between each independent variable and the dependent variable after the effects of the other variables have been removed from both the independent and the dependent variables. Figure 6.14 illustrates the partialling out of variance from the dependent variable (DV) and two independent variables (IV).

Figure 6.14 The Difference Between a Partial and Semi-Partial Correlation

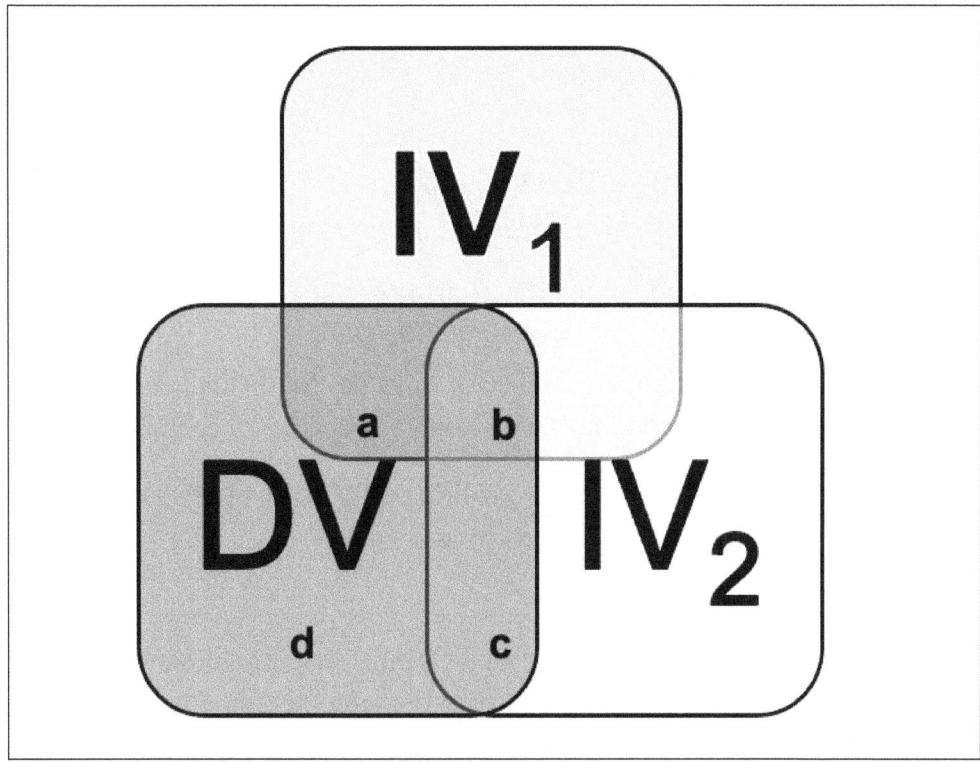

Figure 6.14, adapted from Meyers et al. (2016), demonstrates that in standard multiple regression, the partial correlation of IV_1 (labeled a) is the unique contribution to what is left of the variance in the DV after the contribution of IV_2 has been removed from IV_1 and the DV. Accordingly, in Figure 6.13 after the effects of age and height have been removed from gender and weight, gender is left with a low, negative relationship with weight, rather than the low, positive linear relationship reported in the zero-order column in Figure 6.13. Similarly, after removing the effects of the other variables, age has a lower, positive relationship with weight. Only height maintains a moderate, positive relationship with the dependent variable after the effects of the other independent variables have been removed.

Semi-partial correlations (displayed as **Part** correlations in Figure 6.13), index the correlation between each independent variable (IV) and the dependent variable (DV) after the effects that the other independent variables have on the dependent variable have been removed solely from the independent variable. As Figure 6.14 demonstrates, in standard multiple regression, the semi-partial correlation of IV_2 is the unique contribution of that variable (labeled c) to the total variance of the DVs (a, b, c, and d) after the contribution of IV_1 has been removed from IV_2. When a semi-partial correlation is squared, it indexes the percent of variance uniquely accounted for by each independent variable in the regression model. The squared semi-partial correlations (sr^2) of gender ($-.091^2$), age ($.110^2$) and height ($.455^2$) reveal that the independent variables uniquely account for 1%, 1.2% and 20.7% of the variance, respectively, in weight. Totaling the semi-partial correlations results in a total of 22.9% of the variance in weight uniquely accounted for by the independent variables. Why is the total semi-partial correlation different from R^2 in Figure 6.12 which reports .342 or 34.2% of the variance in weight? The difference is due to the discounting of the shared variance among the independent variables (Meyers et al. 2016).

6.7 Solving the Regression Equation

Now that the constant and regression coefficients for the raw-score equation have been computed, those values may be added to the regression equation:

$$\hat{Y} = a + b_1X_1 + b_2X_2 + \cdots + b_nX_n$$
$$Weight_{pred} = b_1(gender) + b_2(age) + b_3(height)$$
$$Weight_{pred} = (.567) + (-.024)(gender) + (.007)(age) + (.007)(height)$$

To check the accuracy of the equation, add in the actual values for the independent variables from 14-year-old Rose, one individual in the study sample who is 150.5 centimeters in height.

$$Weight_{pred\,Rose} = (.567) + (-.024)(0) + (.007)(14) + (.007)(150.5)$$
$$Weight_{pred\,Rose} = (.567) + 0 + .098 + 1.054$$
$$L10(Weight_{pred\,Rose}) = 1.719$$

Now, of course, Rose does not weigh 1.719 kilograms. Original weight values were transformed to Base 10 logarithm of the weight values in order to normalize the score distribution and meet the normality assumption. Hence, Rose's predicted weight is in base 10 or common logarithm metric. Converting weight back to kilograms, using a calculator, yields the following:

$$\text{Converting } L10(Weight_{pred\,Rose}) \text{ to } (Weight_{pred\,Rose}) = 52.36 \text{ kg}$$

Rose's predicted weight is 52.36 kg which is slightly higher than her actual weight, 50.3 kg—a difference of 2.06 kg. The difference between actual and predicted weight is measurement error.

The predictors selected for a study are not necessarily the only set of variables available to predict the dependent variable. At times researchers may overlook better predictor variables in the literature or discard good candidates due to other limitations, such as the availability of appropriate data collection instruments. Nevertheless, a regression equation is specified and interpreted based on available variables. Important omitted variables may be investigated in future research. However, researchers can preclude omitting important variables by selecting variables based on a thorough literature review and using underlying theory to justify inclusion.

6.8 Sample Report (Standard): The Case of the Weight of U.S. Adolescents

Standard multiple regression was employed to investigate whether gender, age (in years), and height (in centimeters) of students in Grades 6 to 12 were statistically significant predictors of weight (in kilograms). SPSS Explore and SPSS Regression were employed to evaluate whether the variables met the assumptions for multiple regression. As a result of those analyses, the variable weight was transformed, using a log 10 transformation, to reduce skewness.

Table 6.1 Correlations of Weight, Gender, Age and Height (N = 559)

	1	2	3	4
1. Weight	-			
2. Gender	.156	-		
3. Height (cm)	.562	.460	-	
4. Age in years	.333	.006	.377	-

Note: All correlations except that between gender and age are statistically significant ($p<.001$)

Correlations of the variables (Table 6.1) indicate that variables were significantly intercorrelated ($p<.001$) with the exception of gender and age ($p = .445$). Of all the independent variables, gender had the lowest correlation with the dependent variable, weight ($r = .156$).

The prediction model was statistically significant $F(3, 555) = 96.031$, $p < .001$, and accounted for 33.8% of the variance in weight ($R^2 = .342$, Adjusted $R^2 = .338$). The raw and standardized regression coefficients of the independent variables, their correlations with weight, and their squared semi-partial correlations are presented in Table 6.2. Weight was primarily predicted by height, followed by age and gender. Height was the most important of the three predictors, followed by age. Gender, though a significant predictor, was the least important of the three predictors of weight. With the exception of height, the unique variance explained by each of the independent variables was relatively low. Height explained 20.7% of the variance of weight, followed by age, 1.2% of the variance, and gender which explained 1% of the variance.

Table 6.2 Standard Regression Results of the Prediction of Body Weight

Model	B	Std. Error	Beta	Pearson's r	sr^2
Constant	.567	.077			
Gender	-.024	.009	-.104	.156	.008
Age	.007	.001	.565	.562	.012
Height	.007	.002	.121	.333	.207

Note: The dependent variable is weight. $R^2 = .342$. Adjusted $R^2 = .338$. sr^2 is the squared semi-partial correlation. Coefficients significant at $p < .01$.

6.9 SPSS Specifications for Statistical (Stepwise) Multiple Regression

This demonstration uses IBM SPSS to conduct a stepwise regression of the data from The Case of the Weight of U.S. Adolescents. The same dataset used in standard multiple regression (NHANES.sav) will be used in this exercise.

After opening the data file in IBM SPSS, select **Analyze → Regression → Linear**, which opens the **Linear Regression** main dialog window displayed in Figure 6.15. From the variables list panel, select and move the dependent variable **BMXWT_L10** to the **Dependent** panel; then click over **Sex** (Gender), **BMXHT** (height) and **RIDAGEYR** (Age in Years) to the **Independent(s)** panel. To specify stepwise regression, select the down arrow next to the **Method** panel, and select **Stepwise** as the regression method.

Figure 6.15 Main Dialog Window for Linear Multiple Regression

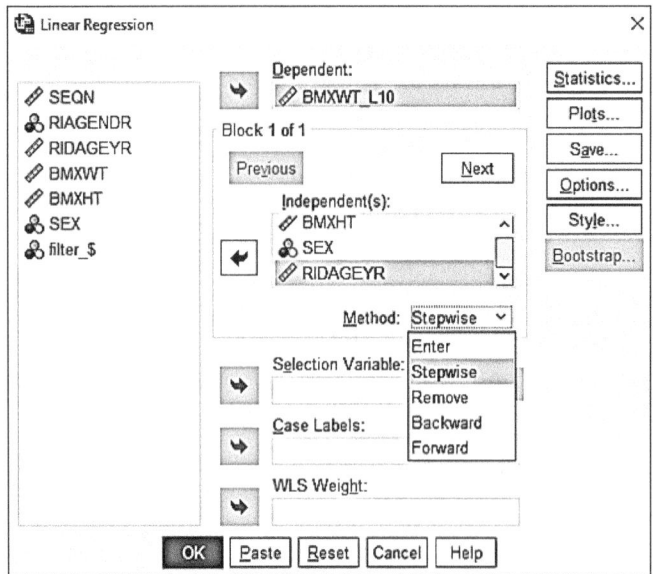

Next, select the **Statistics** button in the upper right corner of the **Linear Regression** dialog window to open the **Linear Regression: Statistics** dialog window displayed in Figure 6.16a. In this dialog window, the defaults **Estimates** (statistics on each independent variable) and **Model fit** (statistics related to how well the entire model fits the data) have been preselected. In addition, select **R square change**, which is useful in reporting how the variance accounted for in the dependent variable changes when variables are entered in multiple steps rather than in a single step (as in standard regression). Leave **Descriptives** unchecked since those statistics will be the same as they were in the standard regression procedure. Select **Part and partial correlations**, which produces partial and semi-partial correlations (Section 6.6), the statistics used to determine the sequence with which variables enter the analysis. **Collinearity diagnostics** and options under **Residuals** statistics will remain unchecked; these statistics were used in completing assumptions testing for multiple regression prior to these analyses (See Chapter 3, Section 5.6 and 6.5 for linear regression assumptions details).

Finally, select the **Options** button to open the **Linear Regression: Options** dialog window displayed in Figure 6.16b. Notice that by contrast to standard regression, **Stepping Methods Criteria** are needed in statistical regression. Setting specific significance levels for **Entry** and **Removal** prevents variables from continually exiting and re-entering the model during the stepwise data analysis process (Section 6.3.3).

In the stepwise procedure, it is more difficult for a variable to be entered into the model than to be removed: hence, the difference in the default **Entry** and **Removal** probability levels in **Linear Regression: Options**. According to default levels, to be entered, a variable must make a significant contribution to the model at or less than an alpha level of .05. To be removed, a variable must be significant at an alpha level greater than .10. Retain entry and removal settings at the default levels. Click **Continue** to return to the **Linear Regression** window, and click **OK** to run the analysis.

Figure 6.16 The Linear Regression Statistics and Options Windows

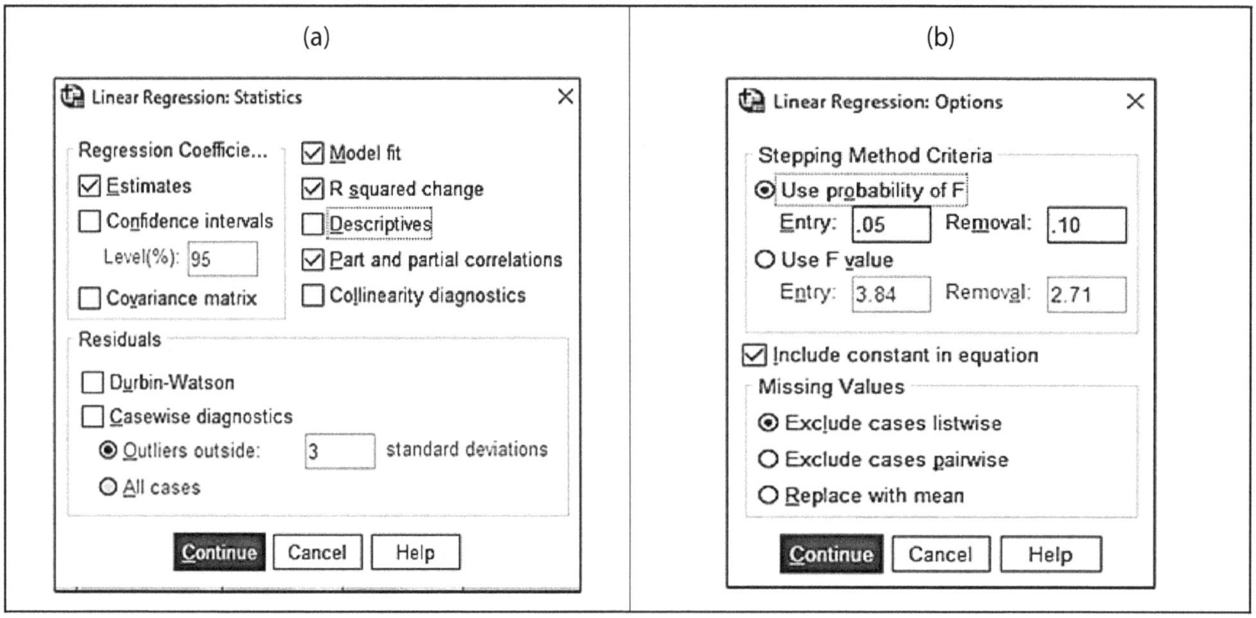

6.10 Statistical (Stepwise) Results: The Case of the Weight of U.S. Adolescents

6.10.1 Results: Estimation of the Overall Model in Stepwise Regression

The process of estimating the overall model in stepwise multiple regression is similar to standard multiple regression. One-way ANOVA is used to evaluate the significance of the overall prediction model, with the combined set of predictors entered as a composite independent variable. However, there are differences between the two regression methods. One key distinguishing feature of statistical stepwise regression is that typically more than one model is tested. The number of models tested is dependent on the number of steps required to build the final model. The number of steps required to build the final model is dependent on the number of variables in the study, the number of variables that meet the requirements for entering the model, and the number exiting after entry. For example, imagine that a researcher is investigating what best predicts the customer-based quality rating of the popular Caribbean and Central American snack, patties. The researcher identified the following independent variables: price, level of spiciness (very spicy, moderately spicy, mildly spicy), flakiness of crust (very flaky, moderately flaky, not flaky), quality of filling (poor, moderate, high).

With four independent variables available in the study, only three meet the entry requirements (price, levels of spiciness, and quality of filling) in a total of three steps. One meets the exit criteria before the process stops (one additional step), resulting in a total of four models and four ANOVAs in the analysis.

In the Case of the Weight of U.S. Adolescents, three models are tested, one for each independent variable. Each model is associated with its own set of hypotheses for significance testing.

Null hypothesis: Variable coefficients at each step of the analysis are not significantly different from zero.
Step 1: $H_o: b_1 = 0$
Step 2: $H_o: b_1 = b_2 = 0$ given b_1 already in the model
Step 3: $H_o: b_1 = b_2 = b_3 = 0$ given b_i already in the model

Alternative hypothesis: At least one variable coefficient is significantly different from zero.

Any step: $H_a: b_j \neq 0$ given b_i already in the model

Figure 6.17 (**Variables Entered/Removed**) summarizes each step of the stepwise procedure for the case. According to the **Model** column, statistical stepwise regression produced three models. Height (**Model 1**) is the first variable to enter because it is the variable most highly correlated with the dependent variable ($r = .562$, $p < .001$, Figure 6.10).

Figure 6.17 Statistical Stepwise Regression Results: Variables Entered\Removed

		Variables Entered/Removed[a]	
Model	Variables Entered	Variables Removed	Method
1	Height (cm)	.	Stepwise (Criteria: Probability-of-F-to-enter <= 0.050, Probability-of-F-to-remove >= 0.100).
2	Age	.	Stepwise (Criteria: Probability-of-F-to-enter <= 0.050, Probability-of-F-to-remove >= 0.100).
3	Gender	.	Stepwise (Criteria: Probability-of-F-to-enter <= 0.050, Probability-of-F-to-remove >= 0.100).

a. Dependent Variable: Weight

After the entry of height, stepwise entry rules change somewhat to the independent variable that contributes the greatest additional unique predictive variance to the dependent variable (semi-partial correlation or sr^2). Based on these conditions, the variable, age, entered the model at the second step, followed by gender, the final variable to enter the model. All of these variables entered because they met entry and removal criteria for contributing to a significant regression model.

Figure 6.18, **Excluded Variables**, provides additional detail about what drove the sequence of entry. Results in the first major row show that after the variable, height, entered the model, gender and age were left behind. Of those two remaining variables, age, which had the larger t value appeared to be the next candidate for entry into the equation. The squared semi-partial correlation for age may be used to verify that decision. It is not in Figure 6.18, but it can be calculated:

$$\begin{aligned} sr^2 &= (Beta\ In * sqrt[Tolerance])^2 \\ sr^2 &= (.142 * \sqrt{.858})^2 \\ sr^2 &= (.142 * .926)^2 \\ sr^2 &= .132^2 \\ sr^2 &= .017 \end{aligned}$$

Compared with gender, which registered a sr^2 of .008, age registered greater unique predictive power than gender, making it a better contributor to the prediction of weight than gender. Hence, age entered the model at the second step of the procedure. Gender entered at the third step because it still explained enough unique variance ($p =< .05$) to enter at Step 3.

Figure 6.18 Stepwise Regression Results: Excluded Variables

Excluded Variables[a]						Collinearity Statistics
Model		Beta In	t	Sig.	Partial Correlation	Tolerance
1	Gender	-.130[b]	-3.329	<.001	-.140	.788
	Age	.142[b]	3.792	<.001	.159	.858
2	Gender	-.104[c]	-2.636	.009	-.111	.756

a. Dependent Variable: Weight
b. Predictors in the Model: (Constant), Height (cm)
c. Predictors in the Model: (Constant), Height (cm), Age in years

Figure 6.19 presents results of the omnibus test for each of the three models (steps) in the procedure. **Model 1** includes the constant as well as the independent variable, height, which was a significant predictor of weight ($p < .001$ noted in the column, **Sig**.).

Figure 6.19 Tests of Significance for Each Step in Stepwise Regression Analysis

ANOVA[a]						
Model		Sum of Squares	df	Mean Square	F	Sig.
1	Regression	2.356	1	2.356	257.605	<.001[b]
	Residual	5.094	557	.009		
	Total	7.450	558			
2	Regression	2.484	2	1.242	139.084	<.001[c]
	Residual	4.965	556	.009		
	Total	7.450	558			
3	Regression	2.546	3	.849	96.031	<.001[d]
	Residual	4.904	555	.009		
	Total	7.450	558			

a. Dependent Variable: Weight
b. Predictors: (Constant), Height (cm)
c. Predictors: (Constant), Height (cm), Age in years
d. Predictors: (Constant), Height (cm), Age in years, Gender

The second major row of the table reports the significance testing results for **Model 2** in which age was added to height (Footnote c). This model was also significant ($p < .001$). Gender was the final variable meeting requirements for entry into the procedure (footnote d) in **Model 3**. This three-predictor model was significant as well ($p < .001$).

Once a third independent variable is in the equation, the stepwise method may remove an independent variable if it no longer merits retention due to variance shared with other independent variables. When independent variables share explanation of the dependent variable, that may result in a reduction of one of those variables' unique predictive contribution and removal from the equation. The default removal standard in SPSS is a variable that explains a nonsignificant ($p > .10$) quantity of unique variance in the dependent variable.

In stepwise regression, the stopping rule governs when the model is complete. If a variable meets the requirements of the rule, the analysis ceases. Hypotheses for the stopping rule follow:

Stopping Rule Null Hypothesis: Given independent variables already in the model, the squared semi-partial correlation between the dependent variable (Y) and the independent variable (X) being considered for entry is not significantly different from zero ($\alpha = .05$).
H_0: $sr_i^2 = 0$, given at least one X_i already in the model

Stopping Rule Alternative hypothesis: Given independent variables already in the model, the squared semi-partial correlation between the dependent variable and the independent variable being considered for entry is significantly different from zero ($\alpha = .05$).
H_a: $sr_i^2 \neq 0$, given at least one X_i already in the model

The stopping rule does not go into effect after Model 1 or 2. It goes into effect after Model 3. If one of the variables in the SPSS demonstration (e.g., gender), did not meet the entry or retention criteria the final stepwise prediction solution would have included two independent variables, rather than three. However, none of the independent variables met the removal criteria; they each maintained a significant, unique contribution to the prediction model that merited retention.

If a fourth independent variable existed and did not meet the requirements necessary for entry into the equation, that variable would have appeared in the **Excluded Variables** table (Figure 6.18), showing a nonsignificant t value. Based on the ANOVA results, the final stepwise regression model confirms the standard regression results that height, age and gender are significant predictors of the dependent variable, weight (at $F(3, 555) = 96.031$ $p < .001$).

6.10.2 Results: Variance Explained in the Dependent Variable

In stepwise regression, each model tested is accompanied by an estimation of the variance in the dependent variable that the model accounts for. If the addition of a new independent variable does not account for a significant amount of unique variance, it is excluded from the model. In the **Excluded Variables** table in Figure 6.17, the 'Beta In' value for age in Model 1 and for gender in Model 2 may be treated as 'Beta' values (unique contribution) if the variables are entered into the equation. Notice that t values for "Beta In" for both entry candidates are significant.

Model Summary is presented in Figure 6.20. This summary includes additional information that is not needed for the model summary for standard regression. In standard regression, all independent variables enter the model simultaneously in a single step. As a result, there are no changes in the model or the statistics. In comparison, stepwise regression is designed to have independent variables enter in the regression procedure at different times (steps), based on whether they are likely to make a significant contribution to the prediction model. Therefore, the model summary table includes statistics provided in standard multiple regression for each step as well as Change Statistics that track the changes in R-square (essentially the value of the squared semi-partial correlation or sr^2 as well as changes in the statistical significance of the models.

Figure 6.20 Model Summary for Stepwise Linear Regression

					Change Statistics				
Model	R	R Square	Adjusted R Square	Std. Error of the Estimate	R Square Change	F Change	df1	df2	Sig. F Change
1	.562a	.316	.315	.09563	.316	257.605	1	557	<.001
2	.577b	.333	.331	.09450	.017	14.377	1	556	<.001
3	.585c	.342	.338	.09400	.008	6.948	1	555	.009

a. Predictors: (Constant), Height (cm)
b. Predictors: (Constant), Height (cm), Age in years
c. Predictors: (Constant), Height (cm), Age in years, Gender

A review of the statistics for each model discloses that the R^2 and Adjusted R^2 values changed with the addition of each new variable. The **R Square Change** column reports growth in the variance explained (in the dependent variable) with each new model. For example, in the first step (Model 1), the variable, height, entered the model resulting in an R^2 and R^2 Change of .316, which is the square of the correlation (R column) between height and weight ($.562^2 = .316$). The addition of the variable, height, to the model added a significant unique contribution ($p < .001$) as indicated by the p value in the **Sig. F Change** column.

In Model 2, the variable, age, was added, and that resulted in an increase of 0.017 in variance explained, noted in the **R Square Change** column for a total R^2 of 0.333 (.316 + .017). The addition of age is a significant, unique contribution ($p < .001$), as is the addition of the variable, gender ($p = .009$). The total variance accounted for is .342, which is the same variance reported for the standard regression analysis (Figure 6.12).

6.10.3 Results: Independent Variables' Contributions to the Model

As with the standard regression method, stepwise regression also reports estimates of the coefficients in each significant model. Accordingly, in the **Coefficients** table in Figure 6.21, there are three sets of coefficient estimates, one for each model (step) in the stepwise regression solution. The null and alternative hypotheses for the coefficient estimates are the same for each model in the analysis:

Null hypothesis: For each model, the regression coefficient of each independent variable is not significantly different from zero after accounting for all other independent variables.
$H_0: b_j = 0$

Alternative hypothesis: For each model, the regression coefficient of each independent variable is not significantly different from zero after accounting for all other independent variables.
$H_a: b_j \neq 0$

Although the coefficients table for statistical and standard regression contains the same statistics, there are some differences that reflect the nature of the regression method. One distinguishing feature of stepwise regression is that the values of the coefficients change with each new model to reflect the additional variables in new models and adjustments in unique contribution to the dependent variable

Figure 6.21 Estimates of Regression Coefficients

		Coefficients^a							
		Unstandardized Coefficients		Standardized Coefficients			Correlations		
Model		B	Std. Error	Beta	t	Sig.	Zero-order	Partial	Part
1	(Constant)	.660	.071		9.249	<.001			
	Height (cm)	.007	.000	.562	16.050	<.001	.562	.562	.562
2	(Constant)	.650	.071		9.213	<.001			
	Height (cm)	.006	.000	.509	13.618	<.001	.562	.500	.472
	Age in years	.008	.002	.142	3.792	<.001	.333	.159	.131
3	(Constant)	.567	.077		7.375	<.001			
	Height (cm)	.007	.001	.565	13.204	<.001	.562	.489	.455
	Age in years	.007	.002	.121	3.196	.001	.333	.134	.110
	Gender	-.024	.009	-.104	-2.636	.009	.156	-.111	-.091

a. Dependent Variable: Weight

due to the new additions. Therefore, as Figure 6.21 illustrates, the unstandardized b coefficient for height in Model 1 is .007, which is significant at $p < .001$. When the variable, age, enters **Model 2** with a b coefficient of .008 ($p < .001$), the b coefficient for height declines to .006 ($p < .001$). When gender enters **Model 3** at $b = -.024$ ($p = .009$), the coefficients already in the model once again adjust, with the b coefficient for height returning to the original value of .007 and age declining to .007. However, both variables continued to report significance levels of $p < .001$ and $p = .001$, respectively.

Notably, the third and final model reported represents the best linear combination of independent variables that significantly predict the dependent variable, weight. The regression coefficients, and indeed the other statistics in **Model 3**, are exactly the same as those in the coefficients table for the standard regression method displayed (Figure 6.12); they are the same model. These equivalent results are attributable to the inclusion of all the independent variables in the stepwise regression solution. If one or more variables had been excluded from the final model, it is likely that the coefficients would have changed in the final model.

6.11 Sample Report (Stepwise): The Case of the Weight of U.S. Adolescents

Height, age, and gender were the independent variables in a statistical, stepwise multiple regression analysis to identify the best linear combination of predictors for the body weight of students in Grade 6 through high school in the United States. The correlations of the variables, presented in Table 6.3, show that height is most highly correlated with weight ($r = 0.562$), followed by age ($r = 0.333$) and gender ($r = .156$). The significant correlational relationship between gender and weight ($p < .001$), in spite of the small correlational magnitude, may be attributable to the large sample size (N=559). Height is also significantly intercorrelated with the other two independent variables, age ($r = .377$) and gender ($r = .460$). Age and gender were not significantly correlated ($r = .006$, $p = .445$).

Table 6.3 Correlations of Gender, Age, and Height (N = 559)

	1	2	3	4
1. Weight	-			
2. Gender	.156	-		
3. Age in Years	.333	.006	-	
4. Height (cm)	.562	.460	.377	-

Note: All correlations except that between gender and age are statistically significant ($p<.001$)

Based on the results of the stepwise regression procedure, the prediction model contained all three of the independent variables and was accomplished in three steps (Table 6.4). Height entered the model first, resulting in an R^2 of .316 and an adjusted R^2 of .315. Age entered the model next, resulting in an additional 2% of variance in weight explained ($R^2 = 0.017$, $p < .001$). With gender entering the model last, a further 1% of the variance in weight was explained ($R^2 = .008$, $p < .001$). The model was statistically significant, $F(3, 555) = 96.031$, $p < .001$, and explained 34% of the variance in body weight ($R^2 = .342$, Adjusted $R^2 = .338$).

Raw and standardized regression coefficients of the independent variables, their correlations with body weight, and their squared semi-partial correlations are displayed in Table 6.4. Findings of the investigation produced a prediction equation in which the body weight of students in the United States in Grades 6 to high school is predicted by their height ($\beta = .546$, $sr^2 = .207$). Therefore, the taller students grow, the greater their body weight. To a lesser degree, student weight is also predicted by students' age ($\beta = .121$, $sr^2 = .012$) and their gender ($\beta = -.104$, $sr^2 = .008$). Therefore, as students age, their weights increase, and boys (coded 1) are likely to record more body weight than girls.

Table 6.4 Stepwise Regression Results of the Case of the Weight of U.S. Adolescents

Step	Model	R^2	Adj. R^2	B	Std. Error	Beta	Pearson r	sr^2
1	Constant			.660	.071			
	Height	.316	.315	.007	.000	.562	.562	.333
2	Constant			.650	.071			
	Height	.333	.331	.006	.000	.509	.562	.225
	Age in Years			.008	.002	.142	.333	.017
3	Constant			.567	.077			
	Height			.007	.001	.546	.562	.207
	Age in Years	.342	.338	.007	.002	.121	.333	.012
	Gender			-.024	.009	-.104	.156	.008

Note: The dependent variable is body weight.

6.12 SPSS Specifications for Hierarchical Regression

Based on preliminary analyses to determine whether the data met the assumptions required for multiple regression, a total of nine outlier cases were removed from the procedure (ID: 17, 52, 72, 80, 161, 217, 218, 234, 272). A total of 354 cases remained for analysis.

The central research questions are:

1. To what extent does the combination of independent variables predict employee salary?
2. To what extent do independent variables add to the prediction of employee salary over and above the independent variables already in the model?

Assume that related research literature and theory guided research decisions for these analyses. Based on those decisions, the following represents blocks of variables entered into the analysis and the sequence in which they entered.

- The demographic variables, age and gender, were entered in the first block. Researchers have consistently identified gender as a key indicator of salary levels.
- While controlling for the effects of age and gender, education level entered the model in the second block of variables. Education level has also been identified as significant predictor of current salary.
- With previously named variables as covariates in the model, beginning salary and months on the job entered the model as the third and final block of variables.

To complete the analysis using the hierarchical entry sequence described, open the data file, **Employee_Fin.sav**. Select **Analyze → Regression → Linear** from the main menu to open the **Linear Regression** main dialog window displayed in Figure 6.22a. From the variables list panel, select **Salary** (annual salary), and move it to the **Dependent** panel. Select **Age** (age) and **Gen** (gender), and move both to the **Independent(s)** panel. This is the first "block" of variables (**Block 1 of 1**) in the hierarchical analysis. Leave the **Method** drop-down menu below the Independent(s) panel at the **Enter** default setting. Like the standard method of regression, this indicates that whatever is in the independent variable(s) panel in Block 1 will be entered into the model simultaneously. Within blocks with two or more variables, a researcher may specify the stepwise method of entry, requesting that the software select the sequence of entry within the block of variables. For this demonstration, however, use simultaneous entry, and to confirm that specification, check that the **Method** drop-down menu remains at **Enter**.

In fact, the regression method will remain at **Enter** for every block of variables entered. Check to ensure that "**Block 1 of 1**" is visible below **Dependent**, signaling that this represents the first block of independent variables. Click the **Next** button above the **Independent(s)** panel to reset the window so that the **Independent(s)** panel is empty and ready for the next block of variables.

Once the **Independent(s)** panel is empty, check to ensure that **Salary** is still visible in the **Dependent** panel (Figure 6.22b), then select the independent variable, **Educ,** and move it to the **Independent(s)** panel below **Block 2 of 2**. This is the second block of variables in the regression analysis. Click the **Next** button to reset the window for the final block of independent variables.

For the third and final block of variables, highlight and move **Salbegin** and **Jobtime** to the **Independent(s)** panel (Figure 6.23a). Leave the method of analysis at **Enter** which specifies that both variables in Block 3 will be entered simultaneously. Confirm that that "**Block 3 of 3**" is visible below the **Dependent** panel.

After entering all the variables in the model, specify the statistical output for the analysis. To do that, click on the **Statistics** pushbutton to open the **Linear Regression: Statistics** window (Figure 6.23b).

Figure 6.22 The Linear Regression Dialog Window: Block 1 and 2 of the Hierarchical Approach

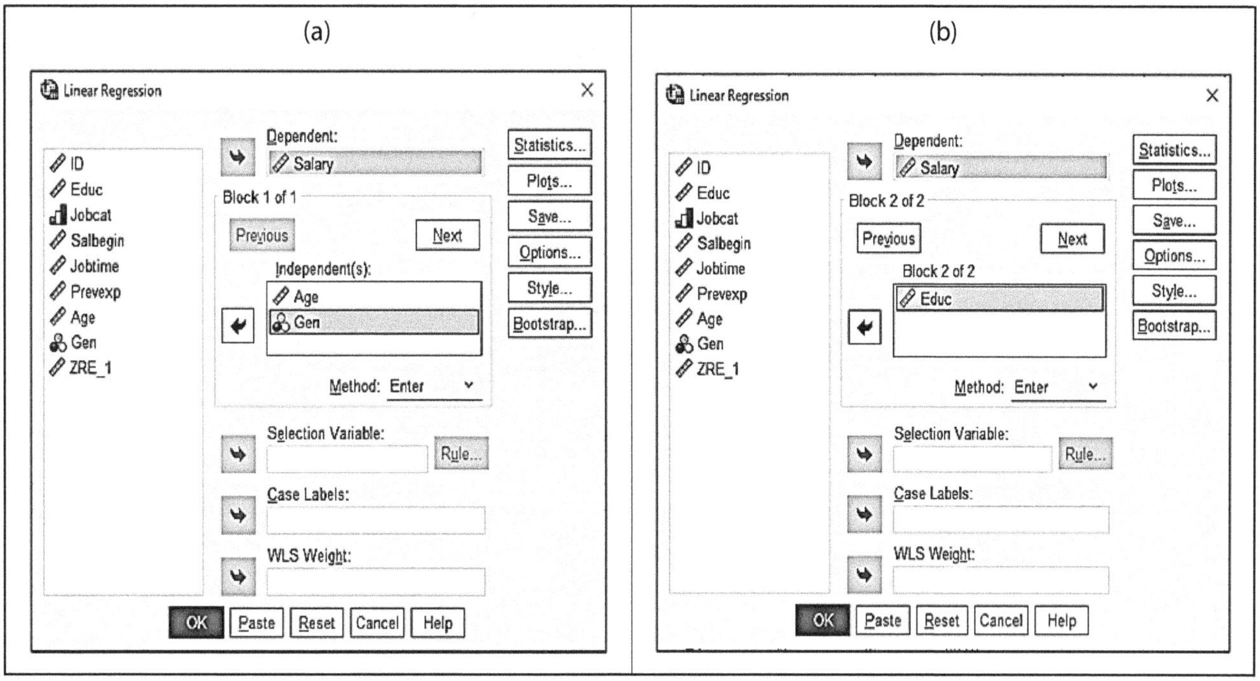

Retain the default selections and, in addition, select **R squared change** (to produce the changes in *R*-square as each block of variables enters the model). In addition, select **Descriptives** and **Part and partial correlations**. Click **Continue** to record the specifications and return to the main dialog window, then click **OK** to run the analyses.

Figure 6.23 The Linear Regression (Block 3 of 3) Window and Regression Statistics Windows

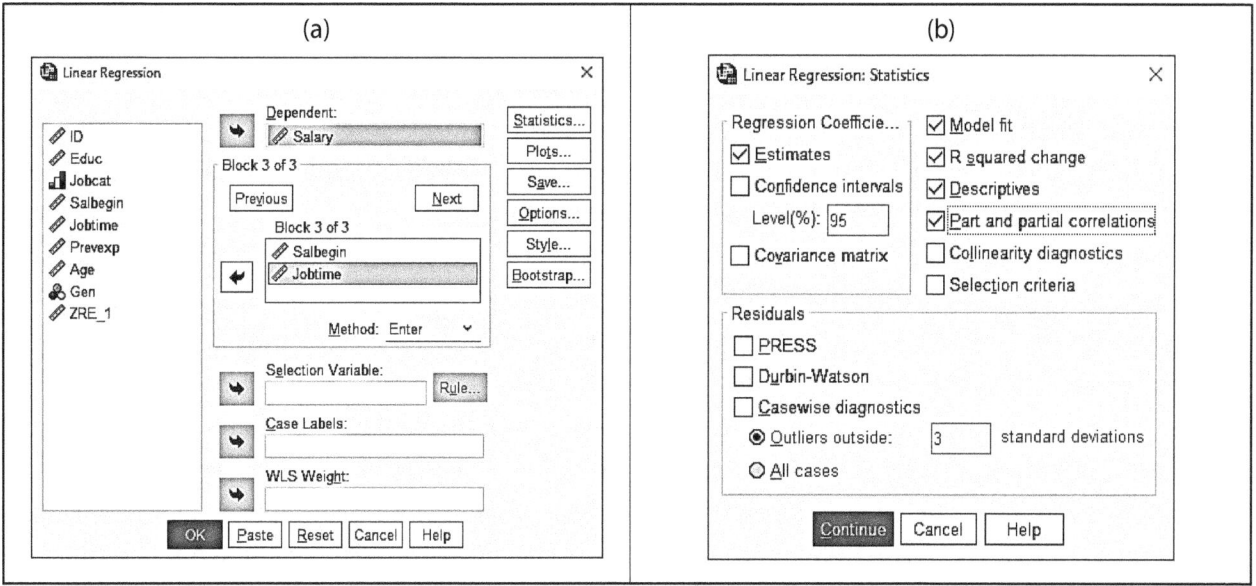

6.13 Results: The Case of Employee Salaries

6.13.1 Descriptive Statistics and Correlations

Figure 6.24 presents **Descriptive Statistics**. These results provide an opportunity to check results for data entry and data analysis specification errors as well as the reasonableness of initial results, indicating that the procedure was completed successfully.

Figure 6.24 Means and Standard Deviations of the Variables

Descriptive Statistics			
	Mean	Std. Deviation	N
Current Salary	$27165.99	$6135.467	353
Age	42.17	12.255	353
Gender	.58	.495	353
Years of Education	12.78	2.307	353
Beginning Salary	$13928.85	$2618.588	353
Months on the Job	80.95	10.118	353

Notably, all the variables in this exercise were scale level (ratio-level) variables, with the exception of gender—a nominal level variable. The values for males and females were 0 and 1, respectively. Accordingly, the mean value for gender is the percentage of subjects who were female (57.6%), which is somewhat higher than the percentage of male clerical employees. As the **N** column displays, a total of 353 cases were included in the analysis. One case was excluded due to a missing value for age. The average beginning salary is about half the average current salary. The average years of education is about one year beyond high school. The average age of the employees is 42 years old and on average employees had been on the job for about seven years (80.95/12).

Figure 6.25 presents correlations of the variables in the analysis. The dependent variable, (**Current Salary**) is in the first row and column of the table. Accordingly, the correlations of **Current Salary** with all independent variables are in the first row of the table and all of the independent variables are significantly correlated with **Current Salary** ($p < .001$ or $p = .01$).

As Figure 6.25 illustrates, the correlation of **Current Salary** with independent variables ranged from a low positive correlation with **Months on the Job** ($r = .135, p = .01$)) to a high moderate correlation with **Beginning Salary** ($r = .709, p < .001$). Therefore, as values for these independent variables increased, so did their salaries. Two independent variables, **Age** and **Gender**, were negatively correlated with **Current Salary**. This means older employees tended to receive lower salaries than younger employees, and the salaries of male employees (coded 0) tended to be higher than their female counterparts (coded 1).

Recall that intercorrelations of independent variables may signal multicollinearity among the variables. Independent variable correlations above .8 are an indication of a violation of the multicollinearity assumption of multiple regression. A review of the intercorrelations failed to find intercorrelations greater than $r = -.55$. Therefore, there is no sign of a violation of that assumption.

Figure 6.25 Correlations of Regression Study Variables

		Current Salary	Age	Gender	Years of Education	Beginning Salary	Months on the Job
Pearson Correlation	Current Salary	1.000	-.354	-.464	.516	.709	.135
	Age	-.354	1.000	.145	-.259	-.065	.036
	Gender	-.464	.145	1.000	-.319	-.550	-.058
	Years of Education	.516	-.259	-.319	1.000	.493	.056
	Beginning Salary	.709	-.065	-.550	.493	1.000	-.063
	Months on the Job	.135	.036	-.058	.056	-.063	1.000
Sig. (1-tailed)	Current Salary	.	<.001	<.001	<.001	<.001	.006
	Age	.000	.	.003	.000	.112	.251
	Gender	.000	.003	.	.000	.000	.137
	Years of Education	.000	.000	.000	.	.000	.147
	Beginning Salary	.000	.112	.000	.000	.	.117
	Months on the Job	.006	.251	.137	.147	.117	.
N	Current Salary	353	353	353	353	353	353
	Age	353	353	353	353	353	353
	Gender	353	353	353	353	353	353
	Years of Education	353	353	353	353	353	353
	Beginning Salary	353	353	353	353	353	353
	Months on the Job	353	353	353	353	353	353

6.13.2 Estimation of the Overall Model

In hierarchical regression, each step in the process represents a different model and, as in stepwise regression, ANOVA is used to test statistical significance of each model. Results of the test of significance for each step in the hierarchical method are presented in Figure 6.26.

The format of the ANOVA table in Figure 6.26 is similar to the stepwise regression results, with each model represented by each block of variables entered in each step of the analysis. The first block includes the variables, age and gender, and results show that this model is statistically significant ($p < .001$). Indeed, each of the models in the analysis provides statistically significant predictive power for current salary, culminating in the third and final model which includes all the independent variables.

Findings of the hierarchical regression analysis result in rejection of the null hypothesis that the slopes of the regression lines are not significantly different from zero when sets of independent variables are entered in the analysis sequentially. With evidence of a statistically significant prediction model, the next step is to assess how much variance individual models or blocks of variables explain; that information is presented in the model summary table (Figure 6.27).

Figure 6.26 Statistical Significance of the Hierarchical Model Predicting Current Salary

Model		Sum of Squares	df	Mean Square	F	Sig.
1	Regression	3970213872.088	2	1985106936.044	74.866	<.001[b]
	Residual	9280459280.887	350	26515597.945		
	Total	13250673152.975	352			
2	Regression	5423533535.182	3	1807844511.727	80.609	<.001[c]
	Residual	7827139617.792	349	22427334.148		
	Total	13250673152.975	352			
3	Regression	8526012203.445	5	1705202440.689	125.238	<.001[d]
	Residual	4724660949.529	347	13615737.607		
	Total	13250673152.975	352			

a. Dependent Variable: Current Salary
b. Predictors: (Constant), Gender, Age
c. Predictors: (Constant), Gender, Age, Years of Education
d. Predictors: (Constant), Gender, Age, Years of Education, Months on the Job, Beginning Salary

6.13.3 Estimation of Variance Accounted for in Dependent Variable

The model summary table in Figure 6.27 displays changes in variance explained in the dependent variable as each block of variables entered the model. Results report increases in R^2 with the addition of each block of independent variables. With the variables, age and gender in Model 1, the R^2 started out at a moderate, but statistically significant $R^2 = .30$ or 30% of the variance in current salary. The addition of years of education in Model 2 increased the variance explained to $R^2 = .429$— an increase of $R^2 = .11$ (or 11% of the variance) over Block 1. Variance explained continued to significantly increase with the addition of the final block of variables—months on the job and beginning salary. The inclusion of these two variables increased the amount of variance in current salary by an additional 23% of the variance.

In summary, the variance accounted for increased from $R^2 = .3$ in the first step to $R^2 = .643$ in the final step or 64% of the variance accounted for in current salary. The final adjusted R-square was .638 or 64% of the variance in salary, which means a rejection of the null hypothesis that $R^2 = 0$.

Figure 6.27 Model Summary Results of the Hierarchical Model Predicting Current Salary

Model Summary									
					Change Statistics				
Model	R	R Square	Adjusted R Square	Std. Error of the Estimate	R Square Change	F Change	df1	df2	Sig. F Change
1	.547[a]	.300	.296	5149.330	.300	74.866	2	350	<.001
2	.640[b]	.409	.404	4735.751	.110	64.801	1	349	<.001
3	.802[c]	.643	.638	3689.951	.234	113.930	2	347	<.001

a. Predictors: (Constant), Gender, Age
b. Predictors: (Constant), Gender, Age, Years of Education
c. Predictors: (Constant), Gender, Age, Years of Education, Months on the Job, Beginning Salary

6.13.4 Estimates of Independent Variable Contribution to the Model

The null and alternative hypotheses for the independent variables' contribution to each model are as follows:

Null hypothesis: For each model, the regression coefficient of each independent variable does not statistically significantly explain unique variance in salary after controlling for previously entered blocks of independent variables.
H_0: $b_j = 0$, where j equals any variable.
Alternative hypothesis: For each model, at least one regression coefficient statistically significantly explains unique variance in salary after controlling for previously entered blocks of independent variables.
H_a: $b_j \neq 0$

Figure 6.28 reports changes in the influence of the independent variables on the dependent variable as new blocks of variables entered the model. The importance of independent variables in the models and changes in importance throughout the process is indexed by unstandardized coefficients (b) and standardized coefficients (beta or β). For example, **Gender** started out as a significant predictor of salary in Model 1 (b = -5,233.114, β = -.422, $p < .001$). However, by Model 3, as more vital independent

Figure 6.28 Independent Variables' Contribution to the Model

		Coefficients[a]							
		Unstandardized Coefficients		Standardized Coefficients			Correlations		
Model		B	Std. Error	Beta	t	Sig.	Zero-order	Partial	Part
1	(Constant)	36371.987	1000.708		36.346	<.001			
	Age	-146.592	22.637	-.293	-6.476	<.001	-.354	-.327	-.290
	Gender	-5233.114	560.884	-.422	-9.330	<.001	-.464	-.446	-.417
2	(Constant)	21787.544	2032.105		10.722	<.001			
	Age	-107.529	21.377	-.215	-5.030	<.001	-.354	-.260	-.207
	Gender	-3954.681	539.729	-.319	-7.327	<.001	-.464	-.365	-.301
	Years of Education	954.136	118.527	.359	8.050	<.001	.516	.396	.331
3	(Constant)	389.281	2448.096		.159	.874			
	Age	-143.291	16.829	-.286	-8.514	<.001	-.354	-.416	-.273
	Gender	-376.275	483.821	-.030	-.778	.437	-.464	-.042	-.025
	Years of Education	297.221	102.204	.112	2.908	.004	.516	.154	.093
	Beginning Salary	1.476	.100	.630	14.741	<.001	.709	.621	.473
	Months on the Job	107.282	19.742	.177	5.434	<.001	.135	.280	.174

a. Dependent Variable: Current Salary

variables entered the model and **Gender** acted as a covariate, **Gender** lost its ability to significantly contribute to the prediction of the dependent variable ($b = -376.275$, $\beta = -.03$, $p = .437$).

Gender is the only independent variable that ceased making a significant unique contribution to the model as the process evolved. This occurred as a result of variance shared with other independent variables in the model. As the process evolved, the unique contributions of the other four independent variables changed as well, but as recorded in the **Sig** column, the variables remained significant contributors to the model throughout the analysis at $p < .001$ for **Age, Beginning Salary** and **Months on the Job** and $p = .004$ for **Years of Education**.

6.14 Sample Report (Hierarchical): The Case of Predicting Employee Salaries

A three-step hierarchical linear regression was performed to determine whether age, gender, years of education, beginning salary, and months on the job improve the prediction of current salary beyond that provided by an unimproved model. In the first block, age and gender simultaneously entered the model. In the second block, years of education entered. In the third and final block of predictors, beginning salary and months on the job simultaneously entered.

Assumptions were evaluated using SPSS regression and SPSS Explore. Of the original 363 cases in the study, nine were significant outliers, and one had missing data. All ten cases were removed from further analysis, leaving a total 353 cases for analysis.

Table 6.5 displays correlations of the variables. All predictor variables are significantly correlated with current salary. However, the variables most strongly correlated with current salary were beginning salary ($r = .709$) and years of education ($r = .516$). Months on the job recorded the lowest correlation with current salary ($r = .135$).

Table 6.5 Correlations of the Variables in the Case of Employee Salaries (N = 353)

	1	2	3	4	5	6
1. Current Salary	—					
2. Age	-.354**	—				
3. Gender	-.464**	.145**	—			
4. Years of Education	.516**	-.259**	-.319**	—		
5. Beginning Salary	.709**	-.065	-.550**	.493**	—	
6. Months on the Job	.135*	.036	-.058	.056	-.063	—

*$p < .05$. **$p < .01$

Results of the hierarchical regression analysis (Table 6.6) indicated that the variables, age and gender, which were added to the model in the first block were statistically significant predictors, $F(2, 350) = 74.866$, $p < .001$, $R^2 = .3$. Age and gender accounted for 30% of the variance in current salary. When education level entered the model, it remained statistically significant, $F(3, 349) = 80.609$, $p < .001$, $R^2 = .409$. Education level explained an additional .11 (11%) of the variance, increasing the predictive power of the model to account for 41% of the variance in current salary.

In the third and final block, beginning salary and months on the job entered the model, adding to the significant predictive power of the independent variables, $F(5, 347) = 125.238$, $p < .001$, $R^2 = .643$, Adj $R^2 = .638$. Beginning salary and months on the job explained an additional .234 (23%) of the variance in current salary. Overall, the model accounted for 64% of the variance in salary.

The best predictors providing the greatest unique contribution in the final model were beginning salary ($sr^2 = .473$), followed by age ($sr^2 = .273$ and months on the job ($sr^2 = .174$). Years of education was also a significant contributor to the model ($sr^2 = .093$). However, gender did not explain significant unique variance in salary ($b = -376.275$, $\beta = -.03$, $p = .437$).

Table 6.6 Hierarchical Regression Results Summary

Block	Model	R^2	Adj. R^2	B	Std. Error	Beta	Pearson r	sr^2
1	(Constant)			36371.987	1000.708			
	Age	.300	.296	-146.592	22.637	-.293	-.354	-.290
	Gender			-5233.114	560.884	-.422	-.464	-.417
2	(Constant)			21787.544	2032.105			
	Age	.409	.404	-107.529	21.377	-.215	-.354	-.207
	Gender			-3954.681	539.729	-.319	-.464	-.301
	Years of Education			954.136	118.527	.359	.516	.331
3	(Constant)			389.281	2448.096			
	Age			-143.291	16.829	-.286	-.354	-.273
	Gender	.643	.638	-376.275	483.821	-.030	-.464	-.025
	Years of Education			297.221	102.204	.112	.516	.093
	Beginning Salary			1.476	.100	.630	.709	.473
	Months on the Job			107.282	19.742	.177	.135	.174

Note: Dependent variable—Current Salary

6.19 Key Terms

Collinearity
Hierarchical regression
Multicollinearity
Multiple regression
Partial correlation
Predictor variable
Semi-partial correlation
Statistical regression
Stepwise regression
t-statistic
Zero-order correlation

7
One-Way Analysis of Variance

Key concepts from previous chapters or courses:
   degrees of freedom (df): the number of values left over after estimating one or more parameters. This number is needed to interpret a statistic (e.g., t score, F ratio, chi-square, H statistic).
   effect size: numerical value that is a measure of the strength of the outcome of a study. Hypothesis testing indicates whether there is a relation of difference larger than zero. Effect size helps you to quantify the difference. The larger the effect size, the greater the magnitude of associations between variables or the magnitude of differences between group means.
   t **distribution**: known as Student's t distribution, this is a probability sampling distribution of the t statistic used in hypothesis testing. It is especially important for interpreting statistics based on data from small samples of fewer than 120 cases.

THIS CHAPTER INTRODUCES

- differences between a t test and one-way ANOVA
- characteristics of a one-way ANOVA
- when to use a one-way ANOVA
- how to use variances in scores to find differences
- how to hand-calculate a one-way ANOVA
- how to calculate the effect size of a one-way ANOVA outcome
- how to use SPSS to conduct a one-way ANOVA.

7.1 Introduction

Attending secondary school is an important turning point for young people in many parts of the world. According to the World Bank (2005), secondary education has been shown to contribute not only to individual earnings and nations' economic growth, but also improvements in health, equity, and social conditions. Yet, despite the efforts of Latin American and Caribbean nations, gaps in access to quality secondary education persist throughout the region (United Nations International Children's Emergency Fund [UNICEF], 2021).

For students throughout the Caribbean, securing a secondary education is a high-stakes enterprise. Secondary school placement examinations, like the Primary Exit Profile [PEP] are the key to unlocking limited seats at the secondary school level. In response to this high-stakes testing environment, a cottage industry of private test preparation programs offers curricula to improve student test performance. Despite this, a common public perception is that performance on high school placement tests depends on the type of pre-secondary school that students attend and by implication, the curriculum employed by each school. One way of testing this hypothesis is by applying a research design that includes the use of the one-way analysis of variance test, commonly known as the one-way ANOVA. Researchers may use one-way ANOVA to investigate whether students from diverse types of pre-secondary schools perform significantly different on high school placement tests. In this chapter, we will explore how to use the one-way between-subjects ANOVA to investigate this research problem.

7.2 The Case of Differences in Achievement

For this case, imagine that a researcher wants to know whether there is a significant difference between written communication test scores of students from three types of schools. The small fictitious dataset in Table 7.1 will be used to illustrate how to tackle this question. The dataset represents communication test scores from three groups, each from a different type of school. For ease in computation, each group consists of five students. The average of each group appears different from the other two group means. The goal of the investigation is to determine whether the group means are more or less different than might have happened by random chance.

The outline of an associated research plan may be found in Box 7.1. The outline describes a plan of action for completing a study similar to the Case of Differences in Achievement.

Table 7.1 Fictitious Written Communication Test Scores of Three Groups of Students

Observations	Groups		
	T_a	T_b	T_c
1	8	4	10
2	6	5	9
3	5	3	11
4	9	3	7
5	7	6	8
Mean scores	7	4.2	9

7.3 One-Way, Between Subjects ANOVA Fundamentals

One-way ANOVA is designed to compute whether there are statistically significant differences between two or more unrelated group averages in a study sample. In most cases, however, researchers use one-way ANOVA in situations where there is a minimum of three groups.

One-way ANOVA is a parametric test in that it makes certain assumptions about the target population from which the sample was drawn. The researcher's data must meet the assumptions in order for the results of the test to be representative of the population.

In addition to the term one-way ANOVA, the test is also known as one-factor ANOVA because it works only in situations where there is a single independent variable. Another term for this statistical procedure is one-way, between-subjects ANOVA because it measures differences between groups. Therefore, if a researcher wants to determine whether there are significant differences in writing performance between students from three types of schools, ANOVA is the likely method of choice because there is a single independent variable (e.g., School Type), comprising three groups (School Type A, School Type B and School Type C), and the researcher wants to compare the average performance of all possible combinations of the groups:

School Type A	vs	School Type B
School Type B	vs	School Type C
School Type A	vs	School Type C

Finally, the term one-way, independent groups ANOVA signifies that the groups are unrelated to each other, and participants may only be in a single group. For example, the score of a participant in a study comparing performance at three different schools may only be used in the group representing the school the participant attends. Similarly, in a study comparing males and females, a participant is restricted to the group of males or females. The term one-way ANOVA throughout this textbook refers to one-way, between-subjects ANOVA.

If the term ANOVA sounds familiar, it should because it is also used to evaluate the statistical significance of linear regression models discussed in Chapter 5. ANOVA is another method of predicting or estimating the value of outcome variables. Researchers can use categorical variables, like gender or teaching method, to estimate a predictive relationship with a dependent variable in linear regression, and the same can be achieved using ANOVA.

7.3.1 *T*-Test and One-Way ANOVA

A *t* test is used to identify significant differences between two groups where there is a single independent variable with no more than two categories (e.g., male/female or experimental/control group). While conducting multiple *t*-tests is feasible, there are some disadvantages in choosing that option.

- Conducting multiple t-tests increases the likelihood of rejecting the null hypothesis when it should not be rejected—Type I error.
- Multiple *t* tests reduce efficiency. Researchers may protect against a Type I error by adjusting the alpha level used to determine when a difference in pairs of mean scores is statistically significant, but in doing that, they are likely to conduct far more statistical tests than necessary.

By contrast, one-way ANOVA may be employed to investigate the differences in multiple-group means in a more efficient manner than multiple *t*-tests, by performing an omnibus test followed by multiple comparison tests. The overall (omnibus) analysis identifies whether there are any pairs of group means with significantly different means. If the omnibus test proves to be significant, ANOVA invokes follow-up *t* tests (multiple comparison tests) in the same analysis that are protective of the alpha level, to detect which pairs of means are significantly different.

Because ANOVA is an extension of the *t* test, it is probably one of the most widely used statistical approaches to measure and evaluate differences in groups means. Unlike the *t* test, there is no restriction on the number of means that might be compared in ANOVA.

Box 7.1 Research Plan Outline: The Case of Differences in Achievement

Research Problem
Researchers stipulate that the general public believes that the different types of pre-secondary schools that students attend perform differentially in educating students. Hence, parents have strong preferences for certain pre-secondary schools over others. The research literature and the data suggest one source of evidence is disparity in pre-secondary student performance on the high school placement assessment.

Variables and Measurement Levels
Independent Variable: School Type (nominal level)
Levels of Independent Variable: School Type 1, School Type 2, School Type 3
Dependent Variable: Written Communication Test Score (interval/ratio level)

Research Question
Is there a significant difference in the mean communication task scores of students attending three types of schools?

Hypotheses
The null and alternative hypotheses are nondirectional; that is, they do not reveal the expected direction of any differences found.

Null Hypothesis: There is no significant difference in the communication scores of students in three types of schools.
$H_o: \mu_1 = \mu_2 = \mu_3$

Alternative Hypothesis: There is a significant difference in the communication scores of students in three types of schools.
$H_a: \mu_i \neq \mu_j$ for any pair of ij

Research Design
The hypotheses and research question disclose three intact categories of the independent variable, school type, from potentially three different populations. All students complete the same assessment represented by the dependent variable, communication test score (O). The dependent variable is used to compare the average difference in achievement among school types (Schtype2) which function as the treatment or experience, X. Therefore, the research design is:

$$\begin{array}{ccc} M & X_{T1} & O \\ M & X_{T2} & O \\ M & X_{T3} & O \end{array}$$

This type of design is called a causal comparative design because it seeks to explore reasons for apparent existing differences among three groups of study participants. Causal comparative research is also referred to as *ex post facto* or "after the fact" designs because the study first observes a difference between two or more groups and then looks back in time to determine possible conditions that might have resulted in the observed differences. Both the effect and the alleged cause have already occurred and are studied by the researcher in retrospect (Gay, 1996). All of the above do not involve random assignment, but rather matching on background variables (M). Accordingly, they are nonexperimental or quasi-experimental designs. They may also be used in experimental studies.

Data Analysis
Borg and Gall (1983) describe the first step in causal comparative research as computing descriptive statistics for each comparison group. The next step is to use one-way ANOVA to complete a test of statistical significance to investigate whether there are statistically significant differences in average scores among the three types of schools. Post-hoc multiple comparisons tests will be conducted if significant differences are found in the omnibus test.

7.3.2 Eligible Data Variables and Measurement Levels

One-way ANOVA is restricted to two variables—one independent and one dependent variable. The independent variable in ANOVA is a grouping variable (also known as a factor) that separates study subjects into mutually exclusive groups or levels of data that are measured at the nominal (e.g., gender, type of car or school) or ordinal level (e.g., age groups or educational levels). The independent variable for this case is a nominal level variable.

The dependent variable is the outcome variable and is measured at the interval or ratio level. For example, the dependent variable for the case in this chapter is test scores. In other situations, salaries, height, weight, and temperature might all be candidates for dependent variables.

7.3.3 Research Questions

The process of conducting quantitative empirical research is similar regardless of the field of study, the research design, or the types of statistical analyses involved. In each investigation, research questions and hypotheses are developed early in the process, typically after a research problem surfaces and is undertaken by a researcher

One-way ANOVA is typically used when researchers want to determine whether there is a **significant difference** between groups of their study subjects on a single dependent variable. For the most part, the research question is a re-organization of the stated research problem so that the independent and dependent variables, as well as the type of relationship that will be investigated, are clearly delineated.

For example, there is a single research question for the case in this chapter (Box 7.1):

Is there a significant difference in the mean written communication score of students attending three types of schools?

Some important points about the research question to note, include:
- The independent variable, School Type, includes three intact groups in that the researcher cannot manipulate these groups; they existed prior to the study, and students were not assigned solely for purposes of the study.
- The ANOVA research question refers to more than a difference in average scores. It refers to a significant difference in achievement. Examining the average scores of students in the three types of schools may reveal what appear to be differences in scores. However, these apparent differences may be the result of estimation error rather than true differences in the target population. A more detailed discussion on this topic may be found in Section 7.3.5 and 7.3.6.
- The research question is a nondirectional question that does not identify a hypothesized direction for the differences. A directional research question might have asked: Does School Type A and School Type B have significantly higher scores than School Type C?

The research question is the central question which is often followed by sub-questions. For example, in some studies, it might be sufficient to identify that the three school types had significantly different scores, but not identify the source of the differences. Typically, however, it is assumed that the intention is to identify the source of the significant differences if they exist, using sub-questions like:
- Is there a significant difference in the written communication scores of at least two types of schools?
- Is the mean score of School Type A significantly different from the mean score of School Type B and School Type C?

- Is the mean score of School Type B significantly different from the mean score of School Type A and School Type C?

The first sub-question is nondirectional and implies the need for additional data analyses (e.g., controlled, post-ANOVA, multiple comparison tests) after the central question has been answered. The last two sub-questions are directional and suggest a different type of secondary analysis called contrast. More on this topic in Section 7.8. It is important to point out that the latter two questions might be the basis of more limited research that is restricted solely to these directional concerns.

7.3.4 Research Design: When to Use One-Way ANOVA

Settling on the method of statistical analysis is a fundamental part of designing a research study. As was discussed in Chapter 2, there are other considerations that should precede the identification of a statistical analysis approach, one of which is the type of research being conducted. Although research design falls outside the scope of this book, it is important to point out that one-way ANOVA is applicable in a multitude of research design frameworks. One-way ANOVA may be used in true experimental, quasi-experimental, or causal-comparative studies, where more than two groups are compared on a specific dependent variable.

In a true experiment, students might be randomly assigned to different methods of teaching communication skills. Their achievement may then be compared using the same communication task test. However, due to ethical and practical concerns, a true experiment may not be viable in an educational setting.

One-way ANOVA may not be the ideal choice for a quasi-experiment. Nevertheless, for illustrative purposes, imagine that the communication task scores of three groups of students with similar backgrounds are exposed to three different communication skills teaching methods in their own school settings. The achievement of the groups is later compared. In this situation there is no random assignment; students' existing school situations (e.g., different classes, each using a different method) are used to divide them into groups for analysis.

In the Case of Differences in Achievement, with the independent variable identified as school type, the study fits into the nonexperimental, causal comparison research design category. The case is an ex-post facto retrospective study in that the intent is to examine characteristic groupings (school type) that already occurred to identify potential causes of the study outcome (student achievement). The study is cross-sectional in that study data are collected at a single point in time rather than over a period of time. For more information about research design and the role of statistical techniques, see Campbell and Stanley (1963); for the classic text on research design, consult Shadish et al. (2001); and for an updated discussion of the subject, consult Mertler (2016).

7.3.5 One-Way ANOVA Hypotheses

Chapter 2 discoursed the role of hypotheses in hypothesis testing when employing inferential statistics like one-way ANOVA. The null and alternative hypotheses reflect the boundaries of the research question and indicate the type of statistical analysis that is needed to test the hypotheses. With an objective of testing for significant differences in communication task scores (dependent variable) of students from three different types of schools (independent variable), it is evident that one-way ANOVA should be used to answer the research question. Symbolically, the null and alternative hypotheses might be framed so that they parallel research questions. In this case the hypotheses reflect the nondirectional central research question described in 7.3.3:

Null hypothesis: There is no significant difference in the mean communication scores of students attending the three types of schools.

$H_o: \mu_a = \mu_b = \mu_c$

where μ_a, μ_b, and μ_c represent the population means of three groups of mean scores.

Alternative hypothesis: There is a significant difference in the mean communication task scores of students attending at least two types of schools.

$H_a: \mu_i \neq \mu_j$

where μ_i and μ_j represent any pair of the three mean scores.

As discussed in Section 7.3.4, research questions that are directional in nature will have directional hypotheses as well. For example, instead of the alternative hypothesis noted, the following might be substituted:

$H_a: \mu_a > \mu_b > \mu_c$

In this case, the directional hypothesis would be: School Type A has significantly higher mean scores than School Type B, which has significantly higher scores than School Type C.

7.3.6 Variance and Group Differences

Conventional wisdom suggests that scores of students within each type of pre-high school are unlikely to be identical. If each student in School Type A had the same score of say, 75, the mean score would be the same as each individual's score. If, simultaneously, each student in School Type B had a score of 80, then School Type B would have a group mean of 80. Since there is no variability within each school, there is no measurement error. Therefore, school means would be the best estimates of the target population mean, and no statistical test would be necessary to conclude, for example, that the School Type B had a higher score than School Type A.

In reality, however, scores within a single group are unlikely to be exactly the same because measurement of anything always captures some variance within the same group. For example, as demonstrated in Figure 7.1, one group may have scores of 10, 9, 11, 7, and 8, while another may have scores of 8, 6, 5, 9, and 7.

Variance within a group is measured by the distance of each score from the mean group score. Accordingly, the mean score for School Type A is 9 (45/5). Two scores are one point away from the mean, while the others are farther away from the mean score.

Variance in scores within a group could stem from sources unrelated to the independent variable or experimental treatment, such as:
- unreliable attitude survey or math test
- environmental events (e.g., severe weather, or rivalry between groups), or
- psychological status of study participants.

Figure 7.1 One-Way ANOVA Design and Score Variance

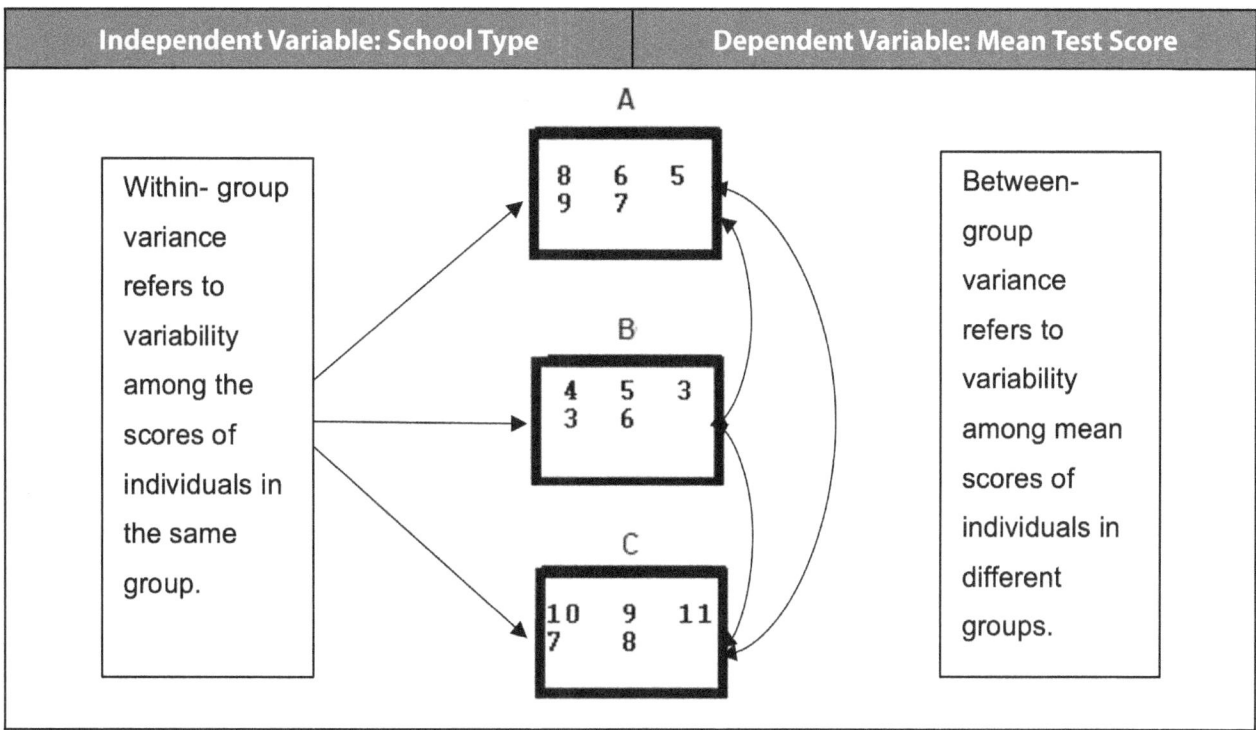

These sources of variability contribute to random error in estimating differences between group means. Steps may be taken when planning and designing research (See Section 2.2) to reduce them (Campbell & Stanley, 1963). Nevertheless, some within-group error is inevitable. If the error in a group is not taken into consideration, the accuracy of study outcomes will be affected. When other groups with within-group variance are added, it becomes increasingly difficult to estimate between-group differences.

As Figure 7.1 illustrates, there is variability in scores within each group, but there is also variability between group scores. This second type of variance is described as between-group or group mean variance. It is the difference between the mean score of each group and the mean score of the entire sample (the grand mean). It is between-group variance that researchers are most interested in and the variance that should primarily affect the dependent variable.

A third type of variance in scores is known as grand variance. This type of variance is a measure of the distance between the score of each individual in the study and the mean of the entire study sample.

Variance has a critical role in comparing group means. Consider two groups of test scores: the variance within groups which is measurement error and the variance between group means. If the variance between two estimated group means is the same or less than the variance within groups (measurement error in the estimate), then the means are not significantly different, and the samples are likely part of the same target population. If the variance between group means is greater than the variance within groups (measurement error), the groups may represent unique target populations. The greater the difference between within- and between-group variances, the greater the likelihood the groups are sufficiently significant to represent different populations. Analysis of variance uses variance in group scores to determine whether systematic differences exist between group means.

Table 7.2 Written Communication Test Scores for Three Groups of Students

Observations	Groups		
	School Type A	School Type B	School Type C
1	8	4	10
2	6	5	9
3	5	3	11
4	9	3	7
5	7	6	8
Sum (Σ)	35	21	45
Mean ($\bar{X}$)	7	4.2	9
Variance (S^2)	2.5	1.7	2.5
Std. Deviation	1.581139	1.30384	1.581139

7.4 Calculations for One-Way Analysis of Variance

Data from Table 7.2 will be used to illustrate how variance is used in one-way ANOVA to compute whether significant differences exist between group means. For ease of computation, the table presents three groups of five communication test scores (a woefully inadequate sample of scores). Although some scores occur in more than one group, assume that participants are restricted to a single group and that each participant's score is denoted by the symbol X_{ik}, where i represents an individual participant's score and k represents that individual's group.

To determine whether significant differences exist between groups, the first task is to compute the following statistics described in Section 7.3.6:

- the sum of the squared differences between each score in a group and the grand mean, also known as the total sum of squared variance (SST), also known as the grand variance
- the sum of the squared variance between group means and the grand mean (SSB), also known as the between-group mean variance or the sum of squared model variances (SSM)
- the mean squared deviations between group mean (MSB) (or the mean of the outcome of the SSB computation) to eliminate the bias introduced by the number of scores involved in the computation
- the sum of the difference between each score in a group and that group's mean score, also called sum of squared error (SSE), sum of squared residuals (SSR) or within-group variance (SSW)
- mean squared deviations from the mean, also known as mean squared error (MSE), mean squared residual (MSR) or mean-square within-group variance (MSW)

7.4.1 Total Sum of Squared Variance (SST)

The total sum of squared variation summarizes how much variance (deviation from the mean of the entire sample) the data represent or the maximum amount of variance that can be explained by the ANOVA procedure. The mathematical formula for the SST is:

$$SST = \sum_{i=1}^{N}(X_i - \bar{X}_{grand})^2$$

1. Calculate the grand mean ($\bar{X}_{grand}$).
2. Calculate the difference between each score (X_i) and the grand mean. Square each difference.
3. Find the sum of the squared differences for the three groups.

The grand mean is the mean of all scores in the entire sample recorded in Table 7.2, school types A, B, and C combined. Therefore, calculation of the grand mean is the first step in finding the SST. One method of finding the grand mean is to do the following:

- Find the total of all the scores regardless of the group each belongs to.
- Divide that number by the number of scores in the sample (15).

Using the sum of scores on the Sum line of Table 7.2, calculate the sum of the scores of the three groups. The study sample includes a total of 15 scores, one for each individual in the sample. Divide the sum of scores by 15 as follows:

$$Grand\ Mean = \frac{(35 + 21 + 45)}{15}$$

$$Grand\ Mean = 6.733$$

Table 7.3 Test Scores and Results of Squared Deviations from the Grand Mean (GM)

Observations	School Type A	(X - GM)²	School Type B	(X - GM)²	School Type C	(X - GM)²
1	8	1.604	4	7.471	10	10.671
2	6	0.538	5	3.004	9	5.138
3	5	3.004	3	13.938	11	18.204
4	9	5.138	3	13.938	7	0.071
5	7	0.071	6	0.538	8	1.604
Sum (Σ)	35	10.356	21	38.889	45	35.689
Mean ($\bar{X}$)	7		4.2		9	

With the grand mean computed, the next step is to calculate the SST. The culmination of Steps 2 and 3 are recorded in the $(X - GM)^2$ column in Table 7.3. The sum of the squared differences for each group appears in the Sum row. To find the sum of the squared differences for all groups:

$$SST = 10.356 + 38.889 + 35.689$$
$$SST = 84.933\ units\ of\ variance$$

7.4.2 Sum of Squared Between-Group Variance (SSB)

To calculate the SSB (amount of variance that can be explained by between-group variance), use the mean score for each group in Table 7.3 as well as the grand mean. Use the following mathematical formula to calculate the SSB:

$$SSB = \sum_{k=1}^{N} N_k (\bar{X}_k - \bar{X}_{grand})^2$$

258 One-Way Analysis of Variance

To apply the formula:
1. Calculate the difference between the mean score of each group ($\bar{X}_k$) and the grand mean ($\bar{X}_{grand}$).
2. Square each of the differences to avoid a negative result.
3. Multiply the result for each group by the number of scores in that group (N_k).
4. Total the results for the groups.

Using the means from Table 7.3, calculate the SSB:

$$\text{SSB} = 5(7-6.733)^2 + 5(4.2-6.733)^2 + 5(9-6.733)^2$$
$$\text{SSB} = 5(.267)^2 + 5(-2.533)^2 + 5(2.267)^2$$
$$\text{SSB} = 0.356 + 32.081 + 25.696$$
$$\text{SSB} = 58.133$$

The SSB value means that of the total amount of variance to be explained (84.933), 58.133 units can be explained by the variance between groups of scores.

7.4.3 The Mean Square Between-Groups Variance (MSB)

The mean squares between groups variance (MSB) eliminates the bias introduced by calculations based on differing numbers of scores. In effect, it is a more stable statistic than sums of squares and represents the average amount of systematic variance explained by the model.

To calculate MSB, take the sum of squared group variance of 58.133 and divide it by the degrees of freedom (df_b) for the three groups in the investigation, 3 – 1 = 2 degrees of freedom:

$$MSB = \frac{SSB}{df_b}$$

$$MSB = \frac{58.133}{2}$$

$$MSB = 29.066$$

7.4.4 Sum of Squared Errors (SSE)

The sum of squared errors statistic summarizes how much of the total variance cannot be explained by the model; this variance represents random error in the model caused by factors unrelated to performance on the Communication Task. The error is computed by totaling the difference between each score in each group and its group mean (Figure 7.3), in effect the sum of the variance from the mean in each group or the sum of the squared errors (SSE).

Here are three commonly used methods of calculating the sum of squared errors:

Method 1
- For each group, find the difference between each score and the group mean, and square the result (variance).
- Total all the variances.

Method 2
- Find the mean variance for each group of scores.
- Multiply each variance by one less than the number of people in the related group (n-1).
- Total the result of step 2.

Method 3
- Subtract sum of squared between-group variances (SSB) (Section 7.4.2) from total sums of squares (SST) (Section 7.4.1): SSE = SST − SSB.

To calculate the SSE using Method 1, use the following mathematical formula and the values in Table 7.4:

$$SSE = \sum N_k (X_{ik} - \bar{X}_k)^2$$

Summing the variances for the three groups (N_k) results in:

$$SSE = 10 + 6.8 + 10$$
$$SSE = 26.8$$

Table 7.4 Test Scores and Squared Deviations from the Group Means

Observations	School A	$(X_{ik}-\bar{X}_k)^2$	School B	$(X_{ik}-\bar{X}_k)^2$	School C	$(X_{ik}-\bar{X}_k)^2$
1	8	1	4	0.04	10	1
2	6	1	5	0.64	9	0
3	5	4	3	1.44	11	4
4	9	4	3	1.44	7	4
5	7	0	6	3.24	8	1
Sum (Σ)	35	10	21	6.8	45	10
Mean ($\bar{X}$)	7		4.2		9	

To confirm the calculated SSE of 26.8, compare with the outcome of Method 3:

$$SSE = SST - SSB$$
$$SSE = 84.933 - 58.133$$
$$SSE = 26.8$$

7.4.5 Mean Squared Error

Like mean square between-group variance, the mean squared error is a more stable statistic than sum of squared errors because it is uninfluenced by the group or sample size involved. It represents the average amount of random variance or error within groups caused by peripheral factors described earlier.

To calculate mean squared error, divide the sum of squared errors (*SSE*) by the degrees of freedom error (df_e), which is the difference between the total sample size (n = 15) and the number of score groups (N_k = 3) or 15 − 3 = 12. Accordingly,

$$MSE = \frac{SSE}{df_e}$$
$$MSE = \frac{26.8}{12}$$
$$MSE = 2.233$$

7.4.6 The *F* Ratio

The analysis of variance test statistic is known as the *F* statistic, named in honor of Sir Ronald Fisher, a 19th century British statistician who created the analysis of variance. The *F* statistic is the ratio of variance between groups (MSB) and variance associated with error (MSE); in other words, it is the ratio of the average variance between groups (systematic variance) and the average variance within groups (unsystematic or error variance).

The numerator of the *F* ratio is the mean square between-group variance that represents difference between the groups. The denominator of the *F* ratio is the mean squared residual that represents the variance within each group. If the value of the ratio is greater than 1, then the differences between the groups are beyond individual differences (unexplained variance) within each group. If the ratio of between and error variance is less than or close to 1, individual differences within the groups outstrip those between groups, indicating any resulting differences may have occurred by chance. The greater the *F* value, the greater the differences between groups are more than might happen by chance.

To calculate the *F* ratio, divide the mean square between-group variance by the mean squared error:

$$F \text{ ratio} = \frac{MSB}{MSE}$$

$$F \text{ ratio} = \frac{29.066}{2.233}$$

$$F \text{ ratio} = 13.015$$

The *F* statistic is the quotient of the *F* ratio, which is 13.015 and well above 1. The size of the *F* statistic indicates that there are more differences between the groups than errors in measurement or within-group differences.

7.4.7 Evaluating the F Statistic

To determine whether differences between groups occurred by chance, the *F* statistic is compared with *F* values in the *F* distribution. The *F* distribution is a distribution of all possible scores (*F* values) used to determine whether the calculated *F* statistic is large enough to represent significant differences between groups rather than a chance occurrence.

To evaluate the *F*-statistic, use the same degrees of freedom used to calculate the mean squared between-group variance (df_b) and mean squared error (df_e) to find the maximum possible *F* value in the *F* distribution. These *F* values are found in the *F* distribution tables in Appendix A.

If the *F* statistic exceeds the critical value (CV) in the distribution table, the *F* statistic reflects a significant difference in the groups being compared. The smaller the degrees of freedom, the larger the value of *F* must be in order to be significant. For instance, if $df_b = 4$ and $df_e = 10$, then an *F* statistic of 3.48 would be needed to be significant at the .05 alpha level. If the df_b were 9 and the df_e were 120, then an *F* statistic of 1.96 would be significant at the .05 level.

In the Case of Differences in Achievement, the degrees of freedom associated with the *F* statistic is 2 (Section 7.4.3) for the numerator value (or the MSB) and 12 (Section 7.4.5) for the denominator value (MSE). To find the critical value, use the *F* distribution table in Appendix A to locate the *F* value associated with the degrees of freedom of 2 and 12. The critical values is 3.89 at an alpha level of .05, 6.93

at an alpha level of .01, and 12.97 at an alpha level of .001. Since the F statistic is 13.0149, it exceeds the critical values at the alpha levels of .05, .01 and .001. This implies that the difference between the groups is significant at the alpha level of .001.

A summary table of the results appear in Figure 7.2. In the first column, the summary table reports the sources of the variance—the between-group variance (model variance), the residual (error variance) and the total variance. Column 2 records the sums of squares that were calculated in Sections 7.4.1, 7.4.2, and 7.4.4. Column 3 records the degrees of freedom from Sections 7.4.3 and 7.4.5. Column 4 reports the mean square statistics that were calculated in Sections 7.4.3 and 7.4.5. The F statistic in Column 5 was calculated in Section 7.4.6. The critical value and the p value follow in Columns 6 and 7. Conducting this statistical analysis using SPSS or another statistical software, such as SAS, STATA or R, produces summary results like those in Figure 7.2.

Figure 7.2 Summary of One-Way Analysis of variance of the Fictitious Communication Test

Source	SS	df	MS	F-Statistic	CV	P
Model	58.133	2	29.067	13.015	6.93	.001
Residual	26.800	12	2.233			
Total	84.933	14				

7.5 Assumptions of One-Way ANOVA

The generalizability and interpretability of statistical outcomes depend on whether the data meet certain statistical assumptions. When the data meet statistical assumptions, researchers are better able to interpret associated statistics without concerns about their accuracy.

For ANOVA, there are three central statistical assumptions that must be met to ensure that statistical estimates accurately reflect the target population:
- independence of observations
- appropriate dependent variable measurement level
- normality
- homogeneity of variance

7.5.1 Assumption of Independence of Observations

In practice, the independence of observations assumption means that each research study participant is independent of other participants in the study sample. Consider the following:
- A researcher is comparing the performance of boys and girls on a spelling test, and Blossom and Carl shared answers to the test questions. Sharing answers can contaminate the data, resulting in measurement error in estimating student performance. The error in estimating Blossom's performance is influenced by the error in estimating Carl's performance due to their communication about the research task.
- Imagine that three intact groups of students are used in a research study of the effect of three different reading teaching methods. However, some of the students are in more than one group. In this case, there may be dependence of observations because some students are in multiple groups. It

is possible that results of the study may be based on some students' exposure to multiple teaching methods rather than resulting from the superiority of a single teaching method. Non-overlapping research study groups reduce the likelihood of dependence of observations.

Violation of the assumption of independence of observations has serious consequences, and chief among them is the increase in the Type I error rate. Fortunately, independence of errors/observations is one assumption that firmly resides within the control of the researcher. Randomly selecting and assigning research participants to comparison groups is one way of avoiding violation of the assumption of independence. Including scrupulous research methods to reduce the likelihood of forbidden communication or cooperation in an investigation is another method of ensuring that comparison groups are independent of each other.

7.5.2 Dependent Variable Measurement Level Assumption

For ANOVA results that are meaningful, the dependent variable must be measured at or higher than the interval level. A nominal or ordinal-level variable leads to means that are difficult to interpret. However, independent variables for ANOVA designs may be nominal or ordinal level variables.

7.5.3 Assumption of Normality

For ANOVA, this assumption refers to the population of scores from which study samples are drawn. For grouped data, it refers to the normal distribution of mean scores across samples of the population. However, the central limit theorem states that with sufficiently large sample sizes, normal distribution of the means of samples of the population is expected. In addition, if there are at least 20 degrees of freedom in univariate ANOVA, the *F* test is robust to violations of normality of data, provided there are no outliers (Tabachnick & Fidell, 2007). When population samples are large, outliers are of more pressing concern than normality.

Since we cannot measure the sampling distribution of the mean directly, if there is suspicion of non-normality in small samples, the normal distribution of the scores of the dependent variable for each group may be assessed as an indication of the underlying distribution. Normal distribution may be assessed by statistical or graphical means; for example, skewness and kurtosis of variables may be assessed. When the values of skewness or kurtosis are zero, the distribution is normal. Other methods of assessing normality include frequency histograms and normal probability plots, discussed in Chapter 3. If the data are not normally distributed, it may be necessary to statistically transform the data for a more normal distribution. Chapter 3 provides more details about data transformation.

7.5.4 Assumption of Homogeneity of Variance

This assumption is related to the variance in each of the groups of a population of scores. When significance testing is involved, as it is with one-way ANOVA, it is assumed that the variance of the outcome variable is the same for all groups (1 through k) of the independent variable as noted in the following null hypothesis. The alternative hypothesis is that variances of the outcome variable are not equal across groups of the independent variable.

Null hypothesis: There is no significant difference in the variance of the communication scores of students in three types of schools.

$H_0: \sigma_1^2 = \sigma_2^2 = \ldots = \sigma_k^2$.

Alternative Hypothesis: There is a significant difference in the variance of the communication scores of students in three types of schools.

$H_a: \sigma_i^2 \neq \sigma_k^2$ for at least two groups of the independent variable

Imagine that a researcher is assessing the homogeneity of variance for the three groups in Table 7.1. In order to accomplish that, the researcher compares the absolute difference between each score in a group with the group mean score. Subsequently, differences across groups are compared to determine whether they are significantly different.

A less labor-intensive method is to use Levene's test to test the assumption of equal score variances across groups of the independent variable. Levene's test produces results that are *p*-values that users can compare to a predetermined alpha level, which is typically $\alpha = 0.05$. The null hypothesis of equal variance is rejected, and the alternative hypothesis must be retained if the Levene's test *p* value is less than the a priori alpha level. Rejection of the null hypothesis means that the assumption of equal variances across groups is violated, and the researcher must conclude that group variances are significantly different from each other. If the *p* value is greater than the alpha level, however, the null hypothesis of equal variances may stand, indicating that the assumption of equal variances across groups has not been violated.

There are alternatives to rectify heteroscedasticity, which include adjustments of the *F*-ratio using the Brown-Forsythe *F* ratio (Brown & Forsythe, 1974) and Welch's *F* ratio (Welch, 1951). Both techniques control the Type I error rate well. However, Welch's test does a better job of detecting an effect when there is one.

7.6 When Group Sizes are Unequal

Independent variables with unequal group sizes are common in research and may be the result of study participant dropouts or planned differences in group sizes. For example, a teacher conducting a study of reading performance may have 30 girls and 37 boys in the study to avoid unethical actions like excluding some students. In other cases, participants may drop out due to illness or loss of interest. Comparison groups of equal sample sizes are known as balanced designs. However, designs with unequal sample sizes are unbalanced designs and may lead to:

- unequal variances across groups, which may result in the violation of the homogeneity of variance assumption in tests like ANOVA. Having both unequal sample sizes and unequal variances can dramatically affect statistical power and Type I error rates (Rusticus & Lovato, 2014). Conversely, studies with equal group size are less sensitive to small departures from equal variance.
- loss of statistical power. The statistical power of tests is best when group sizes are equal.
- difficulties with confounding variables, which are variables that influence independent and dependent variables in a way that has an effect on the dependent variable.

One-way between-subjects ANOVA is robust to problems created by unequal group sizes, especially if conducted using statistical software (Tabachnick & Fidell, 2013). As group sizes become more discrepant, the *F* test can become too liberal, leading to Type I error if the smaller group has the larger variance. More challenges may develop in factorial ANOVA, which will be discussed in Chapter 8.

For one-way ANOVA, one solution to unequal group sizes is to randomly exclude cases from the group with the larger sample size. Otherwise, consider the use of linear regression if the associated research question and hypotheses are not affected by the adjustment. A third approach is to weight each group according to the number of observations when computing the between-group sum of squares. For example, suppose School Type A in the Communication Task example has four rather than five scores as shown in Table 7.1; that is, the score of 9 for School A is missing. The exclusion of one score affects the computation of the grand mean and all subsequent statistical computations which might affect the final outcome.

7.7 Calculating Effect Size

The results of the *F* test in Section 7.4.6 and 7.4.7, suggest that there is a 99% likelihood that a significant difference lies between at least one pair of the school group averages that exist in the target population. But it does not reveal anything about the substance or importance of the difference found. Is the effect of the independent variable on the dependent variable small, medium, or large? In the case of our worked example, is the effect of school type on communication task performance small, medium, or large?

The estimated strength of an effect detected in a research study is called an effect size. Effect sizes are measures of the magnitude or meaningfulness of a relationship between two variables. They are often interpreted as indicating the practical significance of a research finding (VandenBos & American Psychological Association, 2007). Generally, the larger the effect size index, the more important the effect found, and the greater the statistical power to detect true effects in the population.

There are several methods of quantifying the size of a particular effect computed by ANOVA. One of the more widely used indices of effect size in ANOVA is called eta square, symbolized as η^2 (Cohen, 1973). Eta is also known as the correlation ratio (Pearson, 1911). Eta is a measure of the amount of variance in a dependent variable that can be explained by one or more independent variables. In multiple regression, the effect size is called squared multiple correlation, or R^2 (Pearson, 1911; Cohen, 2003). Both effects sizes are interpreted in the same way.

Because eta-squared functions like a correlation coefficient, it takes on values between 0 and 1. The closer the value is to 1, the greater the variance explained by the independent variable.

Eta squared may be calculated by dividing the sum of squares for the group by the total sum of squares—SSB/SST. The subsequent proportion may be converted to a percentage to express the amount of variance accounted for by the independent variable effect. Eta squared may also be converted to a correlation coefficient (*r*) by finding the square root of SSB/SST. Jacob Cohen's (1988) guidelines for interpreting correlation coefficients apply in this situation:

Effect Size	Magnitude of Effect
r = .1 – .29	Small
r = .3 - .49	Medium
r = .5 or more	Large

As an illustration, consider the between-group by total sums of squares from the Case of Differences in Achievement reported in Figure 7.2:

$$\eta^2 = \frac{SSB}{SST}$$

$$\eta^2 = \frac{58.133}{84.933}$$

$$\eta^2 = .637$$

Recall that η^2 is also the squared multiple correlation, R^2. Therefore, to convert to the correlation coefficient (r), calculate the square root of .6374.

$$r = \sqrt{.637}$$

$$r = .798$$

The calculated eta square statistic shows that the independent variable, school type, accounts for 64% of the variance of the dependent variable, communication achievement score. The correlation coefficient (r) of .798 and the interpretation guidelines of this statistic suggests that the impact of school type on communication achievement score represents a large effect.

Some critics of eta-squared assert that this statistic tends to be upwardly biased when applied to the intended target population. It is of especial concern when sample sizes are small, as is the sample size for this case study. For a more unbiased estimate of variance accounted for, methodological researchers (e.g., Kroes & Finley, 2023) suggest that omega (ω^2) squared is a better choice. Omega squared is a more unbiased measure of the strength of association in analysis of variance. Using the results in Figure 7.2, it is calculated as follows:

$$\omega^2 = \frac{SSB - df_b(MSE)}{SST + MSE}$$

$$\omega^2 = \frac{58.133 - 2(2.233)}{84.933 + 2.233}$$

$$\omega^2 = \frac{53.667}{87.167}$$

$$\omega^2 = .616$$

Omega squared leads to a slightly more conservative estimate of the amount of variance school type accounts for—a total of 62% of the variance in the communication task score. Nevertheless, the reduced estimate still represents a large impact of the fictitious scores.

7.8 When Group Means Are Significantly Different

One-way ANOVA is an omnibus test because the F test assesses whether the overall model is significant. It tests the likelihood that there are significant differences in the means of any of the pairs of independent variable groups in the population that the study sample represents.

If the omnibus ANOVA test results in an F ratio that is not statistically significant, that signifies that none of the pairs of group means is statistically significantly different. With a nonsignificant outcome, the researcher may halt further analyses because the omnibus ANOVA test has provided all the information necessary to draw conclusions about the research hypothesis.

When an F-ratio is statistically significant, that indicates the likelihood that significant differences exist between the means of two or more of the independent variable groups in the population that the research sample represents. If only two groups are being compared, the source of the difference is undisputable—the two comparison groups.

If, however, the independent variable includes more than two groups of participants, a statistically significant omnibus ANOVA test tells us there is a significant difference between the means of two groups at a minimum, but it does not disclose which pairs of groups have significantly different means. To discover where the differences lie, it is necessary to carry out additional post-ANOVA statistical procedures commonly described as post-hoc (Latin for "after this") multiple comparison tests. Post-hoc tests are statistical analyses that are specified at the conclusion of an omnibus test. They are designed to identify which group means are significantly different while controlling the Type I error rate—a central objective of the ANOVA technique.

Besides post-hoc tests, researchers may focus on specific hypotheses that exclude the omnibus ANOVA test or go beyond the omnibus test to test a limited set of hypotheses. These tests are described as a priori planned comparisons or contrasts because the associated hypotheses were created before data collection began. For instance, researchers may wish to compare the means of sets of independent variable groups; for example, they may wish to investigate whether the pooled average of four medical treatment groups differs significantly from the average of the control (no treatment) group in an experiment.

If the independent variable is a continuous variable (e.g., traffic noise level in your neighborhood, hours of sleep each night or anxiety levels before exams) that has been categorized (e.g., low, medium, high, and extremely high traffic noise), researchers may examine whether there is a trend in the outcome at levels of the independent variable. These comparisons are called polynomial contrasts or trend analyses.

7.8.1 Post-Hoc Multiple Comparison Tests

Post-hoc procedures compare every possible pair of groups (multiple t-tests), while using a stricter alpha criterion so that the experiment-wise (also called family-wise) error rate does not exceed .05 (or another preselected alpha level). The experiment-wise error rate is the probability of making a Type I error for the set of all possible comparisons. Post-hoc procedures are used following the rejection of the omnibus ANOVA test and when researchers do not have specific a priori hypotheses. To maintain the experiment-wise alpha rate in these situations, divide the a priori alpha level (e.g., .05) by the number of paired group comparisons planned.

$$\text{Alpha level/number of comparisons}$$

If one assumes the chosen alpha level is .05 and the number of comparisons is 15, then .05/15 = .0033. Accordingly, in this instance, post hoc tests of 15 comparisons require an alpha level criterion of .0033 for each comparison pair.

Over the years, statisticians have developed a large number of statistical techniques for comparing group means after the omnibus ANOVA test has been completed. If researchers were to compute post-hoc tests manually, it could be tedious to compute critical values for all possible group comparisons.

However, with the development and popularity of statistical software packages, computing post-hoc procedures are a simple matter regardless of the method selected.

SPSS makes about 18 different methods available for post-hoc tests. Perhaps the best known of the post-hoc techniques are the Tukey method of multiple comparisons developed by Tukey (Braun, 1994), an American mathematician and statistician, and Bonferroni (1936) correction procedures for groups of approximately equal size and equal variance. However, while these tests control the experiment-wise error rate very well, they are conservative tests that lack statistical power when group sizes are unequal, and using these tests are likely to result in outcomes that show differences to be nonsignificant when they are significant.

Tabachnick and Fidell (2007) describe the Scheffé test as one of the most conservative post-hoc tests, but there are no limits to the number and complexity of comparisons that it can handle. Therefore, the Scheffé method is a good candidate for groups of unequal sizes.

Common recommendations for selecting post-hoc tests include the following:

- With equal sample sizes and similar group variances, Ryan, Einot, Gabriel, and Welsch Q (R-E-G-W-Q) or the Tukey procedure has good statistical power and control over the Type I error rate.
- Bonferroni has guaranteed control over the Type I error rate.
- If sample sizes are different, use Scheffé or Gabriel's procedure.
- If variances across groups are expected to be significantly different, use the Games-Howell procedure.

7.8.2 A Priori Planned Comparisons

In a priori planned comparison tests (or planned contrasts), the variance in the entire model is broken down into component parts. Researchers use a priori comparisons when they wish to focus on a few comparisons based on the research literature rather than every possible comparison. When planned comparisons are part of the research design, the choice of which comparisons to explore is specified before data collection and incorporated into the study design, based on an extensive review of the research literature.

Planned comparisons may be orthogonal or nonorthogonal. Orthogonal contrasts are more complex than nonorthogonal contrasts. Therefore, this section will tackle orthogonal, planned contrasts first.

Orthogonal planned contrasts are independent comparisons of two sets of the independent variable groups at a time, where each set may include one or more groups of the independent variable. As a result of conducting independent orthogonal contrasts by twos, there is no need to worry about inflated Type I error. Alpha levels may be the same as those used in omnibus ANOVA tests. For example, imagine a researcher is conducting a study of the effect of three different types of training on young, unmarried fathers' effective parenting skills. Each of three groups is exposed to a different training. The fourth group is the control group receiving no training. After the training, fathers complete an inventory on effective parenting. The researcher focuses the investigation on three planned comparisons to identify any significant differences between:

- the mean of three treatment groups of fathers and the mean of the control group (Contrast 1),
- the mean of treatment Groups 1 and 2 versus treatment Group 3 (Contrast 2),
- the mean of treatment Group 1 and treatment Group 2 (Contrast 3).

Notice that with this research design, the researcher accomplishes two things:

- Contrasts exclude comparisons between individual experimental groups and the control group.
- No individual group is singled out more than once. If a group is singled out in one comparison (e.g., the control group), it is not be used again by itself.

In orthogonal contrasts, there are as many comparisons as there are degrees of freedom for the procedure, which is calculated as k – 1, where k = the number of groups or levels in the independent variable. Therefore, if there are a total of four groups in a research study, there are 4 – 1 or a total of 3 orthogonal comparisons possible. The maximum number of comparisons is one less than the number of groups in the independent variable in the study.

Figure 7.3 visually illustrates the orthogonal comparisons design. With four independent variable groups in the design (Type 1 training, Type 2 training, Type 3 training, and the control group), there are three degrees of freedom and three orthogonal comparisons or contrasts as illustrated.

Figure 7.3 Example of Planned Orthogonal Comparisons

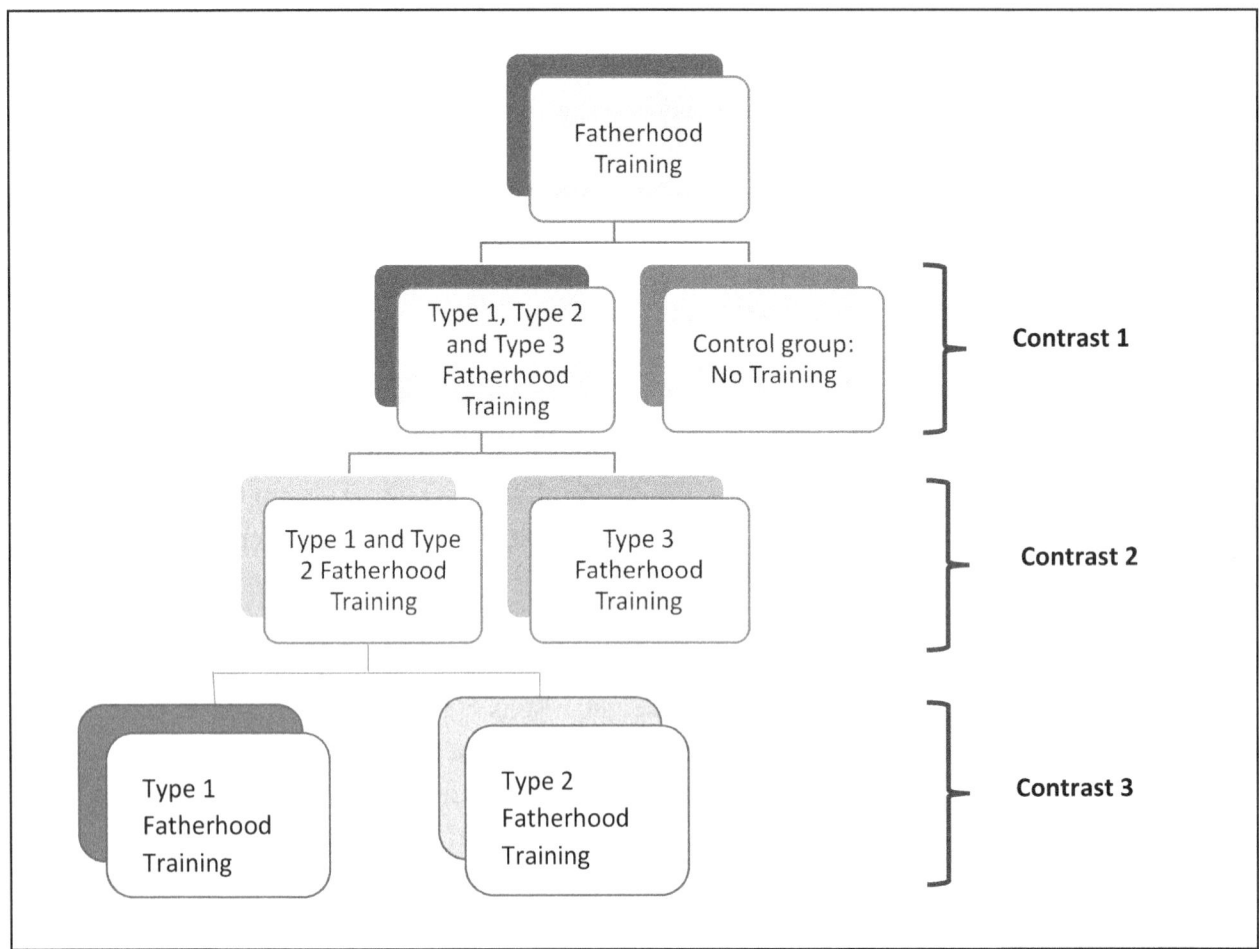

Typically planned contrasts of one-way ANOVA are carried out when a minimum of three groups are involved. Planned contrasts are conducted using statistical software, and researchers must tell the software which groups to compare and when. Those instructions take place by assigning weights to the means of at least two sets of groups which will be compared. Weights of zero are assigned to group means that are left out of the comparison. Group means that are compared (at least two) are assigned weights with opposite signs (plus or minus) so that the weights sum to zero when the study outcome fails to reject the null hypothesis. Imagine you wish to test the null hypothesis that the difference between the means of Group 1 and Group 3 is zero (no significant difference in the training effect). You might assign the weights of 1, 0, -1, 0 to the group means $\bar{X}_1$, $\bar{X}_2$, $\bar{X}_3$, and $\bar{X}_4$, respectively resulting in $1\bar{X}_1$, $0\bar{X}_2$, $-1\bar{X}_3$, $0\bar{X}_4$. Notice two important points about assigning weights:

- The two means (Group 2 and Group 4) that are excluded from the analysis have been assigned zero weights ($0\bar{X}_2$ and $0\bar{X}_4$). Excluded group means are always assigned zero weights.
- All weights add up to zero (1 + 0 -1 + 0 = 0).

Here are some additional guidelines to carry out planned orthogonal contrasts:

1. When designing a research study in which there are specific hypotheses to be tested, use planned contrasts.
2. To avoid uncertainty about where the differences lie, each contrast must be limited to two sets of independent variable levels. Each set may include one or more groups.
3. If a control group is in the research design, compare it with the other groups first to satisfy the prediction that it is different from the treatment groups.
4. After that repeat the process. If there are any sets in a previous comparison that contained more than one group that has not been broken down into smaller sets, create a new contrast that breaks down that set into smaller chunks.
5. Once a group or level of the independent variable has been singled out, it cannot be used in another comparison.
6. The orthogonal contrasts most often used in SPSS are the Helmert contrast and the difference contrast.

In the case of nonorthogonal contrasts, comparisons are not independent of each other. They are related in some way. One distinctive characteristic is that groups may be re-used in a series of comparisons. In the example of the Fatherhood investigation shown in Figure 7.3, when a nonorthogonal method is used, the control group (which did not receive training and only occurs in one contrast in an orthogonal comparison) may be re-used in multiple comparisons. The control group may be compared to the Type 1 or any other training group. There is nothing wrong with this approach. However, with nonorthogonal contrasts, analyses are related, and therefore probability levels for rejecting the null hypothesis should be more stringent. Nonorthogonal contrasts available in SPSS include the deviation, simple, and repeated contrasts.

7.8.3 Trend Analysis

Trend analysis is a special case of orthogonal contrasts. When the independent variable in an investigation is a categorized quantitative variable, the researcher might want to go beyond examining the differences between groups to examine discernible patterns or trends in dependent variable results. Polynomial contrasts are used to test for trends in the outcome data at different levels of the independent variable (e.g., age levels or anxiety levels). At the most fundamental level, this contrast tests for patterns that differ from a linear (or straight line) trend. In a linear trend, as the levels of the independent variable increase, the dependent variable increases concomitantly (positive linear). Of course, the dependent variable values may decline proportionately as the independent levels increase, resulting in a negative, but still linear trend. As contrasts diverge from a linear trend, they may reveal polynomial trends.

For example, the World Bank investigates average life expectancy in countries and geographical regions around the world. An examination of the life expectancy trend in Latin America and the Caribbean for 2020 exposed a slightly nonlinear, quadratic trend across country income groups (World Bank, 2020). Countries are categorized as belonging to one of the following four income groups: low income, lower middle income, upper middle income or high income. Life expectancy trends disclosed an increase in average life expectancy as income increased. The lowest life expectancy rate occurred at the lower middle-income level (no countries were categorized as low income). The highest life expectancy occurred in countries with the highest income. However, the life expectancy trend begins to increase at a slightly lower rate between upper middle and high, compared with lower middle and upper middle income.

Figure 7.4 Latin America and Caribbean Region Life Expectancy at Birth by Income Group

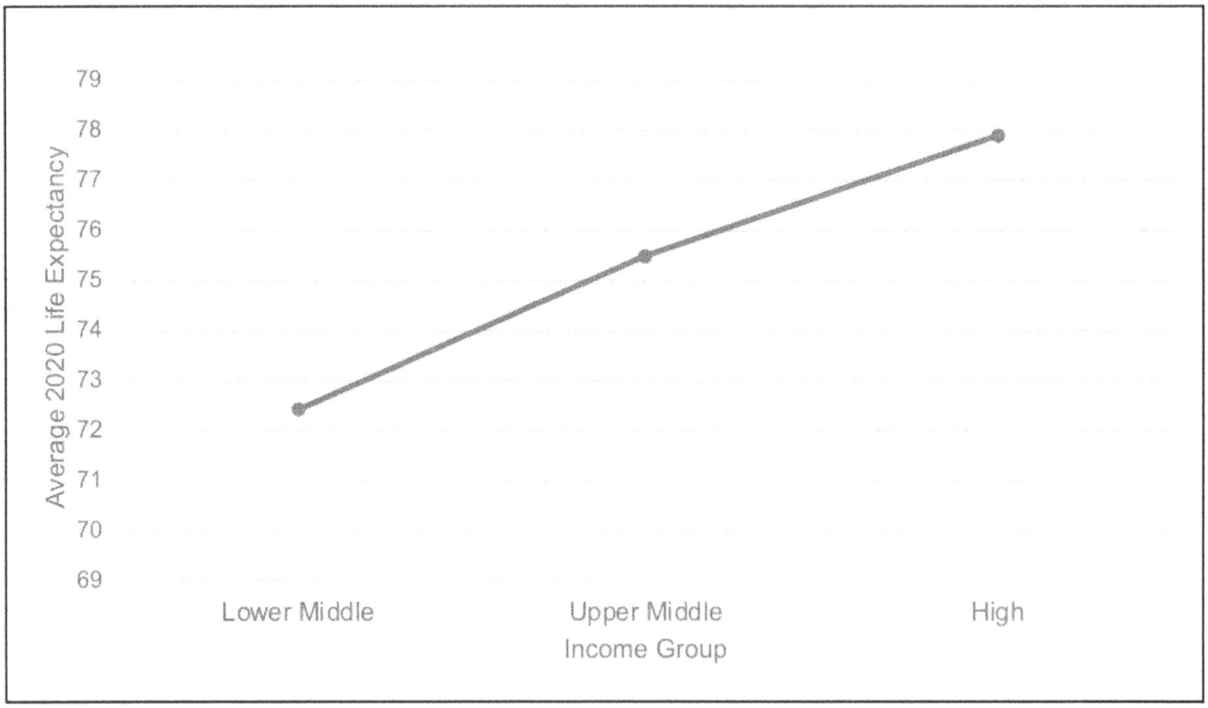

Source: World Bank, World Development Indicators (2020). Life Expectancy at Birth. https://data.worldbank.org/indicator/SP.DYN.LE00.IN.

A clearer, nonlinear trend was found in a 2018 analysis of how many minutes, on average, the main activity of fathers in the United States involved caring for their children. The U.S. Bureau of Labor Statistics (2021) found that the amount of time varied by the age of the child as illustrated in Figure 7.5. Notice that the amount of time fathers spent caring for children declined from children aged less than six years to children aged 6-12, then increased from that point to children aged 13-18. This type of trend is also a quadratic trend where the outcome starts out high, then declines and finally increases again. Like the trend in Figure 7.4, this type of pattern requires a minimum of three points to uncover the trend; limiting the outcome results to two points in this analysis could have resulted in lost information by incorrectly depicting the trend as linear rather than quadratic.

Figure 7.5 Time Fathers Spent Caring for and Helping Children as their Main Activity

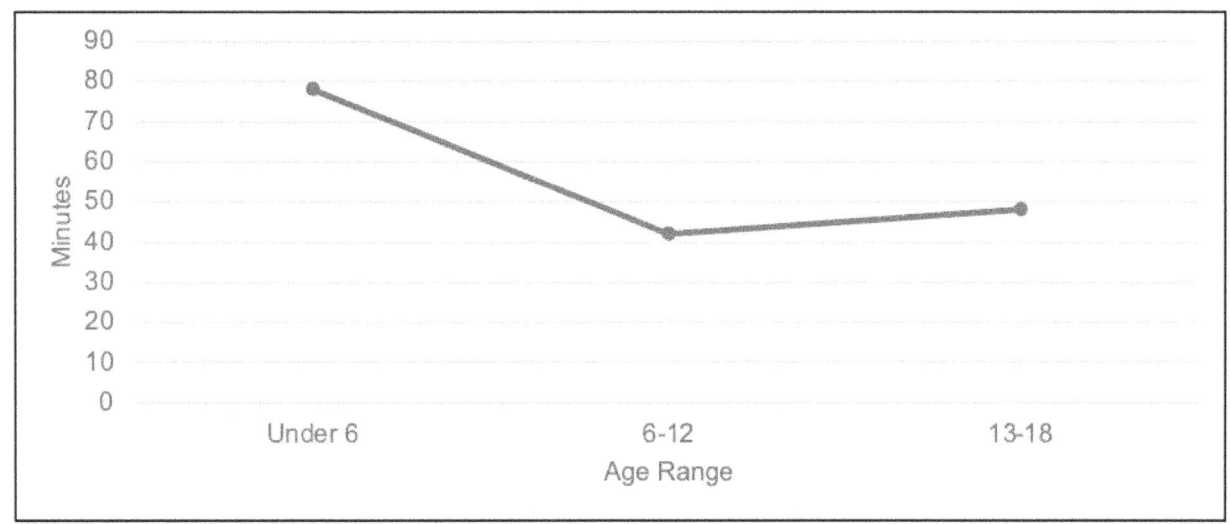

Source: U.S. Bureau of Labor Statistics, U.S. Department of Labor. April 7, 2021) https://www.bls.gov/charts/american-time-use/activity-by-parent.htm.

Other polynomial trend patterns are presented in Figure 7.6. They include quadratic, cubic, quartic and quintic trends.

Figure 7.6 Examples of Polynomial Trends

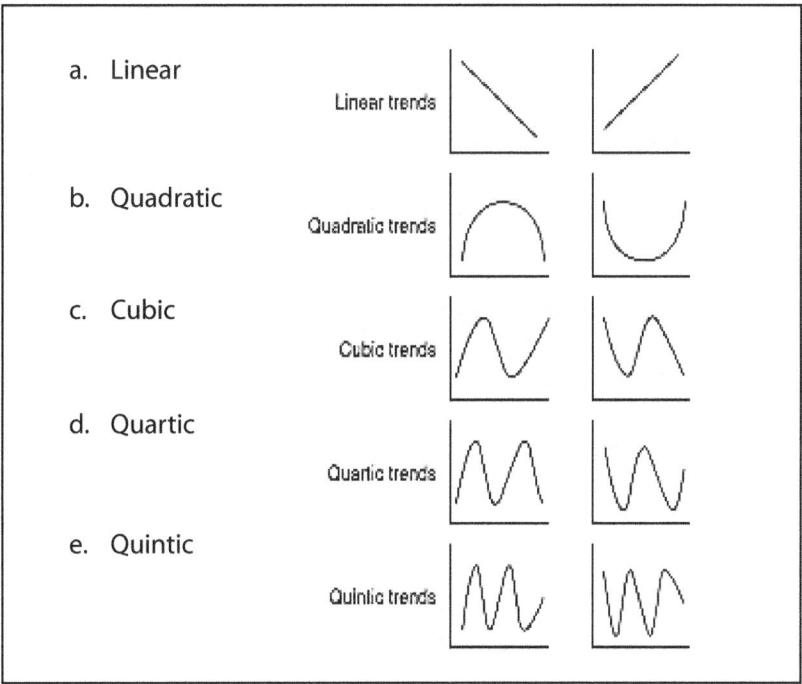

The following guidelines should assist in selecting the appropriate trend analysis for research investigations.

- Figure 7.6a depicts a **linear pattern.** The first illustrates the independent variable (x-axis) increasing as the dependent variable (y-axis) declines (negative trend). The second pictures a trend of the independent variable increasing as the dependent variable also increases (positive trend). For example,

as discussed in Chapter 4, there is a positive, linear trend between height and weight. Three independent variable groupings are needed to confirm a linear trend.
- Patterns displayed in Figure 7.6b illustrate a **quadratic trend** in which more than three levels of the independent variable are needed for plotting the outcome, revealing that the direction of the trend changes once.
- Figure 7.6c shows a **cubic trend** where there are two changes in the direction of the trend. Accordingly, the mean of the dependent variable (e.g., anxiety levels) may increase initially (before a test), followed by a decline (after the test) and a subsequent increase across the levels of the independent variable (immediately before test results). To reveal two changes in the direction of the trend, a minimum of four levels are needed for the independent variable.
- Figure 7.6d shows a **quartic trend** in results, and in this case a total of five levels of the independent variable are needed to reveal this trend since there are three changes of direction.
- Finally, the **quintic trend** in Figure 7.6e exhibits four changes in direction. A minimum of six levels of the independent variable are necessary to expose this pattern in research results. Researchers rarely test for a quintic trend. Regardless of the number of independent variable levels, researchers typically limit trend analysis examinations to quartic or lower trends levels.

7.9 SPSS Procedures for One-Way Analysis of Variance

To recap, a one-way ANOVA focuses on a single between-subjects independent variable (e.g., reading instructional method). The independent variable for a *t* test comprises two groups, but the independent variable in a one-way ANOVA typically comprises more than two groups (e.g., computer-based instruction, teacher directed instruction, and both).

For this exercise, a random sample of data from a national secondary school placement test in Jamaica will be used to demonstrate the procedures for completing a one-way analysis of variance using SPSS. The Communication Task subtest is the focal point of this exercise. The communication task is an extended writing task based on a single prompt. The maximum score possible for this task is 12 and the minimum is 2. Student performance on the communication task (CommTask) is the dependent variable.

This dataset resides in the data file, Equal_N_One_Way.sav. In this exercise, the independent variable is school type (Schtype2) which comprises three types of pre-secondary schools (School Type 1, 2, and 3); these are the levels of the independent variable. School Type 1 and 2 indicate two types of government-supported elementary schools. School Type 3 represents private elementary schools. Each school type group includes 69 study participants. Accordingly, this is a balanced one-way ANOVA test design.

7.9.1 Research Question and Hypotheses

The purpose of the exercise is to determine whether there is a significant difference in the communication task scores of students in three types of schools—School Type 1, 2 and 3. Accordingly, the research question is as follows:

Is there a significant difference among the communication task scores of students in three types of schools—1, 2, and 3?

Given the sample of scores in this exercise, the research hypotheses are as follows:

Null Hypothesis

There is no significant difference among the communication task scores of a sample of students in three types of Jamaican schools: 1, 2 and 3.

$H_0: \mu_1 = \mu_2 = \mu_3$

Alternative Hypothesis

There is a significant difference between the communication task scores of students in two types of schools.

$H_a: \mu_i \neq \mu_j$

7.9.2 SPSS Specifications for Omnibus One-Way ANOVA

To complete a One-Way ANOVA of the communication task performance, locate the dataset, **Equal_N_One_Way.sav** and open it in the **SPSS Statistics Data Editor** window. Select **Analyze → Compare Means → One-Way ANOVA** (Figure 7.7) to open the **SPSS One-Way ANOVA main dialog window** displayed in Figure 7.8.

Figure 7.7 Selecting One-Way ANOVA

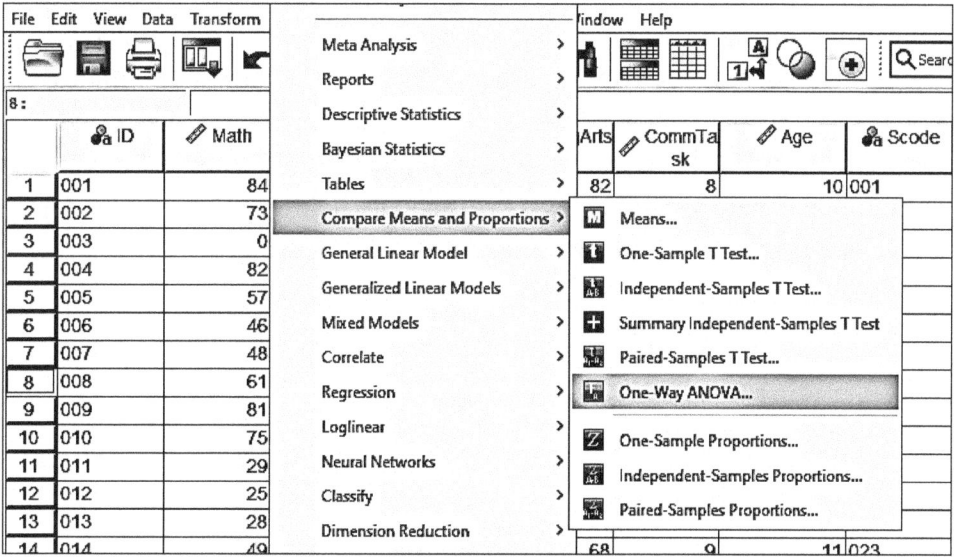

From the variables list on the left, select and move **CommTask** (communication task) to the **Dependent List** box in the top right section of the window. This is the variable that will be used to find the average score for each school type. It is possible to complete multiple one-way ANOVAs in a single run by adding another dependent variable to the **Dependent List** box (e.g., math) if you wish to see how different types of schools perform on more than one variable. For this demonstration, the analysis will be restricted to a single One-Way ANOVA.

Select and click over **Schtype2** (school type) to the **Factor** box in the lower right of the main **One-Way ANOVA** dialog window. The levels or categories of the independent variable define which groups' performance on the communication task the statistical technique will compare.

Figure 7.8 One-Way ANOVA Main Dialog Window

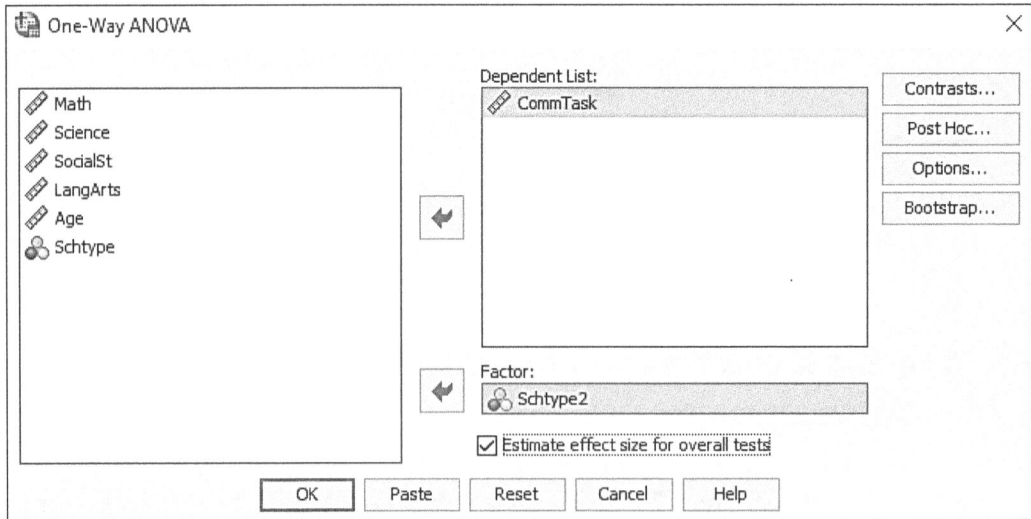

Two assumptions of one-way ANOVA, independence of observations and measurement level of dependent variable, may be addressed using the research design. For example, the assumption that the dependent variable should be measured on at least an equal interval scale is satisfied by the scale (interval or ratio) level measure of the dependent variable. Due to the large sample size (n = 207) and the balanced design (equal sample sizes of each group of the independent variable), normality is unlikely to be violated. However, the assumption of homogeneity of variance should be tested.

To secure descriptive statistics and to test the assumption of homogeneity of variance, select the **Options** pushbutton in the main **One-Way ANOVA** dialog window to arrive at the **Options** window. Check **Descriptive** and **Homogeneity of variance test** (Figure 7.9).

Figure 7.9 One-Way ANOVA Options Window

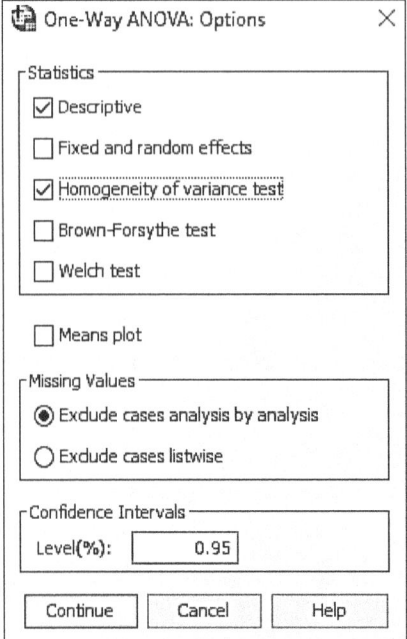

Note that by default, the **Missing Values** option is **Exclude cases analysis by analysis**. This means that a case with a missing value for either the dependent or the independent variable for a given analysis is not used in that analysis. Also, a value outside the range specified for the independent variable is not used. Retain the default option. Click **Continue** to return to the main dialog window.

No a priori planned contrasts will be identified for this exercise. (These would have had to be completed at the same time as your research questions and your hypotheses and based on the research literature.) Accordingly, the **Planned Contrasts** window will not be needed. However, options are available in that window for polynomial contrasts.

Although it is best to wait to review the outcome of the omnibus test before determining whether post-hoc tests are necessary, for purposes of convenience for this demonstration, identify post-hoc tests concurrent with the omnibus test. To do that, select the **Post Hoc** pushbutton in the main **One-Way ANOVA** window to open the **One-Way ANOVA: Post Hoc Multiple Comparisons** window (Figure 7.10).

Figure 7.10 One-Way ANOVA Post Hoc Window

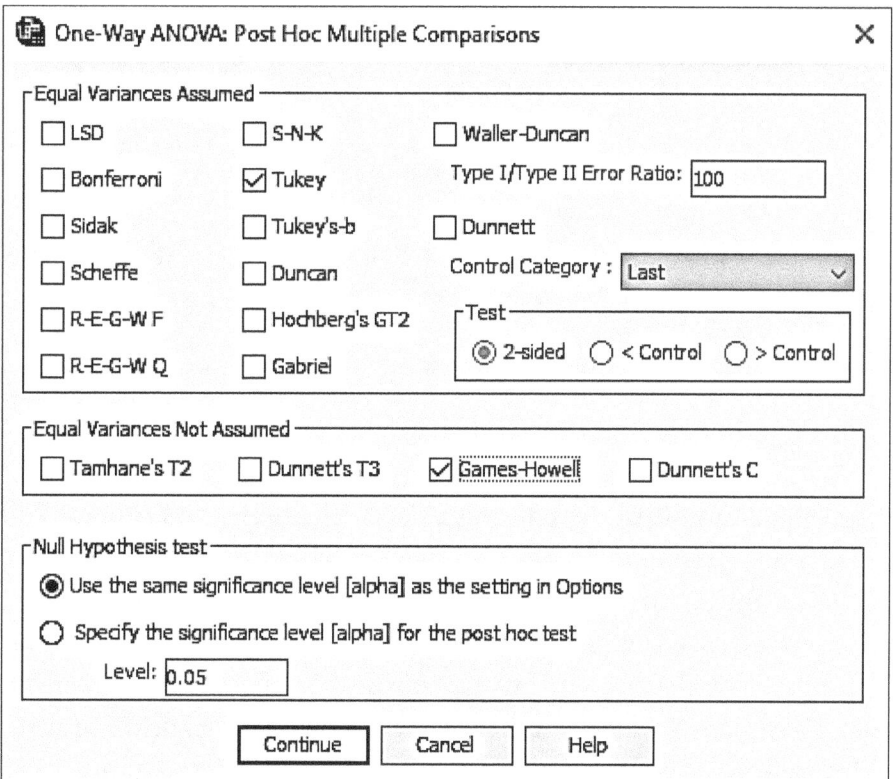

As the **One-Way ANOVA: Post Hoc Multiple Comparisons** window displays, a large number of multiple comparisons tests are available to minimize the Type I error rate as all possible pairs of independent variable group means are tested to identify which pairs of group means are significantly different. Without these post-hoc approaches, the significance level would increase dramatically, and the likelihood of obtaining false-positive results would increase as the number of pairs of groups are tested. As noted at the bottom of the box, some approaches are designed to stabilize or enhance statistical power if assumptions (e.g., equal variance assumption) are violated.

Select **Tukey** in the box labeled, **Equal Variances Assumed**. In addition, in the event that the equal variance assumption is violated, select **Games-Howell** under **Equal Variances Not Assumed**. Retain the significance level of 0.05 in the **Significance level** box at the bottom of the window. Click **Continue** to close the window and return to the main dialog window, then click **OK** to run the analyses.

7.9.3 Omnibus One-Way ANOVA Analysis Results

The descriptive statistics, displayed in Figure 7.11, provide a general picture of the data and the associated samples before examining the results of inferential statistical analyses. The communication task is a continuous measure of students' writing performance. The first two columns show that the student groups are of equal size (n = 69) and that the average score for the groups range from 7.97 for students at Public 2 schools to 9.0 for students at private schools. Public 2 schools recorded the lowest mean score, and private schools recorded the highest mean score. The **Total** row shows that the overall average score for all students is 8.45 based on 207 valid cases. The last two columns reveal that the minimum mark is 0 for each school type and the maximum score is 12.

Figure 7.11 Descriptive Statistics of Students' Communication Task Performance

					95% Confidence Interval for Mean			
	N	Mean	Std. Deviation	Std. Error	Lower Bound	Upper Bound	Minimum	Maximum
Public 1	69	8.38	2.414	.291	7.80	8.96	0	11
Public 2	69	7.97	2.107	.254	7.46	8.48	0	12
Private	69	9.00	2.282	.275	8.45	9.55	0	12
Total	207	8.45	2.299	.160	8.13	8.76	0	12

Figure 7.12 displays the results of the one-way ANOVA omnibus test. Figure 7.12a presents the results of the **Test of Homogeneity of Variances**. The null hypothesis of Levene's test is that there is no significant difference among the variances of the groups being compared.

The results of the test are based on four different types of underlying distributions of the dependent variable. Results are based on the mean, median, median with adjusted degrees of freedom (df), and the trimmed mean. According to Brown and Forsythe (1974), when underlying distributions of the dependent variable are skewed, the assumption of equal variances is best assessed using the median, while normal distributions are best assessed using the mean.

In this case, the significance values (p-values) in the **Sig**. column are larger than the alpha level of .05, regardless of the nature of the underlying distribution, as indicated by results based on the mean, median and trimmed mean. Therefore, the null hypothesis cannot be rejected (Section 7.5.4). Accordingly, there is no statistically significant difference among the variances of the **schtype** groups at the .05 alpha level; the assumption of equality of variances has not been violated, and variances can be treated as if they are essentially equal.

Figure 7.12b presents the results of the ANOVA test and reports a statistically significant F value of 3.594 (**F** column), based on 2 and 204 degrees of freedom noted in the **df** column. The significance level of $p = .029$ in the **Sig.** column means that differences among the average communication task scores of the three types of schools are significant at the a priori alpha level of .05; that is, the significance level is less than the a priori alpha level.

Figure 7.12 One-Way ANOVA Omnibus Test Results

(a)

	Test of Homogeneity of Variances				
		Levene Statistic	df1	df2	Sig.
Communication Task	Based on Mean	.312	2	204	.732
	Based on Median	.031	2	204	.970
	Based on Median and with adjusted df	.031	2	195.010	.970
	Based on trimmed mean	.112	2	204	.894

(b)

ANOVA

Communication Task

	Sum of Squares	df	Mean Square	F	Sig.
Between Groups	37.072	2	18.536	3.594	.029
Within Groups	1052.145	204	5.158		
Total	1089.217	206			

Computing the proportion of variance accounted for in scores by the three school types is one way of estimating the effect size of the differences found. How important are the differences found among the student scores in different types of schools? To compute this statistic (specifically η^2), divide the value for the **Between Groups Sum of Squares** in the first row of the ANOVA table in Figure 7.12b by the **Total Sum of Squares** in the last row. That computation results in 37.072/1089.217, or 0.034. Accordingly, based on Cohen's effect size guidelines in Section 7.7, the school type accounts for a very small amount of the variance in communication task performance (3%). Therefore, other variables are likely to have a larger role in the differences in student performance than the type of school students attend. SPSS results of eta-squared and other effect sizes, epsilon squared and omega squared are displayed in Figure 7.13.

Figure 7.13 ANOVA Effect Sizes

ANOVA Effect Sizes[a,b]				
			95% Confidence Interval	
		Point Estimate	Lower	Upper
Communication Task	Eta-squared	.034	.000	.090
	Epsilon-squared	.025	-.010	.081
	Omega-squared Fixed-effect	.024	-.010	.080
	Omega-squared Random-effect	.012	-.005	.042

a. Eta-squared and Epsilon-squared are estimated based on the fixed-effect model.
b. Negative but less biased estimates are retained, not rounded to zero.

7.9.4 Results of Multiple Comparisons Tests

The omnibus ANOVA test results reported a significant difference in performance of students at three different types of schools. However, the omnibus results do not reveal which pairs of school types have significant differences in scores. Accordingly, to determine the sources of the differences, it is necessary to conduct post-hoc multiple comparisons tests, and in this case, Tukey is used; it tests every possible pair of school type to ascertain whether they have statistically significantly different average scores while controlling the Type I error rate so that it does not rise above the a priori alpha level of .05.

The **Tukey HSD** post-hoc test results are displayed in Figure 7.14a. In the reporting structure, there is one variable group for each major row of the table (e.g., **Public 1** results occupy the first major row where those schools are compared with the other two variable groups). Accordingly, the first major row reports on comparisons between **Public 1** schools and **Public 2** schools, followed by **Public 1** schools and **Private** schools. The second major row displays results of comparisons between **Public 2** schools and **Public 1** schools, followed by **Public 2** schools and **Private** schools. The third and final major row presents results of **Private** schools compared with **Public 1** schools first and then with **Public 2** schools.

In the first major row, in the **Mean Difference** column, the reported difference between the average scores for **Public 1** and **Public 2** school types is 0.406 (8.38 – 7.97). The difference between **Public 1** and **Private** schools is -0.623 (8.38 – 9.0). The **Sig.** column reports the significance level of the average differences between the groups ($p = .547$ and $p = .243$, respectively). Since the significance levels are greater than the a priori alpha level of .05, the differences are statistically nonsignificant at the .05 alpha level.

The second major row compares **Public 2** schools with **Public 1** and **Private** schools, and the outcomes are reported in the same way they were reported in the first major row. In addition, the structure of the table builds in some redundancy in that the first line repeats the nonsignificant results of the comparison between **Public 2** and **Public 1** schools found in the first major row (mean difference = -.406 and $p = .547$). However, the second row introduces the outcome of the performance when **Public 2** schools and **Private** schools are compared. Here, the mean difference in performance (7.97- 9.0 = -1.029) is statistically significant at the .05 alpha level ($p = .023$). Note the table footnote; the asterisk indicates that the mean difference is significant at the .05 alpha level.

The third major row repeats test results presented earlier, comparing **Private** schools and **Public 1** schools as well as **Private** schools and **Public 2** schools. In effect, each combination of school type and associated comparison results is presented twice in the table.

Figure 7.14 Post Hoc Multiple Comparisons Test Results

(a)

Multiple Comparisons

Dependent Variable: Communication Task

	(I) School Type	(J) School Type	Mean Difference (I-J)	Std. Error	Sig.	95% Confidence Interval	
						Lower Bound	Upper Bound
Tukey HSD	Public 1	Public 2	.406	.387	.547	-.51	1.32
		Private	-.623	.387	.243	-1.54	.29
	Public 2	Public 1	-.406	.387	.547	-1.32	.51
		Private	-1.029*	.387	.023	-1.94	-.12
	Private	Public 1	.623	.387	.243	-.29	1.54
		Public 2	1.029*	.387	.023	.12	1.94
Games-Howell	Public 1	Public 2	.406	.386	.546	-.51	1.32
		Private	-.623	.400	.267	-1.57	.32
	Public 2	Public 1	-.406	.386	.546	-1.32	.51
		Private	-1.029*	.374	.018	-1.92	-.14
	Private	Public 1	.623	.400	.267	-.32	1.57
		Public 2	1.029*	.374	.018	.14	1.92

*. The mean difference is significant at the 0.05 level.

(b)

Communication Task

	School Type	N	Subset for alpha = 0.05	
			1	2
Tukey HSDa	Public 2	69	7.97	
	Public 1	69	8.38	8.38
	Private	69		9.00
	Sig.		.547	.243

Means for groups in homogeneous subsets are displayed.
a. Uses Harmonic Mean Sample Size = 69.000.

In the second half of Figure 7.14 are the **Games Howell** results (useful when equal variances cannot be assumed). However, due to the findings of the Test of Homogeneity of Variance (Figure 7.12a), equal variances could be assumed. Therefore, the Tukey HSD rather than the Games-Howell procedure is preferable. Nevertheless, the results are very similar; only scores for Public 2 schools and private schools were found to be significantly different ($p = .018$).

Figure 7.14b summarizes the Tukey HSD results previously described. Means are placed in the same column when they are not significantly different. Accordingly, **Public 1** schools and **Public 2** schools are in **Subset 1**; **Public 1** schools and **Private** schools are in **Subset 2**. The non-significant p values are in the last table row. **Public 2** schools and **Private** schools are in different columns because they have significantly different average scores.

7.9.5 Sample Report: The Case of Differences in Achievement

The Communication Task performance of students in three different school types was compared using one-way ANOVA. The analysis found significantly different performance in the three types of schools, $F(2, 2014) = 3.954$, $p < .05$. The effect size calculated as eta-squared (η^2) was .034, indicating a small effect.

Results of the Tukey post-hoc, multiple comparison test revealed that students at private schools had a significantly higher average performance (M = 9.0, SD = 2.28) than students at Public 2 schools (M = 7.97, SD = 2.11), $p = .023$. There were no significant differences between the mean performance of students at Public 1 schools (M= 8.38, SD = 2.41) and private schools ($p = .243$) or Public 1 schools and Public 2 schools ($p = .547$).

7.10 Analysis of Variance and Linear Regression

It is often said that ANOVA is a special case of linear regression. In fact, they are members of a family of statistics referred to as the general linear model (GLM), all of which investigate how independent variables affect outcome variables. In addition to analysis of variance and linear regression, other statistical methods that are based on GLM include the t test, analysis of covariance (ANCOVA), repeated measures ANOVA (RMANOVA), as well as several multivariate statistical methods.

Certainly, for one-way ANOVA, the outcome of the analysis is the same in linear regression. For illustrative purposes, consider the Communication Task scores in Table 7.2. In that case, the omnibus test shows that the means of the score groups are significantly different from each other ($p < .001$) as reported in Figure 7.2. The same data may be analyzed using linear multiple regression, with a few modifications.

In order to use the linear regression model, replace the three-level independent variable in the ANOVA model with two dummy variables, as discussed in Chapter 3. Independent categorical variables having any number of levels may be substituted with dummy variables equal to one less than the number of categories of the independent variable. With three categories in the independent variable group, two dummy variables are needed. In each case the dummy variable codes are restricted to 1 or 0. The code 1 represents the presence of what is being measured and 0 represents the absence of the measured quality. For example, for Dummy Variable 1, participants in Group A are coded 1, and all other groups are coded 0. For Dummy Variable 2, Group B participants are coded 1, and all others coded 0.

The third category is known as the reference category. The group chosen to be the reference category is dependent on the study's hypothesis. In an experiment, the reference category might be the control group. In other studies, the reference category may be the group used for comparison purposes. In this case, Group C will be used as the reference category, which is always coded 0. Using the 1/0 coding scheme, each group can be described as illustrated in Table 7.5.

Table 7.5 Dummy Coding for a Three Group Design

Group	Dummy Variable 1 (Group 1)	Dummy Variable 2 (Group 2)
Group C (Reference Category)	0	0
Group B	0	1
Group A	1	0

The data entered into **SPSS Regression** appear in Table 7.6. ID 1-5 represent scores for Group A. ID 6-10 represent scores for Group B. ID 11-15 represent scores for Group C. The variable Group identifies each group that was used in the manual computation one-way ANOVA exercise.

Table 7.6 Data Set for Communication Test

ID	Group	Group 1	Group 2	Score
1	A	1	0	10
2	A	1	0	9
3	A	1	0	11
4	A	1	0	7
5	A	1	0	8
6	B	0	1	8
7	B	0	1	6
8	B	0	1	5
9	B	0	1	9
10	B	0	1	7
11	C	0	0	4
12	C	0	0	5
13	C	0	0	3
14	C	0	0	3
15	C	0	0	6

Like one-way ANOVA, multiple linear regression analyses produced the same significant results, an F statistic of 13.0145, significant at $p < .001$. In ANOVA, the means of Group A, B and C are: 9, 7 and 4.2, respectively, as noted in Table 7.3. Linear regression analyses produced the following regression coefficients:

Constant/Intercept 4.2
Group 1(A)   4.8
Group 2 (B)   2.8.

The intercept is the mean of the reference group, Group C. The coefficients for the other two groups are the differences in the mean score of the reference group (Group C) and the other groups. Accordingly,

Group A coefficient = Group A – Group C = 9 – 4.2 = 4.8
Group B coefficient = Group B – Group C = 7 – 4.2 = 2.8.

The ANOVA analysis reports each average and a *p*-value that says at least two averages are significantly different. The regression method reports only one average (constant/intercept), as well as the differences between that average and all other averages, and the *p*-values are used to evaluate those comparisons. It is the same linear model originally independently developed by two branches of social science methodology.

7.11 Other Types of ANOVAs

As there are different types of *t* tests (e.g., independent and dependent *t* tests), so it is that a wide range of ANOVA designs have been developed to evaluate differences in group means. This chapter focuses on one-way ANOVA which is used when there is a single independent variable. Later chapters will investigate other types of ANOVA designs:

Factorial ANOVA: compares the means of two or more categories or groups that belong to two or more independent variables.
Analysis of Covariance (ANCOVA): compares means of one or more independent variables while controlling for one or more other independent variables (covariates) that influence the dependent variable.
Repeated Measures ANOVA (RMANOVA): an extension of the dependent *t*-test that compares the means of the same set of study participants that are measured repeatedly over time on the same dependent variable.
Nonparametric ANOVA (e.g., the Kruskal-Wallis *H* Test and Friedman's Test): alternatives to the one-way and repeated measures ANOVA when data fails to meet the assumptions for parametric tests or when the dependent variable is measured on an ordinal measurement scale.

7.12 Key Terms

Analysis of variance
Between group variance
Cubic trend
Difference contrast
Eta squared (η^2)
Experiment-wise error rate
Family-wise error rate
Grand mean
Grand variance

Helmert contrast
Mean square
Multiple comparisons test
Omega squared (ω^2)
Orthogonal contrast
Pairwise comparisons
Planned contrasts
Polynomial contrast

Post-hoc tests
Squared deviations
Squared residual
Sum of squares
Quadratic trend
Quartic trend
Quintic trend
Within group variance

8
Factorial Analysis of Variance

Key concepts from previous chapters or courses:

between-group variance: the variance in scores attributable only to membership in different groups and exposure to different treatments. In ANOVA it is the degree of group mean differences. It is divided by within-group variance to obtain an F ratio.

eta squared (η^2), omega-squared (ω^2): correlation-based measures of effect size.

mean square: an estimate of variance, typically used in ANOVA. It is calculated by dividing the sum of squares by the degrees of freedom. Different types of mean square are calculated in ANOVA; for example, mean square error or within-group mean score is the average distance of a score from the mean score of its group.

multiple comparisons: also known as paired comparisons. A set of comparisons of pairs of group means to determine whether there are significant differences between them. Multiple comparisons are conducted after an omnibus ANOVA test finds a significant difference and is designed to keep the Type I error rate at a prespecified level.

post hoc test: a test conducted on the basis of an overall ANOVA test which indicated a statistically significant effect or difference across groups that should be examined further.

sum of squares: the value resulting from computing the deviation of each point in a data set from some value (such as a mean), squaring each deviation, and adding the squared deviations. Different types of sums of squares are calculated in ANOVA (e.g., sum of squares between groups).

within-group variance: variance or error in scores among individuals within the same group who experienced the same treatment conditions. It is frequently used in ANOVA and compared with between-group variance to obtain an F ratio.

THIS CHAPTER INTRODUCES

- differences between one-way ANOVA and factorial ANOVA
- characteristics of a factorial ANOVA
- when to use a factorial ANOVA
- how to use variances in scores to find differences
- how to calculate the effect size of a factorial ANOVA outcome
- how to use SPSS to conduct a factorial ANOVA.

8.1 Introduction

Secondary school placement examinations in the Caribbean are high-stakes assessments that often require multiple attempts to meet the standard for admissions. Students often take the test for the first time at age 11, sometimes at age 10. Nevertheless, age 12 is frequently the final opportunity for students to be successfully transported through the secondary school placement test process to the high school of their choice (e.g., Thomas, 2015; Jornitz et al., 2021). In Chapter 7, a comparison of student performance on a written communication secondary school placement assessment revealed that students in some types of schools performed significantly better than those in other types of schools. Could age (representative of academic experience) also contribute to differences in performance on the high school placement test?

The practice exercise in Chapter 7 examined the performance of students attending three different types of pre-secondary school—Public 1 (T_1), Public 2 (T_2), and Private (T_3). In that one-way ANOVA design, types of pre-secondary schools represented a single independent variable with three levels (T_1, T_2, and T_3). The dependent variable was performance on the written communication subtest. Analysis of the data resulted in a finding of significant difference between the performance of two types of schools—Public 2 (T_2) and Private (T_3).

Considering the age range of students completing the placement examination, a researcher could conceivably expand on the first test performance exploration by conducting another one-way between-subjects ANOVA to investigate whether statistically significant differences in performance existed among students of different ages. However, findings of performance differences due to school type in a one-way ANOVA study and findings of differences due to age in a second one-way ANOVA study, provide no insight into the combined impact of the independent variables, age and school type, on performance. To isolate potential differences in performance due to the combined impact of age and school type, researchers must select another type of ANOVA design. This type of design is called a factorial ANOVA, which is the focus of this chapter.

8.2 The Case of Differences in Achievement: The Factorial Effect

After securing the finding of significant differences in communication task performance among students attending three types of schools, researchers wish to determine whether there are significant differences in performance due to the combined impact of school type and an additional independent variable, age. As in the case in Chapter 7, there is a single dependent variable—communication task score. Due to the retrospective nature of the study, investigators decide to use a causal-comparison (or ex post facto) research design and to use factorial ANOVA as the primary statistical vehicle to analyze students' performance. The research plan for the case and associated details may be found in Box 8.1.

Before analyzing actual data, Section 8.3 will present fundamentals of factorial ANOVA. Table 8.1 presents the data to be analyzed using factorial ANOVA. Table 8.1 includes data for a total of 36 cases across three types of schools (Private—T_1, Public 1—T_2, and Public 2—T_3), two groups from each school for a total of six groups. Each school type comprises six cases of 11-year-olds (A_1) and six cases of 12-year-olds (A_2). Table 8.1 also includes descriptive statistics (total, mean, and variance) for each group.

Table 8.1 Fictitious Communication Task Achievement Data and Descriptive Statistics

	Scores			
Age	Private (T$_1$)	Public 1 (T$_2$)	Public 2 (T$_3$)	Age Group
11	10	8	4	
	11	9	5	
	12	9	5	
	11	7	6	
	9	10	4	
	10	8	5	
Age 11 Total	63	51	29	
Age 11 Mean	10.5	8.5	4.833	7.944
Age 11 Variance	1.1	1.1	0.567	6.644
12	9	6	6	
	10	5	6	
	7	7	5	
	8	6	7	
	9	7	6	
	9	8	7	
Age 12 Total	52	39	37	
Age 12 Mean	8.667	6.5	6.167	7.111
Age 12 Variance	1.067	1.1	0.567	2.105
School Type Total	115	90	66	
School Type Mean	9.58	7.5	5.5	
School Type Variance	1.902	2.091	1.0	

8.3 Factorial ANOVA Fundamentals

In one-way ANOVA there is a single independent variable, also called a factor. The independent variable typically comprises more than two levels or groups of people or things (e.g., public, private, and semi-private school types). It is a part of the univariate family of statistics in which there is a single dependent variable. The purpose of one-way ANOVA is to discover whether there is a significant difference among groups of people or things.

Factorial ANOVA is an analysis of variance univariate test in which there is more than one independent variable (e.g., school type and student age) and a single dependent variable (e.g., written communication score). The simplest factorial ANOVA test is a two-way ANOVA that includes two independent variables (e.g., school type and age), each of which has two or more levels or groups (e.g., ages 11 and 12).

Factorial ANOVA designs accord investigators important benefits over one-way ANOVAs. Three of those benefits are: generalizability, economy/efficiency, and interaction.

By augmenting a study with other relevant variables, researchers succeed in increasing generalizability and authenticity. For example, the exercise in Chapter 7 used one-way ANOVA to examine whether differences in student written communication performance were due to the type of pre-secondary school students attended. However, the study failed to consider other variables, like age, that might impact performance, thereby limiting our understanding of contributors to student performance. Including age allows researchers to examine whether performance differences between students attending different types of schools also apply to students of varying ages. Researchers may also learn whether findings related to age differences in performance apply to all or only some types of schools. With the inclusion of additional variables, factorial ANOVA allows for broader interpretations than might occur with multiple one-way ANOVA designs. In addition, factorial designs provide a more authentic context for more meaningful interpretations.

The second advantage of factorial designs is their efficiency. Researchers can compare student performance in separate investigations, each including a different variable; that is, two studies might be conducted, one in which age is the only independent variable and the other in which school type is the sole independent variable. Two studies require two separate samples of participants or at a minimum, two separate designs and data analyses for the same dependent variable (communication task performance). A two-way design requires fewer participants and less time. In effect, researchers get two studies in one by testing all the independent variables at the same time using the same sample. An additional benefit of using a more comprehensive analysis like a factorial design is that it limits the likelihood of a Type I or Type II error by incorporating an initial omnibus test that determines whether there is any significant difference among any of the variables before digging into details about the sources of the differences.

Besides greater data collection and processing efficiency, testing the effect of multiple independent variables in a single analysis has the benefit of allowing researchers to examine whether there are any interactions of the independent variables. Factorial designs allow researchers to investigate whether the effect of one variable is independent of the effects of another; for example, whether the effect of the variable school type is independent of student age, or whether the effect of a specific school type depends on a particular student age. Developmental psychology might suggest that older, more mature students perform better than younger ones, or that schools which are typically regarded as having more rigorous curricula than other schools tend to produce higher performing students. However, what is perhaps most informative is the effects of the interplay between variable groupings that might advance an important finding.

8.3.1 Types of Factorial ANOVA

Factorial ANOVA designs are based on the number of independent variables involved as well as the number of independent variable groups. Factorial ANOVA is designed for a minimum of two independent variables, each split into a bare minimum of two groups, resulting in at least a total of four groups. Factorial ANOVAs are described based on the number of independent variables and the number of incorporated groups associated with each variable.

8.3.1.1 Two-Way Factorial ANOVA Designs

Chapter 7 outlined the essentials about the one-way ANOVA design. The design for factorial ANOVA is an extension of that blueprint.

Imagine that a researcher wants to determine whether there is a significant difference in the income of individuals based on gender (male/female) and high school graduation (yes/no). With a single dependent variable (income) and two independent variables, each of which includes two groups or subdivisions, this is the most basic factorial design. In this two-way factorial design, each group of the independent variable is compared with the other group within the same independent variable as well as with groups within the other independent variable. The objective is to determine whether significant differences exist between any of the variable pairs.

Two-way factorial designs are also known as a two-way between-subjects designs:
- "two-way" because there are two independent variables, and
- "between-subjects" because researchers compare subject groupings.

The term two-way, between-subjects design provides some information about the type of analysis that will be performed. However, researchers sometimes use a more explicit notation system to describe a factorial design. Accordingly, the numerical notation for the income example is known as a 2 x 2 factorial design. The number of digits in the design (2 digits) describes the number of variables in the design, and the values of the digits tell how many groups are in each variable (e.g., male/female for gender). Therefore, in a 2 x 2 design, there are two variables, and each variable is subdivided into two groups. To find the total number of groups in the design, multiply the numbers in the notation; for example, multiplying 2 x 2 equals a total of four groups in the research design.

The Case of Differences in Achievement: The Factorial Effect has a 2 x 3 design, which is a step up from the simplest 2 x 2 factorial ANOVA design. The 2 x 3 factorial design includes two variables, one of which subdivides into two groups (ages 11 and 12 for age) and the other which includes three groups (T_1, T_2, and T_3 for school type). The order of the numbers in the notation system makes no difference. Accordingly, a numerical notation of 3 x 2 is the same as one described as a 2 x 3 design.

Figure 8.1 A 2 x 3 Factorial ANOVA Design

		School Type (T)		
		T_1	T_2	T_3
Age (A)	A_1			
	A_2			

Figure 8.1 illustrates the 2 x 3 factorial design for the variables, age (A) and school type (T). The design includes a total of six cells which represent six groups of study participants.

Figure 8.2 exemplifies two central characteristics of between-subjects factorial designs:
- They are characterized by participants who are limited to a single, nonoverlapping treatment condition; for instance, one group of participants, $\bar{X}_{11}$, represents the group of students aged 11 who attend a single type of school (T_1). A second group of students, $\bar{X}_{21}$, are aged 12, but attend the same type of school as the $\bar{X}_{11}$ group.
- Since each study participant is assigned to one cell, each has a single score for the communication task test.
- The score for each cell group is a group average of the outcome variable, communication task score.

Figure 8.2 Expanded Two-Way Factorial Design

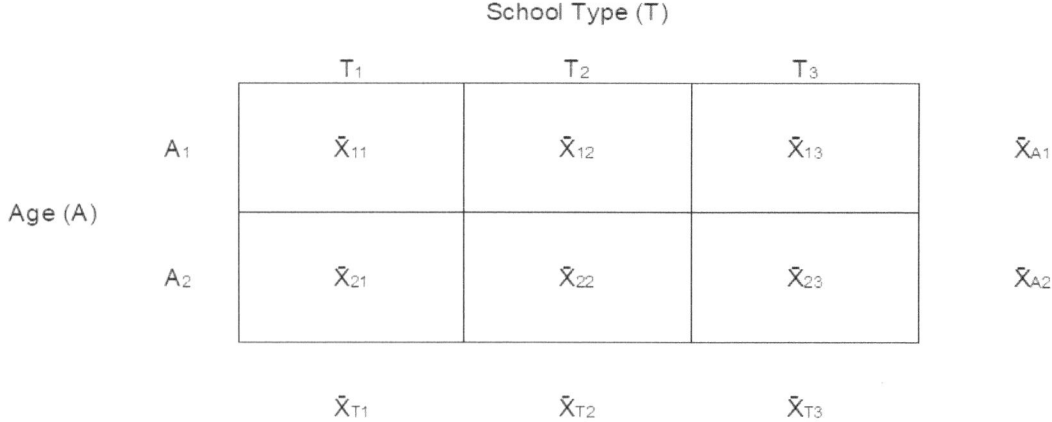

In factorial ANOVA, two types of central analyses take place:
- The examination of whether there is a differential treatment effect for each independent variable. For example, average scores of the three school types represented in Figure 8.2 ($\bar{X}_{T1}$, $\bar{X}_{T2}$, and $\bar{X}_{T3}$) are statistically compared for significant differences among them. In addition, average scores of the two student ages ($\bar{X}_{A1}$, and $\bar{X}_{A2}$) are also evaluated for significant differences. These are known as main effects analyses.
- The investigation of whether there are any interaction effects between the independent variables. For this analysis researchers investigate whether there are significant differences between the average score of one variable against the average scores of each group of another variable. For example, average scores of 11-year-olds ($\bar{X}_{A1}$ in Figure 8.2) may be compared with each group for the other variable ($\bar{X}_{T1}$, $\bar{X}_{T2}$, and $\bar{X}_{T3}$).

Interaction effects become clearer with the help of plots of group means illustrated in Figure 8.3. These plots are based on fictitious data and are being used solely to facilitate a conceptual understanding of interactions in factorial ANOVA. With that said, take a break from the student achievement case study and imagine collecting data on ice cream ratings of males and females. Those ratings are plotted in Figure 8.3. In the plots, the dependent variable values (ratings) are on the vertical axis and independent variable values (ice cream flavors) are on the horizontal axis. The plotted lines represent mean group ratings of the other independent variable, gender.

The first three plots, Figure 8.3(a) demonstrate three examples of plots that show no interaction among ice cream preferences of males (M) and females (F). One common characteristic of the plots that do not display any interaction is that in all cases, the lines are parallel, even when the lines are not straight. In other words, the plots show that the differences in ice cream preferences between males and females are the same for all ice cream flavors. If an interaction is nonsignificant in ANOVA, plotting cell means is unnecessary. Mean ratings are plotted here only for illustrative purposes. If, however, there is a significant interaction in preferences, plots of group means would look like those in Figure 8.3(b).

The second set of plots display three possible types of interactions of ice cream preferences. One consistent indication of interaction is plotted lines that are not parallel. In the first plot, female ratings of ice cream flavors increase with changes in the flavor while the ratings of males remain unchanged for the same flavors. In the second plot, the lines actually cross, indicating that female participants

Figure 8.3: Illustration of interaction effects in ANOVA: Ice Cream Ratings

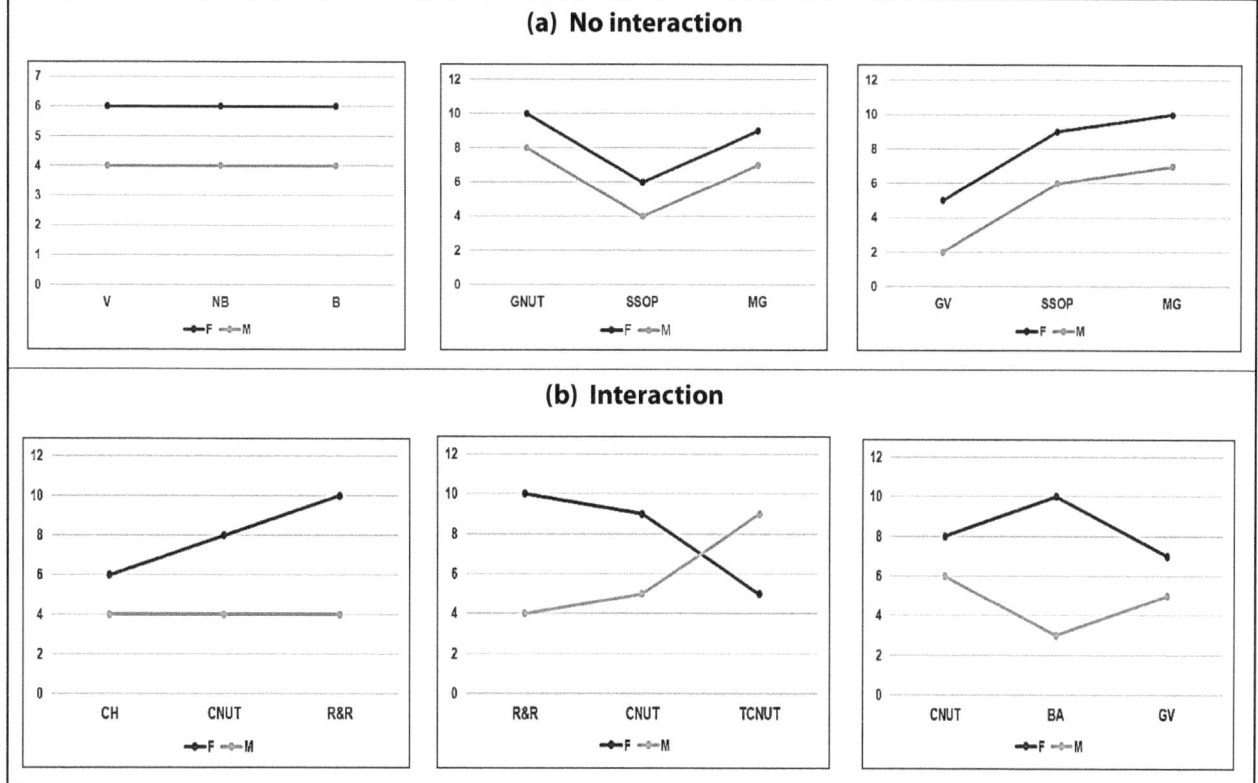

Ice Cream Flavors: B = banana, BA = butter almond, CH = chocolate, CNUT = coconut, GNUT = grape nut, GV = guava, MG = mango, NB = naseberry, R&R = rum and raisin, SSOP = soursop, TCNUT = toffee coconut, V = vanilla.

rate the first two flavors higher than males, but assign the third flavor a rating that is lower than males. In the third plot, the lines do not cross, but they move in opposite directions, reflecting the differing directions of the participants' ratings; in every case the effect of gender is not the same for every type of ice cream flavor.

Lines that are significantly nonparallel reflect an interaction of the independent variables on the dependent variable. When lines cross, the interaction effect is said to be disordinal. Lines that do not cross are said to reflect an ordinal interaction.

8.3.1.2 Factorial Designs Beyond Two-Way ANOVA

So far, the discourse has been restricted to two-way factorial designs. However, there are factorial ANOVA designs beyond two-way ANOVA. For example, adding the variable, poverty level, with two groups (high poverty and low poverty) to the two-way designs in Section 8.3.1.1, expands the designs to three-way factorial ANOVAs. The 2 x 2 design becomes a 2 x 2 x 2 design with gender, high school graduation and poverty level as the independent variables. Here each variable has two levels. The 2 x 3 design becomes a 2 x 3 x 2 design with age, school type, and poverty level as independent variables. Here age and poverty level each has two levels, while school type has three levels.

Two- and three-way ANOVAs are commonly-used research designs. However, four-way ANOVAs or more complex designs are rarely employed due to the difficulty of interpreting very complex results.

8.4 The Research Plan: The Case of Differences in Achievement

The two-way ANOVA design described in Section 8.3.1.1 is an extension of the research problem described in Box 7.1, which included a single independent variable—school type. The two-way design includes an additional variable—age. Details of the new case study are presented in Box 8.1.

8.4.1 The Research Problem

The research problem in the new case once again involves how students in different types of pre-secondary schools perform on the Communication Task in the secondary school placement test. In this case, the problem is expanded to the impact student age has on student performance in a school environment where most students sit the exam between the ages of 11 and 12.

8.4.2 The Research Questions

The first two questions are similar to those in Box 7.1. However, the third question signifies that the research inquiry targets three different effects on test performance: school type, age, and interaction effects. Researchers could choose to concentrate on the main effects of the independent variables or solely on interaction effects of the variables. Study objectives are driven by research interests and pre-existing investigations. Nevertheless, most investigations involving more than a single independent variable examine all possible effects.

8.4.3 Research Hypotheses

The first two hypotheses refer to the tests of main effects and center on the two independent variables. The first hypothesis refers to the row population means (for age) averaged across all columns (school type means). The second describes column population means (for school type) averaged across the rows (age means). In both cases, the null hypotheses suggest that group means will be the same; for example, the score means for students aged 11 and 12 will be equal.

Another way of stating the third null hypothesis for the interaction effect is: There is no significant interaction between age and school type after the main effects have been removed. The third hypothesis refers to whether the effect of one independent variable depends on the effect of the second independent variable. Operationally, the hypothesis suggests that if a researcher plotted the mean communication task scores for the groups of both independent variables, the plotted lines for the variables would be parallel; the pattern of effects on communication task scores would be the same for both variables. More details about these effects in Section 8.4.5.

8.4.4 The Research Design

Like one-way ANOVA, factorial ANOVA may be used to operationalize multiple research designs. However, for this case, with the use of intact variables, random assignment and selection of study participants are infeasible for both treatments (school type and age). These limitations restrict the research design to a nonexperimental causal comparison study.

8.4.5 Data Analysis: Hypothesis Testing

In the Case of Differences in Achievement, the target population could be all students aged 11 and 12 in all types of pre-secondary schools in the nation of Jamaica. Due to accessibility limitations, researchers

Box 8.1: Research Plan: The Case of Differences in Achievement

Research Statement
It is hypothesized that students who attend different types of pre-secondary schools perform dissimilarly on a high-school-placement test. In addition to the availability of different types of pre-secondary schools, students are free to take the high school admissions test at varying ages (e.g., age 11 and 12). Accordingly, based on cognitive development research literature, it is hypothesized that student performance changes with age.

Central Research Question
Is there a significant difference in the performance of students on a high school placement test at different ages in different types of pre-secondary schools?

Research Sub-questions
1. Is there a significant difference in the test performance of students at ages 11, and 12?
2. Is there a significant difference in test performance of students in different types of schools?
3. Is there a significant interaction between the performance of students at different types of schools at different ages?

Variables and Measurement Levels
Independent Variable: School Type (nominal level), Age (ordinal level)
Dependent Variable: Communication task score (interval level)

Hypotheses
Null Hypothesis 1: There is no significant difference in the performance of 11- and 12-year-old students:
$H_o: \mu_{A1} = \mu_{A2}$.

Null Hypothesis 2: There is no significant difference in the performance of students at three types of schools:
$H_o: \mu_{T1} = \mu_{T2} = \mu_{T3}$.

Null Hypothesis 3: The population cell means are equal after the main effects of age and school type have been removed.
$H_o: \mu_{11} - \mu_{21} = \mu_{12} - \mu_{22} = \mu_{13} - \mu_{23}$.

Research Design
This design shares some similarities with the research design in Box 7.1. It is a causal comparative design with average scores for three groups of school type ($X_1 - X_3$) and two groups of age (X_1 and X_2) representing two independent variables (or treatments) in the design notation. The dependent variable is communication task score, represented as O in the design notation.

$$
\begin{array}{ccc}
M & X_{11} & O \\
M & X_{12} & O \\
M & X_{13} & O \\
M & X_{21} & O \\
M & X_{22} & O \\
M & X_{23} & O \\
\end{array}
$$

With the inclusion of two independent variables—age and school type, the research design meets the requirements for a factorial causal comparative design. As the design notation indicates, in this factorial design, one independent variable has two groups of participants while the other has three groups, resulting in 2 x 3 factorial design or six groups of communication task scores to compare (e.g., $\bar{X}_{11}$ and $\bar{X}_{12}$).

Data Analysis
The central statistical analysis for this research study is to conduct hypothesis testing by using two-way ANOVA to calculate the main effects for Hypotheses 1 and 2 and the interaction effects for Hypothesis 3. A significant interaction takes precedence over significant main effect results. Post-ANOVA, simple effects tests will identify the pairs of average scores that contribute to an interaction between school type and age. Upon completing simple effects tests, multiple comparisons may be used to identify the average scores that contribute to a significant main effect for school type. No post hoc test is necessary for the two-group variable, age.

292 Factorial Analysis of Variance

use inferential statistics, in this case two-way ANOVA, to estimate the statistical likelihood that events described in the null hypotheses, and tested using a selected sample, would occur in the target population. This process is also called significance testing.

Box 8.1 includes three null hypotheses (H_0) to be tested in this case study. An alternative (or research) hypothesis (H_a) is adopted if a null hypothesis is rejected. The null hypotheses for the Case of Differences in Achievement are designed to test three distinct effects of interest on the dependent variable: the main effect of age, the main effect of school type, and the interaction effect between age and school type. Each will be discussed below. Each of the three effects is included in what is called an omnibus or global statistical test that is designed to test the overall null hypothesis: (1) There are no significant differences between the groups of study participants; (2) There is no significant interaction between the independent variables. The overall alternative hypothesis is that there is at least one significant difference or interaction among the groups.

8.4.5.1 Main Effects

The first two null hypotheses are designed to separately investigate differences in the effect of each independent variable: Null Hypothesis 1, the row variable (age); and Null Hypothesis 2, the column variable (school type). Main effects tests are like two one-way ANOVAs. Each variable has its own main effect omnibus test. Accordingly, the case in Box 8.1 has two independent variables and two main effects omnibus tests that correspond with Null Hypothesis 1 (for age) and Null Hypothesis 2 (for school type). Both effects are computed in a single ANOVA analysis.

The main effect for each independent variable is conducted as if the other variable does not exist. Testing the column variable (school type) involves comparing the performance of three types of schools to determine the probability that the mean scores are equal in the target population. Here, the main effect of school type involves calculating differences between the performance of students in School Type 1 ($\bar{X}_{T1}$), School Type 2 ($\bar{X}_{T2}$), and School Type 3 ($\bar{X}_{T3}$) while holding the second independent variable, age, constant. Similarly, testing the row variable (age) involves investigating whether there are significant differences in the performance of students aged 11 ($\bar{X}_{A1}$) and 12 ($\bar{X}_{A2}$) while holding school type constant.

8.4.5.2 Interaction Effects

The third null hypothesis is designed to examine the probability of significant differences in mean performance of one independent variable (e.g., mean performance of students aged 11 and 12) in different groups of the second independent variable (e.g., School Type 1, School Type 2, and School Type 3). For example, there may be a significant difference in performance between 11-year-olds in School Type 1 and School Type 3, but not in any other combinations. This would be described as a significant interaction (Section 8.3.1.1).

8.4.6 Post ANOVA Tests

Section 8.3 described the hypotheses and associated statistical effects (main and interaction effects) computed by the omnibus ANOVA analysis. What it did not discuss is what happens when there are significant differences in how the two independent variables affect the dependent variable. In these situations, researchers search for the source(s) of any significant differences detected, using post-ANOVA tests. Researchers' priorities are dependent on whether significant main effects or interactions are found.

Continuing the example of the ice cream ratings in Section 8.3.1, a finding of significant differences in ice cream preferences of males and females averaged across ice cream flavors is a terminal finding; with only two participant groups; there is nothing more that can be said about that main effect result. A second potential main effect is differences in the ratings of the ice cream flavors averaged across gender. With more than two types of ice cream flavors, a main effect for that variable means that follow-up tests may be conducted to identify the source of the difference. Significant differences may be between flavor A and B, B and C, A and C, etc. Under these circumstances, post hoc testing described in Chapter 7 may be considered.

A finding of significant interaction means that differing relationships exist among groups of participants, and therefore, restricting follow-up analyses to investigating significant main effects may result in a limited picture of study findings. Indeed, Howell (2014) declares that "in many cases, the interaction … may be of greater interest than the main effects" (p. 449). The analysis of simple effects helps researchers to fully investigate the causes of significant interactions.

8.4.6.1 Simple Effects

Conducting simple effects is a technique used to examine the effect of one independent variable in one group on each group of another independent variable. For example, female ice cream ratings may be significantly higher than males for some flavors but not for others (e.g., grape nut vs. chocolate). In the Case of Differences in Achievement, a simple effects analysis may reveal differences in test performance among three types of pre-secondary schools for 12-year-old students, but not for 11-year-olds.

In general, testing all possible simple effects after a significant interaction is not recommended; it increases the probability of a Type I error. The preferred practice is to conduct simple effects that are supported by the research literature and are incorporated in research hypotheses.

8.5 Assumptions of Factorial ANOVA

The assumptions of factorial ANOVA and one-way ANOVA described in Chapter 7, are identical:
- Samples are independently selected from a defined population in that each research study participant is independent of other participants in the study sample.
- The dependent variable is measured on at least an equal interval scale.
- Scores for the dependent variable are normally distributed in the target population.
- The variance of scores of the outcome variable is the same for all groups of scores (homogeneity of variance).

The consequences of violating one or more of the above assumptions were discussed in Section 7.5. However factorial ANOVA is robust to violations of the assumptions, especially when there is an equal number of participants in each study group.

8.6 Unequal Sample Sizes

Unlike one-way ANOVA which is robust to problems associated with unequal sample sizes (Tabachnick & Fidell, 2018), unequal sample sizes (unbalanced designs) in factorial ANOVA present more challenges. Factorial ANOVA with unequal cell or group sizes can lead to misleading results where greater emphasis is given to larger groups of observations. In addition, disproportionate group sizes may lead to correlation between the independent variables and confusion about the extent to which each affects the outcome variable. The simplest solution is to ensure group sizes are equal by randomly deleting cases from groups until all groups are equal. Preferred practice is to use the unweighted means approach in factorial ANOVA procedures (Tabachnick & Fidell, 2018). If both of those solutions are untenable, Howell (2014) recommends using the default solution offered by commonly used statistical software, which usually provides the most meaningful results in this situation.

8.7 Calculations for Factorial ANOVA

Section 7.4.2 provided an insider's view of how we arrive at one-way ANOVA results, which helps unveil what contributes to one-way ANOVA computer results. Two-way ANOVA hand calculations are an expansion of the one-way ANOVA procedure, and a variety of statistical software is available that will make quick work of analyzing data for a two-way ANOVA. Nevertheless, hand computation may provide a conceptual understanding of how computerized results are derived to facilitate the accurate interpretation of results.

8.7.1 Hypotheses

Before using SPSS to conduct a factorial ANOVA, the small dataset in Table 8.1 will be used to hand calculate a two-way ANOVA. The hand calculation is based on small sample of fictitious communication task scores from three types of schools with twelve students in each group, for a total of 36 scores. The maximum score possible is 12 and the minimum score is 2. Each school sample includes six students aged 11 and six aged 12 for a total of 18 students aged 11 and 18 students aged 12. With three types of schools and two age groups, as well as equal numbers of students in each group, this is a 2 x 3 balanced factorial design. The data for this example is displayed in Table 8.1. The null hypotheses in Box 8.1 may be used for the hand-calculation.

Null Hypothesis 1

There is no significant difference in the performance of 11- and 12-year-old students:

$H_o: \mu_{A1} = \mu_{A2}$

Null Hypothesis 2

There is no significant difference in the performance of students at three types of schools:

$H_o: \mu_{T1} = \mu_{T2} = \mu_{T3}$

Null Hypothesis 3

Population cell means are equal after the main effects of age and school type have been removed.

$H_o: \mu_{11} - \mu_{21} = \mu_{12} - \mu_{22} = \mu_{13} - \mu_{23}$

Calculation of the two-way ANOVA is computationally similar to the one-way ANOVA described in Chapter 7. Computing the one-way ANOVA includes:
- calculating the total sum of squares (SST); total variance between all scores
- splitting the total sum of squares into variance that can be explained by independent variables (sum of squares between groups or SSB) and variance that cannot be explained (sum of squares error or SSE), also known as sum of squares within groups (SSW).

Figure 8.4 Separating the Variance in Two-Way ANOVA

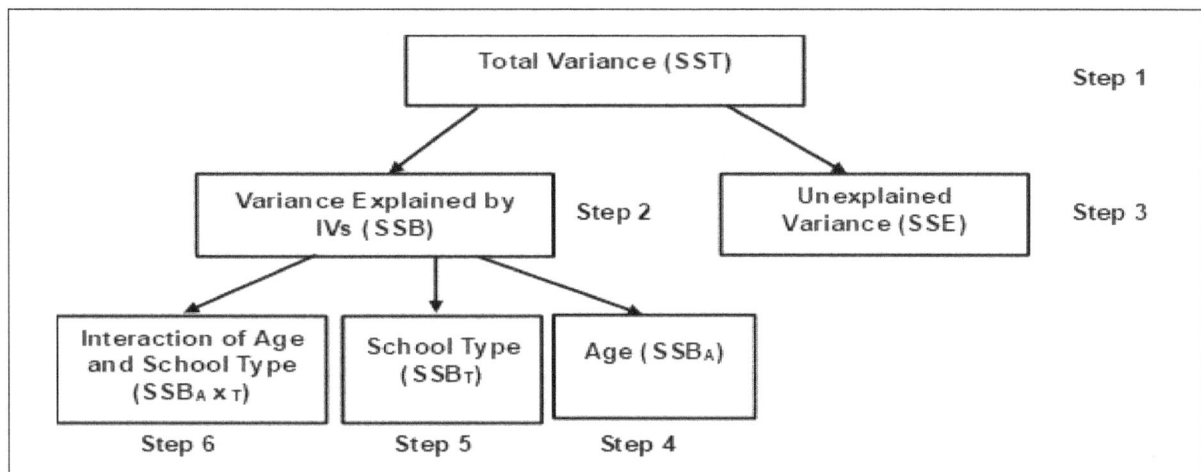

To compute a two-way ANOVA, add the following to the one-way ANOVA computation:
- Since the variance explained by the sum of squares between groups is made up of three between-group components (Figure 8.4) rather than one component in the one-way ANOVA, the between-group sum of squares is divided into variance explained by:
  o the first independent variable, age (SSB_A)
  o the second independent variable, school type (SSB_T), and
  o the interaction between both variables ($SSB_{A \times T}$) depicted in Figure 8.4.

8.7.2 Calculating Total Sum of Squares

The procedure for calculating the SST for two-way ANOVA is identical to the computation for SST for one-way ANOVA:

$$SST = \sum_{i=1}^{N}(X_i - \bar{X}_{grand})^2$$

- Calculate the grand mean ($\bar{X}_{grand}$) or overall average score by finding the sum of all 36 scores and dividing that sum by the total number of scores in the dataset ($N = 36$); that is the same as summing the total for each group (Group Total) shown in Table 8.1 and dividing by the total number of scores: $(115 + 90 + 66)/36 = 7.528$
- Compute the difference between the grand mean and each of the 36 scores in the data set (X_i). Then square the result.
- To derive the SST, find the cumulative total of the squared differences by summing all the squared values.

Adding the squared values will produce a total sum of squares of 154.972 units of variance. As illustrated in Figure 8.4, the total sums of squared variances will be broken down into variance explained by the independent variables age, school type, the interaction of age and school type as well as unexplained variance or error.

8.7.3 Between-Group Sum of Squares (SSB)

Integral to in computing two-way ANOVA is the calculation of the total between-group sum of squares, which is also identical to the method used for the one-way ANOVA. Recall that there is a total of 154.972 units of variance (SST) in the model. Before pulling out the variance explained by specific groupings, it is necessary to calculate the variance explained by all groups:

$$SSB = \sum N_k (\bar{X}_k - \bar{X}_{gm})^2$$

Section 7.4.2.2 showed the process for calculating the sums of squares between groups. In that computation there were three groups of study participants. Here, there are a total of six groups of six participants each. To derive the between-group sums of squares:

- Find the average score for each group ($\bar{X}_k$).
- Subtract each group average from the grand mean ($\bar{X}_{gm}$). (Recall that the grand mean equals 7.528.) Square the result and multiply that by the number of scores in the group (6).
- To compute the SSB, find the total outcome for all the groups.

The final computation looks like this:

$$SSB = 6(10.5 - 7.528)^2 + 6(8.5 - 7.528)^2 + 6(4.833 - 7.528)^2 + 6(8.667 - 7.528)^2$$
$$+ 6(6.5 - 7.528)^2 + 6(6.167 - 7.528)^2$$
$$SSB = 6(8.833) + 6(0.945) + 6(7.263) + 6(1.297) + 6(1.057) + 6(1.852)$$
$$SSB = 52.997 + 5.669 + 43.578 + 7.784 + 6.341 + 11.114$$
$$SSB = 127.483$$

As the computation illustrates, the independent variables account for 127.483 units of variance. The model comprises a total of six groups of scores. Therefore, the degrees of freedom associated with the between-group sums of squares is the number of groups (k) minus 1 or 6 – 1 = 5 df.

8.7.4 Error Sums of Squares (SSE)

Before separating out the variance in the between-group sums of squares for each independent variable, calculate the error sums of squares. The error sum of squares represents the variance that cannot be explained or individual within-group differences in performance.

Section 7.4.2.4 described three methods of calculating the error sum of squares for one-way ANOVA. The same methods work to calculate the error sum of squares for two-way ANOVA. Perhaps the simplest method to find the estimated error variance for the six groups of participants is to subtract SSB from SST:

$$SSE = 154.972 - 127.483$$
$$SSE = 27.489$$

8.7.5 Sums of Squares Main Effects for Age (SSBA)

Going back to the between-group sums of squares, with the value of the error settled, it is time to split the between-group sum of squares into the variance accounted for by each independent variable and the interaction between the variables.

Calculate the between-group variance for the variable, age. To do this, regroup scores according to student age groups, and ignore the school type variable for the moment. Age has two groups—18 students aged 11 (the top half of Table 8.1) and 18 aged 12 (the bottom half of Table 8.1).

Second, apply the between-group sums of squares equation in Section 8.7.3 for the between-group sums of squares for age:
- Find the age 11 variance by subtracting the group average from the grand mean. (Recall that the grand mean = 7.528.) Square the result.
- Do the same for the age 12 group.
- Multiply the outcome for each group by 18 (the number of students in each group)
- Total the result of the products.

$$SSB_A = 18(7.944 - 7.528)^2 + 18(7.111 - 7.528)^2$$
$$SSB_A = 18(0.173) + 18(0.174)$$
$$SSB_A = 3.115 + 3.13$$
$$SSB_A = 6.245$$

Based on the solution for the equation, the main effect of age accounted for 6.245 units of variance.

8.7.6 Sum of Squares Main Effect for School Type (SSBT)

To find the variance accounted for by the independent variable, school type, first regroup the sample of scores into three school type groups, ignoring the ages of students. All students in School Type 1 are in one group; students in School Type 2 are in a second group, and students in School Type 3 are assigned to a third group. There are still a total of 36 students, but this time there are 12 students in each group.

Second, apply the sums of scores between-groups equation for the school type between-group sums of squares, applying the same grand mean used before—7.528.
- Subtract each school type average from the grand mean (both from Table 8.1) and square the result.
- Multiply the variance for each school type by 12 (the number of students in each school type sample).
- Total the three resulting products.

$$SSB_T = 12(9.58 - 7.528)^2 + 12(7.5 - 7.528)^2 + 12(5.5 - 7.528)^2$$
$$SSB_T = 12(4.211) + 12(0.001) + 12(4.113)$$
$$SSB_T = 50.528 + 0.009 + 49.353$$
$$SSB_T = 99.89$$

The main effect of school type accounts for 99.89 units of variance.

8.7.7 The Interaction Effect (SSBA x T)

Calculate how much variance is accounted for by the interaction of the two independent variables, age and school type. One method of deriving the interaction effect is using this two-step process:
- Sum the variance accounted for by age and school type.
- Subtract that total from the between-group sums of squares.

$$SSB_{A \times T} = SSB - (SSB_A + SSB_T)$$
$$SSB_{A \times T} = 127.483 - (6.245 + 99.89)$$
$$SSB_{A \times T} = 127.483 - 106.135$$
$$SSB_{A \times T} = 21.348$$

8.7.8 Computing the *F* Ratios

As described in Chapter 7, the *F* ratio is the analysis of variance measuring stick to determine whether significant differences exist between the average variance between groups (age or school types) and the average variance within groups (error). It is also used to measure the average variance due to interaction between independent variables. This statistic measures the ratio between mean variance that has been accounted for (systematic or between-group variance) and mean variance that has not been accounted for (error or unsystematic variance). In hand calculation, the variance accounted for that will be used to compute *F* ratios includes sum of squares for the main effect of age, school type, and the interaction between age and school type.

Recall that calculating the *F* ratio for the one-way ANOVA involved, first, finding the average between-group sum of squares (MSB) and the average error sum of squares (MSE). The *F* ratio was the ratio of those means—the mean sum of squares to the error sum of squares.

Finding the averages for the one-way ANOVA required four items:
- the between-group sums of squares,
- the error sums of squares, and
- the degrees of freedom for each.

Similarly, the averages needed for the *F* ratios in the two-way ANOVA, require the following:
- sum of squares for age
- sum of squares for school type
- sum of squares for the interaction effect
- error sum of squares
- degrees of freedom for each of the sum of squares.

8.7.8.1 Computing the Degrees of Freedom for F Ratios

Calculate the mean squares by taking the sum of squares for each effect and dividing each by their respective degrees of freedom (df). Start by finding the degrees of freedom for the main effects of age and school type. The degrees of freedom for those main effects are k – 1, where k = the number of groups involved.

- The independent variable, age, has two age groups (age 11 and 12), so the degrees of freedom for Age (df_A) works out to be $df_A = 2 - 1 = 1$.
- The independent variable, school type, includes three groups (School Type 1, 2, and 3), so the degrees of freedom for school type (df_T) becomes $df_T = 3 - 1 = 2$.

The calculation of the degrees of freedom for the interaction effect ($df_{A \times T}$) is somewhat different from the previous two; it is the product of the degrees of freedom for the main effects of age and school type or $df_A \times df_T$. Therefore, outcome is $df_{A \times T} = 1 \times 2 = 2$.

As with the error degrees of freedom (df_E) for one-way ANOVA, the df_E for two-way ANOVA can be computed by taking the total sample size (36) and subtracting from it the total number of groups (6). Accordingly, the computed degrees of freedom for the error sum of squares becomes $df_E = n - k$ or $df_E = 36 - 6 = 30$.

8.7.8.2 Computing the Mean Squares and F-Ratios

With the degrees of freedom computed, continue by calculating the mean square for each effect as well as the error variance. The calculated mean squares for age, school type, the interaction between age and school type as well as the mean square error are as follows:

$$MSB_A = \frac{SSB_A}{df_A} = \frac{6.245}{1} = 6.245$$

$$MSB_T = \frac{SSB_T}{df_T} = \frac{99.89}{2} = 49.945$$

$$MSB_{AxT} = \frac{SSB_{AxT}}{df_{AxT}} = \frac{21.348}{2} = 10.674$$

$$MSE = \frac{SSE}{df_E} = \frac{27.489}{30} = 0.916$$

Finally, calculate the F ratio for each main effect and interaction. To do that, use the same method employed for the one-way ANOVA: for each effect, divide the mean square by the mean square error.

$$F_A = \frac{MSB_A}{MSE} = \frac{6.245}{0.916} = 6.818$$

$$F_T = \frac{MSB_T}{MSE} = \frac{49.945}{0.916} = 54.525$$

$$F_{AxT} = \frac{MSB_{AxT}}{MSE} = \frac{10.674}{0.916} = 11.653$$

8.7.9 Evaluating and Interpreting the F Ratio

With the three F statistics calculated, use them to make decisions about whether the apparent differences between the average scores in the sample displayed in Table 8.1 are likely to occur in a population of similar students. To determine whether to reject the null hypotheses, use the method described in Section 7.4.2.7.

Find the critical value for each effect (age, school type, and the interaction), which are the values of F statistics beyond which researchers may reject the null hypothesis. The critical value for each F statistic is based on the degrees of freedom (Section 8.7.8.1) for the numerator and denominator of the related means square ratio. After identifying the critical values, compare each F statistic against the associated value. The a priori alpha level is .05. Critical values for the effects are as follows:

$$F_A = 4.17, where\ df_A = 1\ and\ df_E = 30$$

$$F_T = 3.32, where\ df_T = 2\ and\ df_E = 30$$

$$F_{AxT} = 3.32, where\ df_{AxT} = 2\ and\ df_E = 3.32$$

Each *F* statistic in 8.7.8.2 exceeds its related critical value. As a result, reject the null in each case at the alpha level of .05 and conclude that there is evidence of significant differences between the average scores of students aged 11 and 12. In addition, there are significant differences between the average scores of students in school type 1, 2, and 3. Finally, there is a significant interaction between age and school type.

Figure **8.5** presents the results of the omnibus test. As the summary table reports (in the **Sig**. column), there are significant main effects for age and school type; there are significant differences between the scores of students age 11 and 12 ($p < .05$) and there are significant differences between School Type 1, 2 and 3 ($p < .05$). There is also a significant interaction between school type and age ($p < .05$).

Figure 8.5 Summary Table: Omnibus Analysis Results

Source	df	Sum of Squares	Mean Square	F	Sig.
School Type	2	99.89	49.945	54.525*	.05
Age	1	6.245	6.245	6.818*	.05
School Type * Age	2	21.348	10.674	11.653*	.05
Error	30	27.489	0.916		
Corrected Total	35	154.972			

What Figure 8.5 does not show is the sources of the differences for school type as well as the interaction between school type and age. Post hoc multiple comparison tests uncover the sources of the differences in variable groups by comparing average scores of all possible pairs of the groups (e.g., differences between School Type 1 vs School Type 2) while controlling the family-wise error rate. Simple effects may be used to find the source of the significant difference in the interaction of the independent variables. Both of these tests may be calculated manually, especially with a small data set. Nevertheless, the next section will use SPSS to replicate the omnibus test and conduct the post-hoc tests.

8.8 SPSS Specifications for Factorial ANOVA

The process for conducting a factorial analysis using SPSS is similar to the method used for one-way ANOVA. Certainly, for every analysis it is important to ensure that the data meet all the assumptions for the statistical procedure before initiating it. The assumptions are the same as those for the one-way ANOVA. The SPSS General Linear Model (GLM) procedure to be used for this analysis includes Levene's test of homogeneity of variance. The data meet the remaining assumptions, including acceptably normal distribution of scores in each group, independence of observations across groups of participants, and appropriate variable measurement levels.

To begin, retrieve the dataset, **Factorial_ANOVA_Demo.sav** (Figure 8.6). Although the data are the same as the dataset displayed in Table 8.1, they have been reconfigured to facilitate analysis. For example, in order to distinguish between school types, the variable, **Schtype**, was created with the values, 1, 2, and 3 representing the three types of schools in the analysis. The variable, **ID**, was created to distinguish between the 36 cases in the dataset (especially helpful when testing for outliers). Finally, the variable, **Age**, was created. The variables in the dataset are: **Schtype** (School Type) and **Age** (Age) as well as **Score** (Communication Score). Each observation exactly corresponds with the data in Table 8.1. For example, the first observation (ID=1) is in Private T_1, is age 11, and has a score of 10.

To conduct factorial ANOVA, select **Analyze → General Linear Model → Univariate** (Figure 8.6) to open the **Univariate** window for factorial ANOVA. In the **Univariate** dialog window (Figure 8.7), highlight the dependent variable, **Score**, in the variables list on the left side of the window, and click the arrow next to the **Dependent Variables** panel to move the variable to that space. There is space for a single variable in this box because univariate statistics is limited to statistics designs, like this one, with a single dependent variable.

Figure 8.6 First 25 Cases of Data Set in SPSS Data Editor

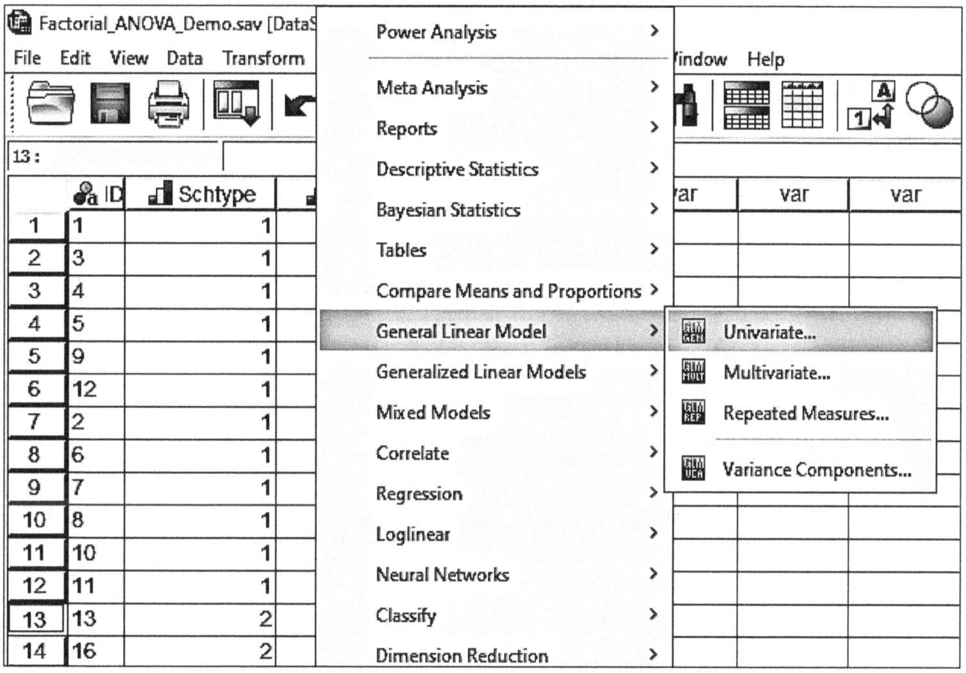

Select the independent variables, **Age** and **Schtype,** by highlighting them and clicking the arrow next to the **Fixed Factor(s)** panel below the **Dependent Variable** panel. Although univariate analyses are limited to one dependent variable, more than one independent variable (fixed factor) is permitted. The remaining spaces in this dialog window remain empty for this analysis.

After selecting the variables for analysis, click on the **Options** pushbutton to open the **Univariate Options** window (Figure 8.8). Check **Descriptive Statistics** and **Homogeneity tests**. The **Homogeneity Tests** option will produce the results of Levene's test of equality of error variances, which assesses the ANOVA assumption that the variance of the values of the dependent variable, **Score**, is not significantly different across the three school type groups and the two age groups. If differences in the group score variances are nonsignificant, the study data have met the ANOVA assumption. Click **Continue** to return to the main Univariate dialog window.

The specifications so far are sufficient to run factorial ANOVA and produce results that include descriptive statistics, results of the homogeneity tests of variance, and the omnibus test for factorial ANOVA. It is conventional to review omnibus test results to assess whether there is a need for post-ANOVA tests. If the omnibus test reveals that there are significant differences between independent

302 Factorial Analysis of Variance

variable groups and/or there is a significant interaction between the independent variables, then appropriate post-ANOVA tests may be conducted. The nature of omnibus test results governs whether there is a need for post-ANOVA tests and which ones should be conducted. However, for this demonstration, previous hand-calculated results (Figure 8.5) resulted in statistically significant F ratios for both independent variables and the interaction of the two independent variables. Therefore, for purposes of efficiency, specifications for post-ANOVA tests will be added to those for the omnibus test.

Figure 8.7 Main Dialog Box for Univariate ANOVA

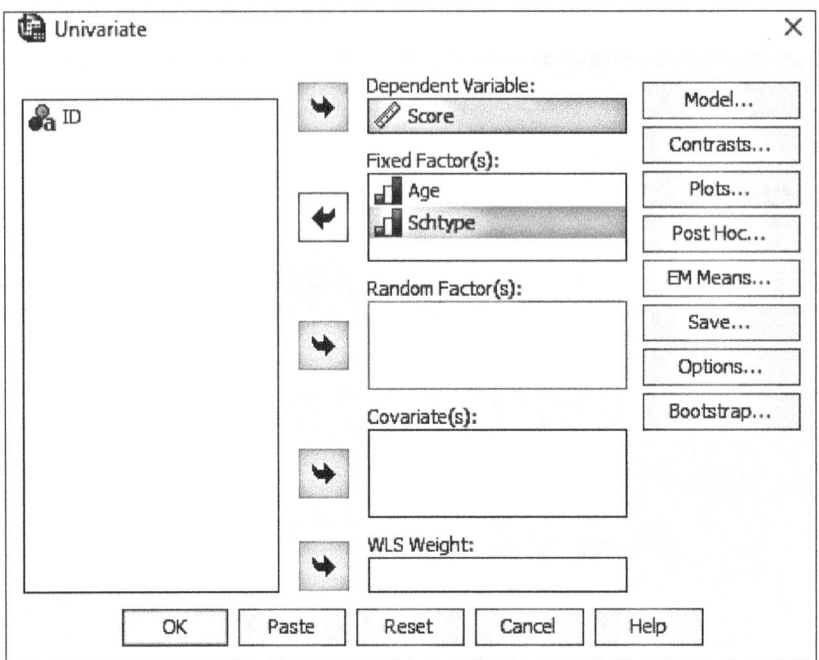

Figure 8.8 Options Window for Univariate ANOVA.

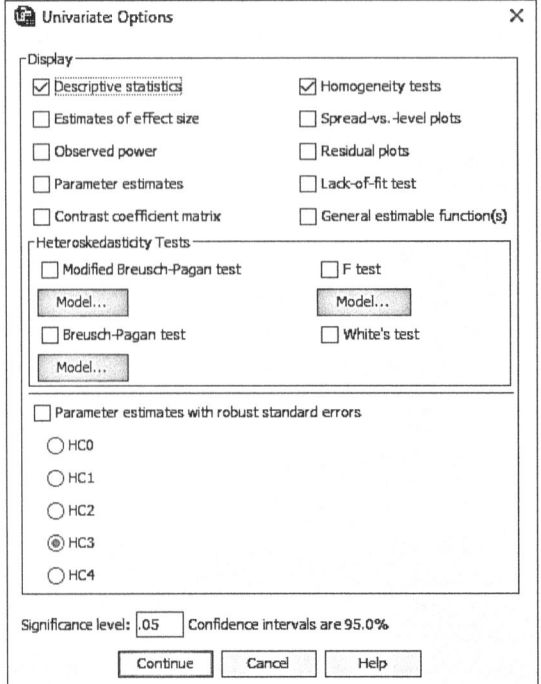

The variable, age, includes two groups (age 11 and age 12), as a result there is no need for post-hoc tests following the omnibus test. Manual computation of the main effect for age (Figure 8.5) showed that students aged 11 on average, have statistically significantly higher scores than students aged 12. On the other hand, the variable school type includes three groups—Private (T_1), Public 1 (T_2), and Public 2 (T_3). Post hoc tests are needed to identify the source of the differences for that variable.

To specify post-hoc tests for school type, click on the **Post Hoc** pushbutton in the Univariate dialog window to open the **Univariate: Post Hoc Multiple Comparisons for Observed Means** dialog window (Figure 8.9). Transfer the variable **Schtype** from the **Factors** box to the **Post Hoc Tests For** panel. Click **Tukey** under **Equal Variances Assumed**, and click **Continue** to return to the **Univariate** main dialog window.

Figure 8.9 Post Hoc Multiple Comparisons Window for Univariate ANOVA

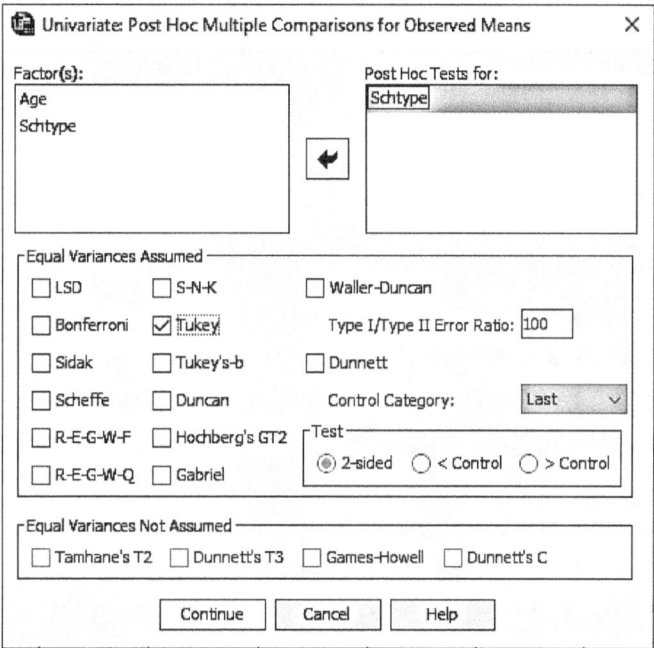

Based on previous calculations (Section 8.7.9) that showed a statistically significant interaction between school type and age, engage two methods to identify the source of the interaction.

- Plot a line graph of the data that will vividly illustrate the combined effect of age and school type on students' scores.
- Specify a simple effects analysis that will detect the effects of one independent variable at different levels of the other independent variable (Section 8.4.6.1).

To produce a two-line graph of the interaction effect, select the **Plots** pushbutton from the main **Univariate** dialog window (Figure 8.7) to open the **Univariate Profile Plots** dialog menu. Click over the variable, **Schtype**, from the **Factors** panel to the **Horizontal Axis** space, and click over the variable, **Age**, to the **Separate Lines** space (Figure 8.10a). Now click on **Add** next to **Plots** to move the line plot (**Schtype*Age**) to the **Plots** panel (Figure 8.10b). This instruction will result in a two-line graph: one displaying the average communication task scores of 11-year-olds in each school type and the other showing the average scores of 12-year-olds. Under **Chart Type**, keep the default selection **Line Chart**. Click **Continue** to return to the main Univariate dialog window.

Figure 8.10 IBM SPSS Univariate Profile Plots Window

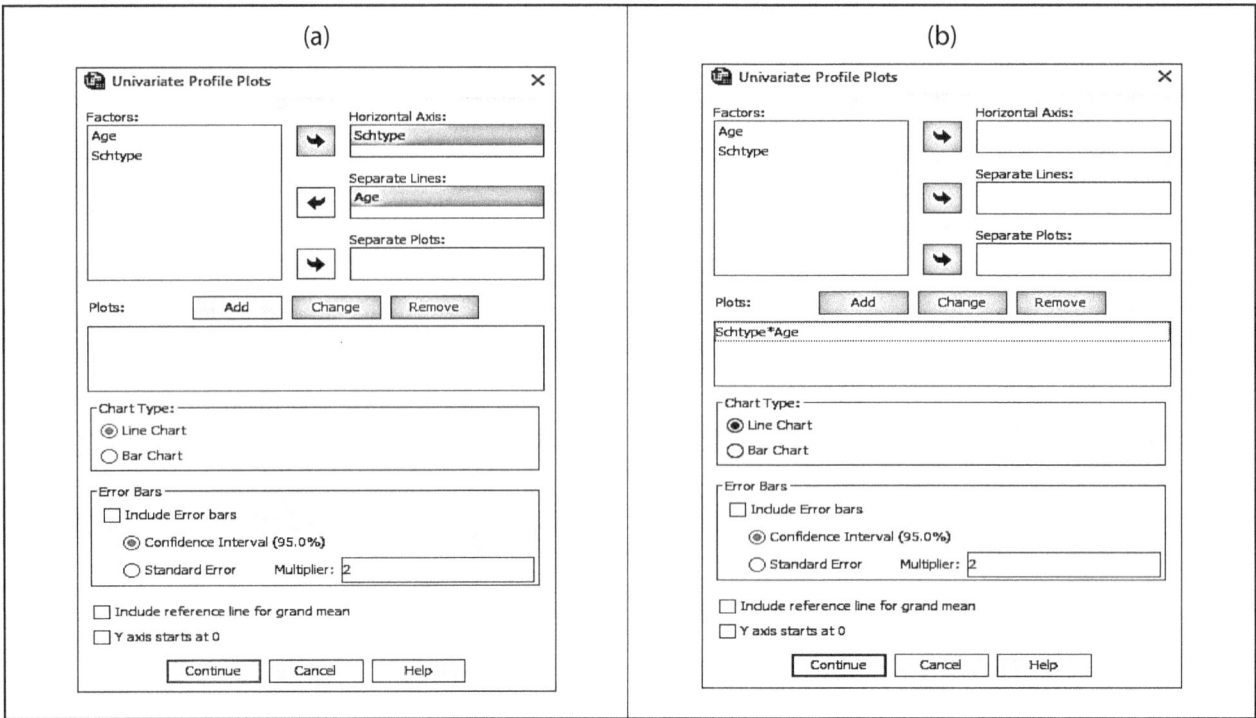

For the tests of simple-effects, click on the **EM** (Estimated Marginal) **Means** pushbutton to open the **Univariate: Estimated Marginal Means** dialog box (Figure 8.11). Highlight the **Age*Schtype** in the **Factors and Factor Interactions** panel, and click on the arrow to transfer it to the **Display means for** box.

To conduct the simple effects test, select the second option below the **Display Means for** panel, which is **Compare simple main effects**. Under **Confidence interval adjustment**, scroll down to select **Bonferroni**, which will maintain the family-wise alpha level at .05, thereby avoiding alpha level inflation that could occur with multiple pairwise comparison of means in this analysis. Click on **Continue** to return to the **Univariate** main dialog window. Then click **OK** to run the analyses.

Figure 8.11 SPSS Univariate Marginal Means Dialog Window

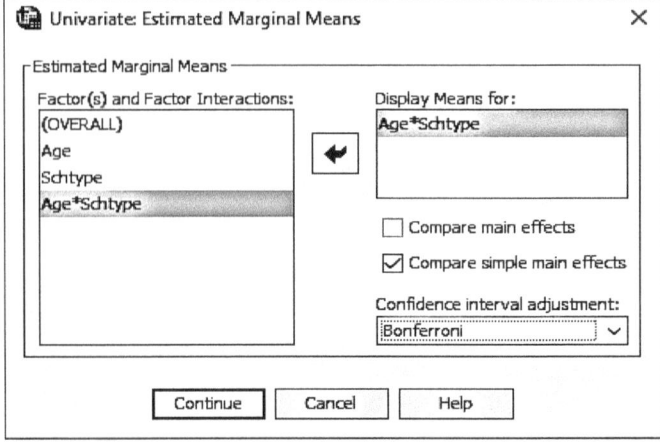

8.9 Factorial ANOVA SPSS Results

The results of the omnibus test appear in Figure 8.12. The traditional Levene's test (differences based on mean) appears in the first row of the first table (Figure 8.12a). The other three tests are more robust to the effects of non-normality and outliers in the dataset. They were added to Levene's test by Brown and Forsythe (1974) to accommodate data with a skewed underlying distribution. Accordingly, error variance significance tests based on the median and trimmed mean could be used as the preferred guideline in interpreting the outcome of error variance tests when statistical techniques robust to non-normal distributions are used. Nevertheless, all the results of the **Levene's Test of Equality of Error Variances** report nonsignificant differences in score variances of the independent variable groupings ($p > .05$). Nonsignificant differences among group scores mean that the data have not violated the ANOVA assumption of equal variances.

Figure 8.12 Results of the factorial ANOVA omnibus analysis

(a)

		Levene's Test of Equality of Error Variances[a,b]			
		Levene Statistic	df1	df2	Sig.
Score	Based on Mean	.447	5	30	.812
	Based on Median	.468	5	30	.797
	Based on Median and with adjusted df	.468	5	25.491	.797
	Based on trimmed mean	.425	5	30	.827

Tests the null hypothesis that the error variance of the dependent variable is equal across groups.
a. Dependent variable: Score
b. Design: Intercept + Age + Schtype + Age * Schtype

(b)

Tests of Between-Subjects Effects

Dependent Variable: Score

Source	Type III Sum of Squares	df	Mean Square	F	Sig.
Corrected Model	127.472[a]	5	25.494	27.812	<.001
Intercept	2040.028	1	2040.028	2225.485	<.001
Age	6.250	1	6.250	6.818	.014
Schtype	100.056	2	50.028	54.576	<.001
Age * Schtype	21.167	2	10.583	11.545	<.001
Error	27.500	30	.917		
Total	2195.000	36			
Corrected Total	154.972	35			

a. R Squared = .823 (Adjusted R Squared = .793)

The results of the omnibus analysis (Figure 8.12b) are similar to the outcomes of the manual calculations displayed in Figure 8.5. SPSS F statistic calculations of the main effects for **Age** and **Schtype** is the same as hand calculations for **Age**, but slightly higher for **Schtype** (difference of 0.051). The outcomes of SPSS and hand calculations show that the main effects of **Schtype** and **Age** are statistically significant. This means that the average scores, averaging across student age, are significantly different across school type, and the average scores for students aged 11 and 12, averaging across school type, are significantly different from each other.

As in one-way ANOVA, to compute the amount of variance each effect accounts for in student scores, divide the school type sum of squares (SS_T) by the corrected total sum of squares (SST). According to the main effect for school type, it accounts for just under an estimated two-thirds of the variance in students' performance (100.056/154.972 = .6456 or 64.56%). By contrast, the main effect for age accounts for approximately 4% of the variance in students' performance (6.250/154.972 = .0403 = 4.03%).

In addition to similar main effect results, the SPSS estimate of the F statistic for the interaction effect is slightly lower than the hand calculation (a difference of 0.108). However, both methods generated statistically significant interactions between school type and age. This means that the effect of school type on test scores is different for 11- and 12-year-olds. The interaction of age and school type accounts for just under 14% of the variance in student performance, (21.167/154.972 = .1366 or 13.66%).

There are two additional statistics in the SPSS table that were not included in the manual calculations and that are typically ignored when interpreting factorial ANOVA results— **Corrected Model** and **Intercept**. These are standard outputs for the General Linear Model, which is the procedure used to conduct factorial ANOVA in SPSS, and they do not add to the understanding of the outcome. The **Corrected Model** includes only statistics for the main effects and interaction. The test of the intercept tests the null hypothesis that the mean of all the scores is zero. It is the estimate of the dependent variable when all the independent variables equal 0.

As in one-way ANOVA, the **Total** line is a total of all sources of variance (both main effects, interaction, error and the intercept). The **Corrected Total**, which is not included in the results of one-way ANOVA, consists of only those sources included in the model (both main effects, interaction and error).

The final statistic of note in the omnibus results is R-Squared and Adjusted R-Squared in the footnote of Figure 8.12b, which reports that 82.3% and 79.3%, respectively, of the variance in student performance (score) is attributable to school type (**Schtype**), age (**Age**), and the interaction of those variables.

Omnibus ANOVA results revealed significant differences between independent variable groups and the interaction of the independent variables, but questions remain about sources of the differences. The results of a post hoc multiple comparisons analysis of school type and simple effects tests for the interaction of school type and age identified sources of the differences found in the omnibus results. However, a significant interaction supersedes significant main effects. Therefore, a discussion of the results of simple effects will precede those of post hoc results.

Figure 8.13 displays a profile plot of the estimated marginal means of the scores of students age 11 and 12. Profile plots graphically illustrate the relationship between independent variables and their effect on the dependent variable. Estimated marginal means are means adjusted for the presence of other variables in the model. In one-way or factorial ANOVA, the estimated marginal means are identical to the observed means. More on this topic in Chapter 9.

Figure 8.13 Two-Way ANOVA Profile Plot: Estimated Marginal Mean Scores

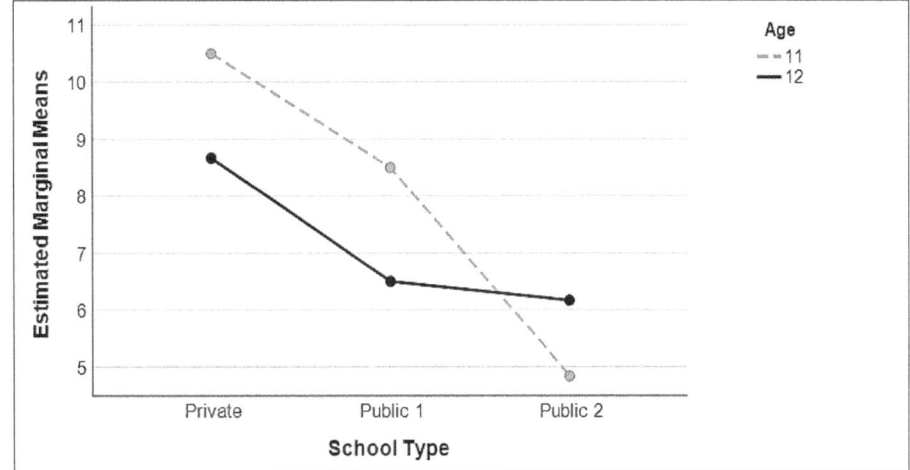

As the plot illustrates, the effect of school type on the pattern of performance is similar across age groups in Private and Public 1, but not in Public 2.
- There is a distinct difference between the scores of 11- and 12-year-olds across school types.
- In the case of Private and Public 1, students age 11 outperform students who are age 12.
- In comparison with Private and Public 1 schools, the pattern of performance changes in Public 2 schools in two ways. First, the scores of students in Public 2 schools are lower than those in Private and Public 1 schools. Second, in Public 2 schools, age group performance flips, and students age 12 perform better than those age 11. This is a result of a more moderate decline in performance in Public 2 schools for students age 12 than for students age 11.

Simple effects test results in Figures 8.14 and 8.15 substantiate findings in the profile plot.
- The results of the **Pairwise Comparisons** in Figure 8.14 show the effect of the variable, age, on student performance for the three types of school. The results show that performance differences between age groups are significant in every school type ($p < .01$ and $p < .05$).

Figure 8.14 Two-Way ANOVA Paired Comparisons for the Main Effect of Age

						95% Confidence Interval for Difference[b]	
School Type	(I) Age	(J) Age	Mean Difference (I-J)	Std. Error	Sig.[b]	Lower Bound	Upper Bound
Private	11	12	1.833*	.553	.002	.704	2.962
	12	11	-1.833*	.553	.002	-2.962	-.704
Public 1	11	12	2.000*	.553	.001	.871	3.129
	12	11	-2.000*	.553	.001	-3.129	-.871
Public 2	11	12	-1.333*	.553	.022	-2.462	-.204
	12	11	1.333*	.553	.022	.204	2.462

Based on estimated marginal means
*. The mean difference is significant at the .05 level.
b. Adjustment for multiple comparisons: Bonferroni.

- The **Pairwise Comparisons** table in Figure 8.15 shows the effect of school type on student performance for each age group. The results support the performance depicted in Figure 8.13, showing significant performance differences between all paired comparisons for students age 11 ($p < .01$).

Figure 8.15 Two-Way ANOVA Pairwise Comparisons for the Effect of School Type

						95% Confidence Interval for Difference[b]	
Age	(I) School Type	(J) School Type	Mean Difference (I-J)	Std. Error	Sig.[b]	Lower Bound	Upper Bound
11	Private	Public 1	2.000*	.553	.003	.598	3.402
		Public 2	5.667*	.553	<.001	4.265	7.068
	Public 1	Private	-2.000*	.553	.003	-3.402	-.598
		Public 2	3.667*	.553	<.001	2.265	5.068
	Public 2	Private	-5.667*	.553	<.001	-7.068	-4.265
		Public 1	-3.667*	.553	<.001	-5.068	-2.265
12	Private	Public 1	2.167*	.553	.001	.765	3.568
		Public 2	2.500*	.553	<.001	1.098	3.902
	Public 1	Private	-2.167*	.553	.001	-3.568	-.765
		Public 2	.333	.553	1.000	-1.068	1.735
	Public 2	Private	-2.500*	.553	<.001	-3.902	-1.098
		Public 1	-.333	.553	1.000	-1.735	1.068

Based on estimated marginal means
*. The mean difference is significant at the .05 level.
b. Adjustment for multiple comparisons: Bonferroni.

- The findings for students age 12 are mixed. The **Mean Difference** column reports that a statistically significant average difference in performance between Private and Public 1 ($M = 2.167$, $p = .001$). Similarly, the mean difference in performance between Private and Public 2 is also statistically significant ($M = 2.5$, $p < .001$). However, the mean difference in performance between Public 1 and Public 2 is comparatively smaller (0.33) than in the other comparisons, and this difference is statistically nonsignificant ($p = 1.0$).

Ordinarily, results of the simple effects pairwise comparisons tests negate the need for post hoc multiple comparisons. Nevertheless, for purposes of discussion, **Multiple Comparisons** in Figure 8.16 presents a significant difference between each pair of school type averaging across age ($p < .001$). Without a significant interaction, investigations would cease with multiple comparisons.

Figure 8.16 Multiple Comparisons for the Main Effect of School Type Ignoring Age

					95% Confidence Interval	
(I) School Type	(J) School Type	Mean Difference (I-J)	Std. Error	Sig.	Lower Bound	Upper Bound
Private	Public 1	2.08*	.391	<.001	1.12	3.05
	Public 2	4.08*	.391	<.001	3.12	5.05
Public 1	Private	-2.08*	.391	<.001	-3.05	-1.12
	Public 2	2.00*	.391	<.001	1.04	2.96
Public 2	Private	-4.08*	.391	<.001	-5.05	-3.12
	Public 1	-2.00*	.391	<.001	-2.96	-1.04

(Multiple Comparisons; Dependent Variable: Score; Tukey HSD)

Based on observed means.
The error term is Mean Square (Error) = .917.
*. The mean difference is significant at the .05 level.

8.10 Differences in Achievement Using Actual Test Scores

Thus far, a small set of fictitious data facilitated a behind-the-scenes peek at the workings of factorial ANOVA. As often happens, analysis of data from an actual assessment may not be as clear-cut as fictitious data designed to illustrate the components of factorial ANOVA.

The data to be used in this example (**Two_Way_Comtask.sav**) are an excerpt of secondary school placement test data randomly selected from the original dataset of scores from schools in the central region of Jamaica. Hypotheses in the research plan in Box 8.1 and SPSS specifications used in Section 8.8, were applied to generate results for this live test data.

8.10.1 Descriptive Summary

As **Descriptive Statistics** in Figure 8.17 reports, the dataset includes a total of 180 observations ages 11 (n = 90) and 12 (n = 90) at three types of schools: School Type 1 (Private) (n = 60), School Type 2 (Public 1) (n = 60) and School type 3 (Public 2) (n = 60). Figure 8.17 verifies the research design is a 2 x 3 balanced two-way ANOVA in that there are 30 student scores in each cell.

Descriptive statistics in Figure 8.17 include the mean and standard deviation (Std. Deviation) for each group of students. The average scores show very little difference between age groups (Age 11 = 8.32) and (Age 12 = 8.16). Scores across school types are somewhat more differentiated; overall scores for Private are higher than those from Public 1 and Public 2 (8.85 versus 8.18 and 7.68, respectively). Public 2 records the lowest scores regardless of student age.

8.10.2 Homogeneity of Variance

Levene's Test of Equality of Error Variances results are presented in Figure 8.18a. Results show nonsignificant differences in the variance across groups with the exception of the significance test based

Figure 8.17 Two-Way ANOVA Data Description Summary

Descriptive Statistics				
Dependent Variable: Communication Task Score				
School Type	Age	Mean	Std. Deviation	N
Private	11	9.10	2.339	30
	12	8.60	1.653	30
	Total	8.85	2.024	60
Public 1	11	8.20	1.846	30
	12	8.17	1.599	30
	Total	8.18	1.712	60
Public 2	11	7.67	3.111	30
	12	7.70	1.878	30
	Total	7.68	2.548	60
Total	11	8.32	2.530	90
	12	8.16	1.735	90
	Total	8.24	2.165	180

on the mean (first row, $p = .044$). ANOVA is robust to somewhat unequal error variances across groups. Nevertheless, a more conservative alpha level (e.g., $\alpha = .01$) would contribute to increased confidence in significance testing and interpretation of analysis results.

8.10.3 The Omnibus Test

The results of the omnibus test, **Test of Between-Subjects Effects** (Figure 8.18b), represent the core of the results of a two-way ANOVA. Here, results in the **Sig.** column show a nonsignificant difference between the average scores of students ages 11 and 12 ($p = .601$), which is much higher than the conventional alpha of $p < .05$ used for rejecting the null hypothesis and certainly higher than the more conservative alpha level of .01. Since the probability is higher rather than lower than .05 and .01, there is insufficient evidence to reject Null Hypothesis 1 that there is no significant difference between the average scores of 11- and 12-year-olds (Box 8.1).

Conversely, the **Sig.** column shows a significant difference between the average scores of the different types of schools ($p = .01$), averaging across age groups. The probability here is the equivalent of the preset alpha level of $\alpha = .01$. Therefore, there is sufficient evidence to reject Null Hypothesis 2 that there is no significant difference between the average scores of the three types of schools. Of course, the results suggest that there is a significant difference in the scores of the three types of schools, but the findings do not reveal the source of the difference. Post-hoc analyses should disclose where the differences lie.

As with the results of Hypothesis 1, the p value indicates that there is not enough evidence to reject the null for Hypothesis 3 that there is no significant interaction between school type and age ($p = .758$). Accordingly, even though descriptive statistics show differences between the average scores of 11-year-olds and 12-year-olds at different types of schools, those differences are likely to have occurred

Figure 8.18 Results of Omnibus Test of Two-Way ANOVA

(a)

Levene's Test of Equality of Error Variances a,b					
		Levene Statistic	df1	df2	Sig.
Communication Task Score	Based on Mean	2.330	5	174	.044
	Based on Median	1.806	5	174	.114
	Based on Median and with adjusted df	1.806	5	124.419	.116
	Based on trimmed mean	2.026	5	174	.077

Tests the null hypothesis that the error variance of the dependent variable is equal across groups.
a. Dependent variable: Communication Task Score
b. Design: Intercept + Schtype + Age + Schtype * Age

(b)

Tests of Between-Subjects Effects					
Dependent Variable: Communication Task Score					
Source	Type III Sum of Squares	df	Mean Square	F	Sig.
Corrected Model	44.894a	5	8.979	1.968	.086
Intercept	12218.272	1	12218.272	2678.118	<.001
Schtype	41.111	2	20.556	4.506	.012
Age	1.250	1	1.250	.274	.601
Schtype * Age	2.533	2	1.267	.278	.758
Error	793.833	174	4.562		
Total	13057.000	180			
Corrected Total	838.728	179			

a. R Squared = .054 (Adjusted R Squared = .026)

by chance and are unlikely to represent a systematic pattern of performance for that age group in the general population. However, there is a difference between at least one pair of school types. The source of the differences may be investigated further using post hoc tests.

8.10.4 Post Hoc Tests

Using the Tukey HSD (Honestly Significant Difference) post hoc test to locate the source of the significant main effect for school type resulted in the outcome displayed in Figure 8.19. Results indicate that the pattern of mean difference noted in Figure 8.17 between Private and Public 2 is statistically significant. Average scores of Private ($M = 8.85$) and Public 2 schools ($M = 7.68$) record a statistically significant mean difference of 1.17 ($p = .009$). All other differences are nonsignificant.

8.10.5 Effect Size

As with the one-way ANOVA, calculating eta squared (η^2) provides an estimate of the effect size or the amount of variance accounted for by the independent variables. The eta squared statistic is probably the most straightforward computation of effect size, although it is a slightly biased estimate of what would occur if a researcher had access to everyone in the target population. Details about how to compute less biased measures like omega squared (ω^2) or Cohen's d are available from Chapter 7 and a variety of statistics texts (e.g., Howell, 2014; Tabachnick & Fidell, 2007). SPSS and other common statistical software also generate these statistics.

Figure 8.19 Results of Multiple Comparisons for Main Effects of School type

						95% Confidence Interval	
	(I) School Type	(J) School Type	Mean Difference (I-J)	Std. Error	Sig.	Lower Bound	Upper Bound
Tukey HSD	Private	Public 1	.67	.390	.204	-.26	1.59
		Public 2	1.17*	.390	.009	.24	2.09
	Public 1	Private	-.67	.390	.204	-1.59	.26
		Public 2	.50	.390	.407	-.42	1.42
	Public 2	Private	-1.17*	.390	.009	-2.09	-.24
		Public 1	-.50	.390	.407	-1.42	.42

Based on observed means.
The error term is Mean Square (Error) = 4.562.
*. The mean difference is significant at the .05 level.

To calculate eta squared for each independent variable and the interaction of the variables, divide the sums of squares for that effect (A = age, T = school type, AxT = interaction of age and school type) by the corrected total sums of squares (SST):

$$\eta_A^2 = \frac{SSB_A}{SST} = \frac{1.250}{838.728} = 0.002$$

$$\eta_T^2 = \frac{SSB_T}{SST} = \frac{41.111}{838.728} = 0.049$$

$$\eta_{AxT}^2 = \frac{SSB_{AxT}}{SST} = \frac{2.533}{838.728} = 0.003$$

Accordingly, using live data from the communication task subtest of the Jamaica secondary school entrance test, the effect size for the variables, age and school type, suggests that these indicators do not account for much of the variance in achievement on the communication task test. School type accounts for most of the variance in test achievement—5% of the variance. Clearly, other variables beyond those available from the database have greater influence on performance on the communication task test component.

8.10.6 Sample Report: The Case of Differences in Achievement: the Factorial Effect

The communication task performance of 180 11- and 12-year-old students in three different types of schools were compared using factorial two-way ANOVA. The analysis found significantly different performance in the three types of schools, $F(2, 174) = 4.506, p < .05)$. The main effect for age ($p = .601$) and the interaction between age and school type ($p = .758$) were found to be nonsignificant. The effect size for school type calculated as eta-squared (η^2) was .049, indicating a small effect.

Results of a Tukey HSD post-hoc, multiple comparison test revealed that students at Type 1 (Private) schools had significantly higher average scores ($M = 8.85$, $SD = 2.42$) than students at Type 3 schools (Public 2) ($M = 7.68$, $SD = 2.55$, $p = .009$).

8.11 Key Terms

Disordinal interaction	Interaction	Simple effects
Factorial ANOVA	Ordinal interaction	Pairwise comparison

9

Analysis of Covariance

Key concepts from previous chapters or courses:
    error variance: any uncontrolled or unexplained variability, such as within-group differences in ANOVA, also known as random error or within-group variance.
    general linear model (GLM): a family of univariate statistical methods, including *t* test, ANOVA, ANCOVA, correlation and linear regression, as well as multivariate models. All of these methods are used to study how one or more independent variables affect one or more continuous dependent variables.
    interaction effect: the joint effect of two or more independent variables on a dependent variable. Interaction effects happen when the effect of one independent variable on an outcome variable depends on the value of another independent variable.
    linear regression: a statistical model used to estimate the linear predictive relationship between the values of one or more independent variables and the values of a single dependent variable.
    main effect: the effect of an independent variable on a dependent variable, ignoring any other variables.

THIS CHAPTER INTRODUCES

- differences between one-way ANOVA and one-way analysis of covariance (ANCOVA)
- characteristics of ANCOVA
- types of ANCOVA
- when to use ANCOVA
- benefits of ANCOVA
- how to calculate the effect size of an ANCOVA outcome
- how to use SPSS to conduct an ANCOVA analysis

9.1 Introduction

According to Altendorf (2019) of the Food and Agriculture Organization (FAO), Latin America and the Caribbean are important exporters of bananas, mangoes, pineapples, avocados, and papayas. Mangoes are also a common snack food in the daily diet of people in this region. Mangoes as a snack food have many documented benefits. They are commonly described as the king of tropical fruit because of their rich flavor combined with multiple health benefits. In addition to being low-fat, mangoes are said to be packed with more than 20 vitamins and minerals (e.g., Fernandez, 2021). In addition, studies of mango consumption as a snack food show that the high fiber content of mangoes produces greater satiety and lower glycemic responses compared to snacks with similar caloric values (e.g., Pinneo et al., 2022). In this chapter, playful, fictionalized data related to snacking will be used to demonstrate the statistical technique of the analysis of covariance (ANCOVA).

ANCOVA is an extension of ANOVA techniques examined in Chapters 7 and 8. Like one-way ANOVA, one-way ANCOVA is designed to compare three or more groups of an independent variable to determine whether there is a significant difference between them in a specific area, like Caribbean nations' gross national income, high school exit examination performance, or appetite for dinner. Factorial ANCOVA is also used to identify group differences and interaction effects of two or more independent variables. However, ANCOVA has more power than ANOVA. The power of ANCOVA comes from its ability to reduce error in the estimate of outcome variables by controlling for effects that are not of primary interest to the study.

9.2 The Case of Appetite After Accounting for Mangoes

Fictionalized data for the Case of Appetite after Accounting for Mangoes appear in Table 9.1. The data include the number of mangoes eaten before a meal, followed by ratings of hunger and fullness, using the hunger and fullness scale based on the Food Satiety Index of Common Foods (Holt et al., 1995). Data are documented for 40 cases to be analyzed. Data file variables are as follows: Gender, ID (Identification Code), M_CV (the Covariate, Mangoes Consumed), and H_DV (Hunger Rating).

Table 9.1 Dataset for the Case of Appetite After Accounting for Mangoes

GENDER											
Female						Male					
ID	M_CV	H_DV	ID	M_CV	H_DV	ID	M_CV	H_DV	ID	M_CV	H_DV
4	2	7	22	2	4	1	0	3	21	2	4
5	2	7	24	3	5	2	0	4	23	2	4
6	2	7	27	3	8	3	0	4	25	3	6
7	3	6	28	4	7	9	3	4	26	3	6
8	3	5	29	4	8	10	1	4	30	4	8
11	1	6	31	4	8	15	2	4	32	4	9
12	1	5	33	5	9	17	3	7	35	5	8
13	2	6	34	5	9	18	3	7	36	5	8
14	2	5	37	1	5	19	4	6	39	1	5

The hunger and fullness scale allows users to rate their level of hunger or fullness, where 0 = empty or uncomfortably hungry, and 10 = uncomfortably full to the point of feeling physically sick. The objective of the investigation is to determine whether the hunger/fullness mean score for males and females are significantly different than might have happened by random chance after accounting for the number of mangoes eaten. The research plan appears in Box 9.1.

9.3 ANCOVA Fundamentals:

Chapter 7, Section 7.6, introduced the use of one-way ANOVA in hypothesis testing in situations where there are three or more groups in a single independent variable. Additionally, Chapter 7 highlighted the special relationship between linear regression and one-way ANOVA by using it to assess differences between groups of study participants. Likewise, linear regression shares a similar relationship with factorial ANOVA; it may be substituted for factorial ANOVA by converting independent variables to dummy-coded variables.

Analysis of covariance also has a special relationship with linear regression. Just as linear multiple regression may include categorical (e.g., gender) and continuous variables (e.g., number of mangoes eaten) as predictors, ANCOVA includes the use of continuous variables, which are known as covariates. However, rather than employing them as predictors, covariates are used to reduce error in the outcome of a study by controlling characteristics that are not of primary interest to study objectives, but might have undue effect on the dependent variable. Removing the unwanted influences of covariates helps to reduce the amount of unexplained variance (also called error variance or noise) and prevent spurious effects on the dependent variable.

In function, ANCOVA is virtually the same as hierarchical multiple regression (Chapter 6). Assume that a hierarchical regression model with the dependent variable, satiety, includes the entry of two blocks of independent variables (Block A and Block B), with one variable in each block (Block A: mangoes eaten, and Block B: gender). The entry of the variable, number of mangoes eaten, followed by the dummy-coded variable, gender (male/female = 1/0), produces the effect gender has on satiety after controlling for the effect of the covariate, the number of mangoes eaten.

9.3.1 Types of ANCOVA

With the exception of the addition of covariates, between-subjects ANCOVA designs are virtually the same as between-subjects ANOVA. Like ANOVA, analysis of covariance is a part of the univariate family of statistics in which there is a single dependent variable. It belongs to the general linear model which is a generalization of the linear regression model for comparing how categorical variables affect one or more continuous variables. An important requirement of this model is that continuous variables must be normally distributed. Linear regression, ANOVA, and ANCOVA are based on this assumption.

As Figure 9.1 illustrates, a one-way ANCOVA has a minimum of three variables – a single independent variable, a single covariate and a single dependent variable. A factorial ANCOVA has a minimum of four variables—two independent variables, one covariate, and one dependent variable.

As in ANOVA, ANCOVA is typically selected when each independent variable has two or more categories. For example, the independent variable, gender, has two categorical groups, male and female. Other variables may have more than two categorical groups. For example, residential location may have three categories— urban, suburban, and rural; and college academic standing may include four categories—freshman, sophomore, junior, and senior. A research design with three independent variables (gender, residential location, and academic standing) would result in a 2 x 3 x 4 design.

Figure 9.1 Types of ANCOVAs: Minimum Characteristics

Type of ANCOVA	IV	DV	COV
One-Way	1	1	1
Factorial	2	1	1

ANCOVA may be mixed with a repeated measures design. In this situation, there may be a single between-subjects variable like gender along with a second independent variable that is a within-subjects variable, like the variable, time. These types of ANCOVAs are called mixed designs. Differences between ANOVA and ANCOVA may be found in Figure 9.2.

Figure 9.2 Similarities and Differences Between ANOVA and ANCOVA.

Characteristics	ANOVA	ANCOVA
General Linear Model	Yes	Yes
Univariate Model	Yes	Yes
Dependent Variable	One	One
Independent Variable	One or more	One or more
Independent Variable Groups	Two or more	Two or more
Covariate	None	One or more
Between Group Design	Yes	Yes
Within Group Design (repeated measures)	Yes	Yes
Mixed Design	Yes	Yes

9.3.2 Characteristics of Covariates

Covariates are characteristics of the participants in a study. ANCOVA designs are not restricted to a single covariate; rather they may include one or more covariates. For example, in the case of Appetite After Accounting for Mangoes, the number of fruits eaten could be the independent variable, with categories of zero, one, and more than one mango. As the data include a diverse group of children, the variable, age, might be one covariate, and the variable, elapsed time since the last meal, might be another. In fact, a good guide for the number of covariates to include in any model is one covariate fewer than the number of groups of the independent variable.

Regardless of the number of covariates, in order for covariates to function as expected, they must have specific characteristics; for example, they must be:
- continuous variables
- observed/measured (as opposed to a manipulated variable)

- control variables
- linearly related to the dependent variable
- unaffected by independent variables

Covariates are typically variables measured at the interval or ratio level. Examples of covariates may include age, time, height and weight, temperature, individual income, driving experience or performance on a pretest.

Covariates are observed or measured, but they are not a part of central analyses, and they are not manipulated by the researcher as are independent variables during data collection.

Covariates are called control variables because they are used to statistically control for influences on the dependent variable that are not the main concern of the research inquiry. This is particularly important when study participants are not randomly selected or assigned; that is, they are not participating in a research experiment.

There should be a linear relationship between covariates and the dependent variable; covariates should be significant predictors of the dependent variable. It is this explanatory relationship that contributes to removing variance (noise) in the dependent variable that is attributable to the covariate, but not to independent variables, which is the primary concern of the research study.

Covariates must be unaffected by the manipulation of independent variables in the study. By using statistics to control for the influence of covariates, researchers increase the precision of the research by reducing the unexplained variance in the DV. Figure 9.3 provides a graphic illustration of the role of a covariate. Figure 9.3a, shows what occurs in a one-way ANOVA. Here, A is the amount of the dependent variable, hunger, which the independent variable gender explains. In Figure 9.3b, the covariate, mangoes consumed, does not affect the independent variable (gender), but as B illustrates, it does account for some amount of hunger/fullness. As a result of the covariate's role, there is a reduced amount of hunger to explain. In addition, error or unexplained variance declines.

Figure 9.3 Role of a Covariate

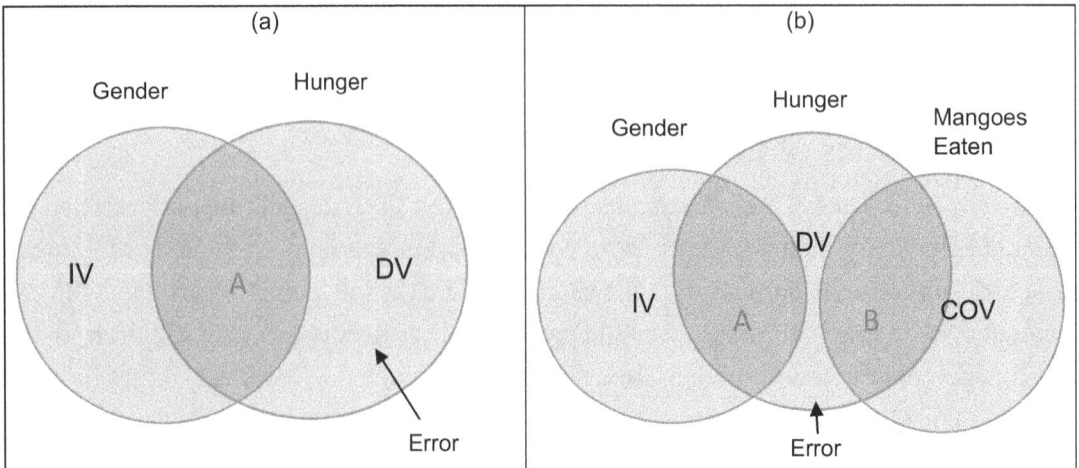

9.4 Benefits of ANCOVA

When a covariate is included that complies with the characteristics outlined in 9.3.2, ANCOVA is superior to its ANOVA counterpart in two distinct respects: (a) increased statistical power, and (b) control.

9.4.1 Increased Statistical Power

An important contribution of covariates is to increase test sensitivity by reducing the probability of a Type II error (a false negative). In the case of ANCOVA, this might occur when a researcher concludes that there is no significant difference in the performance of two groups when in reality there is a difference. Reducing Type II error is particularly important when omnibus main or interaction effects are tested, which will determine whether planned comparison or post hoc investigations are conducted to isolate sources of differences. Since the probability of a Type II error is inversely related to statistical power, ANCOVA is more powerful than its ANOVA counterpart.

F tests associated with a standard ANOVA main effect are computed by dividing the mean squared between variance (MSB) by mean squared error (MSE) (Chapter 8). If MSE is reduced, then the calculated F value will be larger. Subsequently, the calculated p value will be smaller, resulting in an improved probability of rejecting the null hypotheses. This is exactly what happens when an ANCOVA model includes a valid and reliable covariate, thereby increasing the power of the analysis.

9.4.2 Increased Control

In addition to their power function, covariates in ANCOVAs exert increased control over outside influences on the dependent variable. Some researchers refer to covariates of ANCOVA studies as control variables. To bring about the desired control, ANCOVA uses performance on covariates to adjust each group's dependent variable mean score, to place comparison groups in the study on an even footing before conducting comparisons among them. In this way, the researcher accounts for irrelevant influence on the dependent variable before conducting the central analyses in the study.

Suppose a researcher wants to determine whether there is a significant difference in the performance of male and female students in algebra after exposing each group to different instructional treatments or methods (e.g., computer-based, instructor-directed and cooperative learning instruction). The researcher wants to remove the influence of students' previous learnings in algebra. Accordingly, the researcher administers an algebra pretest before introducing the treatments to the comparison groups and uses students' scores on the pretest as the covariate or control variable to adjust posttest scores. If one of the comparison groups had an above-average mean on the covariate compared to the other groups in the study, then that group's average score on the posttreatment algebra test (dependent variable) is lowered. In contrast, any group with a below average score on the covariate will have its average score on the dependent variable raised.

The degree to which any group's average score on the dependent variable is adjusted depends on how far above or below average that group stands on the control variable. By adjusting the average scores on the dependent variable, ANCOVA provides the best estimates of how the comparison groups would have performed if they had all recorded statistically equivalent averages on the control variable before the treatment was instituted. In this way, the covariate is used to exercise control by removing alternative variables that are not the focal point of the study and might otherwise be plausible explanations for the differences between comparison groups.

9.5 Research Plan: The Case of Appetite After Accounting for Mangoes

The research plan outline for the Case of Appetite After Accounting for Mangoes is displayed in Box 9.1.

Box 9.1 Research Plan: The Case of Appetite After Accounting for Mangoes

Research Problem
A researcher wants to help children better manage their eating habits. To accomplish this, researchers investigated primary-aged students' hunger and fullness patterns accounting for snacking on mangoes before dinner. The literature suggests that boys experience greater hunger than girls before meals even after accounting for snacking patterns (e.g., Shomaker et al., 2010). The purpose of the study is to investigate whether these expected differences exist in the Caribbean.

Variables and Measurement Levels
Independent Variables: Gender
Categories of Independent Variables: Male and female
Dependent Variable: Hunger/fullness rating
Covariate: Quantity of mangoes consumed

Research Question
Is there a significant difference in the average hunger/fullness rating of male and female pre-adolescents before meals after accounting for the quantity of mangoes consumed?

Hypotheses
Null Hypothesis: There is no significant difference in the average hunger/fullness ratings of males and females after accounting for the quantity of mangoes eaten.
$H_o: \mu_{1adj} = \mu_{2adj}$
Alternative Hypothesis: There is a significant difference in the average hunger/fullness ratings of males and females after accounting for the quantity of mangoes eaten.
$H_a: \mu_{1adj} \neq \mu_{2adj}$

Research Design
The nonrandomized, causal comparison design that includes three variables: gender comprised of intact comparison groups, male (X_1) and female (X_2) pre-adolescents, hunger/fullness ratings (dependent variable or O), and mangoes consumed as a pre-meal snack, which serves as the single covariate for the study. Participants were matched by a number of factors, including, age, time since last meal, and attitude toward fruit.

$$M \quad X_1 \quad O_{adj}$$
$$M \quad X_2 \quad O_{adj}$$

Data Collection
Data were collected from each study participant (20 males and 20 females), using the hunger/fullness rating scale, and documentation of the quantity of mangoes consumed by each participant.

Data Analysis
When investigating whether statistically significant differences exist between more than two distinct groups of scores on a single continuous, dependent variable and a covariate is present, a one-way analysis of covariance is the appropriate statistical test of significance.

9.5.1 Research Problem and Question

Like most other statistical techniques, ANCOVA generates the answers to specific types of research questions. While ANCOVA and ANOVA share certain functionalities, ANCOVA is designed to do more. ANCOVA removes competing influences on the dependent variable to provide greater precision in the measure of group differences.

The research problem and questions in the research plan embody the objectives for this case: to investigate the differences in hunger ratings between male and female students after accounting for mangoes eaten. The research question identifies three variables: gender (independent variable), mangoes consumed (covariate), and hunger/fullness ratings (dependent variable).

9.5.2 The Variables

The independent variable, gender, is a categorical variable with two intact groups—male and female. Hunger/fullness ratings is the dependent variable, a continuous variable with overall ratings that range from 0 to 10. Ratings and their interpretations may be found in Figure 9.4. Hunger ranges from 0 – 4, 6 is a neutral rating, and fullness ranges from ratings of 7 - 10. The covariate for the study is the quantity of mangoes each participant consumed for their snack before dinner. The covariate is a continuous variable.

Figure 9.4 Hunger & Fullness Scale Ratings and their Interpretations

Rating	Meaning
Hunger	
0	Empty: Uncomfortably hungry.
1	Ravenous: Difficulty concentrating.
2	Very Hungry: Stomach growling.
3	Moderately Hungry: Thoughts about food increasing.
4	Lightly Hungry: Starting to think about food.
Neutral	
5	Neutral: Neither hungry nor full.
Fullness	
6	Lightly full: Satisfied.
7	Moderately Full: Satisfied, comfortable.
8	Full: Comfortably full.
9	Stuffed: Past the point of comfort.
10	Sick: Uncomfortably full.

The scale-level covariate and the two-level categorical independent variable together designate the statistical test as a one-way ANCOVA. The covariate, mangoes consumed, serves the purpose of removing influences that might compete with gender in measuring hunger/fullness differences between male and female subjects.

This case includes a single covariate, mangoes consumed, but others could have been included; for example, age or number of hours since the previous meal. However, in this case, because gender includes only two groups (male and female), adding more than one covariate would violate the assumption, which limits the number of covariates to one less than the number of groups in the independent variable.

9.5.3 Research Design

When ANCOVA is identified as the primary statistical procedure for analyzing data (based on research questions and hypotheses), it is typically part and parcel of one of the following research designs: (a) experiments in which study participants are randomly assigned to treatment and control groups, or

(b) nonexperiments (causal-comparative or quasi-experimental studies) in which subjects cannot be randomly assigned.

Due to the inclusion of intact gender comparison groups, male and female, and the absence of random assignment to treatment groups, the design falls within the confines of a nonexperimental causal comparison design. In these causal comparison cases, covariates are used to adjust dependent variable scores to account for control variables that might influence study outcomes. It is important to remember, however, that using covariates can adjust dependent variable scores and show a larger main effect, or covariates can result in unintended consequences by eliminating the main effect altogether.

The research design for this case is described in Box 9.1 as a causal comparative, nonrandomized group design. It is the same research design used for one-way ANOVA in Chapter 7 (Box 7.1) and factorial ANOVA in Chapter 8 (Box 8.1).

9.5.4 Data Collection

Data collection methods for causal comparative designs are similar to methodology used in other designs. For causal comparison designs, data is typically collected by means of questionnaires, surveys, attitudinal scales, assessments, and behavior observations. Causal comparative research data may be collected live for a specific investigation, or they may be extant or archival data. For a study like the Case of Hunger/Fullness after Accounting for Mangoes, data are often collected via self-reports (mangoes consumed) and the 10-point self-assessment scale (hunger/fullness rating).

9.5.5 Hypothesis Testing and Data Analysis Procedures

Hypotheses determine the type of statistical analysis required. The hypotheses look similar to those for one-way ANOVA, with the exception of "adj", signifying that the means to be compared are adjusted means—adjusted by the covariate to remove any variance in the dependent variable that it accounted for.

The primary data analysis is significance testing using the inferential statistical procedure, one-way ANCOVA. Box 9.1 shows a single null and a single alternative hypothesis because the research design includes a single, two-group independent variable (gender). The purpose of significance testing is to investigate whether there is a main effect for gender. No post hoc test is needed for two-group comparisons.

9.6 ANCOVA Assumptions

Assumptions for ANCOVA are essentially the same as ANOVA assumptions with some variations. Some are study design issues. Others require statistical analysis.

9.6.1 ANOVA Assumptions Required for ANCOVA

ANOVA assumptions shared by ANCOVA include the following:
- Independence of observations: study participants or observations should not be in more than one independent variable group/category.
- Measurement level of variables:
  o Independent variables should be categorical, with each including two or more categories.
  o The dependent variable should be measured on a continuous scale (interval or ratio scale.).
  o Covariates should also be measured on a continuous scale.

- Normal distribution of scores: the sampling distribution of dependent variable data should be roughly normal for each group of study participants. The larger the sample size the more likely the distribution of scores will be normal. In addition, with equal sample sizes in groups, no outliers, and two-tailed tests, ANCOVA is robust to violations of normality and will yield valid results even in the face of some non-normality. However, with small, unequal sample sizes or with outliers, valid outcomes may be difficult to secure. To test for normality in ANCOVA, use the Shapiro-Wilks and the Kolmogorov-Smirnov tests of normality. Both of these tests are sensitive to minor departures from normality (Meyers et al., 2013), especially with small sample sizes. Accordingly, results should be interpreted with caution. Other tests of normality include tests of skewness and kurtosis.
- Homogeneity of variance: data should exhibit homogeneity of variance (equal variance) of dependent variable and covariate scores across all groups. Test this assumption by using the Levene's test of homogeneity of variance.

9.6.2 Additional ANCOVA Assumptions

In addition to ANOVA assumptions, ANCOVA also assumes the following:
- a linear relationship between covariates and dependent variables (linearity of regression)
- homogeneity of regression lines across study groups
- reliability of covariates. Before any ANCOVA analysis occurs, it is important to check that the data meet all ANCOVA assumptions.

Finally, the following issues should be addressed before proceeding with central ANCOVA analyses:
- multicollinearity and singularity among covariates
- outliers in the dataset
- unequal sample sizes and missing data

9.6.2.1 Linearity of Regression

Covariates and the dependent variable (at each level of independent variables) should be linearly related. Covariates should also be linearly related to each other. As with multiple regression, significant departure from linearity reduces the power of ANCOVA (Tabachnick & Fidell, 2018). The error or variance unaccounted for may not be reduced as much as it would be with a linear relationship. Participant groups may not be matched as well as they should be, resulting in limited adjustment of dependent variable scores, which compromise inter-group comparisons. Linearity is especially important when there are substantial differences between group scores on the covariate.

A common method of testing linearity for each covariate is to generate a bivariate scatterplot where dependent variable scores are on the y-axis, and covariate scores are on the x-axis. A visual inspection of the results will reveal whether the relationship is linear. If the relationship is linear the plot should be oval-shaped and symmetrical. If the relationship is curvilinear, consider transforming the variable. Alternatively, consider eliminating the covariate.

9.6.2.2 Homogeneity of Regression

Covariates and independent variables should not interact. A graphic rendering of the regression line for each group in your study should display parallel lines; that is, the slope (or regression coefficient) of the lines predicting the dependent variable from each covariate should be linear for each group in the study. Parallel slopes are necessary to accurately adjust the dependent variable scores for each group.

To test for this assumption, conduct an ANOVA where the sole interest is in examining whether there is a significant interaction between independent variables and each covariate. If the interaction is not statistically significant, the assumption of homogeneity of regression has not been violated. If the interaction is significant, the assumption has been violated and the covariate should not be used. ANCOVA is not an appropriate test in this instance.

9.6.2.3 Reliability of Covariates

Covariates are not always reliable measures, and unreliable ones can lead to less than optimum results. Variables, such as age, height, weight, and other single item variables, are of less concern than scales that measure attitude and other psychological constructs which may fluctuate over time (Tabachnick & Fidell, 2018).

Unreliable covariates may lead to under- or over-adjustments of mean group scores. This could lead to Type I or Type II errors. Unreliable covariates are particularly problematic in nonexperimental studies (Stevens, 2009). To avoid this situation, Tabachnick and Fidell (2018) recommend restricting covariate use to those that can be measured reliably at a correlation coefficient of greater than .8.

9.6.2.4 Multicollinearity and Singularity Among Covariates

A research question requiring the application of ANCOVA may have more than one covariate whose relationship must be evaluated before use. Multicollinearity occurs when two variables are very highly correlated, and singularity occurs when they are perfectly correlated.

If one covariate is highly related to another (at a squared multiple correlation of .5 or greater), it will not adjust the dependent variable over and above the other covariate. Under these circumstances, one of the redundant covariates should be removed from further analysis.

9.6.2.5 Outliers

Univariate outliers are cases with extreme values on one variable, and multivariate outliers are cases with unusual combinations of scores on two or more variables (Tabachnick & Fidell, 2018). For example, a 9-year-old fifth-grader who earns a near-perfect score on the Caribbean Secondary Education Certificate (CSEC) mathematics exam is very likely a candidate for a multivariate outlier on the variables of grade and CSEC score. Age is normal for a fifth-grader, but a near-perfect score is highly unusual.

Multivariate outliers may result in violation of the homogeneity of regression assumption and poor adjustment of the dependent variable, possibly leading to a Type I error. It is important to test for univariate outliers on the dependent variable and all of the covariates individually first, followed by examining the data for multivariate outliers on dependent and covariates combined.

Common methods of screening for univariate outliers prior to ANCOVA include generating boxplots for groups within each independent variable. For example, if gender is an independent variable and basketball score is the dependent variable, boxplots of netball scores should be generated for each gender group. Stem-and-leaf plots and z-scores may also be used to evaluate data for outliers.

For multivariate outliers, bivariate scatterplots for combinations of covariates and dependent variables are a good first step to identify cases that lie outside of the swarm of scores. As a by-product, the scatterplots provide evidence of violations of normality and linearity.

Another approach is to identify the Mahalanobis distance of each case and the group average, known as the centroid. Cases that are significantly distant from the centroid may be candidates for removal. See Chapter 2 for more information about screening and cleaning data before conducting central analyses.

9.6.2.6 Unequal Sample Sizes and Missing Data

Imagine a study in which the independent variables are gender and method of instruction (online or traditional) and the dependent variable is achievement. In this case there are four possible group combinations as illustrated in Figure 9.5.

Figure 9.5 Two-by-two design

	Male	Female
Online	30	30
Traditional	30	30

Ideally, cell sample sizes (e.g., males and females receiving online instruction) should be equal and sufficient to ensure adequate power to reject a hypothesis that should be rejected. However, this does not always happen. If sample sizes are too small, results might be invalid. If sample sizes are unequal, variance might be miscalculated, resulting in computed variances that are larger than the total variance. Additionally, the effects of ANCOVA on one independent variable might be indistinguishable from the effects on the other.

If cell sizes are unequal due to missing data (e.g., unanswered items), the SPSS default specification of SSTYPE3 (Type III sums of squares) assumes that the data was supposed to be complete and the difference in the number of subjects is not meaningful. The Type I sums of square specification in GLM (SSTYPE1) assumes that the difference in number of subjects is meaningful and gives more weight to the values from larger groups.

9.7 SPSS Procedures for One-Way ANCOVA

To recap, this case calls for assessing the hunger/fullness of male and female pre-adolescent students prior to their evening meal (dinner) and after accounting for mangoes consumed. The data file (Fullness.sav) includes a total of 40 cases. Variables include Gender_N (independent variable), Mangoes_CV (covariate) and Hunger_Rating_DV (dependent variable). Imagine that a researcher wants to determine whether the amount of fresh fruit snacking prior to dinner could have an impact on hunger ratings.

Due to the fabricated nature of the data, the reliability of the covariate and dependent variables were not assessed. However, an important prerequisite for using any quantitative measure is to employ instruments that are valid and reliable.

Procedures for conducting a one-way ANCOVA using SPSS are similar to those used for ANOVA. Both procedures are included in this demonstration exercise.

This practicum uses fictitious data from Table 9.1 to conduct the analyses in two steps.
1. To establish the basis for conducting ANCOVA, conduct a one-way ANOVA of the data to investigate whether there is a difference in the average hunger ratings of male and female students. This includes evaluating the data for compliance with ANOVA assumptions.
2. Conduct a one-way ANCOVA to determine whether there is a significant difference in hunger ratings after accounting for pre-dinner mango snacks consumed. Of course, this includes testing for additional ANCOVA assumptions.

9.7.1 ANOVA SPSS Specifications

Two important assumptions of ANOVA are the normal distribution of data and the assumption of homogeneity of variance, which are both detailed in Section 9.6.1. To assess the assumption of normal distribution, consult Chapter 3 for related SPSS procedures. The assessment of the assumption of homogeneity of variance is an integrated component of the SPSS ANOVA procedure which will be conducted first.

One-way ANOVA is a GLM statistical procedure, and SPSS offers two approaches for calculating it. One method is via **Analyze → Compare Means → One-Way ANOVA** and the other method is through **Analyze → General Linear Model → Univariate**. The exercise in Chapter 7 is based on the first method. This exercise uses the GLM procedure.

Start by selecting **File → Open → Data**. Locate the data file, **Fullness.sav**, and double-click on it to bring the file into the **IBM SPSS Statistics Data Editor**. From the **Data Editor**, click on **Analyze → General Linear Model → Univariate** to open the **Univariate** dialog window. In the **Univariate** dialog window, drag or click over **Hunger_Rating_DV** to the **Dependent Variable** panel and **Gender_N** to the **Fixed Factors** panel (Figure 9.6a).

Figure 9.6 SPSS Univariate Window for ANOVA

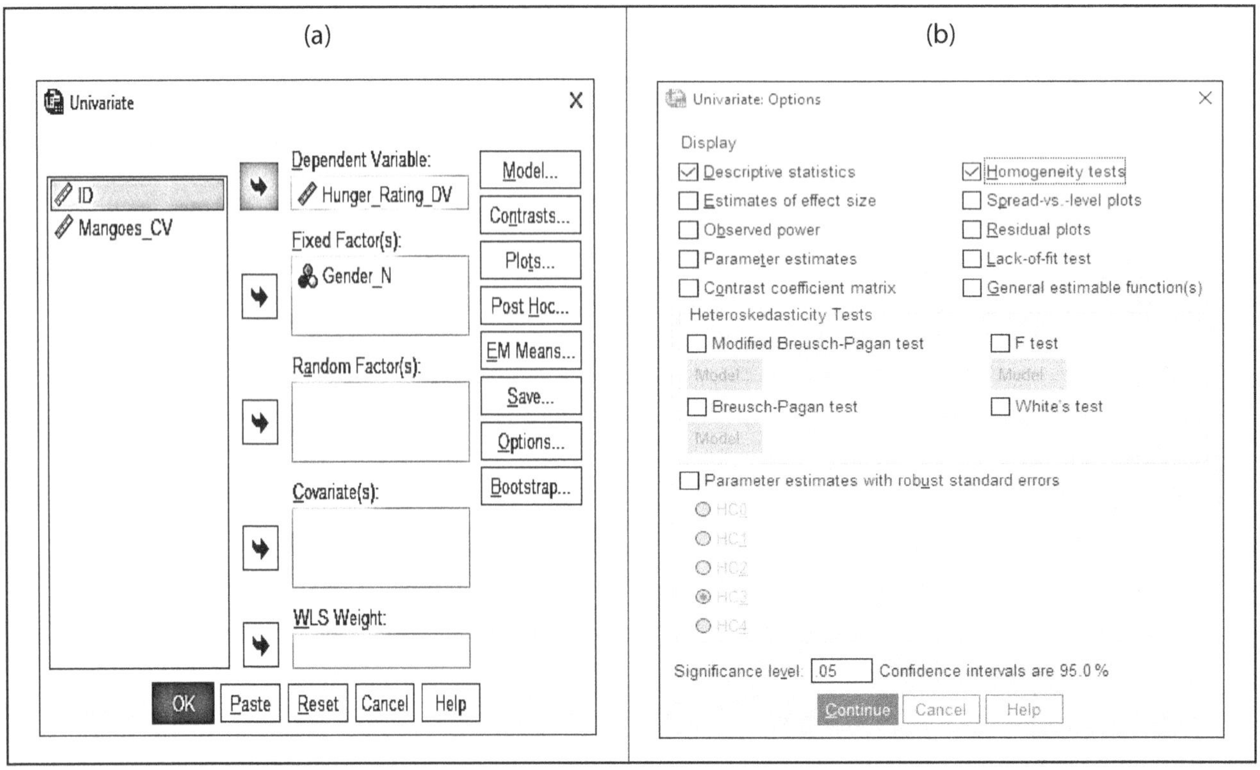

Since **Gender_N** is a two-category independent variable, there is no need for **post hoc** procedures or other post-ANOVA comparisons. Instead, while the **Univariate** dialog window is still active, click on the **Options** pushbutton. Once in the **Univariate Options** dialog window (Figure 9.6b), select **Descriptive Statistics** to generate gender-related descriptives and **Homogeneity Tests** to evaluate the assumption of homogeneity of variance. Click **Continue** to return to the main dialog window, and click **OK** to run the one-way ANOVA analysis.

9.7.2 Results of One-Way ANOVA

The results of the one-way ANOVA omnibus test are presented in Figure 9.7. **Descriptive Statistics**, which are displayed in Figure 9.7a, confirm that an equal number of males ($N = 20$) and females ($N = 20$) participated in the study. The mean hunger rating for the entire group of participants was 6.05, which means that the group as a whole experienced light satiety. The higher the rating the greater fullness participants sensed. The mean self-rating of females ($m = 6.45$) was about one point greater than that of males ($m = 5.65$).

The results of Levene's test (homogeneity of variance) are displayed in Figure 9.7b, **Levene's Test of Equality of Error Variances**. The null hypothesis for Levene's test is that the variances are equal across all groups of the independent variable. As the **Sig**. column reports, the p-values ($p = .222 - .227$) are larger than the alpha level ($\alpha = .05$), so group variances are not significantly different regardless of the underlying distribution of the dependent variable (hunger-fullness rating); hence the data meet the assumption of homogeneity of variances.

With the independent variable comprising two groups, a t test is one option for analysis. However, a one-way ANOVA produces virtually the same results as a t test, and those results are presented in Figure 9.7c. The test results report a statistically nonsignificant F ratio of 2.396 ($p = .130$), based on 1 and 38 degrees of freedom and an alpha of $\alpha = .05$. As was calculated for the one-way ANOVA in Chapter 7, compute eta square (η^2) to generate the percent of variance in the hunger-fullness rating that gender accounts for by dividing the Between-Group Sum of Squares (6.4) by the Total Sum of Squares (107.9). This calculation produces $\eta^2 = .06$, which means that gender accounts for a modest 6% of the variance in hunger-fullness ratings.

9.7.3 ANCOVA Procedures

In an attempt to secure greater precision after a nonsignificant one-way ANOVA outcome, part two of the analysis introduces the quantity of mangoes consumed as a covariate in the exercise. Before conducting the omnibus test, it is necessary to test two assumptions in addition to those required by ANOVA:
- linearity of regression, and
- homogeneity of regression.

9.7.3.1 Evaluating the Assumption of Linearity of Regression

In a scatterplot a positive linear relationship happens when the values of two variables (plotted on the x and y axes) rise at approximately the same rate. A scatterplot of a negative linear relationship results when the values of one variable increase while the values of the other decline at about the same rate. If variables are linearly related, the assumption of linearity has been satisfied.

Figure 9.7 Results of One-Way Between Subjects ANOVA Procedure

(a)

Descriptive Statistics

Dependent Variable: Hunger-Fullness Rating

Gender	Mean	Std. Deviation	N
Female	6.45	1.468	20
Male	5.65	1.785	20
Total	6.05	1.663	40

(b)

Levene's Test of Equality of Error Variances[a,b]

		Levene Statistic	df1	df2	Sig.
Hunger-Fullness Rating	Based on Mean	1.541	1	38	.222
	Based on Median	1.507	1	38	.227
	Based on Median and with adjusted df	1.507	1	37.260	.227
	Based on trimmed mean	1.538	1	38	.222

Tests the null hypothesis that the error variance of the dependent variable is equal across groups.
a. Dependent variable: Hunger-Fullness Rating
b. Design: Intercept + Gender_N

(c)

Tests of Between-Subjects Effects

Dependent Variable: Hunger-Fullness Rating

Source	Type III Sum of Squares	df	Mean Square	F	Sig.
Corrected Model	6.400a	1	6.400	2.396	.130
Intercept	1464.100	1	1464.100	548.136	<.001
Gender_N	6.400	1	6.400	2.396	.130
Error	101.500	38	2.671		
Total	1572.000	40			
Corrected Total	107.900	39			

a. R Squared = .059 (Adjusted R Squared = .035)

To assess the assumption of linearity of regression, first generate a scatterplot of the covariate (**Mangoes_CV**) and the dependent variable (**Hunger_Rating_DV**), then visually examine the plot to evaluate the linear relationship between the two variables. To generate the scatterplot, select **Graphs → Chart Builder** (Figure 9.8) on the data editor page to open the **Chart Builder** window.

Figure 9.8 SPSS Statistics Data Editor Window (Partial View)

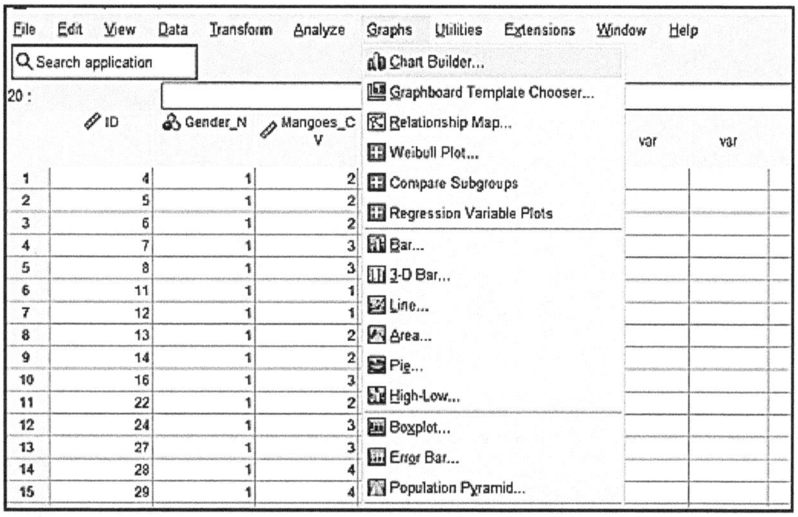

In the first **Chart Builder** window, click **OK** to define the chart. In the second half of the **Chart Builder** definition window, highlight **Gallery** on the left, then **scatter/dot** in the **Choose from:** panel. This will populate the panel on the right with a selection of scatterplots. Select the first scatterplot, and drag it to the **Chart Preview** panel (Figure 9.9)

Figure 9.9 The Chart Builder Dialog Window

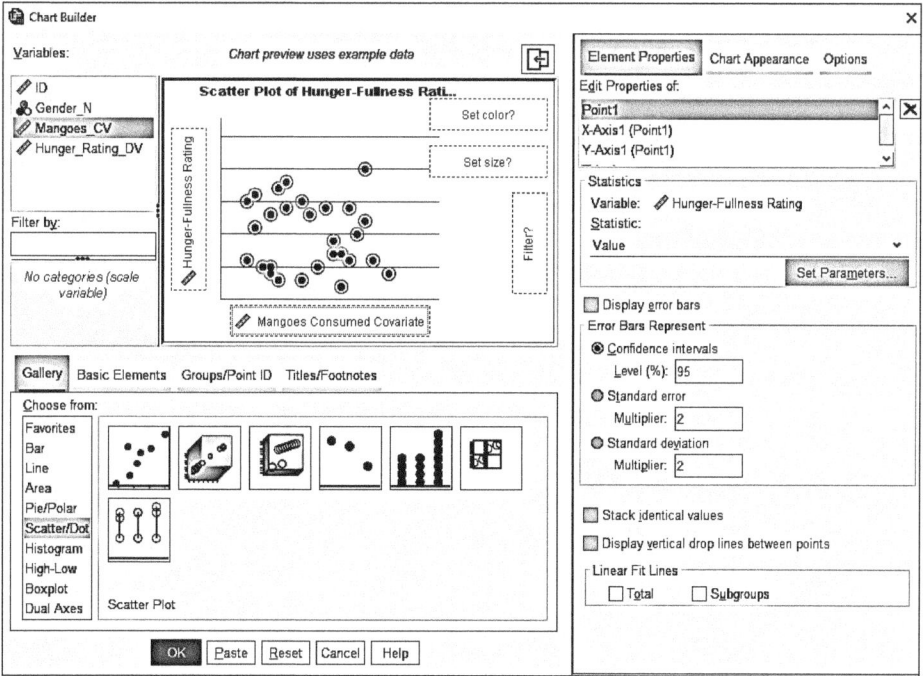

Select the variable, **Hunger_Rating_DV** and drag it to the panel labeled **Y-axis**, then drag **Mangoes_CV** to the panel labeled **X-axis** (Figure 9.9). Click **OK** to run the analysis and generate the scatterplot of the two variables (Figure 9.10). Visual inspection of the plot reveals a direct, positive linear relationship between the two variables. As the values for **Hunger_Rating_DV** increase, so do the values for **Mangoes_CV**.

330 Analysis of Covariance

Figure 9.10 Scatterplot for Assessing the Assumption of Linearity of Regression

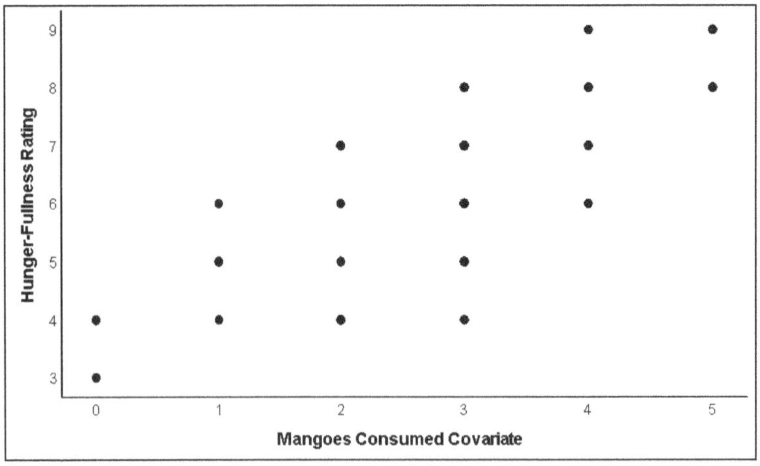

To display the regression line (line of best fit) for the scatterplot, double-click the inside of the scatterplot in the **IBM SPSS Statistics Viewer** window to display the **Chart Editor**. Select **Elements** → **Fit Line at Total** to produce the regression line (Figure 9.11). Note that the **Properties** window also opened automatically. Confirm that the **Fit Line** tab has been selected and that the fit line is set at **linear**, the default setting.

As Figure 9.11 illustrates, there is a linear relationship between the dependent variable, **Hunger-Fullness Rating** and **Mangoes Consumed Covariate**. The note in the upper right corner of the plot notes that R^2 Linear = 0.617, which means that mangoes consumed explain a total of 61.7% of the variance in hunger ratings. The variance explained is comparable to eta square and provides additional evidence that the assumption of linearity of regression has been met.

Figure 9.11 Scatterplot with Line of Best Fit and Properties Window

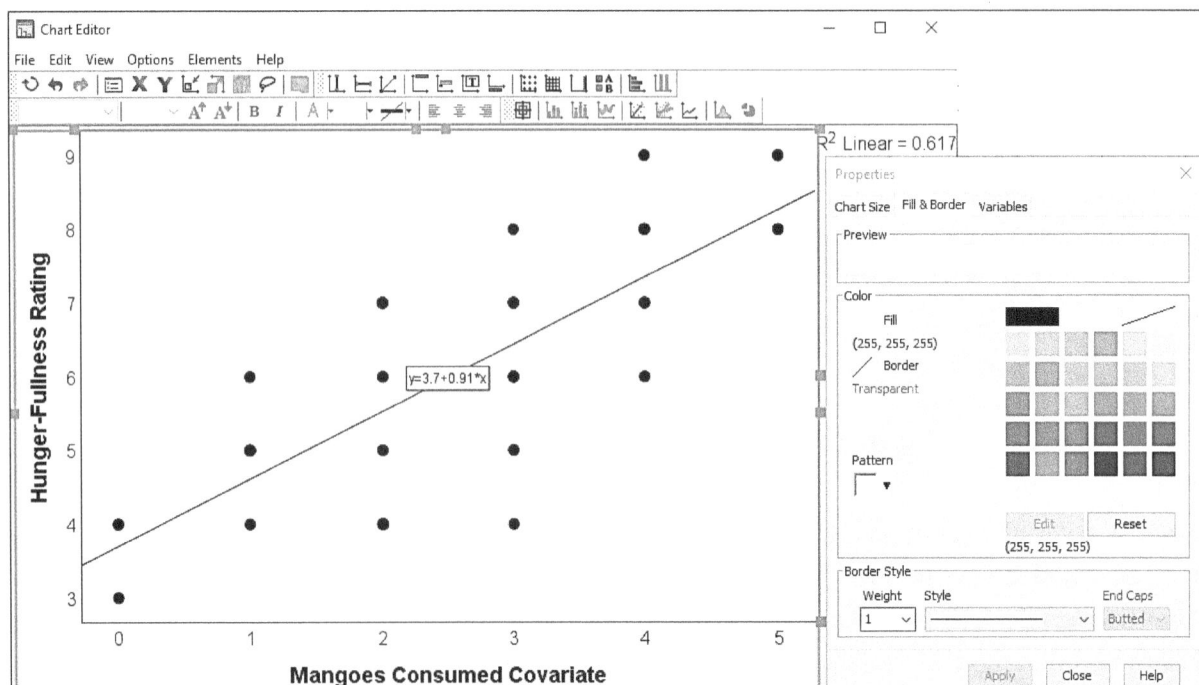

9.7.3.2 Evaluating the Assumption of Homogeneity of Regression

ANCOVA's homogeneity of regression assumption is based on the hypothesis that the regression function and regression line for each independent variable group (male and female) is comparable to the regression line for the entire sample of participants; that is, they have similar predicting power. Therefore, the regression line used to predict hunger-fullness ratings from the mango covariate for each group in the current case should be comparable to the slope of the line in Figure 9.11. Testing this assumption requires conducting an ANCOVA to investigate whether there is an interaction between the covariate, **Mangoes_CV** and the independent variable, **Gender_N**. An interaction between the two variables means that the slopes for males and females are significantly different, and the assumption of homogeneity of regression does not hold. If the interaction is nonsignificant, the assumption has been met.

To test the assumption, select **Analyze → General Linear Model → Univariate** to open the **Univariate** dialog window displayed in Figure 9.12. Drag or click over **Gender_N** to the **Fixed Factor(s)** panel. Similarly, locate **Hunger_Rating_DV** to the **Dependent Variable** panel, and **Mangoes_CV** to the **Covariate** panel (Figure 9.12).

Figure 9.12 Univariate Dialog Window

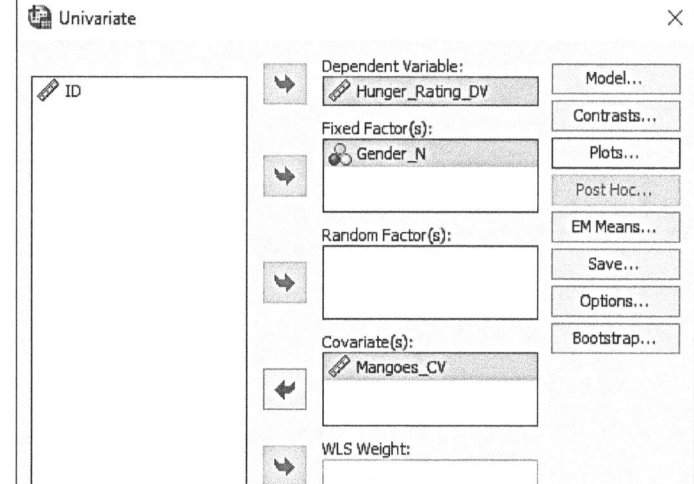

Select the **Model** pushbutton to access the **Univariate Model** window (Figure 9.13). Select **Build Terms** at the top of the window under **Specify Model,** and select **Main effects** by clicking on the down arrow under **Build Terms** in the middle of the window. Now, click over **Gender_N** from the **Factors & Covariates** panel to the **Model** panel, followed by **Mangoes_CV**.

With the main effects identified, specify the interaction between **Gender_N** and **Mangoes_CV**, the primary objective. Replace **Main Effects** below **Build Terms,** by clicking on the down-arrow and selecting **Interaction**. Return to the **Factors & Covariates** panel, and select both **Gender_N** and **Mangoes_CV** by holding down the **Control** button while selecting the two variables. Next, click on the arrow beneath **Interaction** in the middle of the window to move both variables to the **Model** panel. The results should look like the example in Figure 9.14. Click **Continue** to return to the **Univariate** window, and click **OK** to run the homogeneity of regression analysis.

Figure 9.13 The Model Window of the Univariate Model Specifying Main Effects

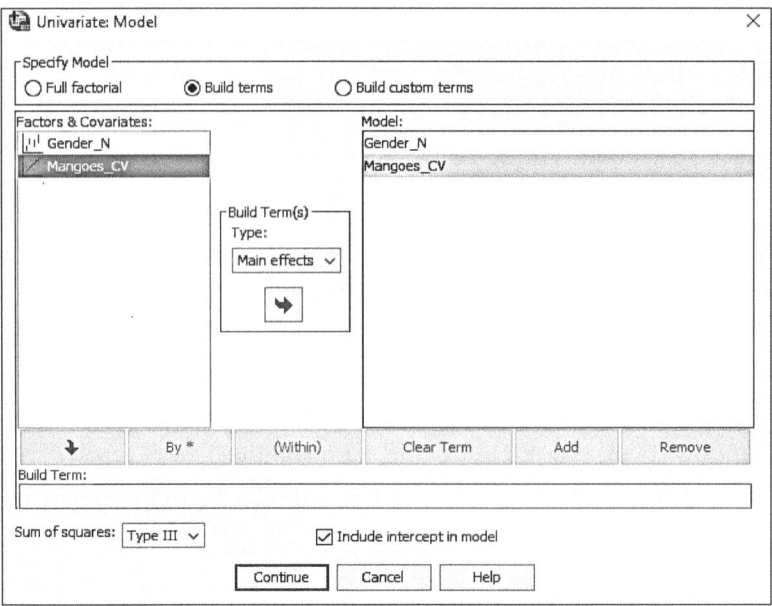

Figure 9.14 The Model Window Specifying the Main and Interaction Effects

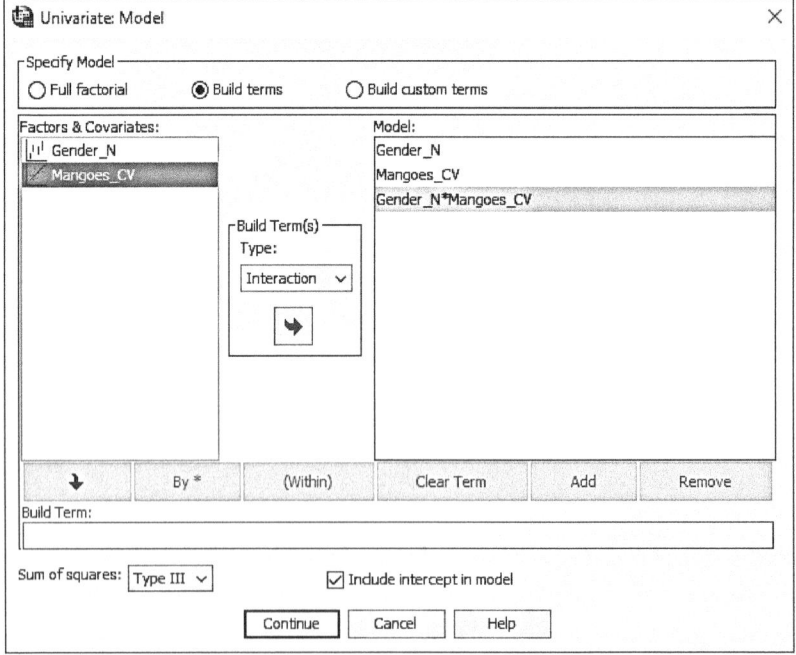

Results of the analysis appear in Figure 9.15. The only part of the summary output that is of interest is the test of significance of the interaction between **Gender_N** and **Mangoes_CV**. Recall that what we are seeking is a nonsignificant interaction effect between **Gender_N** and **Mangoes_CV** to confirm that the regression function and slope for males and females are not significantly different. The p value of .757 in the **Sig.** column for the **Gender_N * Mangoes_CV** means that the probability of the interaction of the variables is greater than the alpha level of $\alpha = .05$. This confirms a nonsignificant interaction effect. Consequently, the homogeneity of regression assumption has not been violated.

Figure 9.15 The Results of the Test of the Assumption of Homogeneity of Regression

Tests of Between-Subjects Effects					
Dependent Variable: Hunger-Fullness Rating					
Source	Type III Sum of Squares	df	Mean Square	F	Sig.
Corrected Model	71.034[a]	3	23.678	23.122	<.001
Intercept	123.308	1	123.308	120.411	<.001
Gender_N	1.591	1	1.591	1.554	.221
Mangoes_CV	59.927	1	59.927	58.519	<.001
Gender_N * Mangoes_CV	.099	1	.099	.097	.757
Error	36.866	36	1.024		
Total	1572.000	40			
Corrected Total	107.900	39			

a. R Squared = .658 (Adjusted R Squared = .630)

9.7.4 One-Way ANCOVA Analysis

Meeting the assumptions of ANCOVA clears the way for the ANCOVA omnibus test to assess whether there is a significant difference between the Hunger ratings of males and females after accounting for mangoes eaten prior to the ratings. To set up the SPSS procedure, select **Analyze → General Linear Model → Univariate** from the main SPSS menu to open the **Univariate** dialog window, and configure it as illustrated in Figure 9.12. Confirm that **Hunger_Rating_DV** is in the **Dependent Variable** panel; **Gender_N** is in the **Fixed Factor(s)** panel, and **Mangoes_CV** is in the **Covariate(s)** panel.

Select the **Model** pushbutton to open the **Univariate Model** dialog window, and confirm that **Full factorial** under **Specify Model,** the default setting, is selected (Figure 9.16a). Click **Continue** to return to the main **Univariate** menu.

Figure 9.16 IBM SPSS Univariate Model and Estimated Marginal Means Windows

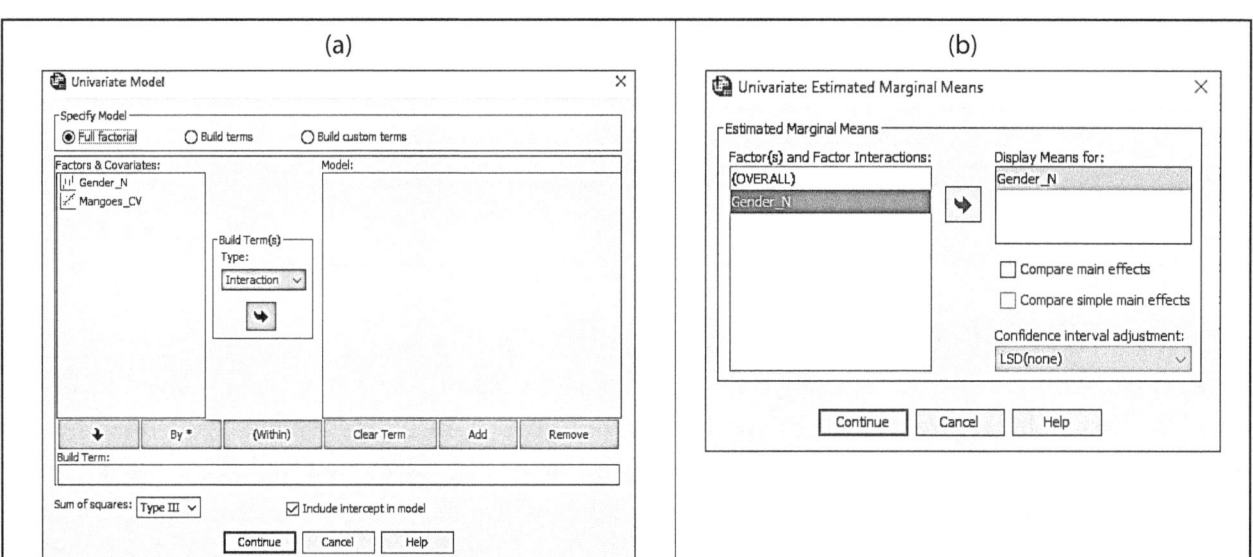

334 Analysis of Covariance

Select the **EM Means** pushbutton in the **Univariate** dialog window to access the **Univariate: Estimated Marginal Means** dialog window displayed in Figure 9.16b. Click over the independent variable, **Gender_N**, from the **Factor(s) and Factor Interactions** panel to the **Display Means for:** panel in order to generate adjusted group means after accounting for mangoes eaten. There are only two groups in the independent variable. Accordingly, there is no need to compare main effects (the first option beneath the **Display Means for:** panel). Click **Continue** to return to the main **Univariates** window.

Finally select the **Options** pushbutton in the main **Univariate** dialog window to open the **Univariate: Options** dialog window. There, select **Descriptive statistics** and **Homogeneity tests** (Figure 9.17) to produce descriptives for the demographic variable and homogeneity of variance results with a covariate added. Click **Continue** to return to the main dialog window, and click **OK** to run the analysis.

Figure 9.17 IBM SPSS Univariate: Options Dialog Window

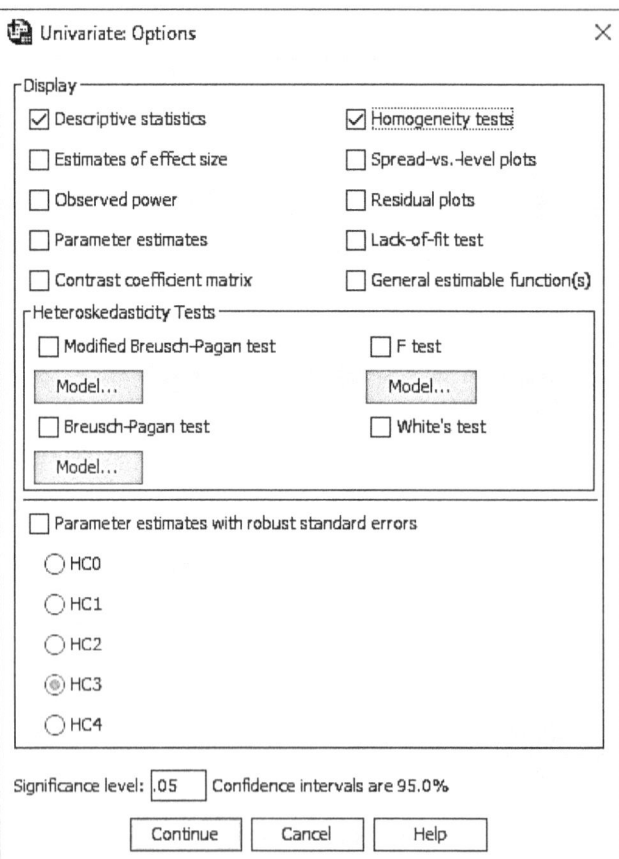

9.7.5 Omnibus One-Way ANCOVA Results

Descriptive statistics and homogeneity of variance test results of ANCOVA are displayed in Figure 9.18. The first table (Figure 9.18a) displays descriptive statistics, including the average hunger ratings for males and females before the number of mangoes eaten were accounted for. The average hunger ratings for males and females differ by 0.8, with females on average rating themselves as experiencing greater satiety ($M = 6.45$) than males ($M = 5.65$). The **N** column reports equal sample size for males and females in the study, making this a balanced, one-way ANCOVA analysis.

The second table (Figure **9.18**b) displays Levene's test of homogeneity of variance which reports a nonsignificant difference in the variance of hunger rating of males and females ($p = .962$), which means that the data met the ANCOVA assumption of equal group variances.

Figure 9.18 Results of One-Way ANCOVA Part One

(a)

Descriptive Statistics

Dependent Variable: Hunger-Fullness Rating

Gender	Mean	Std. Deviation	N
Female	6.45	1.468	20
Male	5.65	1.785	20
Total	6.05	1.663	40

(b)

Levene's Test of Equality of Error Variances[a]

Dependent Variable: Hunger-Fullness Rating

F	df1	df2	Sig.
.002	1	38	.962

Tests the null hypothesis that the error variance of the dependent variable is equal across groups.
a. Design: Intercept + Mangoes_CV + Gender_N

The **Tests of Between-Subjects Effects** and the **Estimated Marginal Means** tables are presented in Figure 9.19. The **Tests of Between-Subjects Effects** (Figure 9.19a) is the summary table for ANCOVA and reports that Mangoes_CV is a statistically significant covariate ($p < .001$). The significant effect of Mangoes_CV means that it significantly adjusted the hunger ratings, thereby cleaning up some of the "noise" in the data introduced by different amounts of snacking, without which differences in hunger would be obscured.

More importantly, by including the covariate to clear up the "noise" in the analysis, the procedure resulted in a statistically significant effect for **Gender_N**. Therefore, there is a significant difference in the hunger ratings of males and females after controlling for the number of mangoes eaten ($p = .043$).

As the table footnote indicates, the model accounts for .657 or 65.7% of the variance in **Hunger_Rating_DV**. The adjusted R^2, which takes into account error variance, is .639, or 63.9% of the variance in hunger ratings. In regression, this is referred to as R^2. In ANOVA it is called η^2 or eta-squared. The eta-square that accounts solely for the contribution made by the independent variable, gender, is computed by dividing the **Gender Type III Sum of Squares** (4.408) by the **Corrected Total Sum of Squares** (107.900), resulting in 4.408/107.900 = .041 or 4.1% of the variance in **Hunger_Rating_DV** that may be attributed to **Gender_N**, which is a relatively weak effect. Moderate and strong effects would be .14 and .22, respectively (Gamst et al., 2013).

Figure 9.19 Results of One-Way ANCOVA Part Two

(a)

Tests of Between-Subjects Effects

Dependent Variable: Hunger-Fullness Rating

Source	Type III Sum of Squares	df	Mean Square	F	Sig.
Corrected Model	70.935[a]	2	35.467	35.500	<.001
Intercept	128.455	1	128.455	128.575	<.001
Mangoes_CV	64.535	1	64.535	64.595	<.001
Gender_N	4.408	1	4.408	4.413	.043
Error	36.965	37	.999		
Total	1572.000	40			
Corrected Total	107.900	39			

a. R Squared = .657 (Adjusted R Squared = .639)

(b)

Estimated Marginal Means: Gender

Dependent Variable: Hunger-Fullness Rating

| Gender | Mean | Std. Error | 95% Confidence Interval | |
			Lower Bound	Upper Bound
Female	6.382[a]	.224	5.929	6.836
Male	5.718[a]	.224	5.264	6.171

a. Covariates appearing in the model are evaluated at the following values: Mangoes Consumed Covariate = 2.58.

Figure 9.19b is the **Estimated Marginal Means** table, which shows the average hunger ratings of the groups after they were adjusted by the covariate **Mangoes_CV**. The mean in Footnote a is the average number of mangoes eaten by study participants. The estimated group means are the predicted values when the covariate is set to the average value shown in Footnote a. Comparing the estimated means with the original means (Figure 9.7a) shows that the covariate adjusted down the mean hunger-fullness rating for females (from 6.45 to 6.382) and adjusted up the mean hunger-fullness rating for males (from 5.65 to 5.718). With the adjusted means, the difference in hunger ratings decline to 0.664. Even with the decline in the difference between the mean ratings (from 0.8 to 0.664), females still record significantly higher hunger-fullness ratings than males. Higher hunger-fullness ratings mean higher satiety. Therefore, females recorded significantly higher satiety than males.

The standard error of the mean (**Std. Error**) in Figure 9.19b measures how far the sample mean of the data is likely to be from the target population mean. The standard error is always smaller than the standard deviation, which measures the amount of variability or mean difference between individual group scores and the mean of the group. The **95% Confidence Interval** columns show that there is a 95%

likelihood that mean hunger ratings for female participants range between 5.929 and 6.836. For males, results show there is a 95% likelihood that their average hunger ratings lie between 5.264 and 6.171.

9.8 Sample Report: The Case of Appetite After Accounting for Mangoes

A one-way between-subjects analysis of covariance was performed to assess the difference in the pre-dinner average hunger ratings of male and female students after accounting for mangoes eaten. The number of mangoes eaten as a snack was used as a covariate in the analysis.

The data conformed to the assumptions of linearity of regression after a linear relationship was observed between the mangoes consumed covariate and the hunger ratings completed by study participants. In addition, the data met the assumption of homogeneity of regression due to the statistically nonsignificant interaction between the independent variable, gender, and the covariate, mangoes eaten, $F(1,36) = 0.097, p = .757$.

The covariate effect was statistically significant, $F(1, 37) = 64.595, p < .001$. In addition, the study found a statistically significant effect for gender $F(1, 37) = 4.413, p = .043, \eta^2 = 0.041$. Estimated marginal means showed that female students indicated a significantly higher fullness rating (adjusted $M = 6.382$, $SE = 0.224$, 95% CI = 5.929 – 6.836) than male students (adjusted $M = 5.718$, $SE = 0.224$, 95% CI = 5.264 – 6.171) after correcting for the quantity of mangoes consumed.

9.9 Key Terms

Adjusted mean
Analysis of covariance (ANCOVA)
Covariate
Estimated means
Homogeneity of regression slopes
Linearity of regression lines
Type III sum of squares

10
Repeated Measures ANOVA

Key concepts from previous chapters or courses:

between-subjects variance: also known as between-group variance in one-way or factorial ANOVA because the designs involve the comparison of the variance in scores of different groups.

effect size: a value measuring the magnitude of an association between two variables. In ANOVA, this refers to mean differences between groups. The larger the effect size, the greater the size of associations.

factorial analysis of variance (ANOVA): a statistical procedure to investigate the effects of two or more independent variables on a single dependent variable by examining whether there are significant differences among the mean scores of the independent variable groups.

multiple comparisons: also known as paired comparisons; comparisons of pairs of group means to determine whether there are significant differences between them. Multiple comparisons are conducted after a significant omnibus ANOVA test and is designed to keep the Type I error rate at a prespecified level.

one-way analysis of variance (ANOVA): also known as one-factor analysis of variance; a statistical procedure to evaluate the effect of three or more groups/conditions of a single independent variable on a dependent variable by investigating whether there are significant differences between the mean scores of the groups.

within-subjects variance: also known as within-group variance in one-way or factorial ANOVA; it refers to variations in scores within the same group that received the same treatment.

THIS CHAPTER INTRODUCES

- differences between repeated measures and between-groups research design
- characteristics of a repeated measures design
- when to use a repeated measures ANOVA
- prerequisite assumptions for repeated measures ANOVA
- how to calculate the effect size of a repeated measures ANOVA outcome
- how to use SPSS to conduct a repeated measures ANOVA.

10.1 Introduction

The previous three chapters investigated three types of ANOVA techniques, all of which were designed to measure differences in score values between groups on a single dependent variable:

- one-way ANOVA: a research technique used to determine whether there are differences in a study using a one-way between-groups design; e.g., differences between the average mathematics scores of two or more groups of a single independent variable, such as teaching/learning methods.
- factorial ANOVA: a research technique used to discover whether differences exist in a study using a factorial between-groups design; e.g., a study of differences in student behavior based on gender and grade point average.
- ANCOVA: a research technique in which a covariate (e.g., depression level) may be added to a one-way or factorial ANOVA design to adjust for external influences on a dependent variable (e.g., mathematics achievement, or student behavior).

The purpose of repeated measures is somewhat different from the previous three statistics techniques. Rather than seeking to detect differences between groups, repeated measures is designed to investigate differences within a group of study participants. Repeated measures models may involve one or both of the following two objectives:

- to investigate differences between points in time, or
- to investigate differences between specific conditions or situations

In between-groups designs, each study participant has a single score on a dependent variable (e.g., math achievement or student behavior). By contrast, in repeated measures each subject has more than one score assigned over time or under different conditions. In each situation, there is a single dependent variable (e.g., essay score).

For example, in a study measuring mathematics achievement due to the implementation of a new pedagogical method, the dependent variable might be mathematics achievement measured before as well as one and two months after the beginning of the pedagogical initiative. Another study might involve the ratings of the corporate responsibility of four social media websites (WhatsApp, Instagram, YouTube and Facebook) by 20 social media users. In this case, each study participant rates each of the four social media sites.

By securing multiple measures from a single group of study participants, repeated measures designs boost the power and efficiency of quantitative research. In all cases, the statistical technique used to analyze repeated measures data is commonly known as repeated measures ANOVA. It is also called RMANOVA, within-subjects/groups ANOVA, or ANOVA for correlated samples.

10.2 The Case of the Sprinting Speeds of Seven Soccer Players

A small dataset taken from Wragg et al. (2000) is the source of the repeated measures ANOVA exercise in this chapter. The dataset consists of the sprint speeds of seven soccer players across six trials (six different occasions) for the Bangsbo endurance sprint test. The objective of the trials was to evaluate the ability of soccer players to perform quality sprints consecutively by determining whether speeds (measured in seconds) decreased significantly over trials. The data analysis for this case involves computing a one-way RMANOVA. The data for the Case of the Sprinting Speeds of Seven Soccer Players appear in Table 10.1. A likely outline for an associated research plan appears in Box 10.1.

Table 10.1 Data for the Case of the Sprinting Speeds of Seven Soccer Players*

ID	Trial (Time)					
	1	2	3	4	5	6
1	7.755	7.743	7.409	7.182	7.263	7.233
2	7.851	7.577	7.364	7.485	7.223	7.199
3	8.048	7.946	7.731	7.554	7.361	7.599
4	8.401	7.906	7.892	8.028	7.685	7.95
5	7.724	7.675	7.394	7.69	7.576	7.867
6	7.672	7.456	7.713	7.589	7.364	7.36
7	8.207	7.918	7.86	8.012	7.747	7.944

*All Speeds measured in seconds

10.3 Repeated Measures ANOVA Fundamentals

Chapter 7 disclosed that one-way ANOVA and the independent samples *t* test have a common feature; they both may be used to investigate the average values of two independent groups to determine whether there are significant differences between the groups. These dependent variable values typically measure a specific characteristic of the groups, such as academic achievement, IQ, or attitude toward a specific sport. One important caveat, however, is that while an independent *t* test is limited to comparing the average differences between two groups on a single characteristic (e.g., average reading achievement score), one-way ANOVA compares the differences between two or more groups.

RMANOVA also has something in common with *t* tests. In this case, the *t* test of note is the dependent samples *t* test, also called the related samples *t* test or the paired samples *t* test. This technique is designed to compare the averages of two sets of scores from a single group of study participants. As Figure 10.1a illustrates, each member of the group is measured twice on the same dependent variable (reading achievement) and has two scores; often the scores are pre- and postscores of an intervention or treatment, such as a new pedagogical reading method. The teacher may compare the two sets of scores to evaluate progress in oral reading achievement.

The dependent samples *t* test has two variables of interest; the dependent variable, which is measured on a continuous scale, and the independent variable, which is a categorical variable. In the reading ability study, the dependent variable is the change in reading achievement. The independent variable has two categories of the variable, time: before and after the instructional treatment.

The dependent samples *t* test is limited to the comparison of two sets of scores within a single group of study participants. RMANOVA may be used to conduct a dependent sample *t* test. However, it is best used to compare more than two sets of scores from a single group of study participants as illustrated in Figure 10.1b. Here, as the graphic illustrates, the same group of seven soccer players engage in a series of three sprint tests during which the researcher records the speed for each participant in each trial. A period of recovery follows Trial 1 and Trial 2. The goal is to compare the average speed of the group in the three trials to estimate the change in speeds from one trial to another.

Figure 10.1 Repeated Measures Designs: Testing for Changes over Time

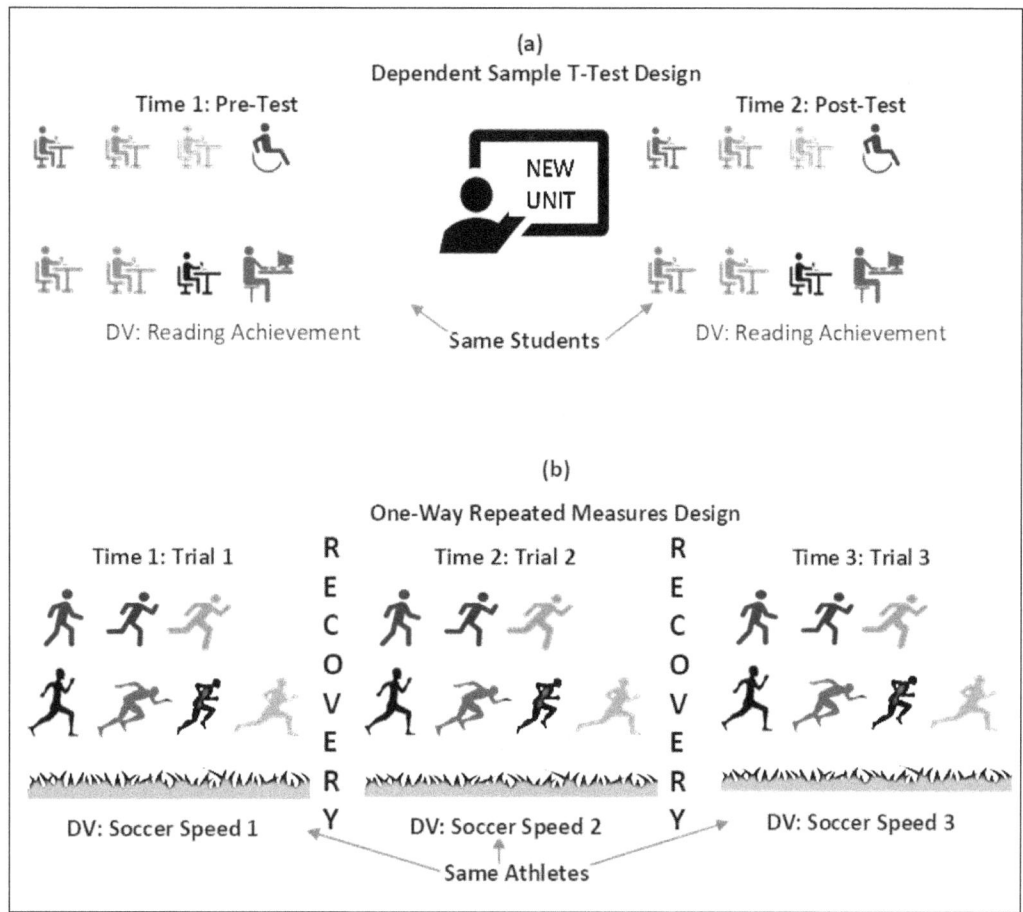

Similar to the dependent samples *t* test, the independent variable for this design is Trial, which has three related categorical groups—Trial 1, Trial 2, and Trial 3; each trial occurs at a specific period in time. The dependent variable is sprint speed, a continuous variable measured on the ratio scale. A statistical design with a single independent variable, a single dependent variable as well as three scores to compare is a one-way RMANOVA.

The foregoing section described how a repeated measures design may be used to analyze changes in average values over time—one of two designs common in repeated measures. The other common repeated measures model is used to investigate differences between a group of average scores from two or more different conditions. Consider a dependent *t* test research design in which a group of 10 trained classroom teachers must score the essays of 30 Grade 6 students, each of which was completed under two different conditions—handwritten and typed on computers as depicted in Figure 10.2a. In one instance teachers rate randomly assigned hand-written essays and then they rate the same essays in random order, but typed on a computer. Each essay receives 10 ratings, and the ratings are compared across conditions—handwritten and typed. In this research design, the independent variable, condition, has two categories—handwritten essay and typed essay. The dependent variable is essay rating. The statistical analysis goal is to investigate differences in the ratings between essay categories.

Figure 10.2b illustrates an expansion of the soccer player speed test study described in Figure 10.1b. The expanded study is intended to investigate differences between the average speeds of the group of soccer players under different conditions: a sandy beach, a grassy field, and a neighborhood street.

Figure 10.2 Repeated Measures Designs: Testing for Differences Across Diverse Conditions

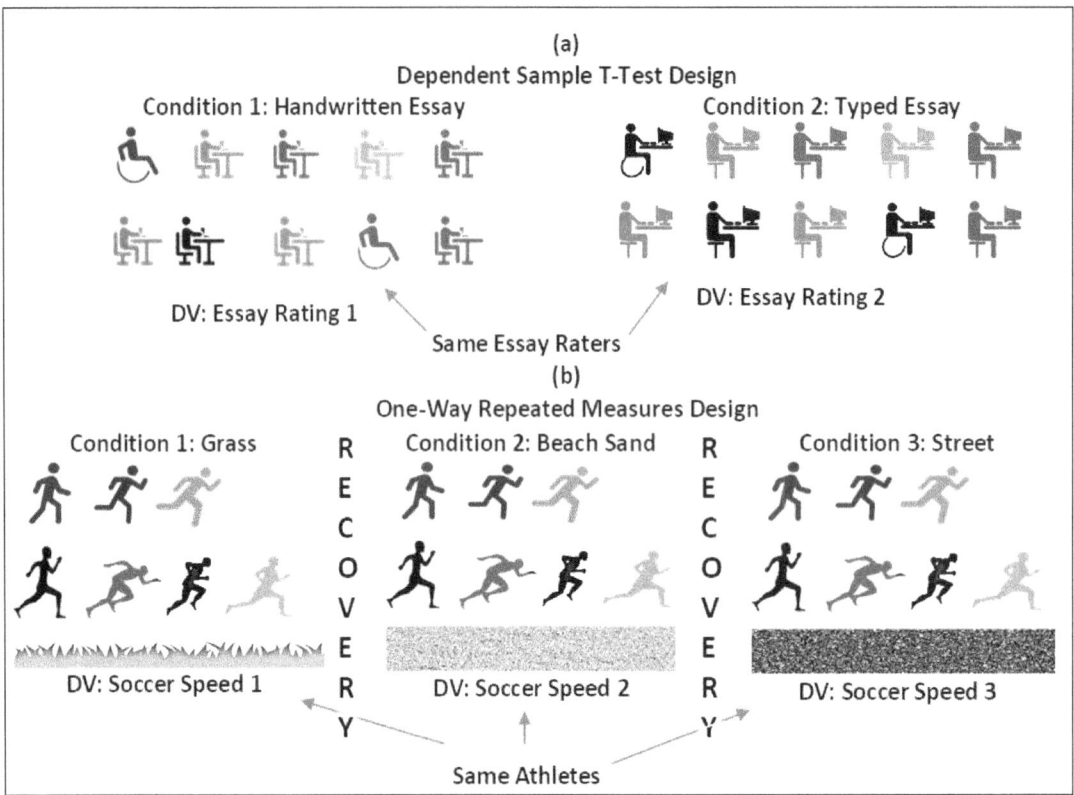

In this research design, the independent variable is condition, and it includes three categories—grass, beach sand, and street. The dependent variable is speed, and the goal is to examine the differences in speed among the three conditions.

A key difference between the independent samples t test and the dependent samples t test is that the former focuses on analyzing differences **between** groups while the latter searches for differences **within** a group of study subjects. The same crucial difference applies to the one-way ANOVA, which is designed to estimate differences **between** groups, and RMANOVA, a statistical technique used to examine differences **within** a single group.

10.4 Characteristics of Repeated Measures Designs

Covered so far are key characteristics of studies that indicate the need to apply RMANOVA:
- Typically, study subjects have more than two scores on the dependent variable (also called repeated or within-group observations). While repeated measures may be used to analyze data with two scores for each subject, the dependent t test is most often used in these situations.
- The scores may represent measures over time points or measures under different conditions. Accordingly, job performance scores for a single individual may be obtained at three different times each year (time points), or a group of 10 students may rate the same support service at four different colleges (different conditions).
- The independent variable is typically the categorical variable, time, or condition. The dependent variable is an interval level or ratio level variable.

- The most common design includes only one group of participants; all group members receive all of the treatments or complete all of the same tasks; for example, all study participants in a chronic pain management clinical trial may receive a placebo, low dose, and high dose treatments, recording the level of pain experienced after each treatment. Here, the study participants are their own control group.

Repeated Measures has some additional unique features.
- Treatments and performances are usually carried out in serial form; that is, one after the other. Therefore, Condition 1 happens prior to Condition 2, which occurs ahead of Condition 3.
- This statistical technique is particularly useful in enhancing power and effectiveness due to its economy in that only a single group is required. It is the only procedure where comparisons of performance are possible without the need for multiple groups. The design of repeated measures makes it more powerful than between-groups models. More details regarding the power of the procedure to be discussed below in Section 10.6.

10.5 Types of Repeated Measures Designs

This section touches on some of the most common repeated measures designs that fall within the univariate statistics domain—the focus of this text.

10.5.1 One-Way Repeated Measures Design

One of the most common designs is the one-way repeated measures design illustrated in Figure 10.1 and 10.2. As in the soccer example, a one-way design has a single independent variable (typically based on condition or time) and a single dependent variable (e.g., speed). The design is labeled repeated measures because a single group of subjects is measured at multiple points in time or under multiple conditions. This is indicated in Figure 10.1b and 10.2b by the repetition of the icons in each time slot or condition. Another way of presenting the design in Figure 10.2b is displayed in Figure 10.3, which indicates that the average speed for each time point derives from the same group of soccer players—S_1, S_2, S_3, S_4, S_5, S_6, S_7. Adding three additional columns to Figure 10.3 results in the design for the data in Table 10.1 where T represents a point in time.

Figure 10.3 Assignment of Subjects in a One-Way Repeated Measures Design (Time)

Subjects	Trials		
	T_1	T_2	T_3
S_1	S_1	S_1	S_1
S_2	S_2	S_2	S_2
S_3	S_3	S_3	S_3
S_4	S_4	S_4	S_4
S_5	S_5	S_5	S_5
S_6	S_6	S_6	S_6
S_7	S_7	S_7	S_7

As implied above, whether the independent variable is time (T) or condition (K), each occasion represents a level of the independent variable. When the independent variable is time, all study participants are assessed at three or more consecutive time points, but not necessarily equidistant from each other (e.g., Winkens et al., 2005). Therefore, time points might be before, in the middle, and at the end of a clinical trial with all subjects measured in that order. Stevens (2009) noted that repeated measures ANOVA is the natural design to use when measuring performance over time.

Figure 10.4 Assignment of Subjects in a One-Way Repeated Measures Design (Condition)

Subjects	Condition		
	K_1	K_2	K_3
S_1	S_1	S_1	S_1
S_2	S_2	S_2	S_2
S_3	S_3	S_3	S_3
S_4	S_4	S_4	S_4
S_5	S_5	S_5	S_5
S_6	S_6	S_6	S_6
S_7	S_7	S_7	S_7

When an independent variable is condition, study participants are all exposed to the same three or more conditions or situations as Figure 10.4 illustrates. For example, if K_1, K_2 and K_3 represent different conditions for playing soccer (grassy field, sandy beach, and street), all players ($S_1 - S_7$) would be exposed to the same pre-established conditions.

10.5.2 Factorial Repeated Measures Design

The factorial repeated measures model shares some of the characteristics of factorial between-groups and one-way repeated measures designs. Like both models, factorial repeated measures ANOVA has a single dependent variable. It may also share the factorial between-groups design characteristic of two or more independent variables. The independent variable could be time: for example, hypertension medication first taken by 10 patients in the mornings over three months, then in the evenings for three months (Figure 10.5). This would be a 2 x 3 factorial repeated measures design with two within-subjects independent variables.

Figure 10.5 Factorial Repeated Measures with Two Time-Related Within-Subjects Variables

Participants	Morning			Evening		
	Month 1	Month 2	Month 3	Month 1	Month 2	Month 3
S_1						
...						
S_{10}						

Alternatively, factorial repeated measures variables may be based on condition/treatment; for example, suppose 10 members of a focus group are asked to rate four different brands of vegetarian patties when they are hungry and when they are full. This two-condition repeated measures design is presented in Figure 10.6.

Figure 10.6 Repeated Measures Design with Two Condition-Related Within-Subjects Variables

	Hungry				Full			
Participants	Brand 1	Brand 2	Brand 3	Brand 4	Brand 1	Brand 2	Brand 3	Brand 4
S_1								
...								
S_{10}								

Finally, repeated measures variables may be a combination of both time and condition. Imagine the soccer trials example with speed tested under three conditions (three kinds of soccer fields), and with sprinting speed measured at each of three points in time (trials), as presented in Figure 10.7. The condition independent variable here includes three types of soccer fields, and the independent variable, time, comprises three levels—Trial 1 - Trial 3.

Figure 10.7 Repeated Measures with Time- and Condition-Related Within-Subjects Variables

	Condition								
	Grassy			Sandy			Asphalt		
Participants	Trial 1	Trial 2	Trial 3	Trial 1	Trial 2	Trial 3	Trial 1	Trial 2	Trial 3
S_1									
...									
S_7									

10.5.3 Simple Mixed Design

In a study by Djaoui et al. (2017), the maximal sprint speeds of professional French soccer players were found to differ significantly by playing position. If this study were replicated over multiple speed trials, researchers would have a mixed design which includes a between-groups and within-groups component.

Figure 10.8 shows what this design might look like if players performed in six speed trials, and two playing positions were involved: wide playing positions (forward, defenders and mid-fielders), and central playing positions (defenders and mid-fielders). Imagine further if there were 10 players in each group. In this study maximal sprint speed would be measured in each of the six trials for each of the 20 participants in the study. The statistical analysis technique in this situation is RMANOVA, specifying the combination of a one-way ANOVA and a one-way RMANOVA.

Figure 10.8 Mixed Design with One Between- and One Within-Group Variable

	Time					
Player Group	Trial 1	Trial 2	Trial 3	Trial 4	Trial 5	Trial 6
Wide S_1 ... S_{10}						
Central S_1 ... S_{10}						

More variables could be added to the design. Additional between-group variables could be added, such as age or performance level (professional versus amateur). Additional time and condition repeated measures variables could be added to the design as well. Mixed Design will be investigated in greater detail in Chapter 11.

10.6 Research Plan: The Case of the Sprinting Speeds of Seven Soccer Players

This section outlines the blueprint for conducting the Case of the Sprinting Speeds of Seven Soccer Players using a one-way repeated measures ANOVA design. Box 10.1 displays the outline for the research plan. The data for the case are in Table 10.1.

10.6.1 Research Objectives and Variables

Sports researchers want to understand more about the physiological mechanism involved in reproducing repeated maximal sprint ability in soccer athletes. To do that, researchers included more authentic features in the treatment, like multidirectional sprinting and recovery that are characteristic of the game. Researchers' desire to investigate changes in serial performance indicates the need for repeated measures analysis.

The research question reflects the variables in the case study which are listed under Research Variables and Measurement Levels in Box 10.1 as well. The dependent variable, sprint speed, is a continuous variable. The independent variable is time, with six levels, Trial 1 – Trial 6, so soccer players sprinted six times with recovery jogging in between the trials. Multiple trials allow researchers to compare speed times within the same soccer player.

Box 10.1 Research Plan for the Case of the Sprinting Speed of Seven Soccer Players

Research Problem
Repeated sprinting ability of soccer players is widely regarded as critical to the outcome in soccer matches. However, there is limited research about this ability. A researcher wants to partially replicate the statistical data analysis conducted for the validation study of the Bangsbo Sprint Test by Wragg et al. (2000), which evaluated soccer players' ability to perform quality sprints in multiple directions, with recovery periods over six trials. The study was also designed to determine whether multidirectional speeds (measured in seconds) with intervening recovery declined significantly over trials.

Research Variables and Measurement Levels
Independent Variable(s): Time (within-subjects variable)
Levels of Independent Variable(s): Trial 1 – Trial 6
Dependent Variable: Speed (measured in seconds)

Research Question
Is there a significant change in the average sprinting speed of seven soccer players across six consecutive trials over time?

Hypotheses
Null Hypothesis: There is no significant difference in the average sprinting speeds of soccer players across six consecutive trials.
$H_o: \mu_1 = \mu_2 = \mu_3 = \mu_4 = \mu_5 = \mu_6$

Alternative Hypothesis: There is a significant difference in the average sprinting speeds of soccer players across six consecutive trials for at least two trials.
$H_a: \mu_i \neq \mu_j$

Research Design

Single Group: X_1 O_{T1} X_2 O_{T2} X_3 O_{T3} X_4 O_{T4} X_5 O_{T5} X_6 O_{T6}

The nonrandomized, single-group repeated measures design includes a total of six random, multidirectional sprint trials completed by each of seven study participants, with each trial interspaced with a period of recovery (jogging). X represents instruction in the sprinting protocol for each trial.

Data Collection
Data were collected from each of seven study participants, all of whom were male and national-level student players. An electronic timing system recorded the time taken between the illumination of the light-emitting diodes (LED) and the breaking of the light at the end of each portion of a 34.2-meter sprinting course. There are four portions of the course altogether, and speed is measured in seconds.

Data Analysis
When investigating whether a significant difference exists among multiple *measures* of the same variable (speed) taken on the same subjects over two or more time periods, a one-way, ordinary repeated measures of analysis of variance (RMANOVA) is the appropriate statistical test of significance.

10.6.2 Research Design

The design used in this case study is a time-course repeated measures design in which there are no experimental interventions. Speed of the soccer sprints are recorded at six time points, with intervening recovery periods of jogging back to the starting position. The research design of the case study is included in Box 10.1. The design notation shows six time periods with $O_{T1} - O_{T6}$ representing the outcomes (speeds) taken at the six time points and $X_1 - X_6$ representing instructions for the sprinting protocol throughout the study.

10.6.3 Data Collection and Hypothesis Testing

Study participants for the original study were volunteers who were national-level student players. Each soccer player underwent the same exercise and followed randomly assigned protocols for the multi-directional sprint test. Using a convenience sample (volunteers) removes the ability to generalize the results to anyone outside of the study sample itself.

Hypothesis testing is designed to be used in experiments and quasi-experimental situations where results may generalize to a wider population. Hypothesis testing here is included in this exercise for demonstration purposes.

The null hypothesis for a research study using a one-way, repeated measures design is that three or more dependent variable scores for the same study participant have the same mean in the target population. That means the assumption in the current case is that within the target population, the means for Trial 1, Trial 2 ... Trial 6 are not significantly different.

The alternative hypothesis is that the speed averages across trials are significantly different in the study sample for at least one pair of averages. A true alternative hypothesis would mean that the averages in the target population are significantly different.

10.7 Calculations for Repeated Measures ANOVA

Table 10.1 serves as the original data source to investigate the primary research question and hypothesis in Box 10.1: Is there a significant change in the average sprinting speed of seven soccer players across six trials? The following statistics will be calculated to answer the research question:

- total sum of squared deviations (SST) or the maximum variance that can be explained by the independent variable
- sum of squares of the scores for each participant, totaled across all participants (SSW).
- sum of squares for time period or treatment (SSM or $SS_{treatment}$). This represents variance explained by changes across the time periods 1 – 6.
- error sum of squares (SSE) or unexplained variance
- mean squares that explain the treatment ($MS_{treatment}$), calculated by dividing $SS_{treatment}$ by the degrees of freedom for treatment ($df_{treatment}$)
- mean squares error sum of squares or unexplained variance (MSE) derived by dividing SSE by degrees of freedom for error df_e
- the F Ratio, which is the ratio of the mean variance explained by treatment ($MS_{treatment}$) and variance explained by extraneous factors (MSE)

The partitioned variance described in the first four steps is illustrated in Figure 10.9. Computation of between-subjects sum of squares (SSB) is excluded from this exercise because it is not needed to determine the significance of the repeated measures model. SSB represents average individual differences between study participants. Data in Table 10.2 will be used for all hand-computations.

Figure 10.9 The Partitioned Variance for Repeated Measures ANOVA

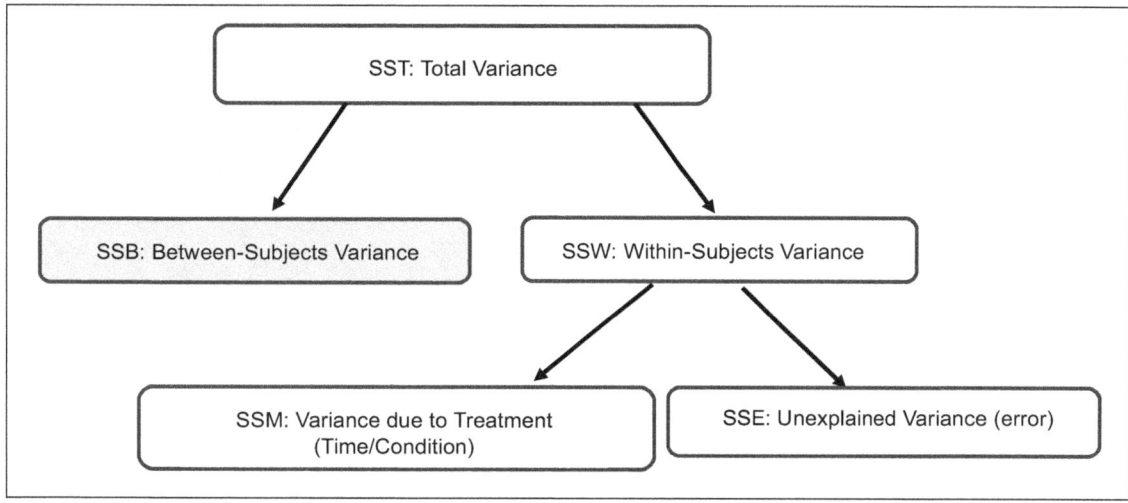

10.7.1 Total Sum of Squares (SST)

The total sum of squares (SST) is the total squared deviations for the entire repeated measures model. The equation is the same as the one for one-way and factorial ANOVA:

$$SST = \sum_{i=1}^{N}(X_i - \bar{X}_{grand})^2$$

Table 10.2 Expanded Data Table for the Case of the Sprinting Speed of Seven Soccer Players

Participant	Trial 1	Trial 2	Trial 3	Trial 4	Trial 5	Trial 6	Participant Mean	Sum of Squares (S²)
1	7.755	7.743	7.409	7.182	7.263	7.233	7.431	0.066
2	7.851	7.577	7.364	7.485	7.223	7.199	7.450	0.060
3	8.048	7.946	7.731	7.554	7.361	7.599	7.707	0.066
4	8.401	7.906	7.892	8.028	7.685	7.95	7.977	0.056
5	7.724	7.675	7.394	7.69	7.576	7.867	7.654	0.025
6	7.672	7.456	7.713	7.589	7.364	7.36	7.526	0.024
7	8.207	7.918	7.86	8.012	7.747	7.944	7.948	0.024
Total	55.658	54.221	53.363	53.54	52.219	53.152	53.692	
Trial Mean	7.951	7.746	7.623	7.649	7.460	7.593	7.670	
(X − GM)²	1.007	0.253	0.329	0.538	0.570	0.707		

To find the Grand Mean (GM), use the data in Table 10.2 and one of the following methods:
1. Sum the sprinting speeds in Table 10.2 for Trial 1 – Trial 6 for Subject 1 – Subject 7 (55.658 + . . . + 53.152), and divide by the total number of speeds (42).

350 Repeated Measures ANOVA

2. Sum the Trial (column) means (7.951 + ... + 7.593), and divide by the total number of trials (6).
3. Sum the subject (row) means (7.431 + ... + 7.948), and divide by the total number of subjects (7).

Regardless of the method selected to calculate the grand mean, it will generate the same outcome—a grand mean of 7.670, which is the total in the last cell in the Mean column/row of Table 10.2. With the grand mean computed, calculate the SST by subtracting each speed from the grand mean (7.670) and squaring the result. The outcome of summing the squared deviations for each trial is presented in the last row of Table 10.2 $(X - GM)^2$. Add the values in the last row to find the SST.

$$SST = 1.007 + 0.253 + 0.329 + 0.538 + 0.570 + 0.707$$
$$SST = 3.404$$

10.7.2 Sum of Squares Within Subjects (SSW)

Calculating the sum of squared deviations within participants is an important departure from between-subjects ANOVA, which includes different participants within each group of the independent variable. The objective of between-subjects ANOVA is to estimate differences *between* groups. Differences within a group of study participants are regarded as error.

Conversely, RMANOVA scores participants repeatedly, and the primary objective is to measure differences in scores within a group from one point in time (or condition) to another. Therefore, the equation for calculating the sum of squared deviations within a group (SSW) becomes an important component of conducting an RMANOVA.

$$SSW = \sum_{i=1}^{N} s_i^2 (k-1)$$

where n = sample size, k = number of treatments/trials, s_i^2 = the variance in scores for the ith soccer player.

To compute SSW:
1. Find the sum of squared sprinting speed deviations for each trial for each player $(s_i^2 = (X_i - \bar{X}_i)^2 / k - 1)$ where X_i = speed for player i, in each trial, $\bar{X}_i$ the mean speed for player i across trials, and k = the number of trials.
2. Multiply the sum of squares for each soccer player by k - 1.
3. Finally, compute the sum of the outcome of Step 2.

To complete Step 1, use Participant 1 in Table 10.2 as an example:

$$S^2_{player\ 1} = ((7.755 - 7.431)^2/5) + ((7.743 - 7.431)^2/5) + ((7.409 - 7.431)^2/5)$$
$$+ ((7.182 - 7.431)^2/5) + ((7.263 - 7.431)^2/5) + ((7.233 - 7.431)^2/5)$$
$$S^2_{player\ 1} = 0.021 + 0.019 + 0 + 0.012 + 0.006 + 0.008$$
$$S^2_{player\ 1} = 0.066$$

The sum of squares for each soccer player is displayed in the last column of Table 10.2. After completing Step 1 for each player, calculate Step 2 and 3 for all players as follows:

$$SSW = (0.066 * 5) + (0.060 * 5) + (0.066 * 5) + (0.056 * 5) + (0.025 * 5)$$
$$+ (0.024 * 5) + (0.024 * 5)$$
$$SSW = 0.33 + 0.3 + 0.33 + 0.28 + 0.125 + 0.12 + 0.12$$
$$SSW = 1.605$$

10.7.3 Treatment (Model) Sum of Squares (SSM or $SS_{treatment}$)

The available variance within the data is quite small as indicated by the similarity in the speeds from one trial to another. It is for that reason that the total variance explained by the data (SST) equals 3.404 units of variance. Of that total, 1.605 units of variance is the explained differences in the speeds of soccer players across the six sprinting trials. The next step is to determine how much variance is explained by the treatment across the trials or points in time in this exercise. The equation for calculating the treatment sum of squares is:

$$SS_{treatment} = \sum_{k=1}^{k} n_k (\bar{x}_k - \bar{x}_{gm})^2$$

To apply the equation for the treatment/trial sum of squares, complete the following:
1. Find the difference between the average for each trial ($\bar{x}_k$) and the grand mean ($\bar{x}_{gm}$).
2. Square the difference for each trial.
3. Multiply the result by the number of participants for each trial (n_k).
4. Total the result for all six trials.

Using the averages in Table 10.2, the calculations are as follows:

$$SS_{treatment} = 7(7.951 - 7.670)^2 + 7(7.746 - 7.670)^2 + 7(7.623 - 7.670)^2$$
$$+ 7(7.649 - 7.670)^2 + 7(7.460 - 7.670)^2 + 7(7.593 - 7.670)^2$$
$$SS_{treatment} = 7(0.281)^2 + 7(0.076)^2 + 7(-0.047)^2 + 7(-0.021)^2 + 7(-0.21)^2$$
$$+ 7(-0.077)^2$$
$$SS_{treatment} = 7(0.079) + 7(0.006) + 7(0.002) + 7(0) + 7(0.044) + 7(0.006)$$
$$SS_{treatment} = 0.959$$

10.7.4 Error Sum of Squares (SSE)

Thus far, sums of squares computations have generated a total of 3.404 units of variance that included 1.605 units of variance, explained by differences within soccer players and slightly less than 1 unit of variance (0.959) explained by the treatment applied (trials). The final sum of squares to be computed in this demonstration is the error sum of squares (SSE) that represents the amount of variance the study cannot explain—extraneous variation outside of the investigator's control.

Recall that the variance within soccer players (SSW) comprises two components: variance explained by the model representing what the investigator manipulates ($SS_{treatment}$), and the error sums of squares (SSE) (Figure 10.9). Therefore, the most direct method of computing the error sum of squares is to subtract $SS_{treatment}$ from SSW as displayed below:

$$\text{SSE} = 1.605 - 0.959$$
$$\text{SSE} = 0.646$$

10.7.5 Degrees of Freedom

The mean squares for treatment and error are important components for calculating the F value for the omnibus repeated measure hypothesis test. Averages in these situations reduce bias that might be introduced using summed values. To find the mean squares, first find the degrees of freedom that will serve as the denominators in the ratios needed to source the mean squares.

Computing the mean squares will necessitate two degrees of freedom; the degrees of freedom for the treatment sum of squares ($df_{treatment}$) and the degrees of freedom for the error sum of squares (df_e).

The degrees of freedom for treatment ($df_{treatment}$) equal $k - 1$, where k is the number of treatments (conditions or time periods). With six trials in the case study, the degrees of freedom equals:

$$df_{treatment} = k - 1$$
$$df_{treatment} = 6 - 1 = 5$$

Calculate the error degrees of freedom by first identifying the within degrees of freedom (df_w) and then subtracting the treatment degrees of freedom (5) from that total.

The within degrees of freedom (df_w) correspond to the sum of $k - 1$ (where k is the number of treatments) for all study participants or the product of $k - 1$ and the total number of participants. There are 7 participants in the study ($N = 7$) and each participant has $6 - 1$ or 5 degrees of freedom. Therefore, the within degrees of freedom is:

$$df_w = N(k - 1)$$
$$df_w = 7(6 - 1)$$
$$df_w = 7 \times 5 = 35$$

To find the error degrees of freedom, use the same method used for computing the SSE,

$$df_e = df_w - df_{treatment}$$
$$df_e = 35 - 5 = 30$$

10.7.6 Mean Squares

As $SS_{treatment}$ represents the amount of variance explained by the treatment across the six trials, so $MS_{treatment}$ represents the average amount of variance explained. This is calculated by dividing $SS_{treatment}$ by $df_{treatment}$:

$$MS_{treatment} = SS_{treatment}/df_{treatment}$$
$$MS_{treatment} = 0.959/5 = 0.192$$

Similarly, to find the average amount of variance explained by extraneous factors, divide SSE by df_e,

$$MSE = SSE/df_e$$
$$MSE = 0.646/(30 = 0.022)$$

10.7.7 The F Ratio

As in one-way ANOVA, the F ratio for one-way repeated measures is a statistic that represents the variance explained by the investigation (systematic variance) and unexplained extraneous (unsystematic) variance. To calculate the F ratio, divide the treatment mean square by the error mean square, a process that is similar to the method used for one-way ANOVA. Therefore,

$$F = MS_{treatment}/MSE$$
$$F = 0.192/(0.022 = 8.73)$$

F value of 8.73 is greater than 1, which means that the repeated trials had some effect above and beyond the effect of extraneous factors (MSE). Accordingly, the changes in at least two sprinting speeds across the six trial periods were significantly different from each other.

As with ANOVA, the F value may be compared with the appropriate critical value to formally determine the significance level of the changes in sprinting speed. To accomplish that, find the critical value for the treatment degrees of freedom ($df_{treatment} = 5$) in the column of the F distribution table (Appendix A), and the degrees of freedom for error ($df_e = 30$) in the corresponding row. The critical value of F, based on 5 and 30 degrees of freedom and an alpha level of .01 is $F(5,30) = 3.70$, evidence that sprinting speeds were significantly different across trials ($p < .01$).

The computation just completed is called an omnibus test for one-way repeated measures ANOVA, which included the F statistic that signifies whether there is a significant difference between at least one pair of sprinting speeds. If the hypothesis test proved to be nonsignificant, all analyses would cease here. However, with a significant effect, the next step is to discover where the effect lies. As discussed in Chapter 7, researchers commonly use one or both of the following post ANOVA tests to identify the source of significant effects:

- a priori planned comparisons, or
- post hoc multiple comparisons tests.

Section 10.11 will investigate how to use SPSS software to conduct post ANOVA tests using the SPSS RMANOVA procedure.

10.8 Underlying Assumptions of Repeated Measures ANOVA

Repeated measures ANOVA statistical assumptions are similar to those for other ANOVA procedures.

10.8.1 Independent and Dependent Variables

The dependent variable should be interval or ratio level variables (e.g., changes in test scores, or ratings of the same product from different sources).

Independent variables must be categorical. In between-subjects ANOVA, independent variables consist of two or more mutually exclusive levels of participants. However, in RMANOVA, independent variables include two or more related levels in which all study participants are measured over the same period of time or under the same conditions. The time periods and conditions are levels of the independent variables (e.g., changes in weight over three or more months, or ratings of patties from four different bakeries).

10.8.2 Outliers and Normality

There should be no significant outlier scores in independent variable groups as these can distort the differences between the related groups and reduce the accuracy of results. Descriptive statistics including frequency counts, stem-and-leaf plots, histograms, and boxplots are commonly used to detect outliers.

The distribution of dependent variable values at each level of an independent variable should be approximately normal. However, like other ANOVA tests, RMANOVA is robust to violations of normality. Accordingly, violation of the normality assumption may not invalidate test results. Test for normality, using the Shapiro-Wilk test of normality as well as descriptive statistics such as, skewness and kurtosis.

10.8.3 The Assumption of Sphericity

The assumption of sphericity in RMANOVA means that the variances of the differences between all possible pairs of levels of the independent are equal (Howell, 2014). Therefore, the variances of the differences between Trial 1 and Trial 3, and Trial 3 and Trial 5 should be approximately equal. The assumption of sphericity is the repeated measures version of the assumption of homogeneity of variance for other types of ANOVA.

Repeated measures ANOVAs are particularly susceptible to violation of the assumption of sphericity, which causes the test to become too liberal and, subsequently, leads to an increase in the Type I error rate. In SPSS, Mauchly's test is used to detect violation of the assumption of sphericity for repeated measures procedures. Sphericity becomes an issue when there are at least three time points or conditions. Therefore, SPSS tests for sphericity only in these situations.

While Mauchly's test is often criticized for failing to find sphericity in small samples, there is little else to test this assumption. As a result, it is commonly used and is the only test of sphericity in SPSS's repeated measures ANOVA procedure.

Mauchly's test evaluates the null hypothesis that the variance of the difference between conditions or time points is equal. If the test statistic, Mauchly's W, is not significant (e.g., $p > .05$), there is not a significant difference between the variances of the differences; in effect, the variances are approximately equal. If, however, Mauchly's W is significant (e.g., $p < .05$), there is a significant difference between the variances of the differences and thus a violation of the assumption of sphericity.

If the assumption of sphericity has been violated, there are corrections to produce F statistics that reduce the likelihood of a Type I error. This is accomplished by estimating the extent to which sphericity has been violated and applying a correction factor to the degrees of freedom used to calculate critical values. The adjustment factor increases critical values so that they are larger than they would be if sphericity had not been violated. Subsequently, F statistics are compared against these larger critical values. Since significance is based on whether the F statistic exceeds the relevant critical value, the likelihood of finding significant differences declines; larger F statistics are needed to find significance.

Three corrections are commonly used in Mauchly's test:
- Greenhouse-Giesser,
- Huynh-Feldt, and
- Lower-bound.

Each correction uses a different method of estimating sphericity, represented by a statistic called epsilon (ε). An epsilon of 1 (i.e., $\varepsilon = 1$) indicates that the condition of sphericity is exactly met. The more epsilon falls below 1 (i.e., $\varepsilon < 1$), the greater the violation of sphericity. Estimates of epsilon appear in Mauchly's test results to guide users in making decisions about the correction to use. Nonetheless, statisticians typically recommend the use of the Greenhouse-Geisser correction if estimated epsilon (ε) is less than 0.75, and the use of the Huynh-Feldt correction if estimated epsilon (ε) is greater than 0.75.

10.9 Benefits and Disadvantages of Repeated Measures ANOVA

10.9.1 Benefits

- Data collected from the same study participants over a period of time often result in reduced or eliminated individual differences (a source of between group error). Stevens (2009) notes that one important goal of systematic research is to enhance statistical power to detect an effect and, therefore, to reject the null by reducing error in the analysis. Repeated measures make this possible by comparing differences within groups.
- Testing is more powerful because the sample size is not divided between groups.
- RMANOVA can be cost-effective and efficient due to a design in which the same participants take part in all conditions of the investigation (e.g., in a study of the sprinting speed of soccer players, all players take part in all of the sprinting trials of the study).

10.9.2 Disadvantages

When test conditions are assigned in a repeated measures study, participants are tested with one condition, then another condition, and so on. In such a design, participants' performance may change due to effects that are not due to treatment conditions for several reasons:

- **Carryover or practice effects**. In repeated measures designs, participants' performance may improve as they progress from one test condition to the next due to situations outside of the effects of treatments in the study. Through practice from one condition to another, study participants are likely to become familiar with the structure of the investigation and related tasks. Consequently, they may learn to do the tasks more effectively as time goes on. Practice effects are a threat to the internal validity of a study.
- **Fatigue**. In some cases, participants' performance may decline during the investigation due to mental or physical fatigue.
- **Order effects**. Participants' performance may improve or deteriorate due to the order of test conditions. Order effects are related to the order in which treatments are administered rather than the treatment itself. Randomization or counterbalancing may correct for this deficit.
- **Missing data**. Different sample sizes due to dropouts can cause an imbalance in the number of repeated responses from each individual. If some study participants have missing data, those individuals are dropped from the study.

10.10 Strategies to Counteract Repeated Measures Disadvantages

10.10.1 Crossover Design

While an ordinary, one-way repeated measures is designed to expose all subjects to the same stimulus in the same order, this can lead to results that might be due to external, nuisance variables that influence study variables. A crossover research design aims to assign groups of study participants different treatments during different time periods of a study; that is, randomly assigned study participants cross over from one treatment to another during the course of the investigation. Therefore, in crossover designs each participant receives all treatments, in random order, reducing the effects of outside influences.

One example of a crossover design is a 12-week study conducted by Nishiwaki, et al. (2014). In the study, they examined the effects of an intervention on physical activity and body composition, using an activity monitor with or without computerized game functions. The researchers used a crossover design with weekly measures to reduce possible nuisance variable effects. An example of the crossover design appears in Figure 10.10.

This represents the most common crossover design, a "two-period, two-treatment" design in which participants are randomly assigned to receive either A and then B, or B and then A. Although the crossover design allows for other analyses, it may also be used for a one-way repeated measures analysis.

One important additional feature that sets crossover designs apart is that the same study participants are in the control group *and* all of the treatment groups, reducing the sample size needed for investigations. Accordingly, subjects are measured by how well they respond to treatments against themselves; the "control" is the original status of the patient. For example, people in the above clinical trial were measured before treatment (the control part) and after treatment (the experimental part).

Figure 10.10 Crossover Design for 12-Week Physical Activity and Body Composition Study

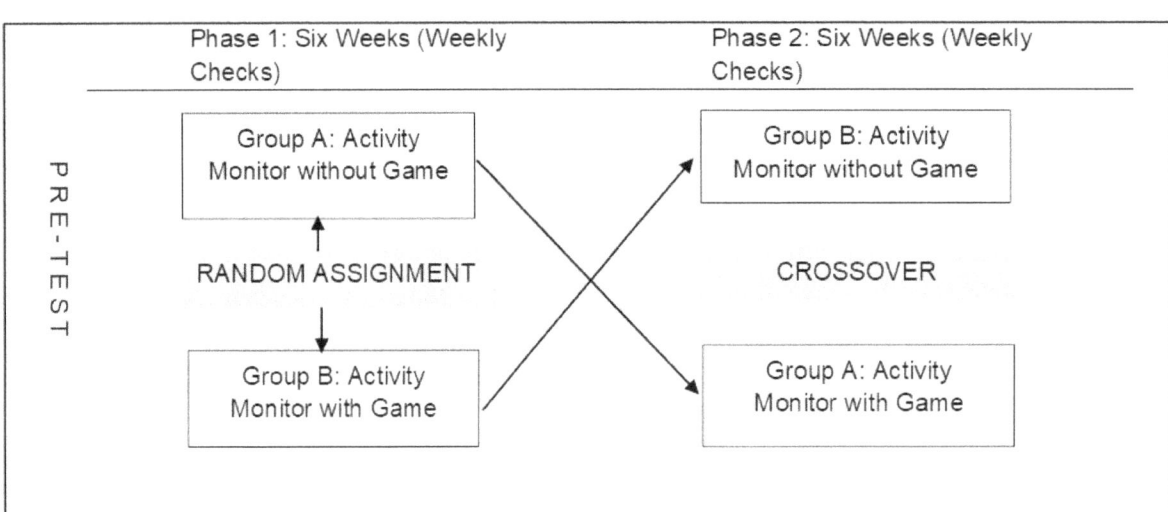

Crossover designs are often employed in pharmaceutical trials in which the goal is to identify whether a new drug (for instance, a new generic drug), results in the same outcomes (the same concentration in the blood stream) as the original drug. In these cases, it is best practice to administer both formulations to each individual in a study, which is a crossover design.

10.10.2 Counterbalancing

Like a crossover design, counterbalancing is one way of dealing with the influence of variables not directly involved in an investigation, such as fatigue and practice effects. This model is designed to guard against **order effects** (the likelihood that the position of the treatment in the treatment order matters) and **sequence effects** (the possibility that a treatment will be affected by the preceding treatment). However, counterbalancing adjusts for biases in different ways from crossover designs. Rather than having participants cross over to a different treatment during the study, counterbalancing involves the administration of a different treatment sequence to each participant. Each treatment sequence is administered an equal number of times in the study.

There are multiple examples of counterbalancing models. The most common models are subject-by-subject, across subjects, complete counterbalancing, and balanced Latin square counterbalancing.

Subject-by-subject counterbalancing. This method of counterbalancing exposes subjects to all treatments in the study more than once. Reverse counterbalancing is one common subject-by-subject counterbalancing method, whereby half of the study participants are exposed to all treatments in the original order, and the other half presented the same treatments in reverse order.

For example, in a cake-tasting competition, half of the cake experts rate the quality of three types of cakes—coconut, pineapple-upside down, and fruit cake—in that order. The remaining half of the experts rate the cakes in the reverse order—fruit cake, pineapple-upside-down, and coconut. However, with more than two conditions as described in the cake example, the likelihood of order effects and carryover effects are still likely to remain prominent.

Across-subjects counterbalancing. Another general approach with many permutations is across-subject counterbalancing. In this approach, each subject receives one sequence of all the possible treatment arrangements. We will examine two across-subjects counterbalancing designs: complete counterbalancing and Latin square design.

Complete counterbalancing. This counterbalancing technique occurs when researchers' designs include multiple combinations of treatments by sequencing them in many different ways. All treatments are used in a study, but in different combinations. Consider the fictitious soccer sprint study (Figure 10.2b and Figure 10.4) in which soccer athletes are timed as they sprint for 34.2 meters on different types of playing fields. The design of complete counterbalancing that might be used in this inquiry is illustrated in Figure 10.11, using six of the soccer players.

As Figure 10.3 illustrates, there are three conditions for the sprinting study—on grass, on sand and on asphalt. With three conditions, there are six possible sequences which Figure 10.11 shows. Here for demonstration purposes, there are six sequences for individuals each of whom is randomly assigned to a different sequence of treatments. If the study includes a larger number of subjects, say 18 participants, they could be randomly assigned to groups of three with each group randomly assigned to one of six treatment sequences (Figure 10.11). The sequences could also be assigned randomly to 6, 12, 18, or more groups as long as they are multiples of six, so that each sequence is run the same number of times in each group. If there are insufficient participants for groups of multiples of six (e.g., 13 groups), the design becomes partial counterbalancing.

Figure 10.11 Across-Subjects Counterbalancing Design in Weight Loss Study

Subject	Treatment: Soccer Fields		
1	Grass	Sand	Asphalt
2	Grass	Asphalt	Sand
3	Sand	Grass	Asphalt
4	Sand	Asphalt	Grass
5	Asphalt	Sand	Grass
6	Asphalt	Grass	Sand

Balanced Latin square counterbalancing. Balanced Latin square counterbalancing (Figure 10.12) also controls for the external influence of nuisance variables that might influence study outcomes, including order effects. However, researchers may only use this method for designs with four or more even-numbered conditions (e.g., 4, 6, 8, 10). This guarantees that a treatment/condition occurs only once in any position. The formula for the design is as follows:
- The first row of the Latin square will follow the formula 1, 2, n, 3, n-1, 4, n-2..., where n is the number of conditions.
- For subsequent rows, add 1 to the item in the same column in the previous row, returning to 1 after you reach n (the number of conditions).

Assume there are four treatments and six study participants (Figure 10.12). With balanced Latin square counterbalancing, every single condition follows every other condition once, allowing researchers to avoid carryover effects during statistical analysis.

Figure 10.12 Example of Balanced Latin Squares Counterbalancing Design

Participants/Groups	Conditions Tested			
	1st	2nd	3rd	4th
A	1	2	4	3
B	2	3	1	4
C	3	4	2	1
D	4	1	3	2
E	1	2	4	3
F	2	3	1	4

10.11 SPSS Procedures for One-Way Repeated Measures ANOVA

Data for this exercise come from the Case of the Seven Sprinting Soccer Players (Table 10.1). The dataset is available in the data file, **Soccer_sprint.sav**. The independent variable in this exercise, trial, includes six levels, Trial 1 – Trial 6, with six measurements of sprinting speed for each soccer player. The dependent variable is speed. This exercise uses SPSS to replicate the omnibus test described in Section 10.7 and compare the outcome with the results obtained by hand-calculations.

10.11.1 SPSS Specifications for One-Way Repeated Measures ANOVA

Select **file → Open → Data** to locate the data file, **Soccer_sprint.sav** and double-click the file to bring it into **SPSS Statistics Data Editor**. Select **Analyze → General Linear Model → Repeated Measures** to reach the **Repeated Measures Define Factor(s)** dialog window (Figure 10.13a).

To define the **Within-Subject Factor Name** (independent variable), click on the panel with the default name, **Factor 1,** and replace it by highlighting **Factor 1** and typing **Trial** to represent the trials during which soccer players' sprinting speeds were recorded. In the box next to **Number of Levels**, enter the number **6** to represent the six levels of the variable, **Trial**. Click the **Add** pushbutton below the number to add the variable level specification to the panel (Figure 10.13b).

Figure 10.13 Define Factor(s) Dialog Window for Repeated Measures

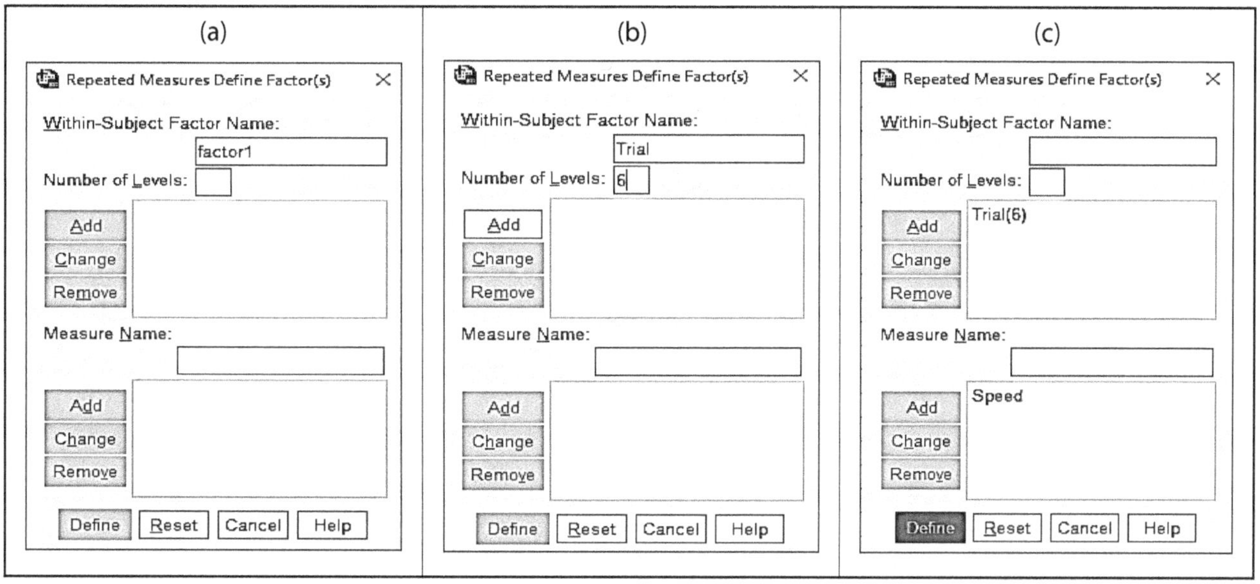

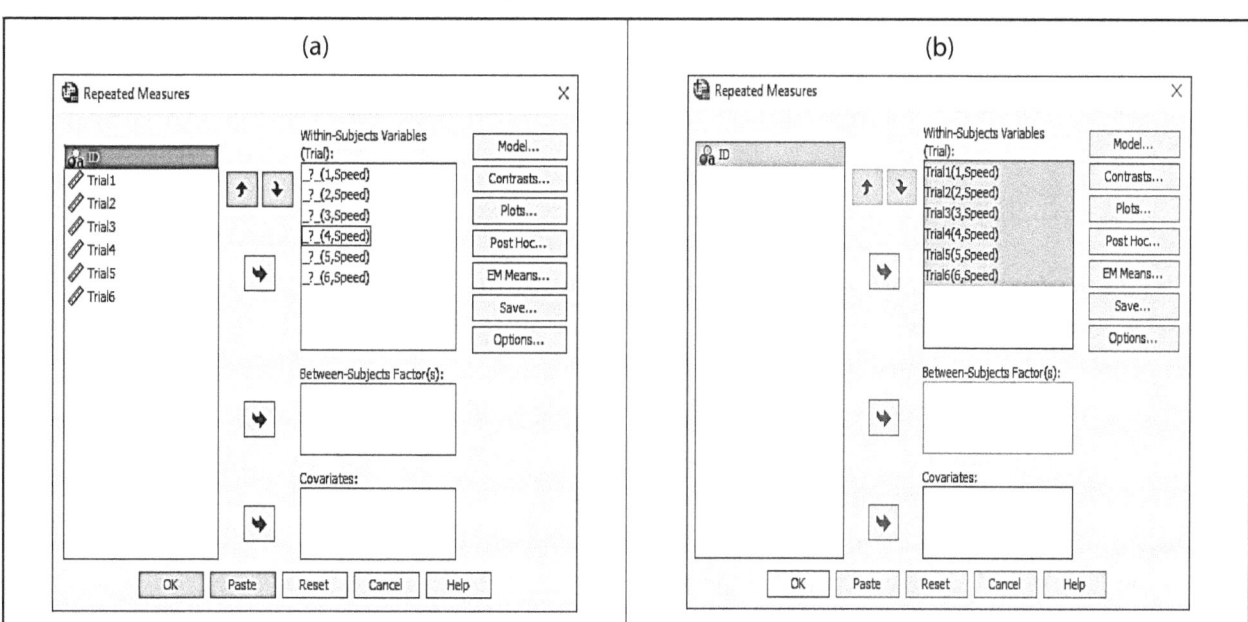

In the **Measure Name** panel, enter the name of the dependent variable, **Speed**. Click the **Add** pushbutton below the **Measure Name** panel to add the dependent variable name to the fully configured dialog box (Figure 10.13c) and to activate the **Define** pushbutton.

Click on the **Define** button to open the **Repeated Measures** dialog window (Figure 10.14a). In the main dialog window are six question marks, followed by the numbers 1 – 6, in the **Within-Subjects Variables** (**Trial**) panel (Figure 10.14a). The six question marks denote the levels of the independent variable, **Trial,** and are placeholders which must be replaced by the variable levels in the panel on the left of the dialog window. To accomplish that, highlight **Trial1** to **Trial6,** and click on the arrow between both panels. Confirm that the variable levels are correctly matched with those in the panel on the right, followed by the dependent variable, **Speed** (Figure 10.14b).

Figure 10.14 Repeated Measures Main Dialog Window

360 Repeated Measures ANOVA

In this research design, there is no grouping variable (no groups of study participants) to add to the **Between-Subjects Factor(s)** panel and no covariate for the **Covariates** panel. Therefore, no additional specifications are required. The fully specified repeated measures design is illustrated in Figure 10.14b. Should you need to reorder the variable levels, highlight the desired levels, and while holding down the mouse button, drag the item to the correct position.

To secure a graphic illustration of the average speeds at each trial, click on the **Plots** pushbutton to access the **Repeated Measures: Profile Plots** dialog window (Figure 10.15). Note that the repeated measures variable, **Trial**, is located in the **Factors** panel on the left.

To specify a profile plot of soccer players' average speed, click over **Trial** to the **Horizontal Axis** panel (Figure 10.15a), then click on **Add** next to the **Plots** panel below the **Factors** panel, and the specified plot should appear in the **Plots** panel (Figure 10.15b). Retain the default specification, **Line Chart**, for **Chart Type**. No additional specifications are needed, so click **Continue** to return to the main dialog window.

Figure 10.15 Repeated Measures Profile Plots Dialog Window

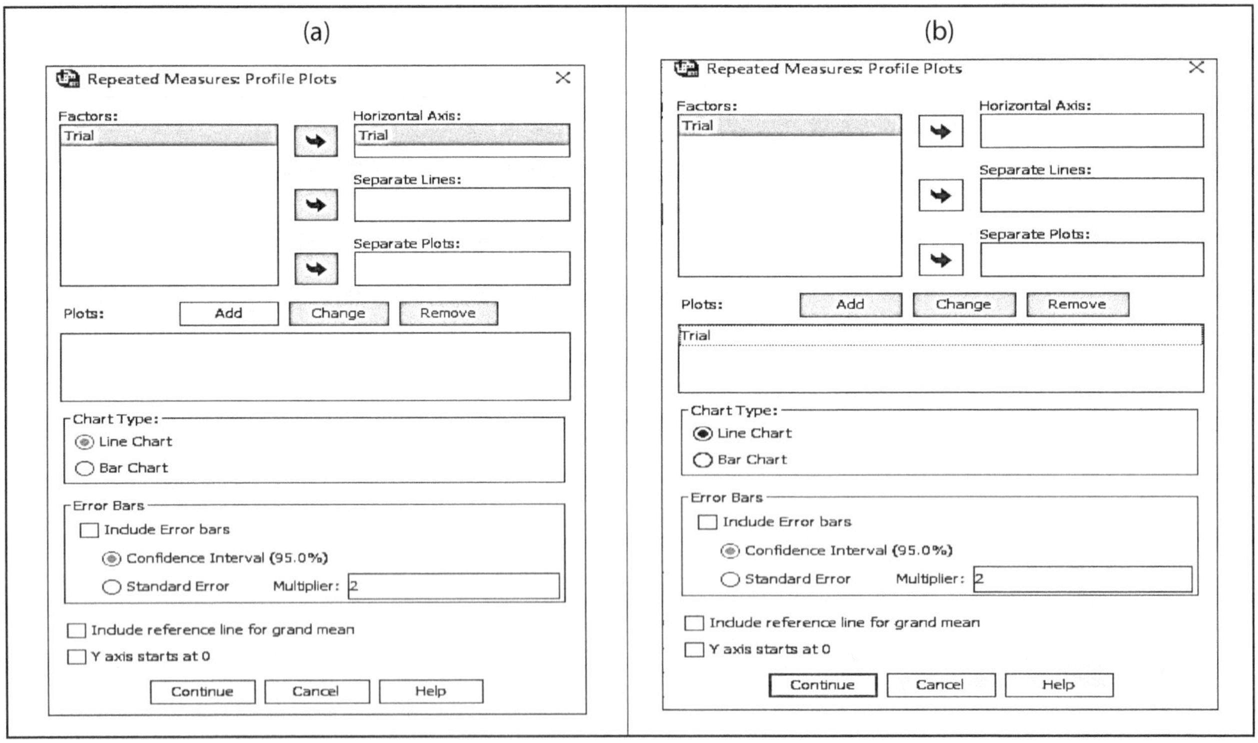

To obtain descriptive statistics for the repeated measures variable, click on the **Options** pushbutton and select **Descriptive statistics** (Figure 10.16). Click on **Continue** to return to the main dialog window.

For purposes of convenience for this exercise, specify the post ANOVA multiple comparisons in the event that the omnibus test finds significant differences across trials. To specify the multiple comparisons, click on the **EM Means** pushbutton in the main dialog window (Figure 10.14) to open the **Repeated Measures Estimated Marginal Means** dialog window.

In the Estimated Marginal Means window, highlight the variable, **Trial** (Figure 10.17a), and click it over to the **Display Means for:** panel on the right (Figure 10.17b). Check the **Compare main effects**

Figure 10.16 Repeated Measures Options Dialog Window

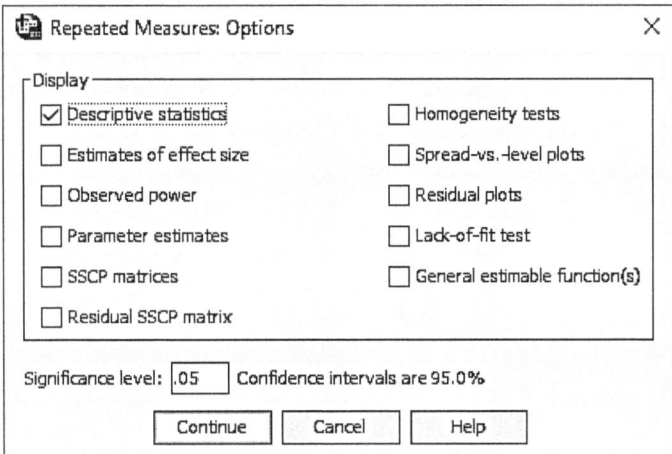

box, and select **Bonferroni** from the drop-down menu below **Confidence interval adjustment** to request pairwise *t*-tests with a Bonferroni adjusted alpha level. This will protect against an inflated Type I error rate in the pairwise comparisons test of the six trials. The completed specification is in Figure 10.17b. Click **Continue** to return to the main dialog window, then Click **OK** to run the procedure.

Figure 10.17 Repeated Measures Estimated Marginal Means Dialog Window

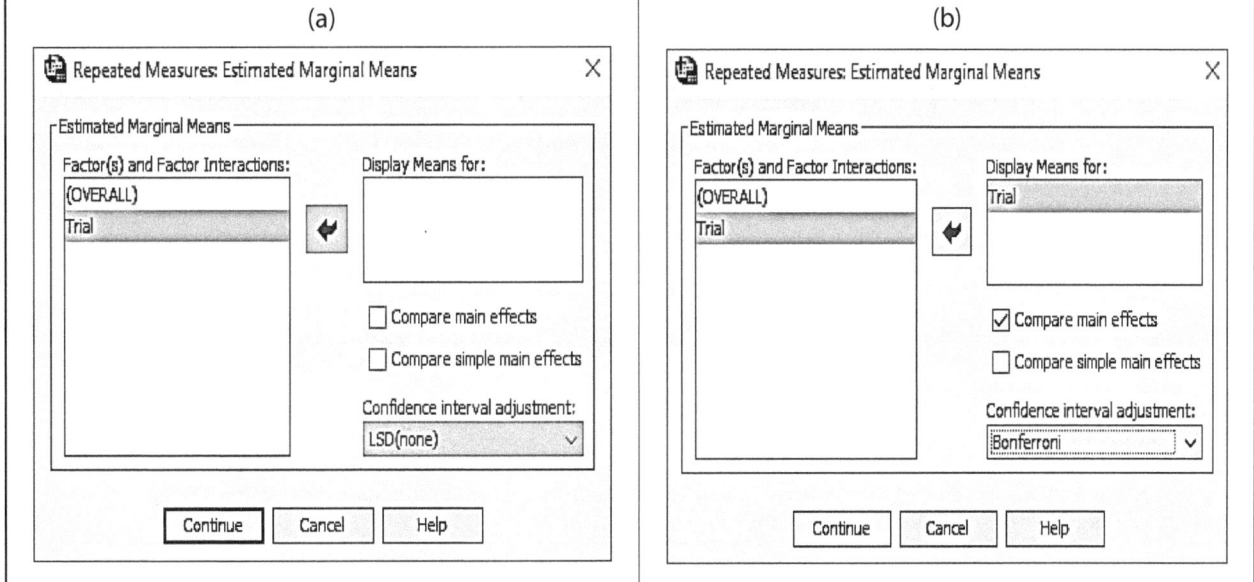

10.11.2 Descriptive Statistics and Mauchly's Test of Sphericity Results

The **Within-Subjects Factors** table (Figure 10.18a) describes the levels of the independent variable and labels the time points 1- 6. These labels will also appear in the pairwise comparisons results. Do not be confused by the **Dependent Variable** column in the **Within-Subjects Factors** table; the time points are not the dependent variable. Rather, the dependent variable, speed, is measured at each time point (trial) noted.

The results of descriptive statistics and Mauchly's test of sphericity analyses are displayed in Figure 10.18b and 10.18c, respectively. The descriptive statistics provide the mean speed and standard deviations

362 Repeated Measures ANOVA

for each of the six trials for all soccer players. The total number of soccer players is recorded in the **N** column. Overall, there is a pattern of declining speed over the trials, with the exception of Trial 4 and Trial 6. As expected, the averages in Table 10.2 mirror those in Figure 10.18b.

The results of the sphericity test (Figure 10.18c) show that Mauchly's test returned a statistically nonsignificant result (**Sig**. column), meaning that the sphericity assumption was retained. Therefore, there is no need to use sphericity-corrected F ratios.

Figure 10.18 Results of the Descriptive Statistics and Mauchly's Test of Sphericity

(a)

Within-Subjects Factors

Measure: Speed

Trial	Dependent Variable
1	Trial1
2	Trial2
3	Trial3
4	Trial4
5	Trial5
6	Trial6

(b)

Descriptive Statistics

	Mean	Std. Deviation	N
Trial 1	7.95114	.275448	7
Trial 2	7.74586	.188368	7
Trial 3	7.62329	.228623	7
Trial 4	7.64857	.298524	7
Trial 5	7.45986	.208345	7
Trial 6	7.59314	.332888	7

(c)

Mauchly's Test of Sphericity[a]

Measure: Speed

Within Subjects Effect	Mauchly's W	Approx. Chi-Square	df	Sig.	Epsilon[b]		
					Greenhouse-Geisser	Huynh-Feldt	Lower-bound
Trial	.045	12.679	14	.627	.555	1.000	.200

Tests the null hypothesis that the error covariance matrix of the orthonormalized transformed dependent variables is proportional to an identity matrix.

a. Design: Intercept
Within-Subjects Design: Trial

b. May be used to adjust the degrees of freedom for the averaged tests of significance. Corrected tests are displayed in the Tests of Within-Subjects Effects table.

10.11.3 Results of the Omnibus Analysis

The results of the omnibus analyses are displayed in Figure 10.19. The **Tests of Between-Subjects Effects** displays the results of the between-subjects portion of the variance discussed in Section 10.7. The **Intercept** row which represents the full model may be ignored for this exercise. Instead, focus on the **Error** row, which is what SPSS labels between-subjects sum of squares (This was not computed manually.). In a one-way RMANOVA analysis, between-subjects sum of squares represent the individual differences between study participants. According to Figure 10.19a, individual differences between study participants accounted for 1.798 units of variance.

The **Tests of Within-Subjects Effects** table (Figure 10.19b) displays the results of the omnibus repeated measures test used to analyze the speed data in Table 10.2. Notice that the sphericity assumed sum of squared deviations (.963) and the mean squares (e.g., .193) are very close to the manually computed values of .959 and .192, respectively, resulting in an F ratio of 8.967 in Figure 10.19b (F column), compared with the manually-computed value of 8.73

Figure 10.19 Results of the Omnibus Test

(a)

Tests of Between-Subjects Effects

Measure: Speed

Transformed Variable: Average

Source	Type III Sum of Squares	df	Mean Square	F	Sig.
Intercept	2471.013	1	2471.013	8247.262	<.001
Error	1.798	6	.300		

(b)

Tests of Within-Subjects Effects

Measure: Speed

	Source	Type III Sum of Squares	df	Mean Square	F	Sig.
Trial	Sphericity Assumed	.963	5	.193	8.967	<.001
	Greenhouse-Geisser	.963	2.776	.347	8.967	.001
	Huynh-Feldt	.963	5.000	.193	8.967	<.001
	Lower-bound	.963	1.000	.963	8.967	.024
Error(Trial)	Sphericity Assumed	.644	30	.021		
	Greenhouse-Geisser	.644	16.657	.039		
	Huynh-Feldt	.644	30.000	.021		
	Lower-bound	.644	6.000	.107		

As the **Sig.** column in Figure 10.19b reports, the effect of the variable, trial, is statistically significant; that is, the changes among the speeds across speed trials are statistically significant ($p < .001$). These results are also in agreement with the original analyses (Wragg et al., 2000).

The results of the omnibus test show a significant change among trial speeds. However, it is not clear where the significant changes in speed occur. Bonferroni-adjusted, pairwise comparisons of Trial levels identified the precise locations of the statistically significant changes. The results are presented in Figure 10.20.

Figure 10.20 Post-ANOVA Pairwise Comparisons

					95% Confidence Interval for Difference[b]	
(I) Trial	(J) Trial	Mean Difference (I-J)	Std. Error	Sig.[b]	Lower Bound	Upper Bound
1	2	.205	.063	.264	-.092	.503
	3	.328*	.068	.044	.008	.648
	4	.303	.077	.118	-.060	.666
	5	.491*	.079	.012	.122	.861
	6	.358	.097	.151	-.096	.812
2	1	-.205	.063	.264	-.503	.092
	3	.123	.077	1.000	-.237	.483
	4	.097	.104	1.000	-.390	.585
	5	.286	.073	.115	-.055	.627
	6	.153	.099	1.000	-.311	.616
3	1	-.328*	.068	.044	-.648	-.008
	2	-.123	.077	1.000	-.483	.237
	4	-.025	.075	1.000	-.379	.328
	5	.163	.069	.846	-.162	.489
	6	.030	.101	1.000	-.444	.504
4	1	-.303	.077	.118	-.666	.060
	2	-.097	.104	1.000	-.585	.390
	3	.025	.075	1.000	-.328	.379
	5	.189	.052	.168	-.057	.434
	6	.055	.062	1.000	-.234	.345
5	1	-.491*	.079	.012	-.861	-.122
	2	-.286	.073	.115	-.627	.055
	3	-.163	.069	.846	-.489	.162
	4	-.189	.052	.168	-.434	.057
	6	-.133	.055	.779	-.392	.126
6	1	-.358	.097	.151	-.812	.096
	2	-.153	.099	1.000	-.616	.311
	3	-.030	.101	1.000	-.504	.444
	4	-.055	.062	1.000	-.345	.234
	5	.133	.055	.779	-.126	.392

Based on estimated marginal means
*. The mean difference is significant at the .05 level.
b. Adjustment for multiple comparisons: Bonferroni.

The first major row in the table compares the speed for Trial 1 to the speed in every other trial. The second major row compares the speed of Trial 2 to the speed of every other trial, and the comparisons continue similarly through Trial 6. The **Mean Difference** column displays the difference between the pairs of relevant sprint speeds. The **Sig**. column reports the significance level of the difference between each pair of sprinting speeds; the greater the difference between the pairs of speeds, the more likely the differences will be statistically significant. Accordingly, significant speed differences ($p < .05$) were found between the paired combinations of Trial 1 and Trial 3 as well as Trial 1 and Trial 5. These pairwise comparison results are somewhat different from the original results. The original investigation used the Tukey procedure and identified two additional pairs—Trial 1 and 6 and Trial 2 and 5—-that were significantly different (Wragg et al., 2000).

10.12 Sample Report: The Case of the Sprinting Speed of Soccer Players

Data from the Bangsbo Sprint Test (Wragg et al., 2000) were re-analyzed to determine whether there was a significant difference in the mean sprint speeds of seven national-level, male soccer players across six sprint trials conducted over a four-week period in random order. No violation of the assumption of sphericity was observed. Therefore, sphericity-corrected F ratios were not needed. Statistically significant differences among the mean speeds in the six time periods were found, $F(5, 30) = 8.967, p < .001$.

The results of Bonferroni-corrected pairwise t-tests showed that the sources of the differences found in the omnibus test were between the mean speeds in Trials 1 and 3 as well as Trials 1 and 5 ($p < .05$). The average speed at Trial 1 ($M = 7.951$, $SD = .275$) significantly declined by Trial 3 ($M = 7.623$, $SD = .229$) and significantly declined further by Trial 5 ($M = 7.46$, $SD = .208$).

10.13 Key Terms

Analysis of variance for correlated samples
Assumption of sphericity
Between-subjects sum of squares
Counterbalancing
Crossover design Factorial repeated measures
Greenhouse-Geisser estimate
Huynh-Feldt estimate
Lower bound estimate
Mauchly's test of Sphericity
One-way repeated measures ANOVA
Treatment sum of squares
Within-subjects ANOVA
Within-subjects sum of squares

11
Mixed Design ANOVA

Key concepts from previous chapters or courses:

between-subjects sum of squares (SSB): in repeated measures ANOVA, the sum of squared deviation scores of each individual's scores.

counterbalancing: a repeated measures design that administers treatments in different sequences to prevent order effects.

crossover design: a research design in which all individuals receive the same treatments but in different sequences; e.g., one group receives Treatment A followed by Treatment B while a second group receives Treatment B followed by Treatment A.

factorial repeated measures ANOVA: a type of repeated measures ANOVA with two or more independent variables with multiple levels. Some variables may consist of non-overlapping groups of subjects. Others may include all subjects in all the variable levels.

one-way repeated measures ANOVA: also called within-subjects ANOVA, a type of ANOVA applied to research data with one independent variable that has multiple levels and with each participant measured more than once.

sphericity: a repeated measures ANOVA assumption that requires equal variances of the differences between all possible pairs of the levels of the independent variable.

total sum of squares (SST): sum of squared deviation between each subject's score and the mean of all scores in a research study.

treatment sum of squares ($SS_{treatment}$): in repeated measures, the sum of squared deviations between the mean of an independent variable level/group (treatment) and the grand mean. It is a component of within-group sum of squares.

within-subjects sum of squares (SSW): in repeated measures, it is the sum of squared deviations between individuals' scores and their group mean.

THIS CHAPTER INTRODUCES

- differences between one-way repeated measures and factorial repeated measures
- characteristics of a mixed design ANOVA
- when to use a mixed design ANOVA
- assumptions for a mixed design ANOVA
- how to use SPSS to conduct a mixed design ANOVA.

11.1 Introduction

Chapter 10 explored the statistical procedure of one-way repeated measures ANOVA (one-way RMANOVA) which is also known as one-way within-subjects ANOVA. Researchers use RMANOVA to statistically analyze data from three or more time periods or conditions. As a result, each subject has scores from multiple time periods or conditions. The purpose of the analysis is to determine whether there is a significant difference in the change in scores among the time periods or conditions.

This chapter introduces an extension of one-way RMANOVA—mixed design ANOVA. Instead of a single within-subject independent variable, mixed design has a minimum of two independent variables—one between-subjects (between groups) and one within-subjects (repeated measures) variable.

Mixed design ANOVA shares characteristics of between-subjects factorial ANOVA and one-way RMANOVA. It may be used to analyze differences in performance between two or more mutually-exclusive groups that are measured repeatedly over three or more time periods, or under three or more conditions.

11.2 The Case of Olympic Track Athletes' Speed

Like other regions around the globe, sports activities and sporting events are an important part of life in Latin American and Caribbean nations. Football (soccer) and track and field are two of the most popular sports activities in the region. The popularity of sports has made it an important research subject.

This case that will be used to demonstrate mixed design ANOVA is based on a small sample of participants in the men's and women's 100-meter events in the 2016 Summer Olympics. The dataset comprises a total of 16 athletes (8 men and 8 women) who competed in the 100-meter dash finals and includes the speeds clocked for each in all three rounds of the event: qualifying heats, semifinals, and final. Speeds are measured in seconds. The dataset includes participants from eight countries: Canada, Cote d'Ivoire, France, Jamaica, the Netherlands, South Africa, Trinidad & Tobago, and the United States.

To demonstrate the analysis of a two-way simple mixed design, the data will be used to investigate whether there is a significant difference in the speeds of male and female athletes across three rounds of performance. Table 11.1 displays the raw data that will be used in the analyses.

The expectation is that speed will increase with each round of the event, and therefore, the time it takes to complete each race will decline as athletes complete each round. Also, male athletes are expected to be significantly faster than female athletes. Review the clocked speeds in Table 11.1 for each 100-meter round in the Summer Olympics of 2016. Are there any patterns of speed that support the expectations described?

11.3 Simple Mixed Design ANOVA Fundamentals

In statistics, when a researcher tests for average differences between groups that have been split on at least two independent variables, a between-subjects variable like gender and a within-subjects variable such as time, the analysis is called a mixed design ANOVA or a mixed factorial design ANOVA. It is also called a split-plot ANOVA design, named after the agricultural setting in which Fisher (1925) first invented the method.

Table 11.1 Raw Data for the Case of Olympic Track Athletes Speed

		Rounds (Time)					Rounds (Time)		
ID	Gender	Heats	Semifinals	Final	ID	Gender	Heats	Semifinals	Final
1	M	10.14	9.98	9.94	9	M	10.01	9.94	9.89
2	M	10.04	9.92	9.91	10	F	11.01	10.94	10.86
3	M	10.03	9.97	9.96	11	F	11.00	10.90	10.92
4	F	11.27	10.96	11.80	12	F	10.96	10.88	10.86
5	F	11.16	10.90	10.90	13	F	11.13	10.90	10.83
6	F	11.21	10.88	10.71	14	M	10.13	10.01	10.06
7	F	11.09	10.90	10.94	15	M	10.07	9.86	9.81
8	M	10.19	9.95	10.04	16	M	10.11	10.01	9.93

Mirshams Shahshahani, P. (2018). *Downloaded IAAF Sprint Results in all Heats for 2004 - 2016 Olympics for both Men and Women*. University of Michigan - Deep Blue. https://doi.org/10.7302/Z20V8B11.

The two-way mixed ANOVA is the simplest of the mixed ANOVA design family—one in which there is a single between-group variable and a single repeated measures variable. Discourse in this chapter will expand later in this chapter to briefly discuss other types of mixed designs.

Chapter 7 investigated the one-way ANOVA, followed by factorial ANOVA in Chapter 8. Chapter 10 explored the characteristics and features of one-way RMANOVA. When comparing one-way and factorial ANOVA with repeated measures, it quickly becomes evident that although all techniques use measures of deviation from the mean to measure differences, the statistical methods have dissimilar objectives. The first two statistical techniques measure differences **between** independent groups of study participants; those groups may represent one variable (e.g., age groups for one-way ANOVA) or more than one variable (e.g., age groups and gender groups for two-way ANOVA). One-way RMANOVA measures changes **within** study participants over time or under different conditions: for example, changes in student track athletes' performance before and after coaching, or changes in performance based on psycho-emotional conditions like anxiety, sadness, happiness. However, all three procedures by themselves are limited in their capacity to investigate trends over time between groups of study subjects. A simple mixed design combines the characteristics and features of the within-design of repeated measures and the between-design of one-way and factorial ANOVA that enhances the authenticity of research and better reflects the complexities of life.

In a study of the impact of emotions on men's and women's (between variable = gender) physical and cognitive performance (dependent variable), researchers using a mixed-design found that different emotional conditions (repeated measures variable) like happiness, anger, and hope had differing impacts on study participants' sports performance (Woodman et al., 2009). An illustration of what the mixed design component of the study might look like is presented in Figure 11.1. As the illustration demonstrates, the independent variable, emotion, is a repeated measures (or within-subjects) variable with three categorical conditions indicated by the facial expressions: happy, angry, and hopeful. The other independent variable, gender, is a between-subjects variable with two categorical groups: male and female. The original study participants consisted of nine males and six females, each of whose physical and cognitive performance (dependent variables) were measured under each of the three emotional conditions.

Figure 11.1 Sample of Mixed Design: Study of Emotions on Sports Performance

Gender	Emotions		
	😊	☹️	😌
Male	👤👤👤👤👤 👤👤👤👤	👤👤👤👤👤 👤👤👤👤	👤👤👤👤👤 👤👤👤👤
Female	👤👤👤 👤👤👤	👤👤👤 👤👤👤	👤👤👤 👤👤👤

As with other factorial ANOVAs, the research question of greatest importance in a two-way mixed design ANOVA is whether there is an interaction between the two independent variables; in the case of the study reflected in Figure 11.1, that refers to an interaction between gender and emotions on cognitive and physical performance. If there is an interaction, the researcher proceeds to identify the source of the interaction, first by conducting the test of simple effects (Chapter 8) to discover whether there are, say, physical performance differences between pairs of the variable, emotion (e.g., emotions of happy versus angry, happy versus hopeful, or angry versus hopeful) in one category of the variable, gender (e.g., females). If there are significant differences in the physical performance of females experiencing different emotions, say, happy versus hopeful emotions, the conclusion might be that hopeful female participants perform significantly better on physical tasks than happy female participants.

If there is not a significant interaction between independent variables, the researcher conducts main effects analyses (Chapter 8) by investigating whether there are significant changes in physical performance among the categories of one independent variable (e.g., emotions) while ignoring the other independent variable (gender). More on these approaches in Section 11.9.

11.4 Characteristics of Simple Mixed Design ANOVA

A mixed design ANOVA is inherently a type of factorial ANOVA in that it has more than one independent, categorical variable. However, this type of factorial design is distinguished by the mandatory inclusion of a minimum of one repeated measures variable as one of the categorical independent variables.

- The most elementary mixed ANOVA design has a single between-subjects/groups variable with two levels or groups (e.g., gender) and a single repeated measures or within-subjects variable, again with a minimum of two levels (e.g., pre-test and a post-test), resulting in a total of two categorical independent variables.
- In addition to the minimum of two independent variables, the dependent variable must be a continuous variable, such as speed, test score, height, weight, elapsed time. If the dependent variable is not measured at an interval level at a minimum, another statistical technique must be used.
- When using mixed design, study participants must be measured in all of the levels of the within-subjects variable. For example, in the Woodman et al. (2009) study, if a study participant had missing data on performance while experiencing one of the emotions, such as sadness, all of that participant's data would be completely removed from the analysis. Because of this limitation, it is

best to use another statistical technique, like linear mixed models if there are missing data for the within-subjects variable. For more information on linear mixed models, refer to texts on multivariate statistics; for example, Tabachnick and Fidell (2018).
- In a mixed design, as in repeated measures, within-subjects variables that are time-based must be completed in the order stipulated in the design; that is, measurement at Time 1 occurs before measurement at Time 2. However, when the within-subjects variable measures conditions, like emotions, subjects need not be measured in a prescribed order.

11.5 Types of Mixed Design ANOVAs

Although mixed design ANOVAs have been limited to two independent variables so far, within the world of univariate statistics, which is the focus in this text, there are probably as many types of mixed designs as there are factorial and repeated measures designs.

11.5.1 One Between- and One Within-Subjects Variable

Perhaps the most basic mixed design includes two independent, categorical variables, but one or both of the variables have more than two levels. In the Woodman et al. (2009) study, and the case study of Olympic track athletes, one independent variable is composed of three categories or levels which represent three conditions. In the Woodman et al. study, participants are measured while experiencing three different types of emotions (happiness, anger, and hope). In the other case, each athlete's speed is measured in three different rounds of the Olympic trials (heats, semifinals, and final). Both designs are described as a 2 x 3 mixed design because there are two independent variables, one of which is a dichotomous variable while the other variable has three categories.

In both examples, only the within-subjects independent variable has more than two categories. However, in a two-way mixed design ANOVA, other permutations may be used as well. For example, in a study of the effect of weight-loss interventions on weight over a period of three months, the between-groups independent variable might comprise three mutually exclusive levels that include two different treatment interventions (exercise and counseling) and a control group. The within-subjects independent variable would be time, and the dependent variable would be weight, measured three times during the course of the three-month study. This would be a 3 x 3 mixed design ANOVA.

11.5.2 Designs with More than Two Variables

Like standard between-group, factorial ANOVA, it is possible to expand beyond two-way factorial models to more complex mixed ANOVA designs that involve three or more independent variables. As the number of variables included in a study increases, the intricacy of the design and analysis also increases.

11.5.2.1 Three-Way Mixed Design

Imagine that the dataset for Olympic track athletes in Table 11.1 was expanded to include performance for the same athletes over two years—Summer Olympics for the years 2012 and 2016 (Figure 11.2b). In that case, researchers would add an additional two-category, within-subjects, independent variable, time, which would grow the original design to a three-way mixed design where A_i represents athletes, and S_j is the sprinting speed of the athletes.

In the first study design, each subject is measured three times in the 100-meter dash. The design in Figure 11.2a reflects the intent to statistically compare the speeds of male and female athletes to see whether there is a significant difference in their performances. Furthermore, the design addresses whether there is a significant difference in speed that is dependent on gender and on (event) rounds (heats, semifinals, final). As stated earlier, this is a 2 x 3 mixed design ANOVA due to the two-category between-groups variable, gender, and the three-category repeated measures variable, rounds.

In addition to elements incorporated by the design in Figure 11.2a, the design in Figure 11.2b adds comparisons across time; that is, it compares athletes across two Summer Olympics (time) in addition to comparisons across (performance) rounds and between male and female athletes. The design also incorporates a more comprehensive investigation of the analysis of interactions among the variables than in the 2 x 3 design, by examining whether differences in athletes' speeds are dependent on the event year (time), (performance) rounds, gender or combinations thereof. The increased analyses garnered by the addition of a single variable illustrate how quickly research designs become very complex. The addition of the independent variable, time, results in a 2 x 2 x 3 mixed design ANOVA with two within-subjects variables. The design notation means that the single between-group variable, gender, has two categories. The next variable is the first repeated measures variable, time, with two categories (2012 and 2016), and the second repeated measures variable is (performance) rounds with three categories (heats, semifinals, and final).

11.5.2.2 Four-Way Mixed Design

Of course, depending on the goal of a research study, more variables may be added to the design. For example, imagine that the between-group variable, diet, is added to the design in Figure 11.2b so that

Figure 11.2 Two-Way (a) and Three-Way (b) Mixed Design

(a) One Between and One Within Variable			
	Performance Rounds		
Gender	Heats	Semi-finals	Final
Male	S_1	S_2	S_3
A_1			
...			
A_4			
Female			
A_1			
...			
A_4			

(b) One Between and Two Within Variables						
	Time					
	2012 Rounds			2016 Rounds		
Gender	Heats	Semi-finals	Final	Heats	Semi-finals	Final
Male	S_1	S_2	S_3	S_1	S_2	S_3
A_1						
...						
A_4						
Female						
A_1						
...						
A_4						

athletes are assigned to one of two additional groups—those who adhere to a vegetarian diet, and those who prefer a meat-based diet. The addition of a fourth variable results in a four-way mixed design ANOVA that includes two between- and two within-subjects variables where the between-subjects variables are gender and diet, and the within-subjects variables are time and (event) rounds. Speed (S_i) remains the dependent variable. Given the number of categories in each independent variable, the design becomes a 2 x 2 x 2 x 3 mixed design ANOVA. Figure 11.3 displays a schematic for the four-way mixed design.

Regardless of the complexity of the design, the analysis of pre-eminent importance is to identify whether a significant interaction is evident among the variables (Section 11. 3). For more advanced designs, the data might be analyzed for several interactions. For example, in the three-way design (Figure 11.2b), the researcher might investigate whether these interactions are significant:
- changes in the speed of athletes across (event) rounds and time, averaging across gender
- changes in the speed of athletes across (event) rounds and gender, averaging across time
- differences in the change in the speed of male and female athletes (gender) across time, averaging across (event) rounds
- differences in the change in speed of male and female athletes (gender) across (event) rounds, averaging across time
- differences in the change in speed of male and female athletes (gender) across (event) rounds and across time.

While it is feasible to create increasingly intricate research designs in an effort to enhance efficiency and authenticity, elaborate designs may result in costly implementation (e.g., time and costs needed to conduct the study) and less than interpretable outcomes. For those intrepid researchers who are interested in exploring more mixed design options, Stevens (2009) provides an extensive discussion of repeated-measures designs.

11.5.2.3 Including a Covariate

Krzysztof et al. (2013) conducted a study of Jamaica's elite sprinter Usain Bolt's three best 100-meter performances between 2008 and 2012 to identify key physical factors that distinguish him from other sprinters and contributed to his success in track and field. Study results revealed that in addition to the advantage of height and lower limb length, stride length and stride frequency allowed him to achieve higher maximal sprinting speed than other elite athletes.

There are other factors (noise) that could affect the speed of a sprinter and might be taken into consideration when assessing an athlete's true performance. In a study of changes in sprinting speeds between races and speed training, Otsuka et al. (2016) identified four such factors and incorporated them in their study as covariates —track conditions, wind speed, air temperature, and gender.

Considering the factors from the Otsuka et al. (2016) study, consider the mixed design in Figure 11.3, and imagine that researchers were interested in accounting for one external condition, wind speed, and its impact on the outcome of the investigation. In this case, as in standard analysis of covariance, the additional variable would be considered a covariate.

Researchers may include covariates, also known as confounding variables or concomitants, in a mixed design ANOVA. This type of design is described as a mixed design analysis of covariance that addresses effects on the dependent variable (e.g., speed in 100-meter sprints) over and above the effect of the covariate (e.g., wind speed). In this situation, the effect of wind velocity on the dependent variable,

Figure 11.3 Four-Way Mixed Design ANOVA

		Time					
		2012			2016		
Diet	Gender	Heats	Semifinals	Final	Heats	Semifinals	Final
Vegetarian	Male A_1 ... A_4 Female A_1 ... A_4	S_1	S_2	S_3	S_1	S_2	S_3
Meat-Based	Male A_1 ... A_4 Female A_1 ... A_4	S_1	S_2	S_3	S_1	S_2	S_3

speed, is removed before the effects of the independent variables, time, (event) rounds, gender, and diet, are investigated. As with ANCOVA, designs may include more than one covariate, thereby, further complicating the research design. Ultimately, researchers must decide whether the interpretability of the outcome is best served by using mixed design ANOVA or whether it would be better to consider other research designs.

11.6 Underlying Assumptions of Mixed Design ANOVA

An important preliminary activity for choosing a statistical technique is to determine whether the data can be analyzed with the selected method. Therefore, selecting mixed design ANOVA as the preferred statistical method should be preceded by associated tests of assumptions to determine whether the data meets mixed design ANOVA assumptions.

11.6.1 Design-Related Assumptions

The dependent variable must be a continuous variable; that is, it should either be at an interval or ratio level. Examples of such variables include scores from 0 to 100, speed in seconds taken to run a 100-meter dash, or weight lost due to weight loss program.

Repeated measures independent variables should comprise two or more related groups or levels, and the same study participants should be in all of these related groups. The related groups might be time points at which all study participants' performance is measured. For example, all Olympic track athletes' speed is measured at prespecified time points (e.g., rounds) of an athletic event. Similarly, related groups might be conditions that all participants undergo. For example, in the study of physical performance under three induced emotional conditions, all participants' physical ability is measured under the same conditions of sadness, happiness, and hopefulness.

As in factorial ANOVA, between-groups independent variables should be composed of a minimum of two categorical mutually exclusive groups. While study participants are measured in each of the levels of the within-group variable, they can only be in one of the groups of each between-groups variable. Frequently, the between-groups variable is a characteristic of study participants, such as gender (male or female), education level (primary, secondary, or college level), or marital status (single, married, divorced, or widowed).

11.6.2 Assumptions in Common with Other ANOVAs

There should not be any outliers in the groups associated with independent variables. Outliers, typically values in a dataset that do not follow the same pattern as the rest of the scores, may distort differences between variable groupings by abnormally increasing or decreasing scores in a way that will reduce the accuracy of results. Detect outliers by using SPSS Explore to generate scatterplots, histograms, stem-and-leaf plots, and boxplots as well as other descriptive statistics described in Chapter 3.

Dependent variable data should be approximately normally distributed for any combination of the independent variables. Therefore, in the study of the impact of emotions on male and female physical performance, the physical performance scores of happy males should be normally distributed as should the scores of happy females and any other combination of groups in that study. Although data used in mixed designs are robust to violations of normality assumptions, it is important to ensure that extreme non-normality is not present in the data. Assess the assumption of normality using the Shapiro-Wilk test of normality or Q-Q plots in SPSS.

The variance of data values should be homogeneous across levels of between-subjects independent variables. This assumption holds that the scores of any two or more groups defined by between-subjects independent variables should have approximately the same error variance. This assumption may be tested using Levene's test of homogeneity of variance in SPSS.

Common remedies for all of the assumptions violations described include larger sample sizes that will result in data that are more robust to assumption violations and data transformations, to improve normality which will also improve homogeneity of variance.

11.6.3 Assumptions in Common with Repeated Measures

As in repeated measures, the assumption of "sphericity" applies only to designs that include repeated measures variables with three or more related levels. This assumption is satisfied if the variances of the difference scores are similar across levels of each repeated measures independent variable for each between-group variable. Use Mauchly's test of sphericity to evaluate this assumption. If the test is significant at $p < .05$, the F and p values of the coefficients in mixed design ANOVA should be adjusted using the Greenhouse-Geisser or the Huynh-Feldt correction.

11.6.4 Assumptions Specific to Mixed Design

The homogeneity of the variance-covariance matrices assumption is satisfied when the pattern of intercorrelations among the various levels of the repeated measures independent variable(s) is consistent across groups of the between-groups independent variable(s). The homogeneity of the variance-covariance matrices assumption is tested using Box's M statistic. If the sample size is large and Box's M returns a p value that is less than .001, then the variance-covariance assumption is violated. However, according to Cohen (2008), with small sample sizes Box's M has little power to achieve significance. Therefore, nonsignificant results may not be tenable.

11.7 Benefits and Disadvantages of Mixed Design ANOVA

11.7.1 Benefits

Because mixed-design ANOVA combines repeated measures and factorial ANOVA, it boasts the advantages of both techniques. Accordingly, mixed design features some of the advantages of repeated measures described in Section 10.9.1.

- Due to the use of repeated measures in mixed design, error is substantially reduced (although not as much as a purely repeated-measures design). Reducing error in the analyses contributes to enhanced power and the increased ability to reject the null hypothesis when it is false.
- Like repeated measures, mixed designs are more efficient because researchers are able to overcome the challenges associated with recruiting study participants by using smaller sample sizes. Measuring study participants repeatedly for the within-subjects component of the design (e.g., measuring sprinters speed over multiple rounds of the Olympics) allows researchers this advantage.

Mixed design also features some of the advantages of between-subjects ANOVA described in Chapters 7 and 8.

- The addition of relevant between-groups variables in factorial ANOVA contributes to the generalizability of study outcomes. Similarly, adding a between-groups factor to a repeated-measures design increases the generalizability of the study.
- Added variables enhance analysis efficiency and design authenticity. Mixed-design ANOVA also allows more flexible inquiries in that they allow researchers to analyze the effects of continuous within- and between-group variables as well as the interaction of their effects.
- Adding a between-group variable may assist in reducing practice effects discussed in repeated measures. In repeated measures, differences may be detected due to study participants learning over time during a study, or due to the order in which study participants are exposed to treatments. For example, envisage a study to determine the extent to which a test-taking skills program helps students who take a secondary school placement exam. In that study, students may complete a practice exam before and after completing the test-taking skills program (treatment group). However, their performance may be due to the program or to practice effects, namely, familiarity with the exam. Comparing the treatment group's performance to the control group's performance allows researchers to identify whether treatment effects occurred.
- A between-subjects variable may also reduce order effects that occur when differences in outcome are due to the order in which they completed a task rather than the task itself. Imagine a study in

which participants are asked to complete a task, first with background noise, followed by completing a similar (parallel) task without background noise. Differences in task performance may be due to the presence or absence of background noise or the order in which the treatment occurred, order effects. Splitting the group in half, so that one half completes the task without background noise "treatment" first and the other half completes the task in the opposite order (with background noise first), would rule out the likelihood of order effects. Adding a between-group variable (e.g., order) in the design identifies the order in which treatments or conditions occur.

11.7.2 Disadvantages

Mixed designs help to reduce the disadvantages of repeated measures that contribute to threats to internal validity: that is, the extent to which you can rule out other explanations for the results of the study. However, there is a special threat to internal validity for mixed designs called differential mortality, which occurs when the dropout rate of the study is different across levels of the between-subjects variable. For example, the design in Figure 11.2b is based on data collected twice over a four-year period—once in 2012 and again in 2016. During the four-year period a lot can happen in the world of elite track and field athletes. Athletes often qualify for a spot on the team for one Summer Olympic Games, but not for another. Alternatively, athletes may qualify for the games, but not for the Olympic finals.

Only complete data may be used in mixed designs. Accordingly, voluntary or involuntary dropouts affect the pattern of data available for analysis. The dropout rate may be much higher for one group (2016 athletes) than the other (2012 athletes). Differential mortality poses a risk because it destroys the assumption of equality or representative samples. For example, differences in outcome may be a result of changes in a target population due to qualifying guidelines or demoralization rather than speed.

Differential mortality is also problematic because mixed designs do not have the ability to accommodate missing data directly. Consequently, researchers must delete incomplete cases, thereby reducing sample size and, therefore, power. Alternatively, researchers may have to impute or estimate missing values.

11.8 Research Plan: The Case of Olympic Track Athletes' Speed

This section supplements the research plan outline (Box 11.1) for the Case of Olympic Track Athletes' Speed. Data for the case is presented in Table 11.1.

11.8.1 Research Objectives and Questions

Mixed-design ANOVA is the factorial version of one-way repeated measures ANOVA. The primary objective is investigating whether there were significant changes in the speed of sixteen athletes in the 2016 Summer Olympics across three rounds of the 100-meter Olympic event. Research Question 1 is similar to the Case of the Sprinting Speeds of Seven Soccer Players in Chapter 10 in that it deals with changes in speed across rounds (time). The second and third questions are similar to those for between-groups factorial designs:
- Is there a significant difference in the speed of male and female track athletes?
- Is there a significant interaction between the variables gender and performance round?

Box 11.1 Research Plan for the Case of Olympic Track Athletes' Speed

Research Problem
The sprinting speeds of elite female Olympic 100 m sprinters have increased substantially over the years. A researcher wishes to determine whether the speeds in different performance rounds of the Olympics (heats, semifinals, and final) differ substantially between male and female athletes as well as whether there is a pattern of change in speed across rounds among the two groups of athletes.

Research Variables and Measurement Levels
Between-subjects, independent variable: Gender Levels: Male and Female
Within-subjects, independent variable: Rounds Levels: Heats, Semifinals and Final
Dependent variable: Speed (in seconds)

Research Questions
Question 1: Is there a significant difference in the average speed of male and female 100 m sprinters?
Question 2: Are there significant changes in the sprinting speed among the 100 m performance rounds?
Question 3: Is there a significant interaction between gender and the change in sprinting speed in performance rounds among the athletes?

Research Hypotheses
Main Effect of Performance Rounds
Null Hypothesis 1: There is no significant difference in the sprinting speed of athletes in three performance rounds.
$H_0: \mu_H = \mu_{SF} = \mu_F$
Alternative Hypothesis 1: There is a significant difference in the sprinting speed of athletes in at least two rounds i and j.
$H_a: \mu_i \neq \mu_j$

Main Effect of Gender
Null Hypothesis 2: There is no significant difference in the average speed of male and female athletes.
$H_0: \mu_1 = \mu_2$
Alternative Hypothesis 2: There is a significant difference in the average sprinting speed of male and female athletes.
$H_a: \mu_1 \neq \mu_2$

Interaction
Null Hypothesis 3: There is no interaction between gender and performance rounds.
Alternative Hypothesis 3: There is a significant interaction between gender and performance rounds.

Research Design
This nonrandomized (NR) 2 x 3 mixed design ANOVA is an ex post facto design, with one repeated measures variable with three levels (rounds) and one between-subjects variable (gender).

NR	X_1	O_h	O_{SF}	O_F
NR	X_2	O_h	O_{SF}	O_F

Data Collection
Publicly-available competition results data were downloaded from the International Association of Athletics Federation (IAAF) (https://www.iaaf.org/results) and reformatted for use (Mirshams, 2018).

Data Analyses
The data will be analyzed to determine whether data assumptions have been met and appropriate steps taken to adjust the data if they fail to meet the assumptions. Subsequently, the omnibus test for a two-way, mixed design, repeated measures ANOVA will be conducted to analyze the data. If the interaction between gender and performance rounds is significant, follow-up tests will be conducted to identify the source of the interaction. In addition, if the main effects test for performance rounds is significant, all possible Bonferroni-adjusted one-way ANOVAs will be conducted for the within-subjects variable, to identify the source of the difference. With two levels comprising the between-subjects variable, gender, no post hoc tests are needed.

11.8.2 Research Study Variables

Research variables include two categorical independent variables: (a) a dichotomous, between-group variable, gender, and (b) a three-group, within-group variable, rounds, which represents the occasions on which speed was measured. The third variable in the exercise is the dependent continuous variable, speed.

11.8.3 Research Design

With one between-groups and one within-group variable, the research design meets the minimum requirements for a mixed design. Data for the exercise were taken from the Summer Olympics archives. Therefore, no researcher manipulation or randomization was involved in the data collection. These types of studies are described as nonrandomized, ex post facto, repeated measures designs.

11.8.4 Hypothesis Testing

Although ex post facto research designs do not support causal interpretations, researchers may yet conduct hypothesis testing to determine the probability that differences observed in a study sample occur in the target population. Therefore, with three research questions, there are three associated null hypotheses. Significant outcomes may serve as the basis for further investigations that may contribute to causal interpretations.

11.9 SPSS Procedures for Mixed Design ANOVA

This exercise introduces procedures for conducting a simple mixed design ANOVA using SPSS and the data in Table 11.1. The dataset is available from the data file, **Olympics_16.sav**. The dataset includes a total of 16 athletes. The data for each study subject occupy a single row in the dataset.

As noted in the research plan (Box 11.1), there is a single repeated measures variable, performance rounds (Rounds), with three levels, heats (Heats), semifinals (Semifinals) and final (Final). Only athletes completing all three rounds are included in the dataset. The athletes include males and females who represent the two levels of the between-subjects variable, gender.

The procedure for analyzing mixed research design is basically the same as that for other ANOVAs, and in particular repeated measures ANOVA.

1. Run descriptive statistics (SPSS Explore) to determine whether the data meet basic ANOVA assumptions: normality and absence of outliers in between- or within-subject variable groups. Take corrective action if data do not meet assumptions (Section 11.6).
2. Conduct Repeated Measures ANOVA. This will produce results of three additional tests of assumptions (Levene's test of homogeneity of variance, Mauchly's test of sphericity and Box's test of equality of covariance matrices) as well as the omnibus test for mixed research design.
3. Conduct follow-up tests for significant interactions first. Then conduct post hoc tests for significant main effects if variables include more than two levels.

11.9.1 SPSS Specifications for Two-Way Mixed Design ANOVA

To begin the exercise, in the SPSS data editor select **File → Open → Data**; navigate to the file that contains the data file **Olympics_16.sav**. Double-click on the file to read it into SPSS (Figure 11.4). The exercise focuses on the following variables: Gender, Heats, Semifinals, and Final.

A preliminary run of SPSS Explore revealed normal distribution of the data, with the exception of data for female sprinters in the final round of the 100-meter dash. Skewness was also most notable in this group. Closer inspection revealed that the outlier was a single female athlete. The outlier in the data, though accurate and reflecting the realities of the race, could be influential especially due to the small sample size. However, this exercise is designed for demonstration purposes only, so the outlier case and the original data were retained unchanged. Remaining assumptions of sphericity, homogeneity of variance and covariance matrices are default components of the procedure.

Figure 11.4 Data Set in SPSS Data Editor

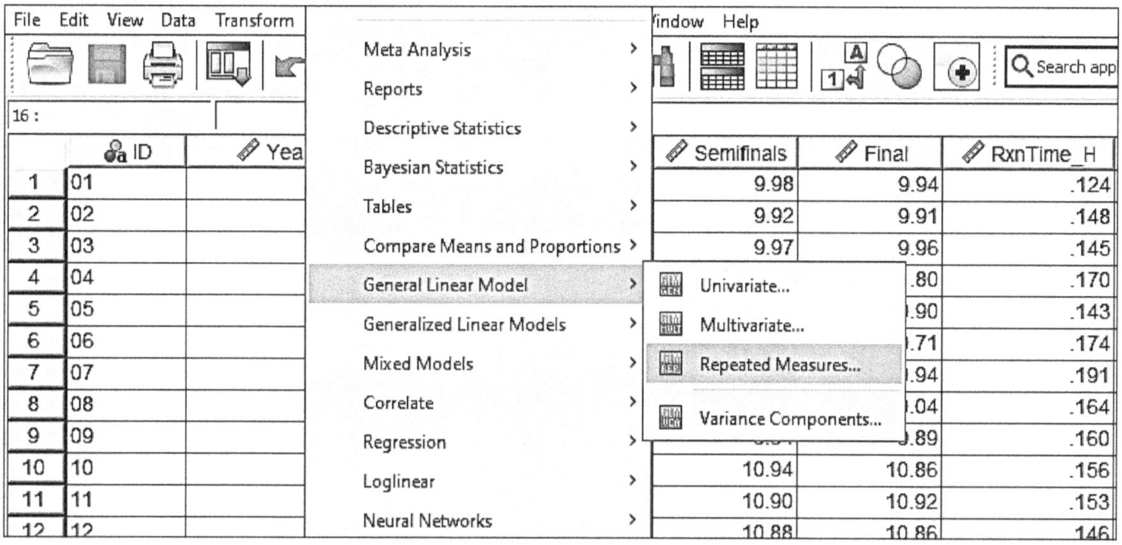

Mixed design is an extension of RMANOVA. Consequently, to specify a two-way, mixed design ANOVA, select **Analyze → General Linear Model → Repeated Measures** to reach the **Repeated Measures Define Factor(s)** dialog window (Figure 11.5a).

Figure 11.5 Configured Repeated Measures Define Factors Dialog Box

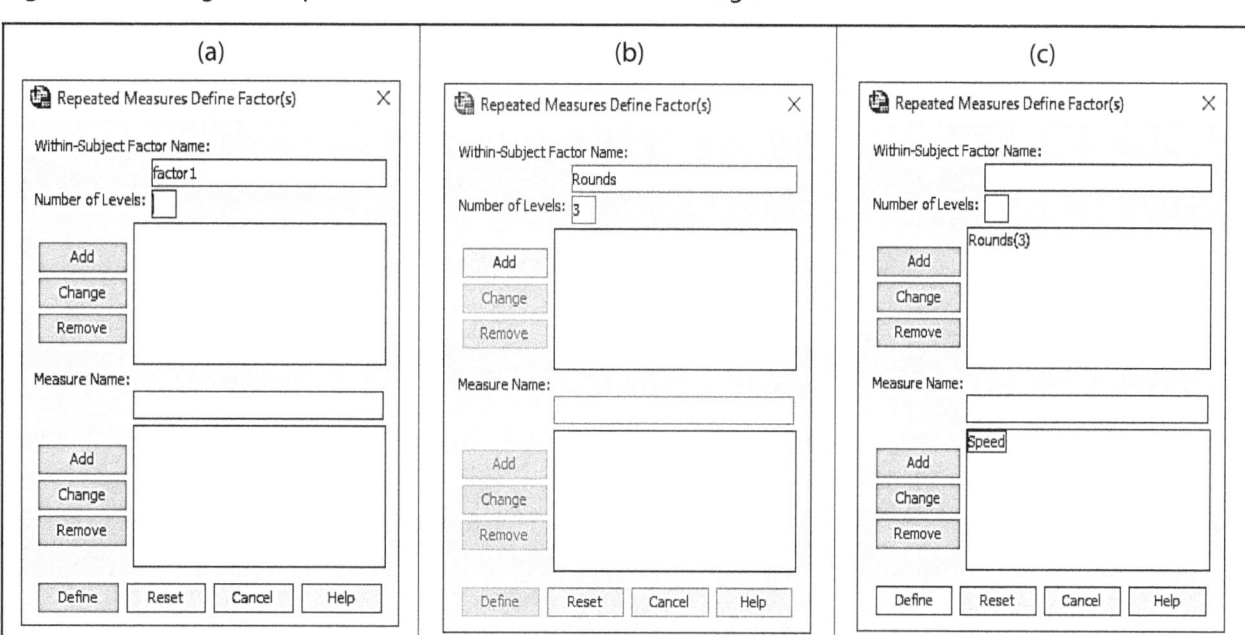

The **Repeated Measures Define Factor(s)** dialog window allows researchers to name the repeated measures variable (the within-subject factor in SPSS) as well as to specify the number and names of the variable levels. To do that, replace the default name **Factor1** in the **Within-Subject Factor Name** panel with the name **Rounds** and type **3** in the **Number of Levels** panel for the number of levels of the repeated measures variable (heats, semifinals, and final) (Figure 11.5b). Click **Add** to move the specifications to the dialog box next to **Add**.

In the **Measure Name** panel, enter the name for the dependent variable, **Speed**. Click on the **Add** pushbutton next to the box below the **Measure Name** panel to add the dependent variable name to the program, and the fully configured dialog box in Figure 11.5c will appear. Also, the **Define** pushbutton at the bottom of the dialog window will be activated.

Click on the **Define** pushbutton to open the **Repeated Measures** main dialog window (Figure 11.6a). Notice that in the main dialog window in the **Within-Subjects Variables (Rounds**, there are three question marks, followed by the numbers 1 – 3, for each of the three levels of the within-subjects (repeated measures) variable. These are placeholders which must be replaced by the variable levels in the panel on the left. To accomplish that, highlight **Heats**, **Semifinal** and **Final** on the left, then click on the single arrow between the two boxes.

Figure 11.6 Main Dialog Window for Repeated Measures

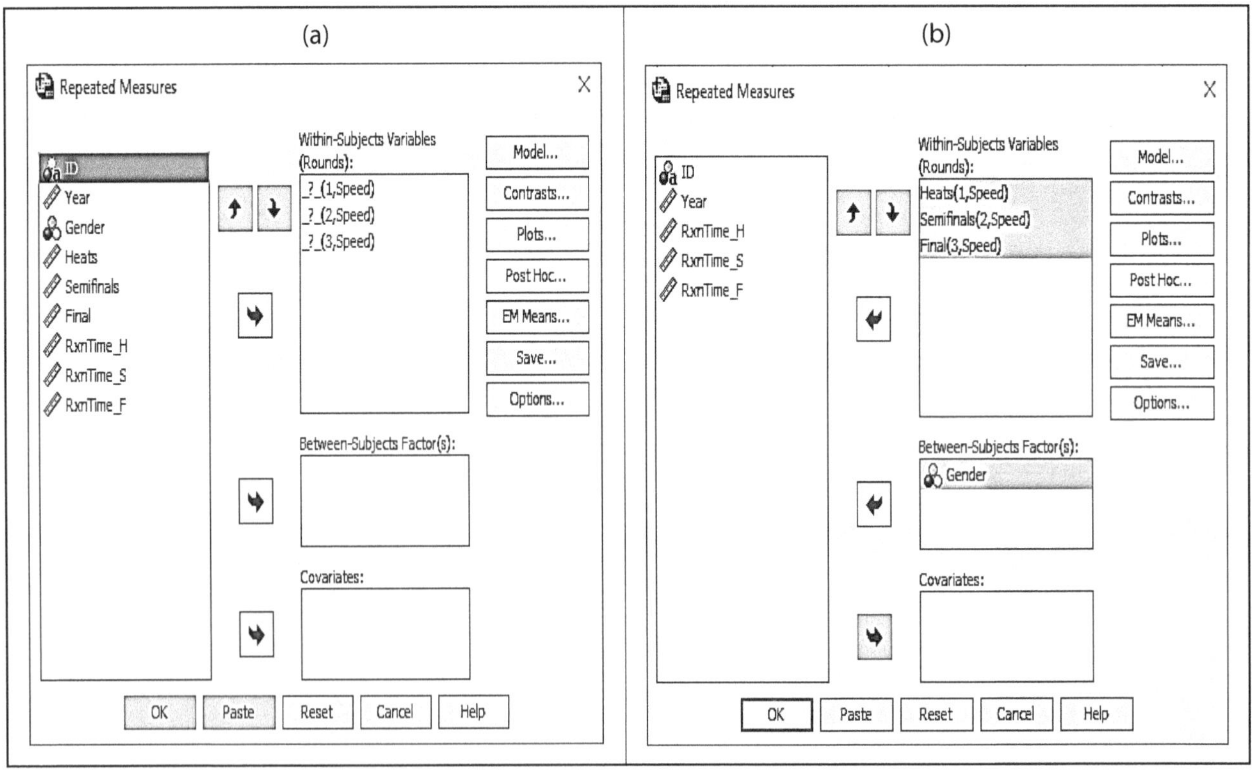

Confirm that the variable levels are matched up, value by value, with those in the panel on the right. Also check that the dependent variable, **Speed**, follows each one (Figure 11.6b).

So far, specifications for mixed design have been similar to the process for repeated measures. However, the next step turns the procedure into one for mixed design. To do that, select the between-subjects variable, **Gender**, in the **Repeated Measures** main dialog window, and click on the arrow next to the **Between-Subjects Factor(s)** panel to transfer the variable to that space.

A plot of the values for **Rounds** and **Gender** will graphically display any interaction that exists between the two variables. To specify the plot, click on the **Plots** pushbutton in the **Repeated Measures** dialog window to activate the **Repeated Measures Profile Plots** window (Figure 11.7). Highlight the variable, **Rounds**, in the **Factors** box and drag it, or click it over to the **Horizontal Axis** panel. Next, highlight **Gender** and transfer that variable to the **Separate Lines** slot.

With the two variables specified (Figure 11.7a), click on the **Add** pushbutton below the **Factors** box to transport the plot instructions to the **Plots** box (Figure 11.7b). This instruction produces two plots showing the speed (in seconds) of male and female athletes in the three (performance) rounds (heats, semifinals, and final) of the 100-meter dash.

Figure 11.7 Repeated Measures Profile Plots Dialog Window

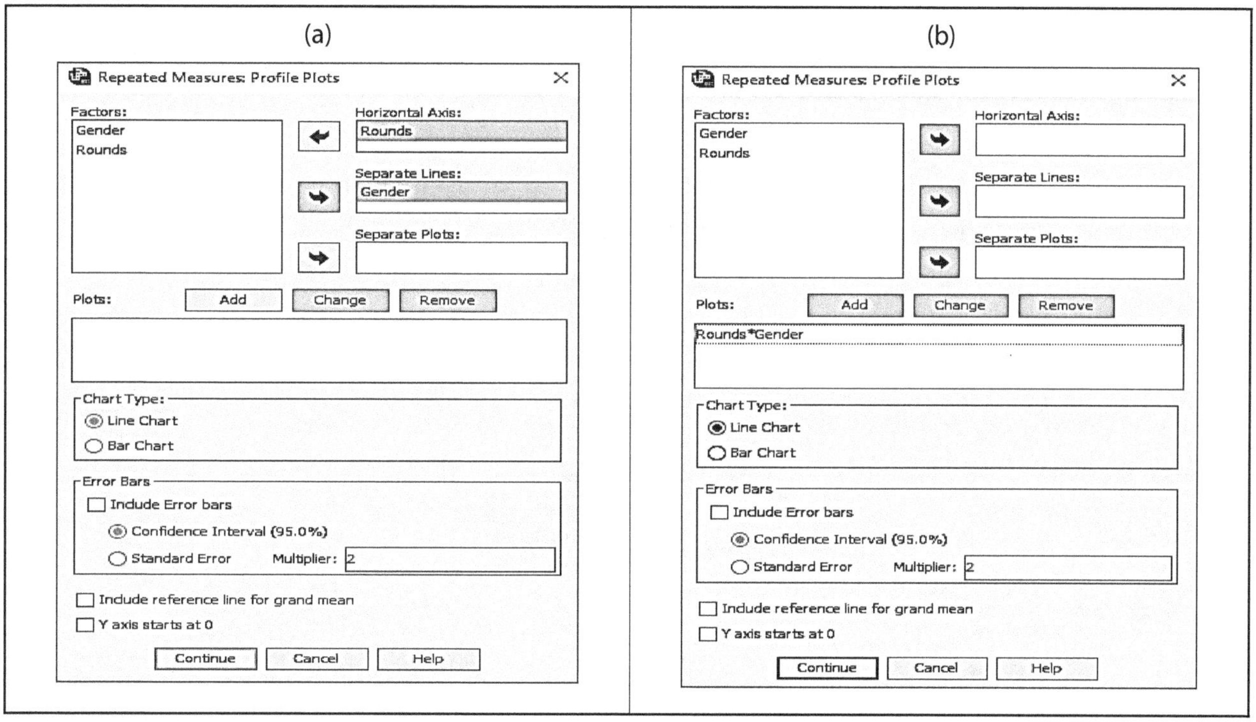

To generate descriptive statistics and test the assumption of homogeneity of variance, click on the **Options** pushbutton in the **Repeated Measures** dialog window (Figure 11.6) to open the **Repeated Measures Options** dialog window (Figure 11.8). Click the boxes next to **Descriptive statistics** and **Homogeneity tests** to specify those procedures.

With three levels in the repeated measures variable, **Rounds** (heats, semifinals, and final), SPSS will automatically execute **Mauchly's Test of Sphericity**. In addition, SPSS will generate results of **Box's Test of the Equality of Covariance Matrices** which tests whether the intercorrelations of the levels of the variable, **Rounds**, are equal across the gender groups (Section 11.6.4). Click **Continue** to return to the main menu.

For purposes of efficiency in this demonstration, specify follow-up tests now in the event that there is a significant interaction between gender and rounds. To complete that task, click on the **EM MEANS**

Figure 11.8 Repeated Measures Options Dialog Window

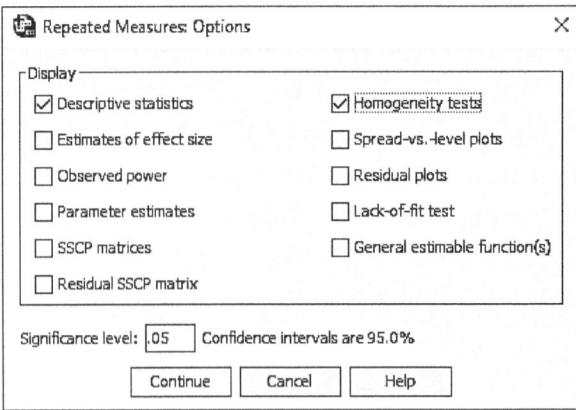

pushbutton in the **Repeated Measures** main dialog window to open the **Repeated Measures Estimated Marginal Means** dialog window (Figure 11.9a).

No post hoc tests are required for **Gender** because it includes only two groups. However, the variable, **Rounds**, has three levels, which necessitates follow-up tests should a significant main effect (significant change in average speed) for **Rounds** occur. Post hoc tests (pairwise comparisons of the athletes' average speeds in the rounds) will identify the source of significant changes in speed between heats and semifinals, heats and final, and semifinals and final.

Figure 11.9 Estimated Marginal Means Dialog Window

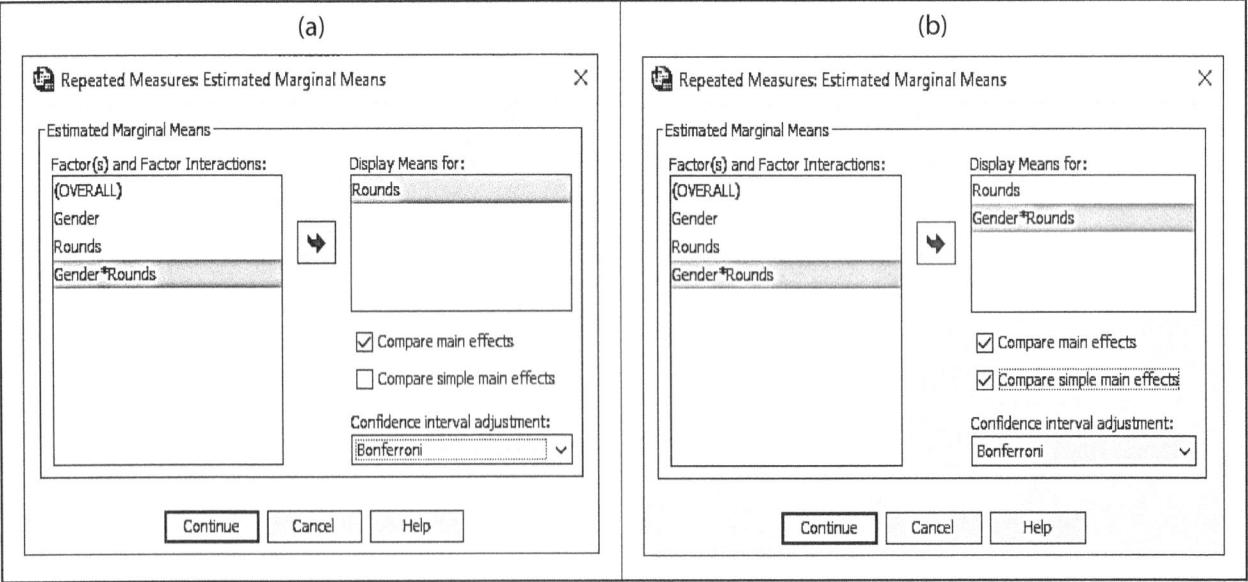

To specify the main effects follow-up test, highlight the variable, **Rounds**, in the **Factor(s) and Factor Interactions** panel on the left, and transport it to the **Display Means for:** panel on the right. Check the **Compare main effects** box below the **Display Means for:** panel. Click on the down arrow in the panel below **Confidence interval adjustment:** and select **Bonferroni** to activate adjustments in the alpha level for the comparisons that prevent inflated alpha levels above .05 (Figure 11.9a).

To specify the test of simple effects, highlight the interaction **Gender*Rounds** in the **Factor(s) and Factor Interactions** panel, and click the arrow to place the interaction in the panel, **Display Means for:** (Figure 11.9b). To conduct the simple effects analysis, select **Compare simple main effects** (Figure 11.9b) below the **Display Means for:** panel. Under **Confidence interval adjustment**, scroll down to confirm that **Bonferroni** is selected. Click **Continue** to return to the main menu, and click **OK** to run the analysis.

11.9.2 Omnibus Results for Mixed Design ANOVA

11.9.2.1 Descriptive Statistics

Figure 11.10 displays descriptive statistics for the two-way mixed design analysis of the Case of Olympic Track Athletes' Speed. Confirm that the **Within-Subjects Factors** table in Figure 11.10a accurately lists the dependent variable (Measure), **Speed**, and the repeated measures variable with its associated levels. The **Between Subjects Factors** table in Figure 11.10b displays the between-subjects variable, **Gender**, together with its two groups, **Female** and **Male** and associated sample sizes. The study included a sample size of 16 subjects—eight female and eight male athletes.

Figure 11.10 Design Features of Mixed Design ANOVA

(a)

Within-Subjects Factors	
Measure: Speed	
Rounds	Dependent Variable
1	Heats
2	Semifinal
3	Final

(b)

Between-Subjects Factors		Value Label	N
Gender	1	Female	8
	2	Male	8

Figure 11.11 displays descriptive statistics (mean and standard deviation) for the groups of each independent variable (rounds and gender) as well as the sample size for each. Some differences in mean speed are evident between the groups for the independent variables. However, without hypothesis testing there is no way of knowing whether the average changes in speed presented are significant or simply reflect errors in measurement.

11.9.2.2 Results of Tests of Assumptions

Figure 11.12 - 11.14 display results of three tests of assumptions for mixed design ANOVA: Mauchly's Test of Sphericity, Levene's Test of Equality of Error Variances and Box's Test of Equality of Covariance Matrices. The results of Mauchly's Test (Figure 11.12) reveal that Mauchly's W is significant ($p = .001$ in the **Sig.** column). Recall that a significant Mauchly's test means that the variances of the changes between the levels of the repeated-measures independent variable are significantly different—a violation of the assumption of sphericity. In this case, heats, semifinals, and final are the levels of the independent variable, **Rounds**. The variances of the changes in speed between the rounds are significantly different. The remedy for this violation is to use the Greenhouse-Geisser correction, which elicits a more accurate significance level of differences in speed than Huynh-Feldt or Lower-bound when variances are unequal.

Figure 11.11 SPSS Descriptive Statistics for Two-Way Mixed Design

	Gender	Mean	Std. Deviation	N
Heats	Female	11.1038	.10901	8
	Male	10.0900	.06256	8
	Total	10.5969	.53049	16
Semifinals	Female	10.9075	.02816	8
	Male	9.9550	.04986	8
	Total	10.4313	.49342	16
Final	Female	10.9775	.33978	8
	Male	9.9425	.08031	8
	Total	10.4600	.58527	16

Descriptive Statistics

Figure 11.12 Results of Mauchly's Test of Sphericity for Mixed Design ANOVA

Mauchly's Test of Sphericity[a]

Measure: Speed

Within Subjects Effect	Mauchly's W	Approx. Chi-Square	df	Sig.	Epsilon[b]		
					Greenhouse-Geisser	Huynh-Feldt	Lower-bound
Rounds	.366	13.078	2	.001	.612	.688	.500

Tests the null hypothesis that the error covariance matrix of the orthonormalized transformed dependent variables is proportional to an identity matrix.
[a]Design: Intercept + GenderN
Within Subjects Design: Rounds
[b]May be used to adjust the degrees of freedom for the averaged tests of significance. Corrected tests are displayed in the Tests of Within-Subjects Effects table.

Figure 11.13 presents the results of Levene's Test of Homogeneity of Variance for all the levels of the variable, rounds. All significance values in the **Sig.** column are greater than .05. Therefore, error variances for speed are not significantly different across male and female athletes. Rather they are equivalent. Accordingly, the equality of variance assumption has not been violated.

The final test of assumption is Box's Test of the Equality of Covariance Matrices (Figure 11.14). This assumption is satisfied when the pattern of intercorrelations among the speed in heats, semifinals and final is consistent among male and female athletes. Recall that a nonsignificant test result means that the assumption has not been violated. Using the recommended alpha level of .001, Box's M returns a nonsignificant outcome $F(6,1420.08) = 3.08$ ($p = .005$).

Figure 11.13 Results of Levene's Test of Homogeneity of Variances

	Levene's Test of Equality of Error Variances[a]				
		Levene Statistic	df1	df2	Sig.
Heats	Based on Mean	2.883	1	14	.112
	Based on Median	2.848	1	14	.114
	Based on Median and with adjusted df	2.848	1	10.423	.121
	Based on trimmed mean	2.880	1	14	.112
Semifinals	Based on Mean	1.828	1	14	.198
	Based on Median	2.263	1	14	.155
	Based on Median and with adjusted df	2.263	1	12.963	.156
	Based on trimmed mean	1.920	1	14	.188
Final	Based on Mean	2.497	1	14	.136
	Based on Median	.892	1	14	.361
	Based on Median and with adjusted df	.892	1	7.398	.375
	Based on trimmed mean	1.522	1	14	.238

Tests the null hypothesis that the error variance of the dependent variable is equal across groups.
[a]Design: Intercept + GenderN
Within Subjects Design: Rounds

The results of the Box's *M* test of the covariance assumption suggest that the assumption has not been violated. However, due to the small sample size ($N = 16$), the test outcome may be questionable.

Figure 11.14 Results of the Box's *M* Test of the Covariance Assumption

Box's Test of Equality of Covariance Matrices[a]	
Box's *M*	24.251
F	3.086
df1	6
df2	1420.075
Sig.	.005

Tests the null hypothesis that the observed covariance matrices of the dependent variables are equal across groups.
a. Design: Intercept + GenderN
Within Subjects Design: Rounds

11.9.2.3 The Interaction Effect Between Gender and Performance Rounds

Results of the omnibus tests, **Tests of Within-Subjects Effects**, are displayed in Figure 11.15. Of primary importance, however, is the test of interaction effects in the second major row of the table, which shows a significance level much greater than $\alpha = .05$. This means that there is a greater than 5% chance that the null hypothesis is true. In fact, the lowest significance (Sig.) level indicates an approximately 53% chance that the null is true. Hence, the null hypothesis must be retained.

Due to the nonsignificant interaction results, no post ANOVA tests are required, and there is no need to review the follow-up tests conducted in the event of a significant interaction. In practice, follow-up tests are not conducted until the results of the omnibus test is reviewed. If interaction effects were significant, the next step would have been to report the results of the test of simple effects as transpired in the factorial ANOVA demonstration (Chapter 8).

11.9.2.4 The Main Effects for Performance Rounds and Gender

With the interaction between gender and rounds nonsignificant, the results of the significance testing for the main effects of rounds and gender take priority. Due to the violation of the sphericity assumption (Figure 11.5), test results for rounds are based on the Greenhouse-Geisser correction (Figure 11.15), and that discloses a significant change ($p = .013$) between the speed recorded in two or more of the three rounds of the 100-meter dash averaging across gender. Huynh-Feldt, the other top option when the sphericity assumption is violated, records a similar significant difference in the average speeds recorded between the various performance rounds ($p = .01$).

Figure 11.15 Results of Tests of Within-Subjects Effects

Tests of Within-Subjects Effects						
Measure: Speed						
Source		Type III Sum of Squares	df	Mean Square	F	Sig.
Rounds	Sphericity Assumed	.251	2	.125	7.079	.003
	Greenhouse-Geisser	.251	1.224	.205	7.079	.013
	Huynh-Feldt	.251	1.376	.182	7.079	.010
	Lower-bound	.251	1.000	.251	7.079	.019
Rounds * Gender	Sphericity Assumed	.015	2	.007	.415	.665
	Greenhouse-Geisser	.015	1.224	.012	.415	.569
	Huynh-Feldt	.015	1.376	.011	.415	.591
	Lower-bound	.015	1.000	.015	.415	.530
Error(Rounds)	Sphericity Assumed	.496	28	.018		
	Greenhouse-Geisser	.496	17.133	.029		
	Huynh-Feldt	.496	19.264	.026		
	Lower-bound	.496	14.000	.035		

The third omnibus test, the **Test of Between-Subjects Effects** (Figure 11.16) records a significant difference ($p < .001$) between the average speed of male and female participants in the 100-meter dash, collapsing across performance rounds. With only two variable groups for gender (male and female), there is no need for a follow-up test. Looking back at the descriptive statistics in Figure 11.11, reveals that male athletes posted significantly faster speeds (fewer seconds) overall than female athletes in every performance round.

Figure 11.16 Results of Tests of Between-Subjects Effects

Tests of Between-Subjects Effects					
Measure: Speed					
Transformed Variable: Average					
Source	Type III Sum of Squares	df	Mean Square	F	Sig.
Intercept	5288.011	1	5288.011	150733.541	<.001
Gender	12.010	1	12.010	342.342	<.001
Error	.491	14	.035		

11.9.2.5 Pairwise Comparisons for Levels of Performance Rounds

The results of post hoc comparisons of sprinting speed in the three rounds are presented in Figure 11.17 and 11.18. The estimated marginal means are presented in Figure 11.17 and pairwise comparisons are displayed in Figure 11.18. A cursory comparison of the estimated means records the slowest average speed in Round 1, heats ($M = 10.597$ seconds) and the fastest in Round 2, semifinals ($M = 10.431$ seconds).

Figure 11.17 Estimated Marginal Means of Speed Across Performance Rounds

Estimates				
Measure: Speed (seconds)				
			95% Confidence Interval	
Rounds	Mean	Std. Error	Lower Bound	Upper Bound
1	10.597	.022	10.549	10.645
2	10.431	.010	10.410	10.453
3	10.460	.062	10.328	10.592

Results of pairwise comparisons in Figure 11.18 show a significant improvement ($p < .001$) in track athletes' average speed from heats to semifinals where the average speed changed from 10.597 seconds to 10.431 seconds, a mean difference in speed of .166 seconds. However, recorded speeds in the semifinals and final round showed no significant increase in speed (from 10.431 seconds to 10.460 seconds, $p = 1.0$) between those rounds. In addition, based on an alpha level of .05, there was no significant difference in athletes' average sprinting speed between the heats and final round (from 10.597 seconds. to 10.46- seconds, $p = .071$).

11.10 Sample Report: The Case of Olympic Athletes' Speed

A study of Olympic track athletes compared the changes in male and female athletes' speed across major performance rounds (heats, semifinals and final) in the 2016 Summer Olympics. Only athletes competing in all performance rounds were included in the analysis. A total of 16 athletes that included eight males and eight females, from eight competing countries were included. Athletes' speeds were taken from the International Association of Athletics Federation (IAAF).

Figure 11.18 Pairwise Comparisons of Speed Across Performance Rounds

Pairwise Comparisons						
Measure: Speed						
					95% Confidence Interval for Difference[b]	
(I) Rounds	(J) Rounds	Mean Difference (I-J)	Std. Error	Sig.[b]	Lower Bound	Upper Bound
1	2	.166*	.022	<.001	.107	.224
	3	.137	.054	.071	-.010	.283
2	1	-.166*	.022	<.001	-.224	-.107
	3	-.029	.057	1.000	-.184	.127
3	1	-.137	.054	.071	-.283	.010
	2	.029	.057	1.000	-.127	.184

Based on estimated marginal means

* The mean difference is significant at the .05 level.

[b]Adjustment for multiple comparisons: Bonferroni.

The main effect for (performance) rounds, the within-subjects independent variable, and the main effect for gender, the between-subjects independent variable, were both statistically significant, $F(1.224, 17.133) = 7.079, p = .013$ and $F(1,14) = 342.342, p < .001$, respectively. However, the interaction between gender and rounds was found to be nonsignificant $F(1.224, 17.133) = .415, p = .569$.

Tests of pairwise comparisons of the main effect for rounds revealed a statistically significant increase in speed between heats (Round 1) and semifinals (Round 2) ($p < .001$). However, changes in speed between other rounds (between heats and final, and between semifinals and final) were found to be nonsignificant ($\rho = .071$ and $\rho = 1.00$, respectively).

11.11 Key Terms

- Box's test of equality of covariance matrices
- Mixed design ANOVA

12

Nonparametric Tests

Key concepts from previous chapters or courses:

omnibus test: any statistical test of significance in which more than two levels (groups) of an independent variable are compared simultaneously or in which there are two or more independent variables.

hypothesis testing: a type of statistical analysis in which assumptions about a population parameter are tested by using data from a sample of the target population to draw conclusions.

degrees of freedom (df): the number of values that are free to vary when computing a statistic. This number is needed to interpret the statistic (e.g., t score, F ratio, chi-square, H statistic).

effect size: a value measuring the degree to which a phenomenon is present in a target population. In ANOVA, this refers to strength of the mean differences between groups. The larger the effect size, the greater the size of associations.

multiple comparisons: comparison tests conducted after finding overall statistical significance (rejection of the null hypothesis) in the omnibus test, to investigate whether differences between specific pairs of group means are statistically significant.

THIS CHAPTER INTRODUCES

- differences between parametric and nonparametric statistics
- choosing between parametric and nonparametric statistical test
- determining whether data meet assumptions for nonparametric tests
- ranking data
- computing nonparametric tests using hand calculations and SPSS
- finding the effect size of a nonparametric test statistic
- performing multiple comparison tests after a statistically significant omnibus test

12.1 Introduction

Memories of the misery of chicken pox, measles, mumps, and whooping cough (pertussis) are nonexistent for many children in the Caribbean today. Childhood vaccinations are now a rite of passage for children throughout the world. Children receive these shots in the blink of an eye—before they realize they have something to cry about.

According to UNICEF (United Nations International Children's Emergency Fund), the rate of childhood vaccinations against the most common childhood diseases is generally high in much of Latin America and the Caribbean region, with most countries attaining vaccination rates of 80-99% of eligible children (WHO-UNICEF, 2018). UNICEF's stated goal is to eradicate childhood diseases across the world. One way of doing that is through monitoring the progress and hindrances in getting children vaccinated, through questionnaires, administered to citizens of Caribbean nations in which many of the items require respondents to specify a level of agreement or disagreement with a statement. These survey response formats are commonly referred to as Likert scales or rating scales named after its inventor Rensis Likert (1932).

Like UNICEF, the World Values Survey Association (WVSA) conducts world-wide research about the values of people around the world using questionnaires with Likert-style survey items as well. Likert scales, which are typically 3- 9-point rating scales (Figure 12.1), are classified as ordinal level or rank-ordered data that frequently do not meet the assumptions of parametric statistics discussed so far (e.g., the assumption of an underlying normal distribution of the data).

Figure 12.1 Examples of Likert Response Formats

Agreement Rating Scale		Frequency Rating Scale	
Agree very strongly	Disagree	Often	Seldom
Agree strongly	Disagree strongly	Sometimes	Never
Agree	Disagree very strongly		
Quality Rating Scale		**Likelihood Rating Scale**	
Extremely poor	Above average	Extremely likely	Slightly unlikely
Below average	Excellent	Very likely	Somewhat unlikely
Average		Somewhat likely	Very unlikely
		Slightly likely	Extremely unlikely
		Neither likely nor unlikely	

How then do researchers analyze these data? The central concern of this chapter is a class of statistical tests known as nonparametric statistics. These techniques are not reliant on normal distribution assumptions and other assumptions required by parametric tests. These tests, while not distribution-free, have assumptions that are less demanding than parametric tests, and therefore, are good alternatives for use when parametric test assumptions have been violated or for investigating rank-ordered data or nominal data (e.g., gender or blood type). This chapter explores common characteristics of nonparametric statistics, specifically four types of nonparametric tests which are alternatives to parametric statistics studied in previous chapters. The nonparametric tests and their parametric cousins are listed in Figure 12.2.

12.2 Nonparametric Statistical Analysis Fundamentals

All of the statistical methods explored in previous chapters belong to the parametric family of statistical tests. In addition to the assumption that research sample data come from a target population with a normal (or at least symmetrical) distribution, ideally parametric tests assume the following as well:

- homogeneity of variance in each group
- equal group sizes for some parametric tests
- no outliers

Figure 12.2 Parametric Tests and Their Nonparametric Equivalents

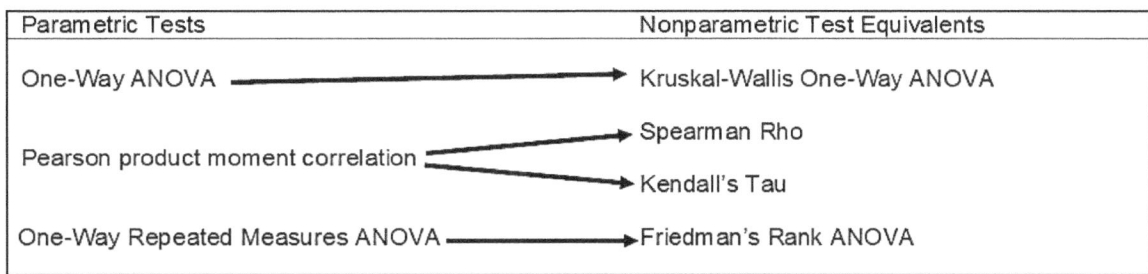

Thus far, parametric statistical tests in this text have assumed that all data have particular underlying characteristics. To avoid inaccurate test results and misinterpretations, statistical tests are routinely preceded by assumptions testing (e.g., test of the assumption of the normal distribution of data) to check for assumptions violations.

There are workarounds when assumptions have been violated. For example, when the equal variance assumption for a one-way ANOVA is violated, researchers may opt for selecting Welch's adjustment in SPSS (Welch's ANOVA) to alleviate some of the problems associated with unequal variances as long as group sizes are equal and normally distributed. In addition, if group sizes are unequal, but groups are normally distributed with equal variances, ANOVA remains a more powerful option that is less likely to result in Type I or Type II errors than nonparametric tests.

There are times, however, when data substantially deviate from parametric test assumptions. Sample sizes may be small and include outliers, or sample data may have severely skewed distributions. In these situations, nonparametric tests, known for having less restrictive assumptions than parameter tests, are statistical test alternatives to consider. Thus, although they are not replacements for parametric tests, nonparametric tests are supplements to parametric tests, and they are so widely used in the world of research today that it is essential that scholars are familiar with those statistical methods in addition to parametric tests. While this chapter does not provide comprehensive coverage of nonparametric statistics, it provides some insight into alternative statistical techniques. For a much more comprehensive treatment, consider Nussbaum (2015), Howell (2014) and others.

John Arbuthnott, a mathematician and physician, was the first to introduce nonparametric test methods in a study of the ratio of males and females at birth (Arbuthnott, 1710; Conover, 1999). He performed a statistical analysis similar to the sign test used today, which is an alternative to the parametric one-sample t-test or paired t-test. However, it was not until the 1940s that additional nonparametric methods were introduced.

Nonparametric tests may be an alternative when any of the following conditions apply:
- The level of measurement of a variable in the analysis is ordinal.
- The study data are best represented by the median score as a measure of central tendency rather than the mean score.
- The sample sizes of study groupings are severely unequal; e.g., a ratio of 4:1.
- Variables are interval or ratio level, but sample sizes are very small.

- The original data were measured at an interval or ratio level, but they violate one or more of the assumptions of parametric tests, consequently reducing the statistical power of the test and reducing the likelihood of detecting an existent significant effect:
  o Instead of exhibiting normal distributions, the data are highly skewed or kurtotic, indicating the presence of outliers (extreme scores) that could influence outcomes.
  o The data substantially violate the assumptions of equal variance (homoscedasticity).
  o The data are no longer interval/ratio data but were collapsed into a small number of categories, which place the data at the ordinal level.

The above conditions support the notion that nonparametric statistics were developed to produce valid results in the following situations:

- Nominal or ordinal data are analyzed.
- Group sample sizes are extremely small or unequal.
- Data depart from parameter test data assumptions to such a degree that the power or accuracy of the tests is in jeopardy.

Nonparametric tests are commonly partially defined as statistical tests used to analyze data at the nominal or ordinal level. In verity, the APA Dictionary of Psychology defines nonparametric tests as "a type of hypothesis test that does not make any assumptions (e.g., of normality or homogeneity of variance) about the population of interest. Nonparametric tests generally are used in situations involving nominal or ordinal data" (VandenBos, 2015, p. 713). Nevertheless, nominal (also called qualitative or categorical data) and ordinal (or ranked) data analysis are handled separately and regarded as having evolved into separate subfields of statistics by many researchers (e.g., Howell, 2014; Nussbaum, 2010) while others continue to collapse both subfields into a single field of nonparametric statistics. Due to the nonparametric tests explored in this chapter, explorations will be limited to those tests used to analyze data at ordinal or higher measurement levels. Still, there is no question that while parametric tests are most frequently used when dependent variables are measured at the interval or ratio level, nonparametric tests are most often used when data are at the nominal or ordinal level.

12.2.1 Distribution and Measurement Level and Nonparametric Tests

Ordinal level or ranked data overcome a key challenge of interval or ratio-level data—dealing with a skewed data distribution. Normal or near-normal distribution is a key assumption of parametric statistics. When sample data are skewed or kurtotic, rank-ordering the data before analysis corrects the problem by assigning the lowest score in the dataset a rank of 1, the next highest a rank of 2, etc., regardless of the differences in the scores. This procedure results in low ranks that represent low scores and high ranks that signify high scores. Subsequently, researchers conduct data analysis using ranks rather than the original scores. In this way the effects of outliers (skewed distributions) are avoided.

Likert-style scales, are a commonly used ordinal response format for attitudinal surveys and questionnaires. As illustrated in Table 12.1, scales may range from 2-point to 9-point response formats. There is continued debate about the efficacy of using parametric tests with Likert-style, ordinal-level data (e.g., Jamieson, 2004), and some argue that parametric tests are sufficiently robust to produce largely unbiased results when there are five or more Likert-style ordered categories, (e.g., Zumbo et al., 1993; Myers et al. 2010), when the data are normally distributed and the sample size is large. However, if the data do not exhibit these characteristics, nonparametric tests are excellent alternatives for data analysis. This is particularly true for research scientists who agree that scale points for Likert-style items are not equally spaced. Therefore, they do not fit the characteristics of an interval level variable.

Consider, for example, a research study in which college students are asked how frequently they watch TV each week, and the response choices are: 0 = Never, 1 = Very Rarely, 2 = Rarely, 3 = Occasionally, 4 = Frequently, 5 = Very Frequently. Each participant's conception of what each descriptor means is likely to vary. Rarely may mean three hours for one person and 6 for another. In this case, one might argue that the scale points are not equally spaced and are rankings. If, however, respondents were asked to specify the number of hours of TV-watching they indulged in each week, the difference between 0 and 2 hours would be the same as the difference between 20 and 22 hours, making the variable, hours, an interval variable and amenable to parametric statistics.

In spite of the disagreement about Likert-style items, research scientists agree that when Likert items are combined to create summative scales, the scale-level variables are interval-level variables. Consider an attitudinal scale that comprises a number of Likert response-style items that measure a single underlying construct like empathy, impulse control, or assertiveness on the emotional quotient inventory (Bar-On, 1997). The Likert response-style items that measure each construct are subscales. Each of these subscales (represented by total subscale scores) would be considered interval-level variables measurable with parametric statistics even though individual items on the scales are ordinal level measures. In effect, the statistical approach selected often depends on a variable's measurement level or by distributional characteristics (skewness and kurtosis). If skewed interval-level data are converted to ordinal-level, then the data are best represented by the median rather than the mean score.

12.2.2 Unequal and Small Sample Sizes

As research sample sizes increase, the central limit theorem says that the sampling distribution will be approximately normal even when the target population has a severely non-normal distribution. However, a small sample size is a key indicator for the use of nonparametric tests. According to Nussbaum (2010), when data are continuous but do not meet the sample size guidelines for parametric tests (e.g., less than 15 cases with 2-15 groups), the assumption of normal distribution is untenable, and it is therefore best to use nonparametric tests.

Severely unequal group size is another marker for choosing nonparametric tests. There are two concerns about unequal group sizes in this regard. They may lead to:

- **loss of robustness of a parametric test.** For example, with unequal group sizes, parametric tests are likely to exhibit greater vulnerability to the violation of the equal variance assumption, especially when sample sizes are small.
- **a violation of the assumption of equal variance in parametric tests**. Having both unequal group sizes and unequal group variances can also impact the statistical power of a test to reveal a true effect when one exists. A loss of statistical power leads to increased Type I error rates.

Parametric tests are likely to retain robustness if the scores of research groups have equal variances but unequal sample sizes. Parametric tests will remain robust if group variances are unequal, but sample sizes are equal. However, if both variances and sample sizes are unequal, nonparametric tests, like Kruskal-Wallis (Section 12.3) are worth considering.

Loss of power due to unequal sample sizes becomes particularly problematic when a research design calls for factorial ANOVA. Imagine a study investigating significant differences in income in which there are two independent variables—educational level (high/low), and age (young/old). However, the groups are unequal. There are twice as many young people as old, and the younger group has a much higher proportion of individuals with a high educational level than the older group. In this situation, due to the larger size of the younger group and the higher educational level of the younger group, the variables (age and educational level) are not independent of each other. Accordingly, the researcher cannot distinguish the effect of age from the effect of educational level. So, data analysis may result in a significant difference in income, apparently due to educational level, but the difference is really driven by age.

Unfortunately, there is no nonparametric test equivalent to factorial ANOVA that is designed to identify significant interaction between independent variables. Researchers are restricted to parametric tests in these situations.

12.2.3 Unequal Group Variance and Nonparametric Tests

As Section 12.2.2 noted, the presence of unequal group variances in a dataset is one of several key considerations in deciding whether to use a parametric or nonparametric test. There is some dispute about the accuracy of that conclusion. For example, Zimmerman (1998) argues that other parametric tests like the Welch or Brown-Forsythe test offer an alternative for parametric tests like a one-way ANOVA when variances are unequal. However, while data that fail to meet the assumption of equal variance do not limit researchers to nonparametric tests, such tests could become the most powerful option if the data fail to comply with additional assumptions. For example, imagine that a researcher plans to use one-way ANOVA to investigate whether differences in mathematics performance among a small sample of students aged 16, 17, and 18 are significantly different. However, the data fail to meet three assumptions for one-way ANOVA:

- The sample size is small—less than 15 in each group.
- The distribution of the data for each age group is notably skewed (nonnormal distribution).
- The score variances of the groups are not equal.

Nonparametric tests like Kruskal-Wallis can be a more powerful option than equivalent parametric statistics when sample sizes are small, and one or more parametric assumptions have been violated (e.g., Nussbaum 2010). However, nonparametric tests also have assumptions about the data to be analyzed, and one of those is the shape of the distribution of the data. While parametric tests require that the data from comparison groups display homogeneity of variance, nonparametric tests simply assume similarity in the shapes of distribution of the data across the groups, as illustrated in Figure 12.3a. This is a requirement of the Mann-Whitney U Test, which is not included in this chapter, as well as the Kruskal-Wallis test, which Section 12.3 will explore in greater detail. The Friedman's rank test, which Section 12.5 will examine in more detail, also assumes that the population distribution across comparison pairs has the same shape although not necessarily normal distribution (e.g., Lomax et al., 2012).

Figure 12.3 Shapes of the Distribution of Math Performance Across Age Groups

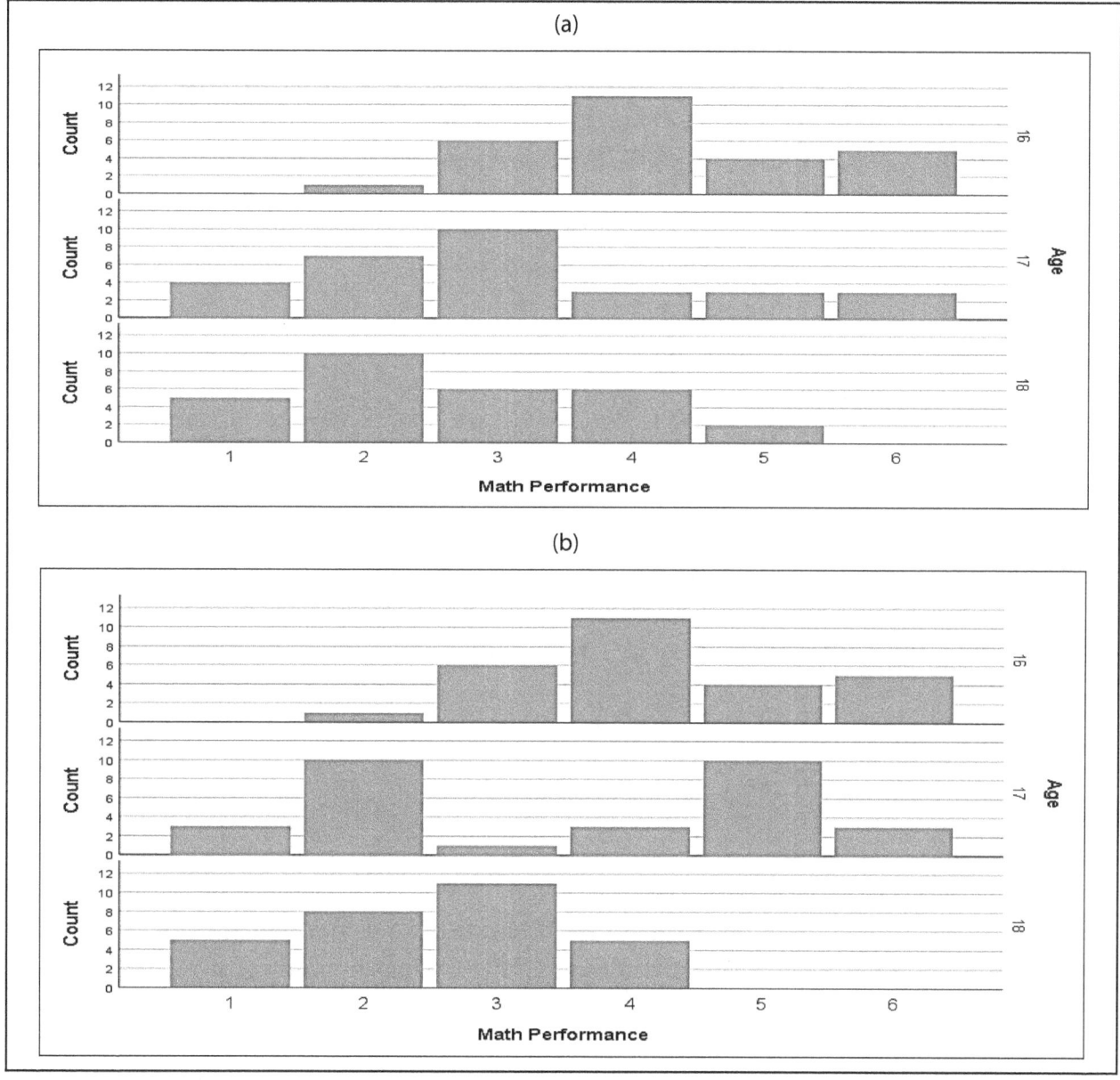

Failure to meet the nonparametric distribution assumption (Figure 12.3b) does not bar researchers from using nonparametric techniques, but it changes the data analysis procedure used (Section 12.3). Fortunately, although heterogeneity has some effect on nonparametric tests like Kruskal-Wallis, the impact is less than occurs with parametric ANOVA (Lomax et al., 2012).

12.2.4 Central Tendency and Nonparametric Tests

When scores are normally distributed, the mean of the distribution of scores is very close to the median of the distribution. However, when scores in a dataset are not normally distributed due to skewed data or a few outliers, the mean and the median scores can be very different. There are methods for addressing skewed distributions; such as transforming variable values so that they are more normally distributed. However, transformation may result in difficulty in interpreting the outcome of the statistical analysis.

In some cases, the median may be a better, more authentic measure of central tendency in datasets that are naturally skewed in the real world, giving nonparametric analyses an advantage over parametric tests. For example, consider the distribution of income. Typically, income tends to exhibit a highly skewed distribution in which most salaries tend to cluster around the median, where half of the salaries are above and half are below the median. However, there is a long tail that stretches off to the right into the very high salary range. (e.g., Figure 12.4). The smaller number of higher incomes pulls the mean far away from the central median value, increasing the mean although most salaries have not changed.

Figure 12.4 Illustration of Central Tendency of Income Distribution in Mexico

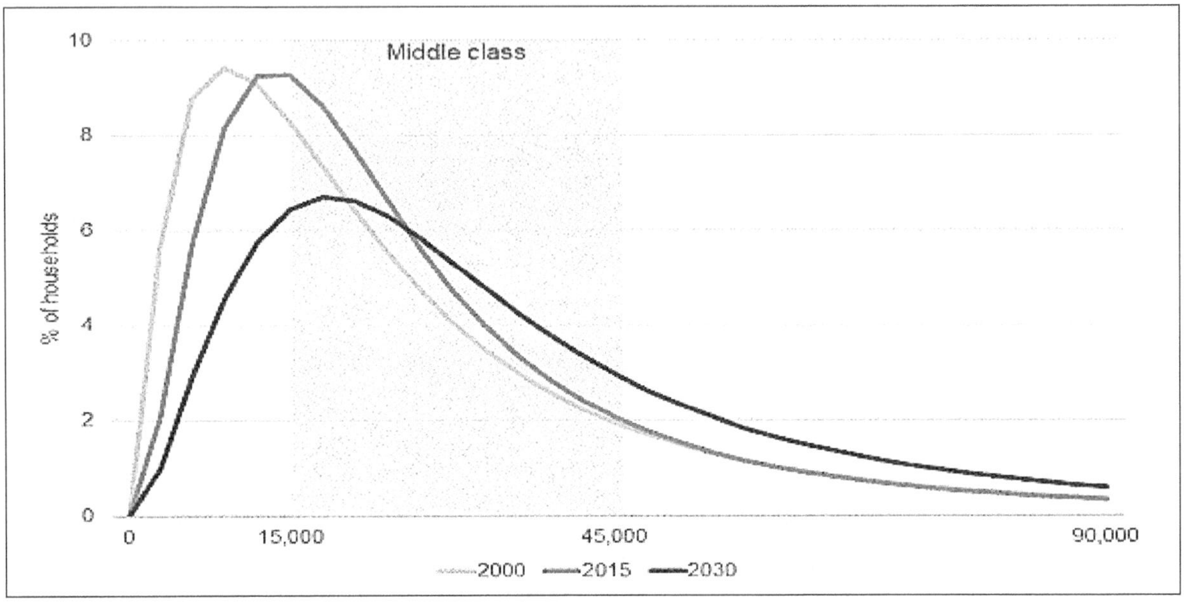

Source: Euromonitor International from national statistics

Skewed distribution can affect the mean, but not the median. Therefore, if outliers are due to skewness or kurtosis and they cannot be removed from a dataset without injecting bias, the median may be a better measure of central tendency, and in such cases, the more conservative nonparametric test equivalents may be more efficacious solutions than parametric ones.

As a rule of thumb, some researchers consider the use of nonparametric tests when the skew or kurtosis of a dependent variables is greater than +/-1.0. Others (e.g., Osborne, 2013) suggest that a skew level of +/- 0.8 should indicate the need to consider a remedy which could include a nonparametric alternative.

12.2.5 Ranks: Key Secret Weapon of Nonparametric Tests

Contrary to parametric tests, nonparametric tests are said to be robust to departures from normal distribution of scores often caused by the presence of outlier scores in a dataset. The key to the robustness of nonparametric tests under these circumstances is the use of ranks rather than raw scores. This applies to dependent variable scores at the ordinal, interval or ratio-level.

In most nonparametric tests, raw data must be converted to ranks because the analysis utilizes the ranks rather than the raw scores for analysis. With ranks being analyzed rather than raw data, nonparametric tests become tests of differences in central tendency which are not affected by a few extreme scores or outliers. Due to the substitution of ranks for raw scores, the shape of the underlying distribution of a set of scores, a key assumption of parametric tests, becomes less important. Normally, extreme scores used in parametric tests inflate the variance (thus the error term) and shift the mean score toward the outlier, resulting in a biasing influence on the outcome of the tests and often reducing the power to identify the occurrence of a true effect. Nonparametric tests neutralize the problem of the loss of power in parametric tests caused by outliers. Here is an example. Consider the following scores:

Normal Distribution

Raw Scores (1):	15	18	19	20	21	24
Ranks (1):	1	2	3	4	5	6

Skewed Distribution

Raw Scores (1):	31	32	33	35	57	58
Ranks (1):	1	2	3	4	5	6

Notice that regardless of the quantities represented, the distribution of the values or the presence of outliers, the ranks remain the same. There are six unique values, and the rankings are 1 – 6 for both lists. In this way, nonparametric statistics reduce the importance of the underlying distribution.

The assignment of ranks follows the same process regardless of the data involved:
1. Order raw scores from lowest to highest as illustrated in the above lists.
2. Assign a rank of 1 to the lowest value, then appoint the next lowest a rank of 2, and so on. The largest value is assigned a rank of N (in the example above, $N = 6$).

A complication may arise when there are duplicate values, the same values for two or more subjects in a dataset. In theory, nonparametric tests were developed for continuous distributions where the probability of duplicate values is zero. In practice, however, duplicate values or ties often occur, and these prohibit the assignment of unique rankings. For example, imagine that the following values replaced the first set of the data above:

Original Raw Scores (1):	15	18	19	20	21	24
Replacement Raw Scores (2):	15	15	19	20	21	24

Notice that the first two values are the same—15. To resolve this tie, the practice is to (1) assign each value consecutive ranks, (2) find the average of the first two ranks $1 + 2 = 3/2 = 1.5$ and (3) assign 1.5 to each of the values. Therefore, the rankings would adjust in the following way:

Replacement Raw Scores (2):	15	15	19	20	21	24
Adjusted Ranks (2):	1.5	1.5	3	4	5	6

What happens if there are three ties in a set of values? Assume that the following values replaced the original dataset:

Replacement Raw Scores (3):	15	18	19	20	20	20

In this case, the ranks would be determined by finding the average of the three consecutive rank assignments: (4 + 5 + 6)/3 = 15/3 = 5. The adjusted ranks are:

Adjusted Ranks (3):   1   2   3   5   5   5

Therefore, if the absolute values of two or more observations are equal, each observation is assigned the average of the corresponding ranks. Using this approach of assigning the average rank when there are ties ensures that the sum of the ranks (21) is the same as it would have been if the values were unique.

12.2.6 The Cost of Converting Continuous Scores to Ordinal Level Values

Sections 12.2.1 to 12.2.3, explored situations that call for considering the use of nonparametric tests, including skewed or kurtotic distributions and very small or unequal sample sizes. While nonparametric tests are valuable alternatives under these conditions, there are also drawbacks.

Converting continuous scores to ranked categories, creating multiple categories or dichotomous variables, creates a discontinuity in outcome variables as category boundaries are crossed. While it may be necessary due to the violation of assumptions, it may reduce the authenticity of the research investigation.

Useful information is often lost when interval/ratio data are reduced to ranks, simply due to the comparative coarseness of the ordinal level scale. This becomes more problematic as the number of categories used declines and when continuous scores are converted to dichotomous scores. This can lead to a reduction in statistical power to detect existent effects between independent and outcome variables—a Type II error. According to Cohen (1983), dichotomizing continuous scores at the median (a median split), which puts scores above and below the median into two categories, reduces power by the same amount as discarding a third of a dataset.

Reducing continuous scores to ranks (especially a small number of ranks) or dichotomizing the data may also lead to a more serious error. Cut-points that determine the categories for conversion may be arbitrarily assigned with little or no empirical justification for category boundaries, making interpretation of research outcomes difficult.

Cut-points can create false positives. Cut-points are also manipulatable so they may be set in a way that results in a desirable outcome. For example, individuals close to a category on either side of a cut-point may be characterized as being very different when in fact they are very similar, resulting in a Type I error.

12.2.7 Overall Advantages of Nonparametric Tests

- Nonparametric tests assess the median of dependent variable scores rather than the mean. The median is a better measure of central tendency when data are naturally skewed.
- Nonparametric tests are most often simpler to use than their parametric counterparts, leaving less room for improper use and interpretation.
- Typically, nonparametric tests have less power than parametric counterparts to correctly reject the null hypothesis. However, nonparametric tests may be readily substituted for parametric techniques if samples are too small to meet parametric sample size requirements. Very small sample sizes (≥ 5 per group) are acceptable when using nonparametric tests.

- Nonparametric tests can effectively analyze a broader scope of data than parametric tests, including continuous (interval and ratio) data, nominal and ordinal level data.
- Outliers in a dataset can impact the outcome of parametric tests (Howell. 2014). Converting data to ranks and using nonparametric tests reduce or remove outlier effects.

12.2.8 Overall Disadvantages of Nonparametric Tests

- Nonparametric tests are less powerful than parametric tests if parametric test assumptions have not been violated. Therefore, rejection of a null hypothesis is less likely when it is false if the data come from a normal distribution.
- Tied ranks can be problematic when they are common in a dataset. When they do occur, corrections to test statistics are necessary to determine whether results are significant. SPSS and other statistical software packages automatically apply the necessary corrections to avoid p-value inflation and the increased likelihood of a Type I error.
- Because nonparametric tests typically answer questions about the median rather than the mean score, the tests do not answer the same questions as the parametric procedures.
- Compared to parametric tests like the t-test and the one-way ANOVA, critical value tables for some nonparametric tests are not included in some statistical packages.

12.3 Kruskal-Wallis One-Way ANOVA (H) Test Fundamentals

The Kruskal-Wallis test is a nonparametric equivalent of the parametric one-way, between-subjects ANOVA. The test is known by other monikers, including Kruskal-Wallis test by ranks, Kruskal–Wallis H test, nonparametric one-way ANOVA, and one-way ANOVA on ranks. The test is named after William Kruskal and Wilson Allen Wallis (1952), statisticians at the University of Chicago.

Kruskal-Wallis is a nonparametric significance test that compares the scores of three or more independent groups on a continuous or ranked dependent variable. Researchers use the test to determine whether there are statistically significant differences among the group scores. If scores are significantly different, that would indicate that the groups are from different populations with different distributions and/or different central tendencies. If scores are not significantly different, the groups are sampled from the same population with similar score distributions and central tendencies (McDonald, 2014).

The Kruskal-Wallis test is an extension of the Mann-Whitney U Test which is used for comparing two groups and is the equivalent of the parametric t test. Like the one-way ANOVA, Kruskal-Wallis may be used to investigate whether there are significant differences between two groups on a dependent variable. However, it was specifically designed for investigations with three or more groups, making it a more efficient method of investigating the performance of multiple groups while controlling the Type 1 error rate inflation. Kruskal-Wallis also has the capacity to conduct pairwise comparison tests should the omnibus test prove to be significant. Pairwise comparison tests identify the pair(s) of groups that contributed to the significant outcome of the omnibus test.

An asymptotic Kruskal-Wallis significance test (using a chi-square distribution) may be used with group sample sizes of five or more subjects each. (An asymptotic test is based on the assumption that a dataset is large.) In hand calculations of the significance test, the hand-calculated test statistic, H, is treated as if it were a chi-square statistic and compared against the appropriate chi-square critical value for k − 1 degrees of freedom at the pre-set alpha level. Therefore, if there are three groups and the pre-set alpha level is .05, the chi-square statistical table (Appendix A) for 2 degrees of freedom (3 − 1) reports a critical value of 5.99. An H statistic that exceeds the critical value leads to rejection of the null hypothesis.

SPSS uses essentially the same procedure to analyze the data. With sample sizes less than 5, Kruskal-Wallis uses a more conservative test that does not make asymptotic test assumptions, but rather, produces an exact probability test for significance testing.

12.3.1 The Case of Performance in Three Instructional Programs

The dataset for this case is a fictitious dataset of 30 subjects based on the mathematics Caribbean Secondary Education Certificate (CSEC) examination, a Caribbean-wide secondary exit examination administered by the Caribbean Examinations Council (CXC) headquartered in Barbados. Students typically sit this exit examination at the end of fifth form in lower secondary school. CXC offers a total of 33 subject-area examinations (CXC, 2021). The exams cover a range of subject areas, and students may take as many subject-area exams as they wish. Students are placed in one of six grades (1 through 6) based on their performance, with 1 − 3 indicating successful performance. Successful completion of assessments results in the issuance of a CSEC certificate for each subject area exam students pass. Regional universities typically require successful completion of 4-5 CSEC subjects for admission.

Of the 30 cases in the hypothetical dataset, 10 were exposed to one of three different types of instructional methods: (1) teacher-directed in-person instruction, (2) computer-based instruction, and (3) a hybrid of both instructional approaches. Table 12.1 displays the data to be included in data analysis. The dataset is also available in **Math_Instr.sav**. Performance levels have been reverse scored to facilitate interpretation. Therefore, a grade of 1 equals a performance level of 6 and a grade level of 6 equals a performance level of 1. Box 12.1 presents the outline of an associated research plan.

Table 12.1 Mathematics Performance of Students in Three Instructional Programs

ID	Method	Performance Level	ID	Method	Performance Level	ID	Method	Performance Level
01	1	6	11	2	4	21	3	5
02	1	5	12	2	3	22	3	3
03	1	5	13	2	3	23	3	3
04	1	5	14	2	3	24	3	2
05	1	5	15	2	2	25	3	3
06	1	4	16	2	2	26	3	2
07	1	6	17	2	1	27	3	5
08	1	5	18	2	2	28	3	4
09	1	3	19	2	4	29	3	2
10	1	4	20	2	2	30	3	1

12.3.2 Research Plan: The Case of Performance in Three Instructional Programs

12.3.2.1 Eligible Data Variables and Measurement Levels

Like the one-way ANOVA, Kruskal-Wallis has a single dependent and a single independent variable. As is typical in one-way nonparametric ANOVA, the independent variable in the Case of Performance in Three Instructional Programs is a nominal-level grouping variable with three groups: computer-based instruction, in-person instruction, and a combination of both types of instruction. The dependent variable, math performance, is measured at the ordinal level, performance levels 1 through 6.

12.3.2.2 Research Question and Hypotheses

Kruskal-Wallis seeks to determine whether there is a significant difference among score distributions of three groups. The hypotheses for the case are:

Null Hypothesis
There is no significant difference in the median mathematics performance of students in the three pedagogical groups.

Alternative Hypothesis
There is a significant difference in the median mathematics performance of at least two pedagogical groups of students.

12.3.2.3 Study Sample

One advantage of Kruskal-Wallis is that it may be used with groups of equal or different sample sizes. However, according to Sawilowsky (1990), Type I error rates rise sharply when unequal group sizes are accompanied by unequal variances across groups. Kruskal-Wallis is best used when groupings have the same distributional shape even when group sizes are small. In this case, group sizes are equal at 10 cases per group. If group sizes were less than five subjects per cell, the exact probability test would be required.

12.3.2.4 Research Design

Kruskal-Wallis is equivalent to one-way ANOVA, and it is used in research designs similar to those described in Chapter 7. Subjects are in groups, matched based on background factors, so they were neither randomly selected nor assigned. Furthermore, a pretest was not administered. Therefore, the study uses a nonrandomized, pre-experimental, one-shot causal comparative research design.

12.3.2.5 Data Analysis

The Kruskal-Wallis H test is essentially the result of applying one-way ANOVA to score rankings. Dependent variable scores are ranked before analysis. An enduring reason for using ranks to run Kruskal-Wallis is that ranks can eliminate or reduce the effects of outlier values. Imagine that the two highest ranks are 24 and 25, but the two highest raw scores are 78 and 150. Ordinarily the score of 150 might be an outlier. However, the rank of 25 which represents the score of 150 is unlikely to be tagged an outlier.

Box 12.1 Research Plan for the Case of Performance in Three Instructional Programs

Research Problem
Researchers have found that pedagogical methods affect student achievement differently. With students' mathematics performance lagging in CSEC, a researcher desires to investigate whether a significant pattern of performance exists in students exposed to one of three different pedagogical methods: (a) in-person, teacher-directed instruction (Group 1); (b) computer-assisted instruction (Group 2); and (c) a combination of computer-assisted and teacher-directed instruction (group 3). Students are demographically and achievement level comparable before the study.

Variables and Measurement Levels
Variable 1: Mathematics Performance Level
Level of Measurement: Ordinal
Variable 2: Instructional Method
Level of Measurement: Nominal

Research Question
Is there a statistically significant difference in the median mathematics performance of CSEC examinees taught in three different types of pedagogical groups?

Hypotheses: Two-Tailed Test of Significance
Assuming all population distributions have the same shape (normal or not), hypotheses are:

Null Hypothesis: There is no significant difference in the median mathematics performance of students in three different pedagogical groups.
$H_0: \tilde{\mu}_1 = \tilde{\mu}_2 \ldots = \tilde{\mu}_k$, where $\tilde{\mu}$ = group median and k = number of groups

Alternative Hypothesis: There is a significant difference in the median mathematics performance of students in at least two pedagogical groups of students.
$H_a: \tilde{\mu}_i \neq \tilde{\mu}_j$, where $\tilde{\mu}$ = group median and i and j = any pair of groups

Sample Size
Total sample(N) = 30
Each Group = 10

Research Design
Kruskal-Wallis is typically used in a causal comparative design in which multiple groups are compared. The groups will be compared using their median performance in mathematics.

M X_1 O
M X_2 O
M X_3 O

Data Analysis
Descriptive statistics as well as tests of assumptions for the Kruskal-Wallis precede the central analysis. Next, with the dependent variable measured at the ordinal level, the Kruskal-Wallis (H test) statistical procedure will be used to determine whether differences among the group are significant at $\alpha = .05$. If the outcome of the omnibus test reveals significant differences, pairwise comparison tests, using Dunn's test, will identify which group pairs are significantly different.

Section 12.2.5 explored the general procedure for ranking scores before conducting a nonparametric test. The ranking method used changes somewhat from one nonparametric approach to another. For example, when ranking scores for Kruskal-Wallis, the practice is to rank all scores without regard to group membership.

Table 12.2 Mathematics Scores and Ranks for Students in Three Instructional Programs

1: In-Person			2: Computer-Based			3: Hybrid		
ID	Score	Rank	ID	Score	Rank	ID	Score	Rank
01	6	29.5	11	4	19	21	5	25
02	5	25	12	3	13	22	3	13
03	5	25	13	3	13	23	3	13
04	5	25	14	3	13	24	2	6
05	5	25	15	2	6	25	3	13
06	4	19	16	2	6	26	2	6
07	6	29.5	17	1	1.5	27	5	25
08	5	25	18	2	6	28	4	19
09	3	13	19	4	19	29	2	6
10	4	19	20	2	6	30	1	1.5
Sum	48	235	Sum	26	102.5	Sum	30	127.5

Consider the Case of Performance in Three Instructional Programs. Scores and associated ranks are displayed in Table 12.2. Note that the ranking process cuts across groups. For example, individuals in Program 2 and Program 3 have scores of 1, which means they are tied ranks. Therefore, as the ranking practice requires, one score of 1 is ranked 1 and the other is ranked 2. The average of the ranks 1 + 2 = 3/2 = 1.5. Therefore, each of the two scores of 1 is assigned a rank of 1.5. The next available rank is 3, but there are seven occurrences of the absolute value of 2—four in Program 2 and three in Program 3. These scores of 2 are ranked 3 + 4 + 5 + 6 + 7 + 8 + 9 = 42/7 = 6. Therefore, each score of 2 is ranked 6, and so on.

Should the Kruskal-Wallis omnibus test find significant differences among median group scores, pairwise comparison tests, using Dunn's test (with Bonferroni correction) will identify which group pairs are significantly different. Other options for post-hoc tests include Mann-Whitney test with Bonferroni correction or the Conover-Iman test.

12.3.3 Calculations for Kruskal-Wallis

Among the advantages of Kruskal-Wallis is the relative ease with which the H statistic can be calculated, especially when small sample sizes are involved. The following demonstration of hand-calculating the Kruskal-Wallis H statistic uses the fictitious dataset in Table 12.2. The scores range from 1 – 6, so there are multiple tied scores/ranks in the dataset. However, each case is unique and assigned to a single group.

12.3.3.1 Test Statistic Computational Procedures

The computational formula for the Kruskal-Wallis H test is as follows:

$$H = \frac{12}{N(N+1)} * \left(\frac{\sum R_j^2}{n_j}\right) - 3(N+1)$$

where

n_j = the number of participants in the jth group.
R_j = the number of ranks in the jth group
$N = \sum n_j$ = total sample size

Consequently, the Kruskal-Wallis formula means that computing the statistic (H) involves the following steps:

1. Divide the value, 12, by the product of the sample size (N) and the sample size plus 1 (12/(N (N + 1)).
2. Find the sum of the ranks for each group (the last row of Table 12.2) and square it (R_j^2). Note that if the sum of the ranks is different across groups, the differences are likely to be significant.
3. Divide the squared sum of each group (Step 2) by the number of participants (10) in that group R_j^2/n_j.
4. Add together the results of the computations in Step 3 ($\sum R_j^2/n_j$)
5. Add 1 to the sample size and multiply that by 3 (3(N + 1))
6. Multiply the result of the computation in Step 1 by the results of Step 4.
7. Subtract the result of Step 5 from the result of Step 6.

Using the above instructions, find the H statistic as follows:

$$H = \frac{12}{30(30+1)} * \left(\frac{235^2}{10} + \frac{102.5^2}{10} + \frac{127^2}{10}\right) - 3(30+1)$$

$$H = \left(\frac{12}{930}\right) * (5{,}522.5 + 1{,}050.625 + 1{,}625.625) - 93$$

$$H = \left(\frac{12}{930}\right) * (8{,}198.75) - 93$$

$$H = 0.013 * 8{,}198.75 - 93$$

$$H = 105.790 - 93$$

$$H = 12.79$$

Although ranks were used in computing the H statistic, note that H is unadjusted for the number of ties in the dataset.

12.3.3.2 Evaluating the Significance of the H Statistic

As stated earlier, the Kruskal-Wallis H statistic is treated as if it were a chi-square statistic. As a chi-square statistic, compare the value of H against the appropriate chi-square critical value for k − 1 (2 degrees of freedom) at an alpha level of .05. According to the chi-square statistical table in Appendix A, the critical value for 2 df at a .05 alpha level is 5.99. The H statistic of 12.790 exceeds the critical value of 5.99, resulting in a rejection of the null hypothesis that states there is no significant difference among the scores of the students in the three types of instructional groups.

12.3.3.3 Effect Size

In Chapter 7, the effect size was computed for the one-way ANOVA example using eta squared (η^2). For this exercise, use the same statistic to calculate the effect size for Kruskal-Wallis to measure the magnitude of the effect being studied. Use the following formula:

$$\eta_H^2 = \frac{(H - k + 1)}{(N - k)}$$

where
H = the value obtained in the Kruskal-Wallis test;
k = the number of groups;
N = the total sample size.

Like ANOVA, eta-squared estimated values range from 0 to 1. Eta-squared multiplied by 100 specifies the percentage of variance in the dependent variable explained by the independent variable. As eta square increases, the percentage of the variance in the dependent variable explained by the independent variable also increases. Guidelines for interpreting eta-squared values for the H statistic are as follows:

Effect Size	Range
Small	.01 - <.06
Moderate	.06 - <.14
Large	≥ .41

To compute the effect size for H, substitute the values for H, k and N into the formula:

$$\eta_H^2 = \frac{12.79 - 3 + 1}{30 - 3}$$

$$\eta_H^2 = \frac{8.79}{27}$$

$$\eta_H^2 = .33$$

Therefore, instructional approaches in the study account for 33% of the variance in the dependent variable, students' math performance level.

12.3.4 Assumptions of Kruskal-Wallis

Although Kruskal-Wallis, like other nonparametric tests, is commonly called the distribution-free option for one-way ANOVA, there are some assumptions that are prerequisites for the test. If the assumptions are violated, accurate interpretations of test results falter. There are five main assumptions summarized below. The first four are incorporated in the design of research studies. The remaining assumption may be evaluated using statistical software.

- **Dependent Variable Measurement Level**. The dependent variable is measured at the ordinal level at a minimum. Examples of ordinal level variables include Likert response formats like agreement or satisfaction ratings, educational level, and income levels. Kruskal-Wallis is also an equivalent statistical method when data are measured at the interval or ratio level but violate other assumptions (e.g., normal distribution or homogeneity of variance) required by parametric one-way ANOVA.
- **Independent Variable Measurement Level**. The independent variable is limited to a single, categorical variable that consists of two or more independent groups. Kruskal-Wallis may be

successfully used to compare two groups that comprise an independent variable. However, it was designed specifically for three or more groups. The Mann-Whitney test is the preferred method when an independent variable comprises two groups.

- **Independence of Observations**. The independence of observations assumption requires that independent variable groups must be mutually exclusive. Accordingly, there should be no relationship between the data values in each group or among the groups. In effect, study participants should not be in more than one group.
- **Sample Size**. One advantage of Kruskal-Wallis over parametric one-way ANOVA is that it analyzes small-sample data with greater accuracy than parametric one-way ANOVA. Where minimum group sizes for parametric one-way ANOVA is 15, the assumption for Kruskal-Wallis is that groups have a minimum of five observations each. In group sizes of fewer than five observations, the exact significance level is computed.
- **Shape of Distribution of Scores**. Kruskal-Wallis assumes that the distributions of group scores have the same shape otherwise known as the same variance. If your distributions have the same shape, you can use Kruskal-Wallis to compare the medians of the dependent variable for the groups you wish to compare. If the shapes of the distributions are different, you may use Kruskal-Wallis to compare the mean ranks of the groups to determine whether they are the same.

12.3.5 After the Omnibus Test

Like parametric one-way ANOVA, the Kruskal-Wallis significance test discloses whether significant differences in one or more pairs of the independent variable groups exist. However, the test does not reveal which groups are significantly different. To identify the pair-wise differences, multiple pair-wise testing must be conducted. To avoid the inflation of a Type I error rate of the a priori alpha level, one of two approaches is adopted:

- all pairwise multiple comparison tests, and
- the stepwise step-down procedure

12.3.5.1 All Pairwise Multiple Comparison Procedure

For Kruskal-Wallis, conducting all pairwise comparisons of the groups using Dunn's test of significance is one way of identifying the source of difference found in the omnibus test. Dunn's test is a nonparametric pairwise multiple comparison procedure that is commonly used after Kruskal-Wallis is found to be significant. The purpose of the test is to automatically conduct all pairwise group comparisons to determine whether there is a significant difference between any of the group pairs, but without inflating the a priori alpha level which could result in a finding that groups are significantly different when they are not—a Type I error. In order to prevent alpha level inflation and falsely rejecting the null hypothesis, Dunn (1961) used the Bonferroni correction to reduce the significance level (p-value) at which a null hypothesis may be rejected. This leaves the a priori alpha level intact while preventing a Type I error.

As an example of how this works, consider a study in which all possible pairs of the mean scores of three groups are compared to determine whether any is significantly different. In the all pairwise multiple comparison procedure, three comparisons would take place:

Group 1 and Group 2   Group 1 and Group 3   Group 2 and Group 3

Each comparison is completed independent of the other two. Three comparisons with a pre-set alpha level of .05, would result in an actual alpha of 15% (0.05 + 0.05 + 0.05) and the likelihood that one of the comparisons will result in a false positive. Dunn's adjustment corrects for the alpha inflation by dividing the pre-set alpha by the number of intended comparisons. Therefore, .05/3 corresponds to 0.017 for each comparison and an overall alpha level of .05.

There have been several modifications of the Bonferroni adjustment over the years (e.g., Sidak, 1967; Holland & Copenhaver, 1988). Nevertheless, Dunn's Bonferroni remains the most commonly used post-ANOVA test. Dunn's Bonferroni is not only convenient in that most statistical software applications will complete all-pairwise comparisons automatically upon a significant finding, but also the conservative nature of the test statistic effectively reduces the likelihood of a false positive.

12.3.5.2 Stepwise Step-down Procedure

Nussbaum (2015) argues that while the use of all pairwise comparison tests are useful and convenient when the number of comparisons is relatively low, it becomes unnecessary and overly conservative when there are more than, say, three comparisons. Other researchers note that the Bonferroni correction is unnecessary when the intent is to conduct one or two a priori planned comparisons. In these instances, the stepwise step-down procedure is a more powerful alternative for conducting planned comparisons, while preserving the a priori alpha level.

Contrary to all pairwise multiple comparison tests, stepwise step-down tests are conducted in a predetermined order and subsequent comparisons are performed only when the previous comparison proves to be significant. In stepwise step-down tests, the two groups most likely to be significant (those with the largest mean rank differences) are tested first using a nonparametric test like Mann-Whitney *U*. Successive comparisons are performed only when the previous comparison proves to be significantly different. This method acquired its name because the extent of the differences between the group pairs declines as the comparisons proceed.

Consider a study in which there are four groups of participants and, therefore, six comparison groups for post hoc testing:

| Group 1 and Group 2 | Group 1 and Group 3 | Group 1 and Group 4 |
| Group 2 and Group 3 | Group 2 and Group 4 | Group 3 and Group 4 |

Imagine that the two groups with the largest difference in mean rank are Group 1 and Group 4, and the groups prove to be significantly different from each other. The pair with the next largest mean difference in rank are Group 1 and Group 3. Testing that pair for significant difference in rank results in a nonsignificant difference. This outcome means that none of the remaining tests could be significantly different because they all have smaller discrepancies in mean rank. Hence, the process stops at the second pairwise test. As a result of the predetermined order in which groups are compared, fewer comparisons are needed, and Bonferroni corrections can be less restrictive.

12.3.6 SPSS Procedures for the Kruskal-Wallis H Test

This demonstration exercise will utilize data from the Case of Performance in Three Instructional Programs and the associated data in Table 12.2. SPSS will be employed to illustrate how to complete the Kruskal-Wallis *H* test. The data file for this exercise is **Math_Instr.sav**.

To recap, the independent variable in this exercise is Instructional Method (Instruction) which comprises three groups each with a different type of instruction: 1 = In-Person, 2 = Computer-Assisted, 3 = Hybrid. The dependent variable is Performance Score (Score), which is an ordinal variable with values of 1 - 6. Kruskal-Wallis is the selected statistical procedure for this case due to the ordinal measurement level of the dependent variable and the small sample size.

A previously conducted Kolmogorov-Smirnov Test (Figure 12.5) reported a significant nonnormal distribution of the dependent variable for In-Person (instruction) ($p = .02$), but the Shapiro-Wilk test reported a normal distribution. Therefore, nonnormality may not be a sufficient concern to choose Kruskal-Wallis. Nevertheless, it will be used for demonstration purposes.

Figure 12.5 Results of Tests of Normality

	Instructional Method	Kolmogorov-Smirnov[a]			Shapiro-Wilk		
		Statistic	df	Sig.	Statistic	df	Sig.
Performance	In-Person	.286	10	.020	.885	10	.149
	Computer-Based	.233	10	.133	.904	10	.245
	Hybrid	.200	10	.200*	.918	10	.344

* This is a lower bound of the true significance.
a Lilliefors Significance Correction

12.3.6.1 SPSS Specifications for the Kruskal-Wallis H Test

To complete the Kruskal-Wallis *H* test, navigate to the location of the data file to open the dataset in the **SPSS Statistics Data Editor** (Figure 12.6). Confirm that Instruction (Instructional Method) is a nominal level variable and Score is at the ordinal level.

Figure 12.6 SPSS Data Editor

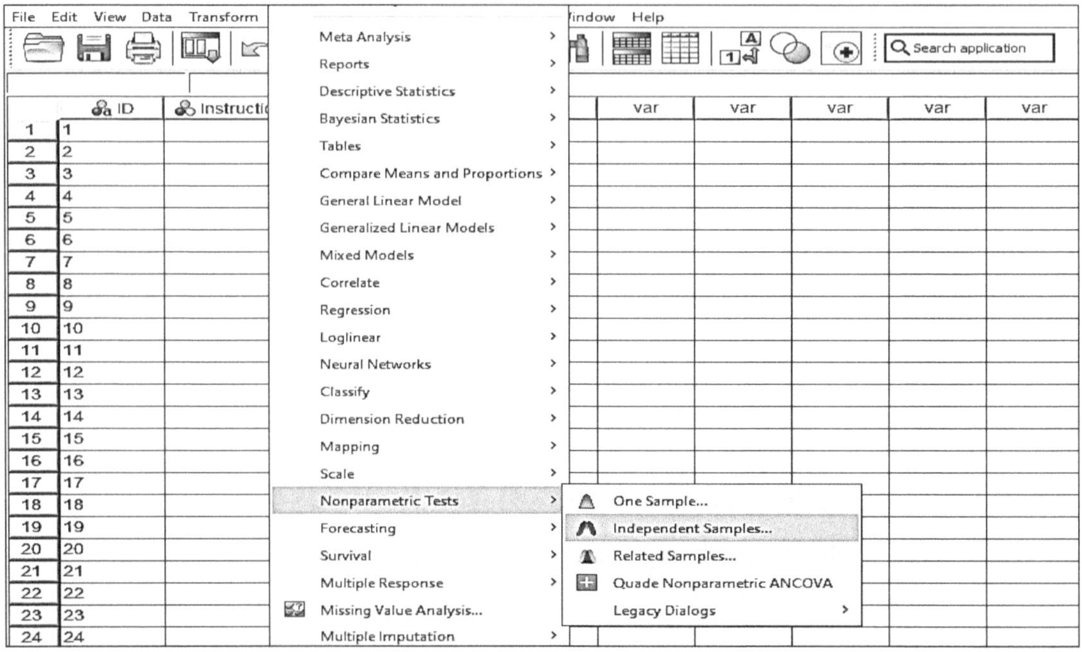

SPSS provides two procedures for conducting the Kruskal-Wallis *H* test:
1. The **Legacy Dialogs** → **K Independent Samples** procedure
2. The **Nonparametric Tests** → **Independent Samples** procedure.

Use the **Nonparametric Tests** → **Independent Samples** procedure for this exercise; it is a more efficient test than the Legacy Dialogs option in that it automatically runs associated post hoc tests. However, the **Legacy Dialogs** procedure is available for earlier versions of SPSS Statistics.

Once the dataset is in the **Data Editor**, click on **Analyze** → **Nonparametric Tests** → **Independent Samples**. In the **Nonparametric Tests: Two or More Independent Samples** dialog box, select the **Objective** tab, then below "**What is Your Objective?**" retain the default selection, **Automatically compare distributions across groups** (Figure 12.7).

Figure 12.7 Kruskal-Wallis Dialog Box: Objective Tab

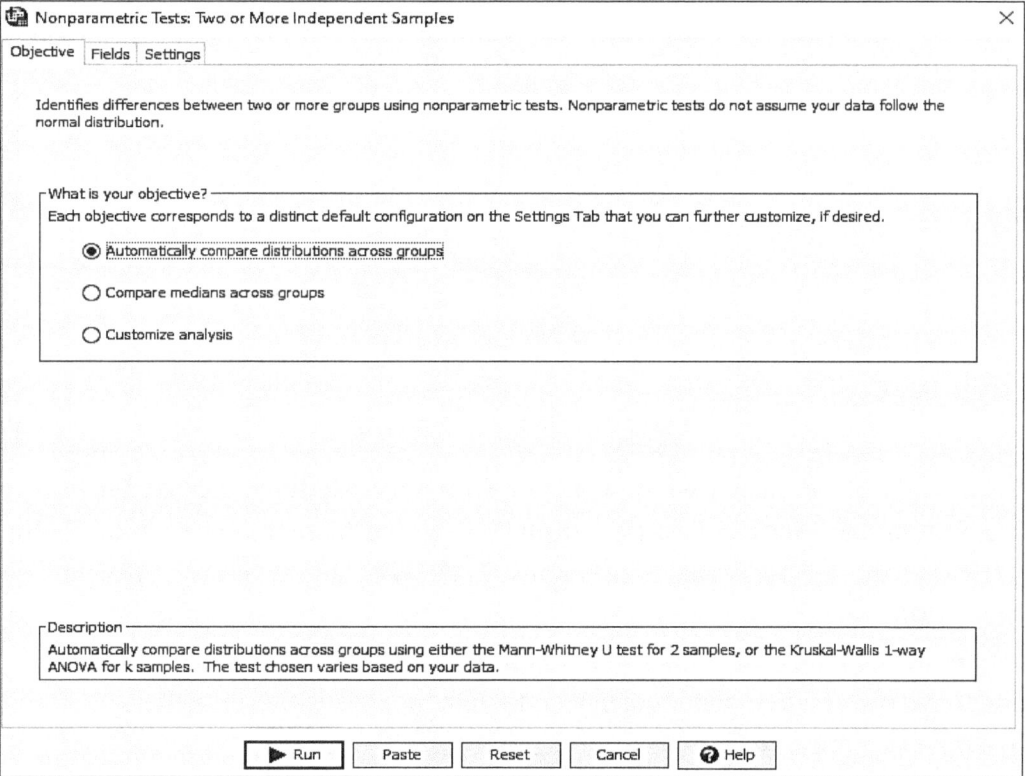

Next, select the **Fields** tab (Figure 12.8). With the selection of **Use predefined roles**, SPSS can select a test for users, but for more options select **Use custom field assignments**. Highlight the dependent variable, **Score**, and click on the arrow to move **Score** to the **Test Fields** panel on the right. Click over the independent variable, **Instruction**, to the **Groups** panel that is below the **Test Fields** panel (Figure 12.8).

Select the **Settings** tab (Figure 12.9). The first option, **Automatically choose the tests based on the data,** allows SPSS to select a significance test. The second option, **Customize tests,** provides the opportunity to customize options. Choose the latter. Below **Compare Distributions across Groups**, select **Kruskal-Wallis 1-way ANOVA (k samples)**. Below that test selection are options for follow-up analyses, should the omnibus test prove to be significant. Click on the down-arrow next to **Multiple Comparisons** to reveal three follow-up analyses to choose from:

1. None (no multiple comparisons)
2. All pairwise, and
3. Stepwise step-down.

Select **All Pairwise** to compare every possible group combination.

Figure 12.8 Kruskal-Wallis Dialog Box: Fields Tab

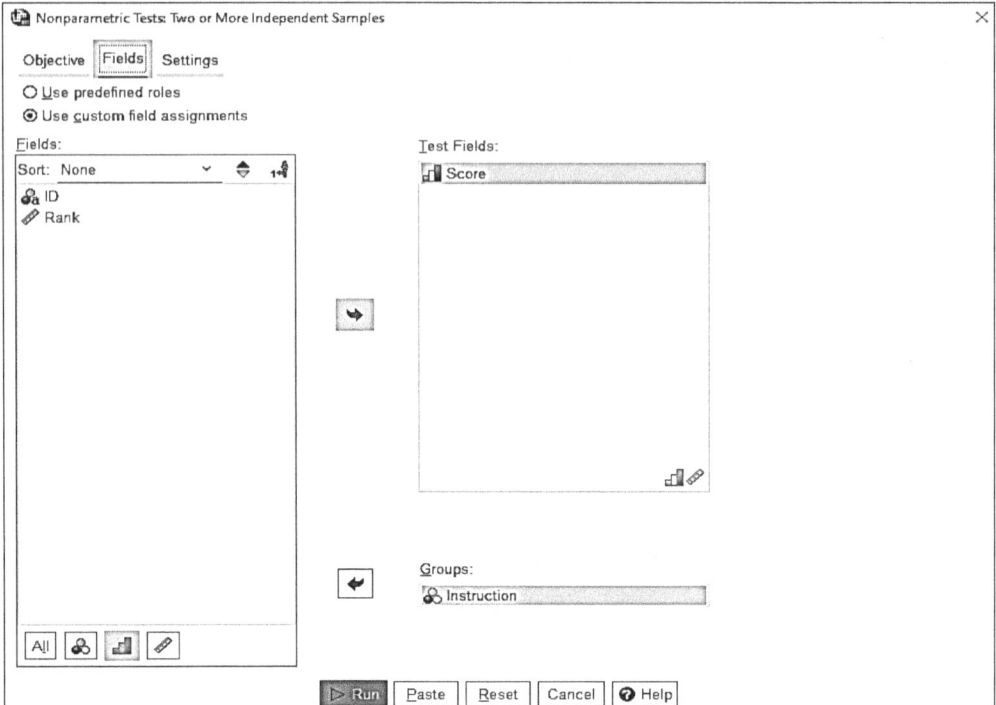

In the lower right of the dialog window, below **Compare Medians Across Groups**, is another omnibus test option, the **Median Test (k samples)**. Both the Median and Kruskal-Wallis tests are nonparametric tests that are equivalent to the one-way ANOVA. However, there are subtle differences.

The **Kruskal-Wallis test**
- investigates actual ranks (converts scores to ranks), and
- is a more powerful test than the Median test, though more sensitive to outliers.

The Median Test (k samples)
- restricts its analysis to whether dependent variable values are above or equal to/below the median. If there are fewer than five cases in more than 20% of the cells, the Median test is likely to generate biased results.
- is more robust to outliers than the Kruskal-Wallis test because it has no equal variance assumption.

Since more than 20% of the cells have fewer than five cases each, retain **Kruskal-Wallis 1-way ANOVA (k samples)** to complete the analysis. Click **Run** to conduct the procedure.

12.3.6.2 Results of the Omnibus Kruskal-Wallis H Test

Figure 12.10 displays the results of the Kruskal-Wallis Test. The decision (noted in the Decision column) is to reject the null hypothesis; the significance levels of .001, is below the a priori alpha level of .05 (Footnote a). This omnibus test result leads to the conclusion that significant differences in score distributions exist among the groups.

Figure 12.9 Kruskal-Wallis Dialog Box: Settings Tab

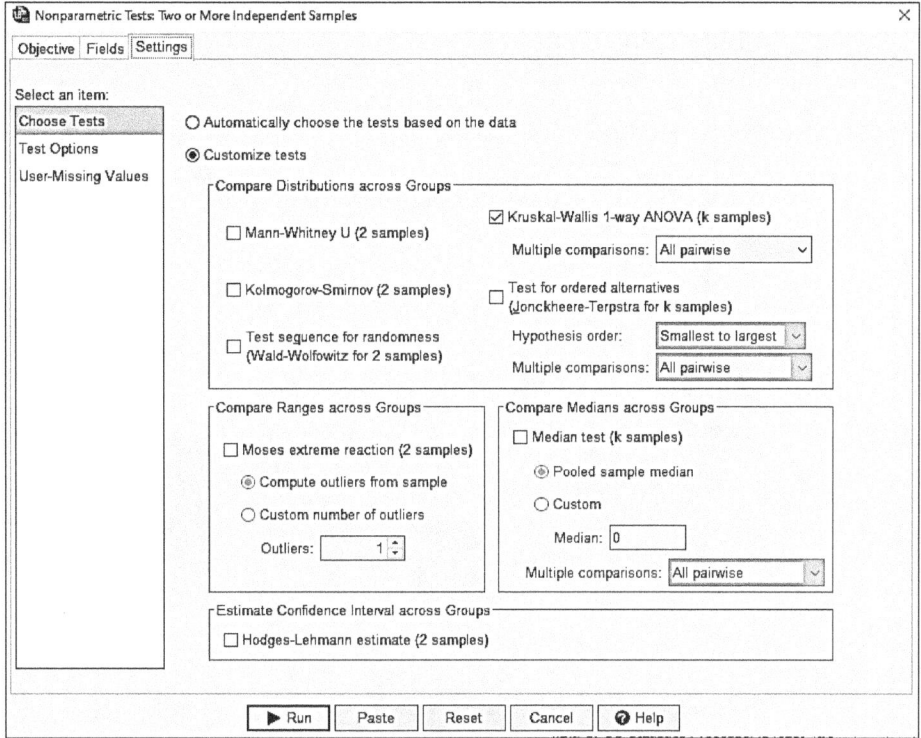

Figure 12.10 Kruskal-Wallis Hypothesis Test Summary

	Hypothesis Test Summary			
	Null Hypothesis	Test	Sig.[a,b]	Decision
1	The distribution of Score is the same across categories of Instruction Method.	Independent-Samples Kruskal-Wallis Test	.001	Reject the null hypothesis.

a. The significance level is .050.
b. Asymptotic significance is displayed.

Figure 12.11 provides the same summary of significant differences among the groups as Figure 12.10, but with a bit more detail. The test statistic, H, of 13.355 is slightly higher than the H statistic computed by hand ($H = 12.790$. The results suggest that there is a significant difference in the scores of students receiving different methods of instruction.

Figure 12.11 Independent-Samples Test Summary

Independent-Samples Kruskal-Wallis Test Summary	
Total N	30
Test Statistic	13.355[a]
Degree Of Freedom	2
Asymptotic Sig.(2-sided test)	.001

a. The test statistic is adjusted for ties.

Although the omnibus test reports that performance associated with the three instructional methods are significantly different, there is no evidence of which pairs of groups differ. Figure 12.12 presents a graphical display of the scores of the three groups.

Figure 12.12 Boxplot of Instructional Method Group Scores

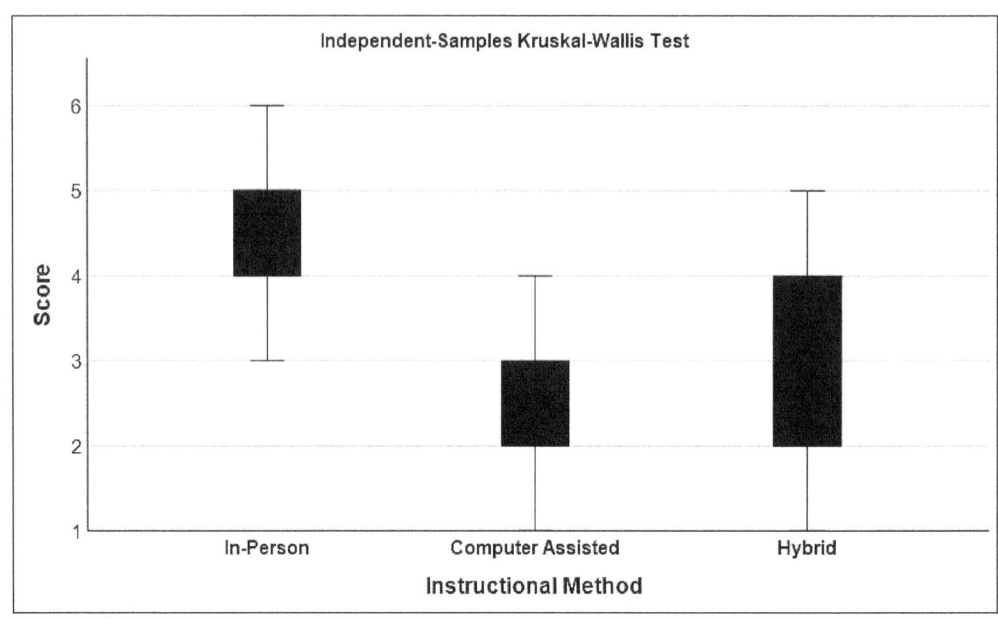

Visual examination of the boxplots, suggest that the medians of the hybrid and computer-assisted groups are relatively similar. However, the median of the in-person instructional method group appears to noticeably exceed the medians of the other two groups. Follow-up analyses provide additional details about the differences observed in the boxplots.

12.3.6.3 Follow-Up Data Analyses

Figure 12.13 offers the results of pairwise comparisons of the three groups to determine whether any of the group pairs have significantly different scores. Column 1 describes all possible combinations of the comparison groups: computer-based (**Comp-Based**) vs. hybrid (**Hybrid**), computer-based (**Comp-Based**) vs. in-person (**In-Person**), and hybrid (**Hybrid**) vs. in-person (**In-Person**). Column 2 presents the test statistic resulting from computing the difference between each pair of groups. Each test statistic is converted into a standardized test statistic or a z score by dividing the test statistic by its standard error in Column 3 (**Std. Error**). The significance level for each z score appears in the **Sig** column, and the associated adjusted Bonferroni-corrected significance level, to prevent alpha level inflation due to comparison of multiple tests, appears in the last column (**Adj. Sig.**).

As the results demonstrate, the scores of two paired comparisons show significant differences: computer-based vs. in-person ($p = .002$) and hybrid vs. in-person ($p = .016$). There is no significant difference between the scores of the computer-based and hybrid groups.

Figure 12.14 is a diagrammatic illustration of the results, which shows the average rank of the scores achieved by each group. The groups that are significantly different are signified by the black lines connecting **In-Person** and the other two groups. Why is there a difference? Students in the **In-Person** course have much higher scores than students in the other two groups, while performance of students in the **Hybrid** and **Computer-Assisted** instructional methods groups record performances that are relatively similar.

Figure 12.13 Follow-Up Pairwise Comparisons

Pairwise Comparisons of Instructional Method					
Sample 1-Sample 2	Test Statistic	Std. Error	Std. Test Statistic	Sig.	Adj. Sig.[a]
Comp-Based-Hybrid	-2.500	3.853	-.649	.516	1.000
Comp-Based-In-Person	13.250	3.853	3.439	<.001	.002
Hybrid-In-Person	10.750	3.853	2.790	.005	.016

Each row tests the null hypothesis that the Sample 1 and Sample 2 distributions are the same.
Asymptotic significances (2-sided tests) are displayed. The significance level is .050.
a. Significance values have been adjusted by the Bonferroni correction for multiple tests.

Figure 12.14 Diagram of Paired Comparisons Analysis

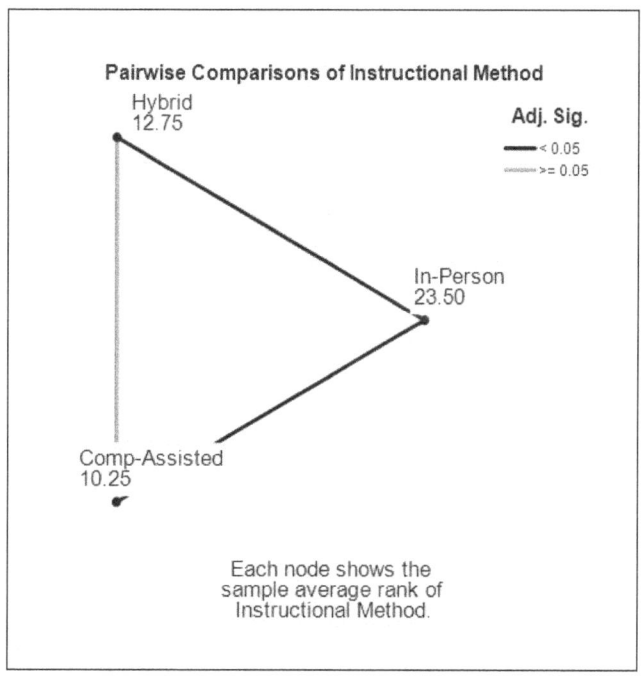

12.3.7 Sample Report: The Case of Performance in Three Instructional Programs

A Kruskal-Wallis test was conducted to determine whether there were significant differences in the mathematics performance of students exposed to one of three different instructional methods:
- In-person instruction
- Computer-assisted instruction, and
- A hybrid of in-person and computer-assisted.

Data assumptions of the Kruskal-Wallis test were evaluated before proceeding to the omnibus test. A statistically significant difference was found, $\chi^2_H(2, 30) = 13.355, p = .001$, signifying that there were significant differences among the performance of students in the three instructional methods groups.

Effect size using eta² (η^2) showed that instructional approaches in the study account for 33% of the variance in students' mathematics performance level.

The Dunn-Bonferroni test, which corrects for multiple test comparisons, was used for all pairwise comparisons of the performance in the three groups to determine which pairs differed significantly. Results revealed significant differences between in-person and computer assisted groups ($z = 3.439$, $p = .002$) as well as between in-person and hybrid groups ($z = 2.790$, $p = .016$). No significant difference was found between hybrid and computer-assisted groups.

12.4 Nonparametric Correlation: Spearman's Rho and Kendall's Tau

12.4.1 The Case of Hypertension in Portland

The following data are a random sample of blood pressure measures taken on a medical service trip to Jamaica in 2019. Using a medical mobile clinic system, health professionals provided walk-in medical screening and primary care to residents in the parish of Portland in Jamaica. More than 300 individuals received service on the first of the five-day mission. A random sample of 16 of those patients is represented in Table 12.3. A research plan outline appears in Box 12.2. The dataset resides in the data file, **Hypertension.sav**.

Table 12.3 Age, Systolic and Diastolic Blood Pressure of Clinic Patients in Portland

ID	Systolic	Diastolic	Age	ID	Systolic	Diastolic	Age
30348	106	74	51	30117	128	84	57
30232	108	64	27	30061	151	81	82
30049	118	58	67	30019	155	80	59
30046	121	74	61	30264	164	92	46
30161	122	84	35	30173	168	89	72
30102	124	80	22	30120	184	99	69
30045	124	62	84	30275	195	88	79
30336	126	77	63	30028	221	108	86

12.4.2 The Case of Trust and the Importance of Religion

The World Values Survey Association (WVSA) is an international group devoted to the study of the values of people around the world. The primary survey instrument is the World Values Survey (WVS) that is administered every five years in countries around the world. Among questions that address social, political, economic, religious, and cultural values are six items that address the question of how much respondents trust people from the following groups on a scale of 1 (Do Not Trust at All) to 4 (Trust Completely):

- their families
- their neighborhoods
- people they know
- people they meet for the first time (FM)

- people of another religion, and (AR)
- people of another nationality (AN)

Delhey et al. (2011) describe the first three items as measures of trust of people that respondents know personally (in-group) and the latter three as measures of trust of those outside of a person's immediate group (out-group or general trust). The purpose of this case is to analyze the association among the out-group trust items and the variable, Importance of Religion (RI), rated on a scale of 1 (Not at All Important) to 4 (Very Important), using a small random sample of the Trinidad and Tobago WVS dataset (Table 12.4).

Table 12.4 Trinidad and Tobago's Responses to World Values Survey Trust Items

ID	RI	Trust			ID	RI	Trust		
		FM	AR	AN			FM	AR	AN
001	2	1	3	1	016	4	2	1	1
002	4	2	3	3	017	3	2	3	3
003	3	2	2	2	018	4	1	3	3
004	1	1	1	1	019	4	3	3	3
005	3	1	2	1	020	3	1	2	2
006	4	3	3	3	021	4	2	3	1
007	4	3	3	3	022	4	3	1	3
008	3	1	3	3	023	4	1	3	3
009	4	2	3	3	024	4	1	4	1
010	4	2	2	2	025	4	2	4	4
011	4	3	3	3	026	4	2	3	2
012	4	2	3	3	027	3	1	2	2
013	4	1	1	1	028	4	2	3	3
014		2	2	2	029	4	2	3	3
015	4	3	3	3	030	4	3	3	1

The data are from the sixth round of the survey (Inglehart et al., 2018). The dataset appears in Table 12.4 and in the datafile **WVS_Trust.sav**. The research plan outline for a study like this one appears in Box 12.3.

12.4.3 Nonparametric Correlation Fundamentals

The Pearson product moment correlation (also known as Pearson's r) is one of the most commonly-used correlation coefficients. However, Pearson's r, like most parametric statistics, is based on data assumptions that must be met in order to accurately portray test results. For example, Pearson's r assumes that the data analyzed are (a) continuous (interval or ratio), (b) normally distributed, and (c) represent a linear relationship between the values of two variables.

Pearson's r is robust to violations of data assumptions. Nevertheless, there are alternatives with fewer restrictive assumptions. This section will discuss two alternatives for Pearson's r: (a) Spearman's rho, also known as Spearman rank (ρ) and (b) Kendall's tau (τ), also known as Kendall's rank correlation. Both statistical tests are nonparametric correlations, and neither have assumptions of linearity or normality required by Pearson's *r*. Both are designed to analyze ordinal or continuous data and are special cases of a more general correlation coefficient.

Spearman's rho and Kendall's tau are especially useful in analyzing ordinal data, which include measurements on Likert scales (e.g., a 7-point happiness scale that ranges from score point 1 that means "extremely unhappy" to score point 7, meaning "extremely happy"). Another example of ordinal data is ordered categories such as a 4-point educational level ranking scale ranging from "elementary" to "graduate level." Ordinal data may also be scores on a rubric (e.g., a 7-point scale used to evaluate performance on a persuasive essay examination).

Pearson's *r* and the nonparametric alternatives share some objectives in common.

- They measure the extent to which two variables change together. For example, they measure the extent to which values of two variables increase or decrease concomitantly.
- They evaluate the extent of the bivariate relationship by virtue of the direction of the association between the variables (positive, negative, or no relationship).
- They quantify the strength of the relationship on a scale of -1 to 1, where the closer the coefficient is to -1 or 1 the stronger the relationship (negative or positive) between the two variables involved. The closer the coefficient is to 0, the weaker the relationship.
- The correlation coefficients may be subjected to significance testing to determine the likelihood that the bivariate association exists in the target population.
- All three correlation coefficients measure the association between two variables. In most cases, neither has the role of predictor or outcome variable. If one variable is an independent variable and the other a dependent variable, a different statistical procedure, such as bivariate linear regression, must be used to measure that relationship.

All three correlation coefficients share important properties. However, they differ in two fundamental respects: (a) the nature of the association that they measure, and (b) the procedure used for assessing the relationship.

12.4.3.1 Is the Bivariate Relationship Monotonic?

Pearson's *r* measures the linear association between two continuous variables. For example, researchers might use Pearson's *r* to evaluate the relationship between weather (temperature) and ice cream sales. One does not cause the other, but as one changes, the other changes proportionately; as the temperature increases, more people tend to purchase ice cream as one delicious method of cooling down.

Like Pearson's *r*, Spearman's rho and Kendall's tau also estimate the association between two variables, but the relationships assessed are not necessarily linear. While both of these nonparametric procedures can successfully analyze linear data, they also evaluate the association between variables represented by nonlinear data.

When nonlinear data points are plotted, the directions of the association might look like the scatterplots illustrated in Figure 12.15. The first two plots are of monotonic associations. In a monotonic relationship, one of the following happens: (a) As the value of X increases, the value of Y increases, and (b) as the value of X increases, the value of Y decreases. The third is a curvilinear relationship, but it is not monotonic because the values of the two variables do not always change in the same direction; as the value of X increases the value of Y may decrease or increase.

Figure 12.15 Examples of Nonlinear Scatterplots

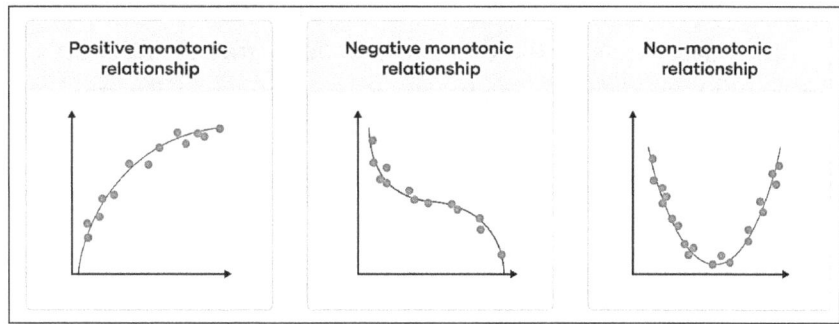

In a linear relationship, the variables move in the same direction at a constant rate. In a monotonic relationship, the two variables tend to move in the same relative direction, but not necessarily at a constant rate. Therefore, linear relationships are monotonic, but monotonic relationships are not always linear.

To determine whether a bivariate relationship is nonlinear monotonic or linear monotonic, generate a scatterplot of the data. If it is linear, a Pearson's r correlation may be used to establish the relationship between the variables. If the scatterplot is nonlinear monotonic, use Spearman's rho or Kendall's tau to assess the relationship.

12.4.3.2 Assigning Ranks

For Pearson's r, raw data are analyzed without further modification. However, for nonparametric correlation, the data must be ranked before analysis. The ranking may be completed by hand or specified in SPSS.

For nonparametric correlation, the general procedure for ranking is identical to that described in Section 12.2.5 where high numbers are given high ranks and low numbers are given low ranks. When two cases are assigned the same rank, the rank is described as a tied rank. There are different methods of dealing with ranks in the presence of ties, but the most common, called mid-ranks (and the SPSS default), is to find the average of the tied rank and assign that average to each case with the tied rank. For more information on tied ranks, see Section 12.2.5.

There is one aspect of the ranking procedure that differs across nonparametric tests, and that is whether cases are ranked within each group or across groups, without regard to group membership. In Kruskal-Wallis, cases are ranked without regard to group membership. However, in the case of Spearman's rho and Kendall's tau, the values must be ranked separately for each variable. Accordingly, if a researcher plans to examine the relationship between job satisfaction and salary range, or the job satisfaction ratings of males and females, each must be ranked separately.

So far, the exploration of Spearman's rho and Kendall's tau has revealed that the two techniques share some characteristics. However, there are some differences between them as well. In addition, they both have characteristics that are distinct from each other and Pearson's r.

12.5 Spearman's Rho

Spearman's rho (ρ or r_s) was developed by Charles Spearman (1904), an English psychologist known for his work in statistics. It was developed as an alternative to Pearson's r to assess the monotonic association between two variables using ranks of the original values.

Because Spearman's rho works with ranked data, it is not subject to some of the assumptions that are characteristic of Pearson's *r*. Rather, it was designed to analyze nonlinear data, making it robust to outliers and subsequent skewness in the data, which can sometimes distort Pearson's *r*. Hence, Spearman's rho is often used when data fail to meet the assumptions of Pearson's r.

A sample research plan outline for studies requiring Spearman's rho (or Kendall's tau) appears in Box 12.2. The data to be used with Spearman's rho is from the Case of Hypertension in Portland, which appear in Table 12.2.

12.5.1 Research Plan: The Case of Hypertension in Portland

12.5.1.1 Eligible Data Variables and Measurement Levels

Spearman's rho is used with ordinal level data. Rankings may be embedded in the values themselves; for instance, Likert-type answer options are already rankings (e.g., happiness ratings where 1 = unhappy, 2 = somewhat unhappy, 3 = happy, 4 = somewhat happy, and 5 = happy).

Spearman's rho is not restricted to the analysis of ordinal level data, but may be used to analyze variables on the interval or ratio level scale. Accordingly, for the Case of Hypertension in Portland, Spearman's rho will be used to analyze the relationships between the interval/ratio level variables, age, systolic and diastolic blood pressure.

When continuous variables are included, a prerequisite for data analysis is to transform the data before analysis takes place. This requires that the researcher rank the values for each variable separately before analysis. Alternatively, when statistical software is used, the researcher must specify the need for the software to rank the data.

12.5.1.2 Research Questions/Hypotheses

The purpose of the case study is to determine whether there is a monotonic association between two sets of variables. Although three variables are included in this case, each hypothesis is related to a single pair of variables. They are analyzed together for purposes of convenience and efficiency only. Therefore, the research questions for the Case of Hypertension are:

Research Question 1
Is there a bivariate, monotonic association between systolic blood pressure and age?

Research Question 2
Is there a bivariate, monotonic association between diastolic blood pressure and age?

12.5.1.3 Research Design

Spearman's rho is used in correlational nonexperimental research designs when data fail to meet the assumptions of its parametric counterpart, Pearson's *r*. In addition, nonexperimental research designs are frequently selected when researchers cannot manipulate the variables involved in the study. In the Case of Hypertension, researchers cannot manipulate blood pressure or age.

12.5.1.4 Data Analysis

In the first demonstration exercise, Spearman's rho will be used to calculate the association between systolic blood pressure and age. The second describes how to calculate Spearman's rho correlation

Box 12.2 Research Plan Outline: The Case of Hypertension in Portland

Research Problem
The Caribbean Public Health Agency (CARPHA), a regional public health agency for the Caribbean, recently reported that the Caribbean region has the highest prevalence of raised blood pressure in the Americas ranging from a high of 27.1% to a low of 20.9%. Increase in blood pressure is also commonly associated with increase in age (e.g., Ostchega et al., 2020). The data in Table 12.3 are a small sample of blood pressure readings of mobile clinic patients that will be used to begin to examine this phenomenon.

Variables and Measurement Levels
Variable 1: Age
Level of Measurement: Interval/Ratio
Variable 2: Systolic BP
Level of Measurement: Interval/Ratio
Variable 3: Diastolic BP
Level of Measurement: Interval/Ratio

Research Questions
Is there a significant bivariate monotonic association between systolic blood pressure and age?
Is there a significant bivariate monotonic association between diastolic blood pressure and age?

Hypotheses: Two-Tailed Test of Significance (Spearman's rho)
Null Hypothesis 1: There is no significant monotonic association in the population between age and systolic blood pressure.
Null Hypothesis 2: There is no significant monotonic association in the population between age and diastolic blood pressure.
$H_0: \rho_1 = 0$
$H_0: \rho_2 = 0$

Alternative Hypothesis 1: There is a significant monotonic association in the population between age and systolic blood pressure.
Alternative Hypothesis 2: There is a significant monotonic association in the population between age and diastolic blood pressure.
$H_a: \rho_1 \neq 0$
$H_a: \rho_2 \neq 0$

Research Design
Nonexperimental correlational research design

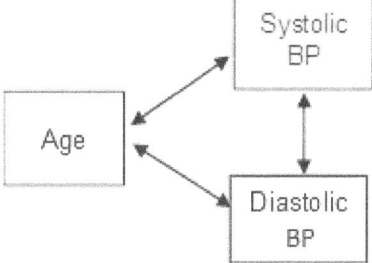

Data Analysis
Descriptive statistics, including means and standard deviation as well as tests of assumptions for nonparametric procedures, are generated first. Preliminary analyses show that the variable values, which are measured on the interval/ratio scale, fail to meet some parametric correlational assumptions. Accordingly, their values will be transformed into ranks, and they will be correlated using Spearman's rho, followed by significance testing to determine whether correlation coefficients are significant at $\alpha = .05$.

between diastolic blood pressure and age. Both could be completed in the same procedure. Nevertheless, for ease of comparisons, the analyses will be completed separately.

The computational formula for Spearman's rho is:

$$\rho = 1 - \frac{6\sum D_i^2}{N(N^2 - 1)}$$

where
N = the number of paired ranks in the data file, and
D_i = the difference between the paired ranks for each subject.

When there are no tied ranks, the formula reduces to Pearson's correlation coefficient (r) applied to the ranks. According to Nussbaum (2015), if there are more than two tied ranks, Kendall's tau (τ) is more suitable for use. In the Case of Hypertension in Portland, there is one tied rank for systolic, three for diastolic, and none for age. Therefore, arguably, Spearman's rho is more suitable for correlating age with systolic, rather than diastolic blood pressure.

The Spearman's rho (ρ) coefficient is used to test the null hypothesis that ρ is not statistically significantly different from zero in the target population. There is no commonly used population parameter for Spearman's rho as there is for Pearson's r. Accordingly, ρ is used as the sample statistic and as the population parameter.

When the sample size equals 10 or more, the table of critical values for Spearman's rho (Appendix A) may be used to determine statistical significance. If the correlation coefficient exceeds the critical value for the given degrees of freedom ($N - 2$) and the alpha level (e.g., .05), researchers may reject the null hypothesis which states that $\rho = 0$. If the correlation coefficient does not exceed the critical value, there is not enough evidence to reject the null. Statistical software conducts the entire analysis. When sample sizes are small (e.g., $N < 10$), researchers may look up the (exact) significance level in the Quantile table for Spearman's rho (Nussbaum, 2015).

12.5.2 Calculations for Spearman's Rho

This exercise will use the formula for Spearman's rho and data from the Case of Hypertension in Portland (Table 12.3) to explore the hand-computation of the Spearman's rho coefficient for age and systolic blood pressure. Ranks for the values of the variables are presented in Table 12.5.

To calculate Spearman's rho for systolic and age, using the formula in Section 12.5.1.4:
1. Rank the values (Table 12.5).
2. Calculate the differences between the pairs of ranks to be analyzed (Table 12.6).
3. Square the differences (Table 12.6).
4. Total the squared differences (Table 12.6).
5. Multiply the outcome of item 4 (307.5) by 6 to find the numerator for the formula.
6. For the denominator, square the number of ranked pairs and subtract 1 from that total.
7. Multiply the outcome of item 6 by the number of ranked pairs.
8. Divide the outcome of item 5 by the outcome of item 7.
9. Subtract the outcome of item 8 from 1.

Table 12.5 Original Values and Ranks for Variables from the Case of Hypertension

	Original Scores			Ranked Scores		
ID	Systolic	Diastolic	Age	Systolic	Diastolic	Age
30348	106	74	51	1	4.5	5
30232	108	64	27	2	3	2
30049	118	58	67	3	1	10
30046	121	74	61	4	4.5	8
30161	122	84	35	5	10.5	3
30102	124	80	22	6.5	7.5	1
30045	124	62	84	6.5	2	15
30336	126	77	63	8	6	9
30117	128	84	57	9	10.5	6
30061	151	81	82	10	9	14
30019	155	80	59	11	7.5	7
30264	164	92	46	12	14	4
30173	168	89	72	13	13	12
30120	184	99	69	14	15	11
30275	195	88	79	15	12	13
30028	221	108	86	16	16	16

Table 12.6 Ranked Scores and Rank Differences for Systolic Blood Pressure and Age

	Ranked Scores		Ranked Differences	
ID	Systolic	Age	Difference	Difference^2
30348	1	5	4	16
30232	2	2	0	0
30049	3	10	7	49
30046	4	8	4	16
30161	5	3	-2	4
30102	6.5	1	-5.5	30.25
30045	6.5	15	8.5	72.25
30336	8	9	1	1
30117	9	6	-3	9
30061	10	14	4	16
30019	11	7	-4	16
30264	12	4	-8	64
30173	13	12	-1	1
30120	14	11	-3	9
30275	15	13	-2	4
30028	16	16	0	0
$\sum D_i^2$				307.5

To solve the equation, beginning at Step 5:

$$\rho = 1 - \frac{6 * 307.5}{N(N^2 - 1)}$$

$$\rho = 1 - \frac{1,845}{16 * 255}$$

$$\rho = 1 - \frac{1,845}{4,080}$$

$$\rho = 1 - .452$$

$$\rho = .548$$

To determine whether to reject the null hypothesis, apply the procedures described in Section 12.5.1.4. For this example, the critical value is based on the degrees of freedom of n − 2 = 16 − 2 = 14. At the a priori alpha level of .05 (two-tailed), the critical value for rejecting the null hypothesis is 0.538. Since the calculated correlation coefficient is 0.548, it exceeds the critical value of 0.538, and therefore, it results in a rejection of the null hypothesis that the correlation coefficient equals zero. Instead, the Spearman's rho coefficient suggests that there is a 95% probability of a significant association between systolic blood pressure and age.

12.5.3 Spearman's Rho Assumptions

Spearman's rho requires that data must meet the following assumptions:
- The two variables to be correlated must be measured at or above an ordinal level. If data belong to either of the latter two categories, the data must be transformed into ranks.
- Each study subject should have a value for each of the variables. For example, in the hand-calculated example in 12.5.2, each subject had a value for age and a value for systolic blood pressure. Accordingly, 16 paired values were analyzed.
- Values of both variables must be monotonically related. A visual inspection of a scatterplot of the values should reveal whether the variables are monotonically related.

Unlike Pearson's *r*, Spearman's rho has no data assumption of linearity or normality, characteristics that make this procedure a useful alternative to Pearson's *r*.

12.5.4 SPSS Specifications and Results for Spearman's Rho

This exercise uses data from the Case of Hypertension in Portland (Table 12.5) that will illustrate how to estimate the association between rankings of variable values using SPSS. The data in this exercise are systolic blood pressure (BP-Systolic), diastolic blood pressure (BP-Diastolic) and age (Age). The objectives are to assess the association between systolic blood pressure and age as well as diastolic blood pressure and age. The dataset resides in the data file **hypertension.sav**.

12.5.4.1 SPSS Specifications for Spearman's Rho: Systolic Blood Pressure and Age

Open the dataset, **Hypertension.sav** and select **Analyze → Correlate → Bivariate** (Figure 12.16) to open the **Bivariate Correlation** window (Figure 12.17).

Figure 12.16 SPSS Data Editor

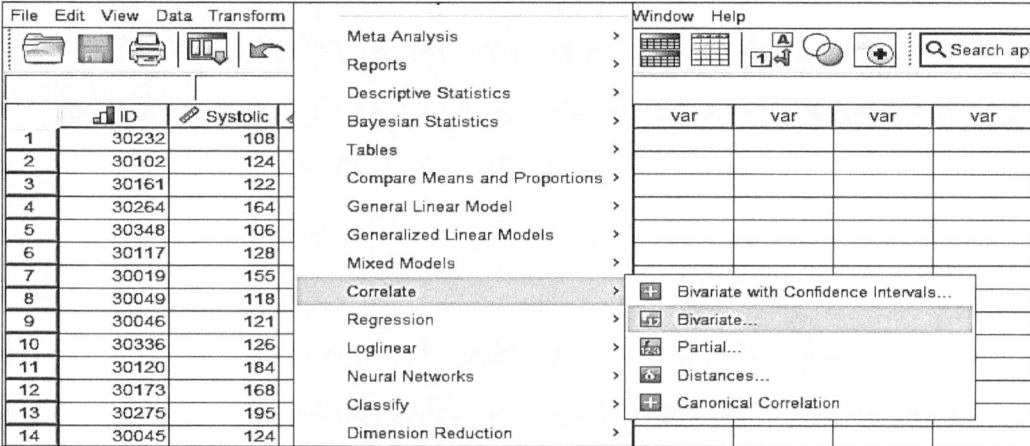

In the **Bivariate Correlation** window, move the variables to be correlated—**Systolic, Diastolic** and **Age**—to the **Variables** panel. Select **Spearman** which is below **Correlation Coefficients** and uncheck the **Pearson** checkbox (the default in SPSS). Leave the remaining defaults in place— **Test of Significance**: **Two-tailed** and **Flag significant correlations**.

Figure 12.17 Bivariate Correlation Dialog Menu

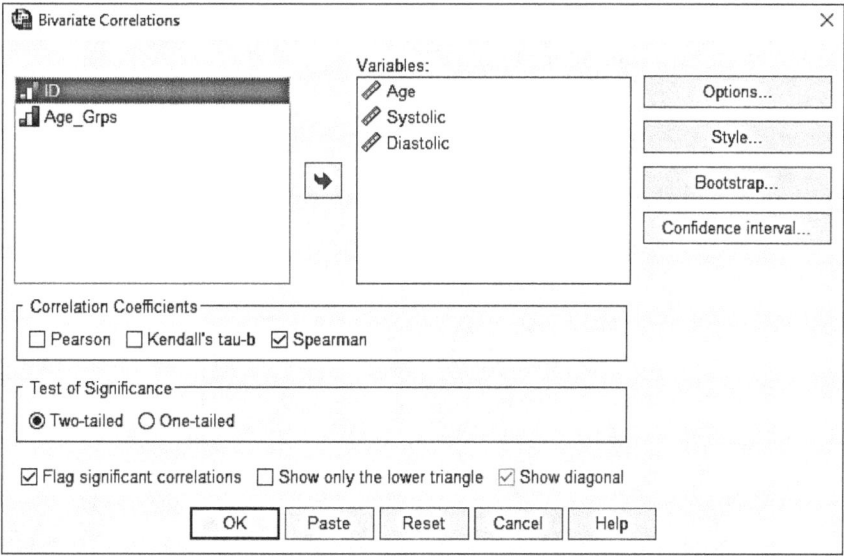

Click the **Options** pushbutton to open the **Bivariate Correlations: Options** window (Figure 12.18). Notice that the options under **Statistics** have been grayed out due to the selection of **Spearman** rather than **Pearson**. Selecting **Pearson** would have resulted in the availability of the **Statistics** options because **Pearson** is based on interval/ratio values. However, **Spearman** is amenable to ranked values which are associated with median, rather than the mean scores, to describe central tendency of a set of values. In this case, specifying **Spearman** in SPSS automatically converts the original raw scores (interval/ratio level data) into ranks.

Figure 12.18 SPSS Bivariate Correlations Options Menu

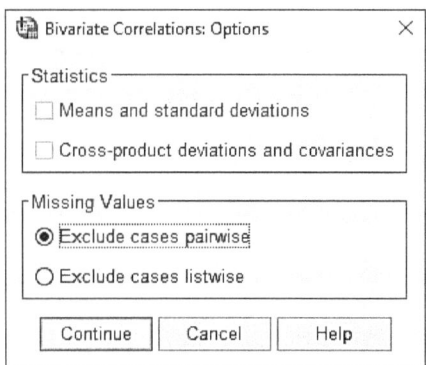

The **Missing Values** option presents two approaches to handling missing values. **Exclude cases pairwise**, the SPSS default, results in the deletion of a case if it has a missing value on one of the two variables correlated; the analysis is limited to complete data. Accept the default of excluding variables pairwise. Click **Continue** to return to the main dialog menu.

In the main dialog window, select **Confidence Interval**. In the **Bivariate Correlations Confidence Interval** dialog menu (Figure 12.19), select **Estimate confidence interval of bivariate correlation parameter** to get the commonly-used 95% confidence interval for the Spearman's rho correlation coefficient. In addition, retain the default variance estimation method, **Fieller, Hartley and Pearson** (1957). Click **Continue** to return to the main menu, then click **OK**

Figure 12.19 Bivariate Correlations Confidence Interval Window

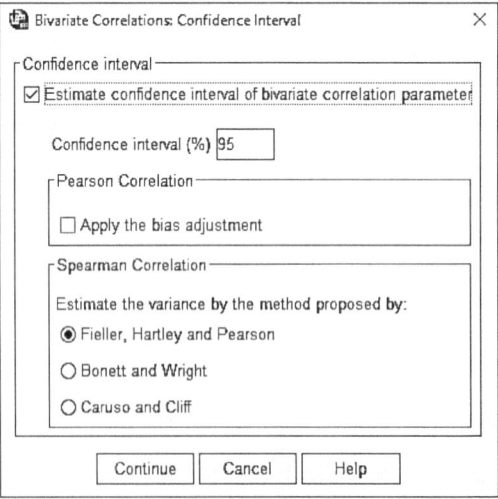

12.5.4.2 Spearman's Rho Results: Blood Pressure and Age

The results of the **Spearman's rho** correlation between systolic blood pressure (**BP-Systolic**) diastolic blood pressure (**BP-Diastolic**) and age (**Age**) are displayed in Figure 12.20.

The sample size (N=16) confirms that no cases were excluded from the analysis (compare with Table 12.5). As expected, there is a significantly, high correlation between BP-Systolic and BP-Diastolic ($r = .849$, $p < .001$). More pertinent to the objectives of the exercise, the estimated correlation coefficient for **BP-Systolic** and **Age** is .547, .001 lower than the hand-calculated coefficient (Section 12.5.2). The

Figure 12.20 SPSS Spearman's Rho Results: BP-Systolic and Age

		Correlations			
			Age	BP-Systolic	BP-Diastolic
Spearman's rho	Age	Correlation Coefficient	1.000	.547*	.212
		Sig. (2-tailed)	.	.028	.430
		N	16	16	16
	BP-Systolic	Correlation Coefficient	.547*	1.000	.849**
		Sig. (2-tailed)	.028	.	<.001
		N	16	16	16
	BP-Diastolic	Correlation Coefficient	.212	.849**	1.000
		Sig. (2-tailed)	.430	<.001	.
		N	16	16	16

* Correlation is significant at the 0.05 level (2-tailed).
** Correlation is significant at the 0.01 level (2-tailed).

correlation coefficient is moderate in magnitude and positive in direction, which suggests that as age increases, so does blood pressure. Results of the two-way significance test reveal a significance level of $p = .028$, which is below the a priori alpha level of .05. Therefore, there is less than a 3% probability that the coefficient does not reflect the population coefficient. This outcome results in a rejection of the null hypothesis.

Unlike the results for **BP-Systolic** and **Age**, the Spearman's rho correlation between **BP-Diastolic** and **Age** suggests a low ($\rho = .212$), positive, but nonsignificant relationship ($p = .430$) between the two variables. Therefore, there is not adequate evidence of a statistically significant correlational relationship between **BP-diastolic** and **Age** to reject the null hypothesis.

Although the analysis generated a significant correlation between systolic blood pressure and age, the confidence interval (Figure 12.21) has a range of .055 (lower bound) to .826 (upper bound). This means there is a 95% likelihood that the true correlation coefficient lies between poor and high correlation. This wide range may be due to the small sample size. With a nonsignificant relationship between **BP-Diastolic** and **Age**, the confidence interval is even is even wider.

Figure 12.21 Spearman's Rho Coefficient Confidence Interval

	Confidence Intervals of Spearman's rho			
	Spearman's rho	Significance (2-tailed)	95% Confidence Intervals (2-tailed)[a,b]	
			Lower	Upper
Age - BP-Systolic	.547	.028	.055	.826
Age - BP-Diastolic	.212	.430	-.331	.650

a. Estimation is based on Fisher's r-to-z transformation.
b. Estimation of standard error is based on the formula proposed by Fieller, Hartley, and Pearson.

12.5.5 Sample Report: The Case of Hypertension in Portland

A Spearman's rho test of significance was conducted to determine whether there were significant correlational relationships between systolic blood pressure and age as well as diastolic blood pressure and age. SPSS Statistics was used to conduct all analyses.

Data assumptions of Spearman's rho were evaluated and met before proceeding to the omnibus test. A positive, moderate, statistically significant relationship was found between systolic blood pressure and age ($\rho(16) = .547$, $p = .028$), signifying that as age increases so does systolic blood pressure. However, the correlation between diastolic blood pressure and age revealed a somewhat different relationship. Here, a low, positive, but nonsignificant correlation was found between the variables ($\rho(16) = .212$, $p = .43$), indicating the coefficient is not significantly different from 0.

12.6 Kendall's Tau

Like Spearman's rho, Kendall's tau (τ) is a measure of association between two ordinal-level variables. Kendall's tau was named after the British statistician, Maurice Kendall (1938). Notably, while Spearman's rho is often regarded as the Pearson's r of rank-ordered data, Kendall's tau is distinctively different from both procedures.

Kendall's tau is an alternative to Pearson's r when continuous data have been converted to ranks or when they are the result of Likert-style response scales. It is also considered an alternative to Spearman's rho when researchers have small sample sizes with multiple tied ranks. Contrary to Spearman's rho, Kendall's tau is designed to produce unbiased correlation coefficients when sample sizes are small. Kendall's tau also adjusts for ties in the data, using a statistic designed for that purpose, the tau-b statistic. The tau-a statistic is used when there are no tied ranks.

The basis for the benefits of Kendall's tau is its computational approach, which is markedly different from Spearman's and Pearson's methodologies. Unlike Pearson's and Spearman's computational formulas, Kendall's tau is not built on the covariance of variable values, but rather on the number of concordant and discordant pairs of points in a dataset.

12.6.1 Hand Calculating Kendall's Tau-a

A pair of points is concordant if Y values increase when X values do. It is discordant if Y declines when X increases. Imagine a study about the association between ratings (on a 6-point scale) of the job satisfaction and happiness of six employees.

To calculate the Kendall's tau coefficient, the researcher assigns each study participant a rating or rank for each of the variables (Figure 12.22). Next, the researcher identifies the number of concordant and discordant pairs in the dataset. To do that, identify the pair of ranks for each participant. Each pair of ranks represents the coordinates for each participant plotted on the x, y axes; the coordinates for Participant 03 is 3,5 which means that Participant 03 gave job satisfaction a rating of 3 and happiness a rating of 5, both on a scale of 1 – 6. Next compare the coordinates for each participant with the coordinates for every other participant. If ranking 3,5 is compared with ranking 5,6 for Participant 05, the pair of rankings would be considered concordant, since as job satisfaction increases from 3 to 5, happiness also increases from 5 to 6. If the coordinates of 3,5 is then matched with the coordinates 4,3 for Participant 04, that pair of rankings would be discordant because as job satisfaction increases from 3 to 4, happiness declines from 5 to 3.

Figure 12.22 Job Satisfaction and Happiness Ratings

Participant	Job Satisfaction (X)	Happiness (Y)
01	1	1
02	2	2
03	3	5
04	4	3
05	5	6
06	6	4

Paired rankings of two participants can also be tied on the x or y coordinate. For example, if a pair of study participants had the coordinates 4,6 and 4,5, they would be considered tied on x. If a pair of participants had the coordinates 4,6 and 2,6, they would be tied on y.

Another method of identifying all of the concordant and discordant pairs is to use a matrix of the two variables involved. A matrix of the variables, job satisfaction and happiness, is displayed in Figure 12.23. For example, consider all the rankings that might be paired with 1,1. All the pairings with the other coordinates are concordant. However, consider the ranking 3,5. When paired with some coordinates, the pairing is concordant (e.g., 2,2). However, pairing with other coordinates (e.g., 3,5 and 4,3) is discordant.

Figure 12.23 Matrix of Coordinates to Identify Concordant/Discordant Pairs

Pairs	1,1	2,2	3,5	4,3	5,6	6,4
1,1	-	Concordant	Concordant	Concordant	Concordant	Concordant
2,2		-	Concordant	Concordant	Concordant	Concordant
3,5			-	Discordant	Concordant	Discordant
4,3				-	Discordant	Concordant
5,6					-	Discordant
6,4						-

12.6.2 Two Computational Examples of Kendall's Tau

With six participants in Figure 12.22, there are 15 total pairs of points—11 concordant and 4 discordant pairs. The computation for deriving the total possible pairs of points is:

$$\frac{N(N-1)}{2} = \frac{6(6-1)}{2} = \frac{6*5}{2} = \frac{30}{2} = 15$$

After computing the number of concordant and discordant pairs of points, use the following equation to calculate the Kendall's tau coefficient:

$$\tau = \frac{n_c - n_d}{N}$$

428 Nonparametric Tests

where

n_c is the number of concordant pairs,
n_d is the number of discordant pairs, and
N is the total number of paired ranks.

Therefore, using the number of concordant pairs (11), the number of discordant pairs (4) and the total number of pairs, find the Kendall's tau coefficient for the data in Figure 12.22 as follows:

$$\tau = \frac{n_c - n_d}{n}$$

$$\tau = \frac{11 - 4}{15}$$

$$\tau = .467$$

A Kendall's tau coefficient of .467 is evidence to reject the null hypothesis that job satisfaction and happiness are independent of each other; rather, they are moderately dependent.

This Kendall's tau equation is best used when there are no ties in the data and is referred to as Tau-a (τ_a). When there are ties in the data, researchers use a different computation that makes a correction for ties—Tau-b (τ_b). In fact, in many statistical software applications, like SPSS, τ_b is the default for Kendall's tau. For information on calculating τ_b manually, see Nussbaum (2015).

Consider the association of age and diastolic blood pressure rankings in Table 12.5. The concordance matrix for that data is in Figure 12.24. Age occurs first in each coordinate pairing. There are 61 concordant pairings in the ranked data (n_c) and 56 discordant pairings (n_d). Additionally, a total of three tied ranks are in the data. Choosing to compute τ_a for this data means the ties are ignored, and the result is a coefficient that's almost zero:

$$\tau_a = \frac{n_c - n_d}{N} = \tau_a = \frac{61 - 56}{120} = .042$$

Figure 12.24 Coordinates Matrix of Concordant/Discordant/Tied Pairs of Age and Diastolic BP

	1,7.5	2,3	3,10.5	4,14	5,4.5	6,10.5	7,7.5	8,4.5	9,6	10,1	11,15	12,13	13,12	14,9	15,2	16,16
1,7.5		C	C	C	D	C	T	D	D	D	C	C	D	D	D	C
2,3			C	C	D	C	C	D	D	D	C	C	D	D	D	C
3,10.5				C	D	T	C	D	D	D	C	C	D	D	D	C
4,14					D	C	C	D	D	D	C	C	D	D	D	C
5,4.5						D	D	T	C	C	D	D	C	C	C	C
6,10.5							C	D	D	D	C	C	D	D	D	C
7,7.5								D	D	D	C	C	D	D	D	C
8,4.5									C	C	D	D	C	C	C	C
9,6										C	D	D	C	C	C	C
10,1											D	D	C	C	C	C
11,15												C	D	D	D	C
12,13													D	D	D	C
13,12														C	C	C
14,9															C	C
15,2																C
16,16																

SPSS-generated Kendall's tau (τ_b) results for diastolic blood pressure and age are slightly higher ($\tau_b(16) = 0.127$, $p = .498$) than the manually-computed results ($\tau_a = .042$). In both cases, the relationship was statistically nonsignificant. However, Kendall's tau correlation coefficients are slightly more conservative than the Spearman's rho coefficient of $\rho = .212$ for the same variables.

12.6.3 Kendall's Tau Assumptions

Beyond the assumptions for all nonparametric techniques discussed earlier in this chapter, there are two assumptions for Kendall's tau:

- The two variables to be correlated should be measured on an ordinal, interval or ratio scale. In the case of the latter, scores must be converted to ranks before analysis. Exercise frequency rated on a 7-point scale is one example of a variable measured on an ordinal scale (where 1 = once weekly to 7 = every day of the week).
- Kendall's tau measures whether there is a monotonic relationship between two variables. Therefore, the data involved should appear to represent a monotonic relationship.

12.6.4 Research Plan: The Case of Trust and the Importance of Religion

Box 12.3 presents an outline of the research plan for this case. Table 12.4 presents the data.

The purpose of this case study is to determine whether there is a bivariate monotonic association between pairs of four variables, three of which measure levels of general trust or trust in individuals that respondents do not know personally, and one that measures the importance of religion. Each hypothesis is related to a single pair of variables. The four variables are analyzed together for purposes of convenience and efficiency. Details of the case are in Section 12.4.2.

The research question and null hypothesis for the case are as follows:

Research Question
Are there bivariate, monotonic associations among the four variables, RI, FM, AR, AN?

Null Hypothesis
There is no significant monotonic association among the four variables.

All variables are ordinal-level variables. One is respondents' ratings of the importance of religion (RI). The remaining three are measures of trust of people: (a) in a first-time meeting (FM), (2) from another religious group (AR), and (3) of another nationality (AN).

All variable values originate from 4-point Likert-type response scales. Data from the trust scales are based on 4-point scales, ranging from 1 (Do not trust at all) to 4 (Trust completely). Ratings of the importance of religion range from 1 (Not at all important) to 4 (Very Important).

This case is a descriptive nonexperimental research design, often used when researchers cannot manipulate the variables in a study. In this case, the study is based on data from one-on-one, in-person surveys, administered by research assistants. The target population is Trinidad and Tobago, and the data are a small random sample (n = 30) of the original sample of the population.

Perhaps due to the small number of points on the response scale, there are a multitude of tied ranks in the data. Consequently, the primary data analysis tool in this case is Kendall's tau-b, which increases the accuracy of the statistic by making a correction for ties in the data.

Box 12.3 Research Plan Outline: the Case of Trust and the Importance of Religion

Research Problem
Delhey et al. (2011) describe three items on the World Values Survey that measure respondents' trust of people they do not know personally as items measuring general trust, based on a confirmatory factor analysis they conducted. A researcher desires to investigate whether there is a monotonic association among these variables, including another, importance of religion.

Variables and Measurement Levels
Variable 1: Importance of religion (RI)
Variable 2: Trust of people meeting for the first time (FM)
Variable 3: Trust of people from another religious group (AR)
Variable 4: Trust of people of another nationality (AN)
Level of Measurement for all Variables: Ordinal

Research Question
Are there significant bivariate monotonic associations among the four variables RI, FM, AR, AN?

Hypotheses: Two-Tailed Test of Significance (Kendall's Tau)
Null Hypothesis: There is no significant monotonic association among the four variables.
$H_o: \tau = 0$

Alternative Hypothesis: There is a significant monotonic association among two or more variables.
$H_a: \tau \neq 0$

Research Design
Nonexperimental correlational research design

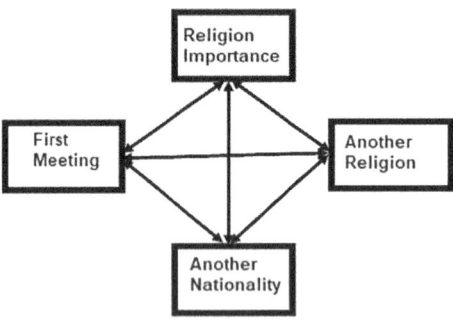

Data Analysis
Descriptive statistics, including means and standard deviation as well as tests of assumptions for Kendall's tau, come first. Next, with the variables measured on the ordinal level scale, the variables will be correlated using Kendall's tau followed by significance testing to determine whether correlation coefficients are significant at $\alpha = .05$.

12.6.5 SPSS Specifications for Kendall's Tau-b

To begin, open the data file, **WVS_Trust.sav** in the **SPSS Statistics Data Editor**. Select **Analyze → Correlate → Bivariate** to open the **Bivariate Correlations** dialog menu (Figure 12.25).

Once in the dialog window, specify a matrix of pairwise correlations of the variables by selecting the four variables to be correlated, and drag or click them over to the **Variables** panel. Next, unselect **Pearson** and select **Kendall's tau-b** to specify the statistical significance test.

Figure 12.25 SPSS Bivariate Correlations Dialog Menu

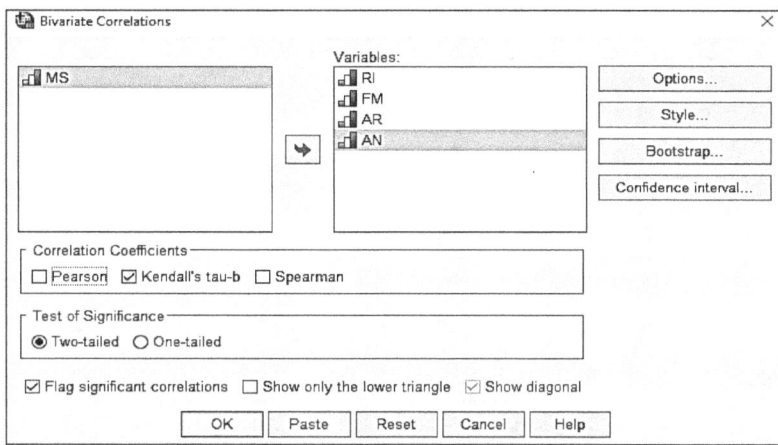

Complete the specifications by clicking on the **Options** pushbutton to confirm that the default missing value specification, **Exclude Cases Pairwise** has been selected. Click **Continue** to return to the **Bivariate Correlations** dialog menu, then click **OK** to run the procedure.

12.6.6 SPSS Results of Kendall's Tau-b

Kendall's tau-b results appear in Figure 12.26. Kendall's tau-b analyses resulted in significant positive monotonic associations between the **importance of religion** and two of the three trust variables: **trust people they meet for the first time** ($p = .004$), and **trust people of another religion** ($p = .048$). In addition, there was moderate correlation between **trust people of another nationality** and **trust people they meet for the first time** ($p = .027$) as well as **trust people of another religion** ($p = .005$). Asterisks denote all significant pairs.

All statistically significant variable pairs are positively correlated, which means that as the ratings for one variable increase, so do the ratings for the other variable. For example, the importance of religion is positively related to the tendency to trust people at the first meeting ($\tau_b(29) = .503, p < .01$).

12.7 Friedman's Rank

Friedman's rank is the nonparametric equivalent of the one-way repeated measures analysis of variance, named after Milton Friedman (1937), an American economist and statistician. The test is also known as nonparametric repeated measures ANOVA, nonparametric Friedman test, the Friedman rank sum test, or the Friedman's ANOVA by ranks test. This text will refer to the test as the Friedman's rank test.

As the name suggests, typically this test is used to analyze rankings or ordinal level data rather than the interval/ratio data that parametric repeated measures ANOVA analyzes. Like repeated measures ANOVA, however, researchers use this test to investigate the existence of significant differences between dependent samples of rankings or ratings from more than two time points or more than two conditions for each study participant. The Friedman's rank test is an extension of the sign test, a nonparametric test that is used to test for significant differences between rankings at two time points or under two conditions.

Figure 12.26 Spearman Rho and Kendall's Tau-b Results

<table>
<tr><th colspan="7">Correlations</th></tr>
<tr><th colspan="3"></th><th>Importance of Religion</th><th>Trust People They Meet for the First Time</th><th>Trust People of Another Religion</th><th>Trust People of Another Nationality</th></tr>
<tr><td rowspan="12">Kendall's tau_b</td><td rowspan="3">Importance of Religion</td><td>Correlation Coefficient</td><td>1.000</td><td>.503**</td><td>.344*</td><td>.335</td></tr>
<tr><td>Sig. (2-tailed)</td><td>.</td><td>.004</td><td>.048</td><td>.054</td></tr>
<tr><td>N</td><td>29</td><td>29</td><td>29</td><td>29</td></tr>
<tr><td rowspan="3">Trust People They Meet for the First Time</td><td>Correlation Coefficient</td><td>.503**</td><td>1.000</td><td>.167</td><td>.367*</td></tr>
<tr><td>Sig. (2-tailed)</td><td>.004</td><td>.</td><td>.314</td><td>.027</td></tr>
<tr><td>N</td><td>29</td><td>30</td><td>30</td><td>30</td></tr>
<tr><td rowspan="3">Trust People of Another Religion</td><td>Correlation Coefficient</td><td>.344*</td><td>.167</td><td>1.000</td><td>465**</td></tr>
<tr><td>Sig. (2-tailed)</td><td>.048</td><td>.314</td><td>.</td><td>.005</td></tr>
<tr><td>N</td><td>29</td><td>30</td><td>30</td><td>30</td></tr>
<tr><td rowspan="3">Trust People of Another Nationality</td><td>Correlation Coefficient</td><td>.335</td><td>.367*</td><td>.465**</td><td>1.000</td></tr>
<tr><td>Sig. (2-tailed)</td><td>.054</td><td>.027</td><td>.005</td><td>.</td></tr>
<tr><td>N</td><td>29</td><td>30</td><td>30</td><td>30</td></tr>
</table>

**. Correlation is significant at the 0.01 level (2-tailed).
*. Correlation is significant at the 0.05 level (2-tailed).

12.7.1 The Case of a Public Health Video Campaign

In this era of the increasing frequency of pandemics, using public health media campaigns to keep the public informed is paramount. Observing the impact of the COVID pandemic, Scott Gottlieb, a former commissioner of the U.S. Federal Drug Administration observed that public health had become part of national security (April 26, 2020). The Case of a Public Health Video Campaign uses a small dataset of five variables and 20 subjects to evaluate four methods for informing the public about a medical condition. The dataset appears in Table 12.7 and an outline of a research plan for the case appears in Box 12.4.

The dataset is the result of responses from 20 individuals to four five-item Likert-style survey subscales about four public health videos —new general video (NG), new medical profession video (NM), the old video (OV) and a demonstration video using props (DV). The objective of the study was to determine whether new methods were more popular than the old ones. Variables for the study include:
- study participant identification code (ID)
- new general video overall rating (NG)
- new medical profession video overall rating (NM)
- old video overall rating
- demonstration video overall rating

12.7.2 Friedman's Rank Test Fundamentals

The Friedman's rank test is an alternative to repeated measures ANOVA when data are measured at the ordinal level, and therefore, it tests for median differences within subjects rather than between groups of study participants. Consider a situation in which 20 male and female students are asked to rank their attitudes toward mathematics on a 6-point scale, at three time points: before a cooperative learning unit is introduced (Time 1), in the middle of the mathematics unit (Time 2) and after the unit ends (Time 3) (Figure 12.27).

Figure 12.27 Friedman's One-Way Repeated Measures Over Time Design

	Attitude Ratings		
Subjects	Time 1	Time 2	Time 3
S_1	S_1	S_1	S_1
...			
S_{20}	S_{20}	S_{20}	S_{20}

In this instance, there are three time periods with 20 subject ratings for each. Friedman's rank tests for significant differences between time periods—from Time 1 to Time 2, Time 2 to Time 3, and Time 1 to Time 3. The sample values in this example are dependent because each study participant has a value for each time period. Investigations of changes among different time periods represent one of two situations in which Friedman's rank is commonly used.

The data represented in Table 12.7 is an example of testing differences under three or more conditions, the second situation in which Friedman's rank is most commonly used. In this study, 20 participants are asked to (a) rank the videos' ability to aid understanding about a medical condition, and (b) rank their impression of each video by responding to five Likert-style items about each. In this case, Friedman's rank is used to test for differences under four different conditions or communication methods. Data for each type of video includes a sampling of video ratings with 20 values per group (no missing values). The ratings of the videos are dependent or related since each study participant produced a rating for each of the three conditions (video) (Figure 12.28).

If Friedman's rank yields significant results, at least two of the samples are significantly different from each other. However, the test does not reveal where the differences lie or the number of paired differences. One option for identifying the location and number of differences between sample pairs, is a post hoc test like the Wilcoxon signed ranks test.

12.7.3 Research Plan: The Case of of a Public Health Video Campaign

The research study for the Case of a Public Health Video Campaign (Box 12.4) is a repeated measures design. It tests for differences among four public health videos by measuring viewers' general impression of the videos using five Likert-style survey items on a 5-point scale.

Table 12.7 Excerpt from Public Health Video Campaign Data

ID	Overall Score				ID	Overall Score			
	NG	NM	OV	DV		NG	NM	OV	DV
01	25	23	13	22	11	24	23	15	21
02	23	23	20	22	12	22	24	14	19
03	20	17	14	23	13	24	23	14	19
04	24	24	23	25	14	25	25	17	18
05	25	22	18	23	15	24	22	14	17
06	24	22	23	25	16	25	23	13	20
07	25	22	9	21	17	25	22	16	21
08	24	23	6	21	18	25	21	11	22
09	25	21	11	23	19	25	23	19	25
10	24	23	5	25	20	24	23	18	25

Figure 12.28 Friedman's One-Way Repeated Measures Over Conditions

	Video Ratings			
Subjects	General	New Medical	Old	Demonstration
S_1	S_1	S_1	S_1	
...				
S_{20}	S_{20}	S_{20}	S_{20}	

As a repeated measures design, the subjects serve as their own control group by rating their general impressions of each type of video. This, together with reducing practice effects by random presentation of the videos, reduces the threat to internal validity and increases the likelihood of rejecting the null hypothesis if it is false (Cook & Campbell, 1979).

Friedman's test statistic (χ^2_F) can be computed manually, using the following formula:

$$\chi^2_F = \frac{12}{NK(K+1)} \sum R_j^2 - 3N(K+1)$$

where
R_j = the sum of the ranks for column or condition, j;
N = the number of subjects (rows); and
K = the number of conditions (columns).

Data analysis procedures for conducting the Friedman's rank test are also widely available in statistical software packages. This exercise will hand-calculate the statistics and use SPSS to generate test results. But before calculating the statistic, the following steps are required:
1. Rank the raw scores for each participant separately, and
2. Sum the rankings for each condition.

Box 12.4 Sample Research Plan Outline for the Case of a Public Health Video Campaign

Research Problem
A researcher at the University of Sheffield wanted to identify the best method for informing the public about a certain medical condition, using four videos—new general video (GV), new medical profession video (NM), the old video (OV), and a demonstration using props (DV). The study sought responses from 20 individuals to Likert-style questions about the public health videos to see whether the new methods were more popular than the old ones. This case uses an excerpt of the original dataset.

Variables and Measurement Levels
Independent Variable: General Impression
Level 1: General Video (GV)
Level 2: New Medical (NM)
Level 3: Old Video (OV)
Level 4: Demonstration Video (DV)
Dependent Variable: General Impression Ratings
Level of Measurement for all Variables: interval/ratio

Research Question
Is there a significant difference in participants' general impression rankings of four public health information videos?

Hypotheses: Two-Tailed Test of Significance (Friedman's Rank)
Null Hypothesis: There is no significant difference in participants' general impression rankings of the four videos.
$H_0: m_{ng} = m_{nm} = m_{ov} = m_{dv}$

Alternative Hypothesis: There is a significant difference in at least one pair of general impression rankings.
$H_a: m_i \neq m_j$

Research Design
Nonexperimental repeated measures research design

$$X_1 \quad O_{R1} \quad X_2 \quad O_{R2} \quad X_3 \quad O_{R3} \quad X_4 \quad O_{R4}$$

The nonrandomized, single-group repeated measures design includes a total of four random exposures to different video (X) methods for informing the public about a certain medical condition and summative ratings of participants' impressions of each video (O), completed by each of 20 study participants.

Data Analysis
Descriptive statistics, including means and standard deviation as well as tests of assumptions for nonparametric procedure, Friedman's rank, come first. Next, with the variables measured on the ordinal level scale, is significance testing, using Friedman's rank to determine whether there are significant changes in the rankings at α = .05. Post hoc testing will follow the omnibus test to identify the location of significant changes in ratings.

When conducting Friedman's rank significance testing, use either the asymptotic p-value or the exact p-value for the test. The asymptotic p-value is calculated based on the estimated distribution of the ratings involved. A p-value calculated using the true distribution is called an exact p-value. For large sample sizes, the exact and asymptotic p-values are very similar. The asymptotic version of Friedman's significance test requires a minimum sample size of 12 (Nussbaum, 2015). Therefore, the case study sample size of 20 is adequate for the asymptotic p-value. The exact p-value is best used when sample sizes are less than 12.

12.7.4 A Computational Example of Deriving the Friedman's Rank Statistic

The data for this demonstration may be found in Table 12.7. These data are summed values, and therefore, ratio-level values. Consequently, before applying the Friedman's rank formula, it is important to rank the values for each participant separately (Table 12.8).

A total of four rows in Table 12.8 are highlighted. Each of these rows includes two or more scores that are virtually the same; that is, they are tied ranks. Therefore, the rankings are averages that represent the duplicated total scores. For example, ID 02 has the same overall score of 23 for Video NG and video NM. Therefore, they both have an assigned ranking of 3.5 (3 + 4 = 7/2 = 3.5).

The Friedman's rank computational formula for data with ties is somewhat more complicated than the formula for values without ties. To simplify the hand calculation, ignore rows with tied ranks for this illustration. Highlighted rows (ID = 02, 04, 14, and 19) are those excluded from the calculation, leaving a total of 16 blocks or rows of ranks (Table 12.9).

Table 12.8 Original Scores and their Rankings of the General Impression Videos

	Overall General Impression Scores				Ranks of General Impression			
ID	NG	NM	OV	DV	NG	NM	OV	DV
01	25	23	13	22	4	3	1	2
02	23	23	20	22	3.5	3.5	1	2
03	20	17	14	23	3	2	1	4
04	24	24	23	25	2.5	2.5	1	4
05	25	22	18	23	4	2	1	3
06	24	22	23	25	3	2	1	4
07	25	22	9	21	4	3	1	2
08	24	23	6	21	4	3	1	2
09	25	21	11	23	4	2	1	3
10	24	23	5	25	3	2	1	4
11	24	23	15	21	4	3	1	2
12	22	24	14	19	3	4	1	2
13	24	23	14	19	4	3	1	2
14	25	25	17	18	3.5	3.5	1	2
15	24	22	14	17	4	3	1	2
16	25	23	13	20	4	3	1	2
17	25	22	16	21	4	3	1	2
18	25	21	11	22	4	2	1	3
19	25	23	19	25	3.5	2	1	3.5
20	24	23	18	25	3	2	2	4

The next step in the hand computation process is to sum the rankings for each condition. Table 12.9 includes the final dataset (rows without tied ranks) as well as the summed rankings for each video (condition).

Use the Friedman's rank formula for data without ties to compute the statistic χ_F^2. Note that R_j^2 equals the square of the sum of the rankings for each condition in Table 12.9.

$$\chi_F^2 = \frac{12}{nk(k+1)} \sum R_j^2 - 3N(K+1)$$

$$\chi_F^2 = \frac{12}{16(4)(4+1)} * (59^2 + 42^2 + 17^2 + 43^2) - 3(16)(4+1)$$

$$\chi_F^2 = \frac{12}{16(4)(5)} (3{,}481 + 1{,}764 + 289 + 1{,}849) - 3(16)(5)$$

$$\chi_F^2 = \frac{12}{320} (7{,}383) - 240$$

$$\chi_F^2 = 36.86$$

Table 12.9 General Impression Rankings for Observations Without Ties

	Ranks of General Impression Scores			
ID	NG	NM	OV	DV
01	4	3	1	2
03	3	2	1	4
05	4	2	1	3
06	3	2	1	4
07	4	3	1	2
08	4	3	1	2
09	4	2	1	3
10	3	2	1	4
11	4	3	1	2
12	3	4	1	2
13	4	3	1	2
15	4	3	1	2
16	4	3	1	2
17	4	3	1	2
18	4	2	1	3
20	3	2	2	4
Sum	59	42	17	43

When computing χ_F^2 manually, use the table of critical values for the chi-square test (Appendix A) to estimate the chi-square critical value at an alpha level of .05. The calculated test statistic must exceed the critical value in order to reject the null hypothesis.

To use the table of critical values, compute the chi square degrees of freedom, which is k − 1 or 4 − 1 = 3 degrees of freedom. The chi-square critical value for 3 degrees of freedom at an alpha level of .05 is 7.82. Comparing the critical value of 7.82 to the obtained test statistic for Friedman's test of 36.86, reveals that the test statistic exceeds the critical value, thereby rejecting the null hypothesis. As a result, there is evidence of a significant difference in participants' general impression ratings of one or more pairs of videos ($\chi^2_F(3) = 36.86, p < 0.05$)

12.7.5 Friedman's Rank Test Assumptions

Like other nonparametric tests in this chapter, Friedman's rank has fewer assumptions than its parametric counterpart. However, like all statistical test assumptions, in order for statistical test results to be accurate, the data must satisfy the data properties described.

With the exception of Assumption 4, all assumptions for the Friedman's test are methodological or research design related. Therefore, ensuring that the data meet the assumptions is almost always within the control of the researcher and must be addressed during research design. Thus, the researcher must ensure:

- Study participants comprise a single group that is measured at least three times or in a minimum of three conditions.
- Research study participants represent a random sample of the target population.
- At a minimum, the dependent variable is measured at the ordinal level.
- Data need not be normally distributed, but the score distributions for the time or condition groups should be similar in shape.
- The data from one participant should not affect or interact with data from another participant; participants' data should be independent of each other.

12.7.6 SPSS Specifications for Friedman's Rank

For purposes of comparison with the manual computation of the statistic, this exercise will employ the amended dataset that excludes tied ranks to estimate the significance of the differences in the general impressions of four videos. The dataset may be found in the data file, **Hlth_Vid_NT.sav**. The independent variable conditions to be analyzed in this exercise are new general (NG), new medical (NM), old video (OV) and demonstration video (DV).

12.7.6.1 SPSS Setup for Friedman's Rank

SPSS offers two methods for conducting Friedman's test:

Method 1: **Analyze → Nonparametric Tests → Legacy Dialogs → K Related Samples**

Method 2: **Analyze → Nonparametric Tests → Related Samples**

Method 1, known as the legacy dialog method, is an older procedure and easier to use than the newer Method 2. However, the new method runs the Dunn-Bonferroni post hoc tests automatically if the omnibus test is significant.

Opting for Method 1 means that a significant omnibus Friedman's test will require the administration of Bonferroni-corrected Wilcoxon signed-rank tests as the post hoc for each pair of variables. When performing multiple paired comparisons, the Type 1 error rate becomes inflated due to an increase in the a priori alpha level. To make the Bonferroni adjustment, divide the a priori alpha level by the number of Wilcoxon tests planned. Use the corrected alpha level to determine whether there are significant differences between variable pairs.

The new method will only run the Friedman tests if all the variables to be entered are classified as scale in SPSS. Accordingly, even though Friedman's test is suitable for testing differences between ordinal-level variables, they must be classified in SPSS as scale-level measures to run. To ensure variables are set as **Scale**, before running the analysis, confirm that the variables are correctly labeled in the Measure column in SPSS's **Variable View** page.

By default, variables are assigned to the **Input** role unless otherwise specified in **Variable View**. This indicates that variables will be used as input variables (e.g., predictor or independent variables). To confirm the input role of the variables, click on the **Variable View** tab and check the column, **Role**, to confirm that each variable is assigned as **input**.

To take advantage of automatic post hoc opportunities, use Method 2 to run the analysis. Select **Analyze → Nonparametric Tests → Related Samples** (Figure 12.29) to open the dialog window.

Figure 12.29 SPSS Statistics Data Editor Menu

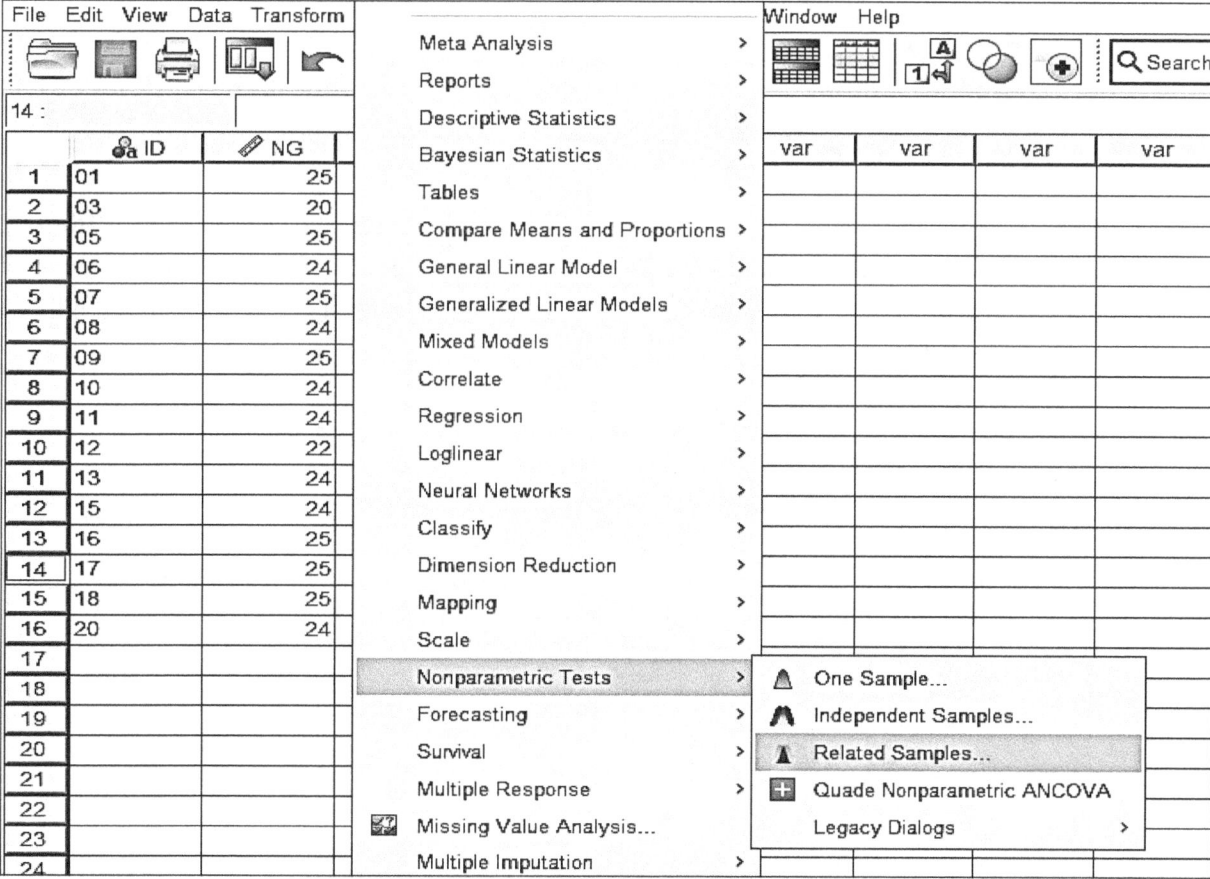

In the **Nonparametric Test: Two or More Related Samples** dialog window, there are three tabs. Clicking on the first tab reveals test objectives: (1) **Automatically compare observed data to hypothesized** (default), and (2) **Customize analysis**. Accept the default setting (Figure 12.30).

Click on the **Fields** tab (Figure 12.31) to choose (1) **Use Pre-defined roles**, or (2) **Use custom field assignments**. The first option applies when researchers assign specific roles for variables in **SPSS**

Statistics Data Editor. With the default setting of **Input** selected in **SPSS Data Editor**, select **Use custom field assignments** (Option 2) on the **Fields** page.

Figure 12.30 Nonparametric Test: Two or More Relate Samples—Objective Dialog Window

Next, move the four conditions, **New General Video**, **New Medical Video**, **Old Video** and **Demonstration**, from the **Fields** panel to the **Test Fields** panel (Figure 12.31). SPSS will automatically conduct Friedman's rank test if there are three or more conditions and a Wilcoxon Signed ranks test if there are two conditions.

Figure 12.31 Nonparametric Test: Two or More Related Samples—Fields Dialog Window

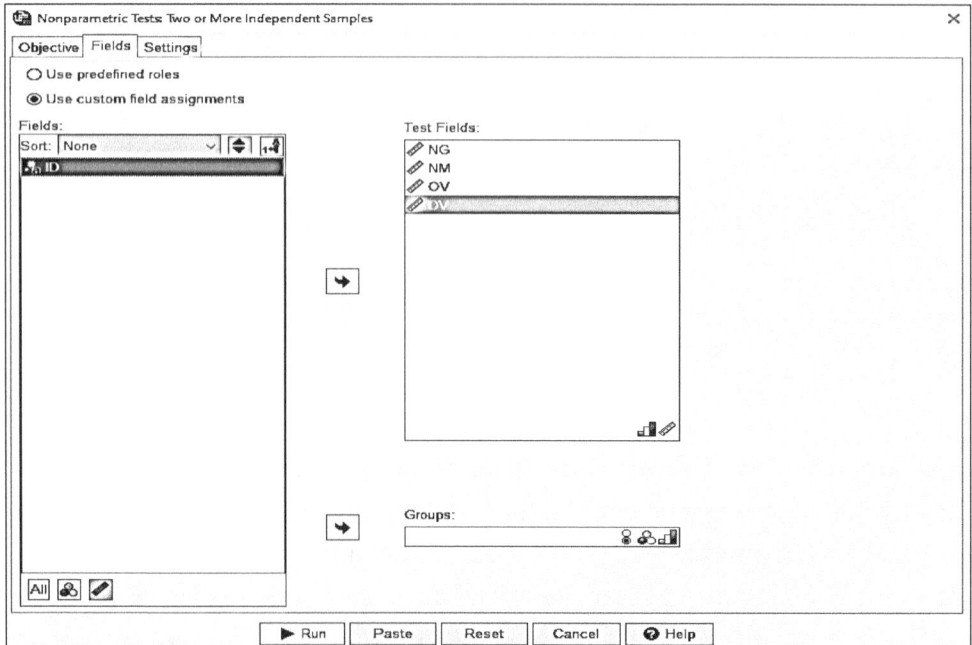

Click on the **Settings** tab to confirm the selection of **Choose Tests** below **Select an Item** on the far left. In addition, confirm that the option, **Automatically choose the tests based on the data** is the designated procedure (Figure 12.32). This means that Friedman's rank will be conducted, followed by the Dunn-Bonferroni post hoc test for all paired comparisons should the omnibus test prove to be significant. The post hoc test will identify the pairs of videos that have significantly different ratings. Click **Run** to perform the analysis.

Figure 12.32 Nonparametric Test: Two or More Related Samples—Settings Dialog Window

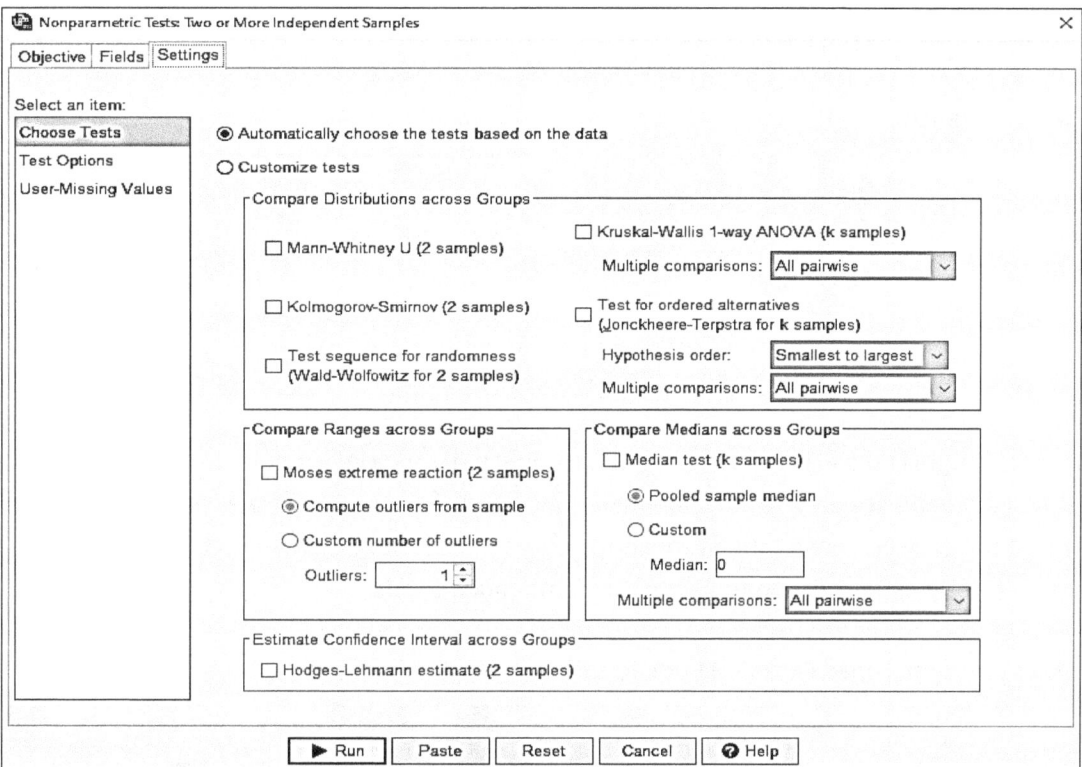

12.7.6.2 Friedman's Rank Test Results

Figure 12.33 displays the summary table for the Freidman's rank test results. The null hypothesis for the significance test appears in the **Null Hypothesis** column, along with a description of the test conducted (**Test column**). The **Sig.** column reports a significant difference between at least one pair of video ratings. Therefore, as reported in the **Decision** column, the test result is to reject the null hypothesis (There is no significant difference among the video ratings).

The **Related-Samples Friedman's Rank Summary** (Figure 12.34) provides additional details about the test results, including the **sample size** (16), **test statistic** (33.75) and the **degrees of freedom** on which the test statistic is based (3). The results of the 2-sided test of significance (**Asymptotic Sig. (2-sided test)**) are also displayed ($p < .001$).

Due to the decision to have SPSS automatically select the significance test based on the data (rather than selecting to customize test specifications), SPSS not only selected Friedman's as the omnibus test, but also the Dunn-Bonferroni test as the post-hoc test. SPSS automatically conducted the post-hoc test after Friedman's test determined that ratings for at least one pair of video methods were significantly different.

Figure 12.33 Friedman's Rank Test Summary Results

	Hypothesis Test Summary			
	Null Hypothesis	Test	Sig.[a,b]	Decision
1	The distributions of New General Video, New Medical Video, Old Video and Demonstration are the same.	Related-Samples Friedman's Two-Way Analysis of Variance by Ranks	<.001	Reject the null hypothesis.

a. The significance level is .050.
b. Asymptotic significance is displayed.

Figure 12.34 Friedman's Test Results Details

Related-Samples Friedman's Rank Summary	
Total N	16
Test Statistic	33.750
Degree Of Freedom	3
Asymptotic Sig.(2-sided test)	<.001

Figure 12.35 reports the results of testing all possible pairs of conditions— unstandardized and standardized test statistics for each pair. The (unstandardized) **Test Statistic** in Column 1 of Figure 12.35 is the difference between each pair of mean rankings of the videos. **Std. Test Statistic** is the z-score of the first set of statistics needed to calculate significance levels.

Figure 12.35 Wilcoxon Signed Ranks Post Hoc Test Results for All Paired Comparisons

Pairwise Comparisons					
Sample 1-Sample 2	Test Statistic	Std. Error	Std. Test Statistic	Sig.	Adj. Sig.[a]
Old Video-New Medical Video	1.500	.456	3.286	.001	.006
Old Video-Demonstration	-1.625	.456	-3.560	<.001	.002
Old Video-New General Video	2.625	.456	5.751	<.001	.000
New Medical Video-Demonstration	-.125	.456	-.274	.784	1.000
New Medical Video-New General Video	1.125	.456	2.465	.014	.082
Demonstration-New General Video	1.000	.456	2.191	.028	.171

Each row tests the null hypothesis that the Sample 1 and Sample 2 distributions are the same. Asymptotic significances (2-sided tests) are displayed. The significance level is .050.

a. Significance values have been adjusted by the Bonferroni correction for multiple tests.

Two types of significance levels are included in Figure 12.35: (1) unadjusted significance levels that do not include a Bonferroni correction for the number of post-hoc tests conducted (**Sig.**) and (2) adjusted significance levels based on the Bonferroni correction for multiple tests; the p-values in the **Sig** column are multiplied by the number of tests being carried out to secure the **Adj. Sig** *p*-values (Bonferroni corrected alpha level = .008).

Dunn's post-hoc test results report that the ratings of the first three video pairs in the table are significantly different (p < .01). The largest difference in ratings occurred between the old video and the new general video (*p* = .000), followed by the old video and the demonstration (*p* = .002), and the old video and the new medical video (*p* = .006). There were no significant differences between the remaining pairs of public health campaign methods.

The diagram in Figure 12.36 presents the mean rank for each public health campaign method. Gray lines represent statistically significant pairs of methods. Remaining lines represent statistically nonsignificant pairs of methods.

Overall, significance testing results using SPSS were similar to those based on hand calculations. In both cases the omnibus test resulted in statistically significant differences between at least one pair of video methods. The hand-calculated test statistic was $\chi^2_{F_{(3)}} = 36.86$, while SPSS yielded $\chi^2_{F_{(3)}} = 33.75$.

Figure 12.36 Mean Ranks for Public Health Campaign Methods

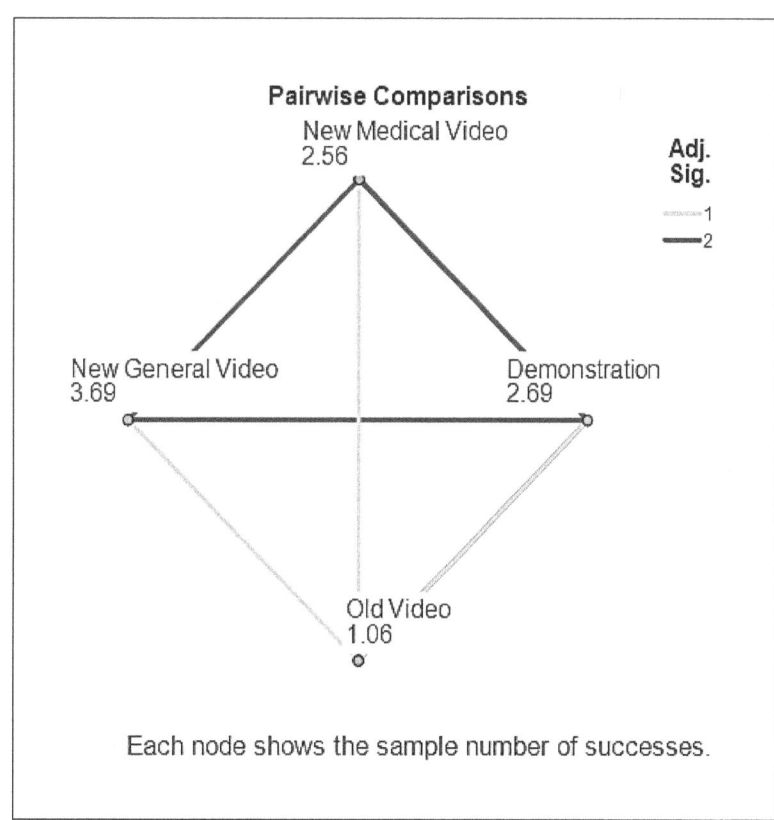

12.7.6.3 Measures of effect sizes

Although reporting the statistical significance of the differences in video ratings is essential in showing whether research findings are due to chance, reporting the effect size is probably equally important since it demonstrates the magnitude of the differences found. As the effect size increases, so does the practical significance of study results. Kendall's coefficient of concordance (Kendall's *W*) may be used to measure the effect size of the Friedman's rank test.

Kendall's *W* evaluates the agreement in study participants' ratings. The Kendall's *W* coefficient ranges from 0 to 1, where 0 means study participants did not rank the video methods in the same way and were in complete disagreement. A coefficient of 1 means subjects' ratings of the videos were virtually the same.

The SPSS specifications for conducting Kendall's *W* are similar to the ones required for Friedman's test in Section 12.7.6.1. Start by selecting **Analyze → Nonparametric Tests → Related Samples**. Use the same specifications in Figure 12.30 and 12.31. However, instead of selecting the default specification in the **Settings** menu (Figure 12.32), select **Customize tests**, then select **Kendall's coefficient of concordance (k samples)**.

Kendall's *W* results are displayed in Figure 12.37 and report a strong effect size of .703. This means that study participants' ratings reflect sizeable agreement on the ratings of public health video methods.

Figure 12.37 Effect Size for the Friedman's Rank Test Ratings Agreement

Related-Samples Kendall's Coefficient of Concordance Summary	
Total N	16
Kendall's *W*	.703
Test Statistic	33.750
Degree Of Freedom	3
Asymptotic Sig.(2-sided test)	<.001

12.7.6.4 Sample Report: The Case of a Public Health Video Campaign

Friedman's test was conducted to assess whether there were significant differences among the mean ranks of the ratings of the four public health campaign methods. Before the omnibus test was conducted, the data were screened for violations of the assumptions of Friedman's rank.

Results of Friedman's rank revealed that there were statistically significant differences among study participants' rankings of the four campaign methods, $\chi^2_F(3) = 33.75$, $p < .001$. Kendall's coefficient of concordance (Kendall's *W*), which was calculated to determine the effect size of the omnibus test, indicated a large effect size of .70.

The Dunn-Bonferroni post-hoc test was conducted to determine which pairs of video methods had significantly different rankings at a Bonferroni-corrected alpha level of .008. Results disclosed that rankings between the old video and the new medical video ($Z = 3.286$, $p = 0.001$), the old video and the demonstration ($Z = -3.56$, $p < 0.001$), as well as the old and new general video ($Z = 5.751$, $p < 0.001$), were all statistically significantly different. However, differences in the rankings between the new general video and the new medical video, the new medical video and the demonstration, as well as between the new general video and the demonstration were all statistically nonsignificant. Overall, results showed that the old video was routinely ranked lower than all other campaign methods. Based on Cohen's (1988) effect size interpretations, the differences in the rankings were of high practical significance ($W(3) = .703$, $p < .001$).

12.8 Key Terms

Friedman's rank
Kendall's tau-a
Kendall's tau-b
Kendall's W
Kruskal-Wallis test
Likert response scales
Median test
Nonparametric tests
Pairwise comparisons
Rankings
Spearman's rho

Appendix A

Statistical Tables

Table A.1 Critical Values of the Chi-Square Test Statistic

df	\multicolumn{9}{c}{Alpha Level}
	0.99	0.98	0.95	0.90	0.10	0.05	0.03	0.01	0.001
1	0.00	0.00	0.00	0.02	2.71	3.84	5.02	6.63	10.83
2	0.02	0.05	0.10	0.21	4.61	5.99	7.38	9.21	13.82
3	0.11	0.22	0.35	0.58	6.25	7.81	9.35	11.34	16.27
4	0.30	0.48	0.71	1.06	7.78	9.49	11.14	13.28	18.47
5	0.55	0.83	1.15	1.61	9.24	11.07	12.83	15.09	20.52
6	0.87	1.24	1.64	2.20	10.64	12.59	14.45	16.81	22.46
7	1.24	1.69	2.17	2.83	12.02	14.07	16.01	18.48	24.32
8	1.65	2.18	2.73	3.49	13.36	15.51	17.53	20.09	26.12
9	2.09	2.70	3.33	4.17	14.68	16.92	19.02	21.67	27.88
10	2.56	3.25	3.94	4.87	15.99	18.31	20.48	23.21	29.59
11	3.05	3.82	4.57	5.58	17.28	19.68	21.92	24.72	31.26
12	3.57	4.40	5.23	6.30	18.55	21.03	23.34	26.22	32.91
13	4.11	5.01	5.89	7.04	19.81	22.36	24.74	27.69	34.53
14	4.66	5.63	6.57	7.79	21.06	23.68	26.12	29.14	36.12
15	5.23	6.26	7.26	8.55	22.31	25.00	27.49	30.58	37.70
16	5.81	6.91	7.96	9.31	23.54	26.30	28.85	32.00	39.25
17	6.41	7.56	8.67	10.09	24.77	27.59	30.19	33.41	40.79
18	7.01	8.23	9.39	10.86	25.99	28.87	31.53	34.81	42.31
19	7.63	8.91	10.12	11.65	27.20	30.14	32.85	36.19	43.82
20	8.26	9.59	10.85	12.44	28.41	31.41	34.17	37.57	45.31
21	8.90	10.28	11.59	13.24	29.62	32.67	35.48	38.93	46.80
22	9.54	10.98	12.34	14.04	30.81	33.92	36.78	40.29	48.27
23	10.20	11.69	13.09	14.85	32.01	35.17	38.08	41.64	49.73
24	10.86	12.40	13.85	15.66	33.20	36.42	39.36	42.98	51.18
25	11.52	13.12	14.61	16.47	34.38	37.65	40.65	44.31	52.62
26	12.20	13.84	15.38	17.29	35.56	38.89	41.92	45.64	54.05
27	12.88	14.57	16.15	18.11	36.74	40.11	43.19	46.96	55.48
28	13.56	15.31	16.93	18.94	37.92	41.34	44.46	48.28	56.89
29	14.26	16.05	17.71	19.77	39.09	42.56	45.72	49.59	58.30
30	14.95	16.79	18.49	20.60	40.26	43.77	46.98	50.89	59.70
40	22.16	24.43	26.51	29.05	51.81	55.76	59.34	63.69	73.40
50	29.71	32.36	34.76	37.69	63.17	67.50	71.42	76.15	86.66
60	37.48	40.48	43.19	46.46	74.40	79.08	83.30	88.38	99.61
70	45.44	48.76	51.74	55.33	85.53	90.53	95.02	100.43	112.32
80	53.54	57.15	60.39	64.28	96.58	101.88	106.63	112.33	124.84
90	61.75	65.65	69.13	73.29	107.57	113.15	118.14	124.12	137.21
100	70.06	74.22	77.93	82.36	118.50	124.34	129.56	135.81	149.45

Table entries computed by author.

Table A.2 Significant Values of Correlation Coefficients

	Apha Level for One-Tailed Test			
	.050	.025	.010	.005
	Alpha Level for Two-Tailed Test			
df	.100	.050	.020	.010
1	.988	.997	1.000	1.000
2	.900	.950	.980	.990
3	.805	.878	.934	.959
4	.729	.811	.882	.917
5	.669	.754	.833	.875
6	.621	.707	.789	.834
7	.582	.666	.750	.798
8	.549	.632	.715	.765
9	.521	.602	.685	.735
10	.497	.576	.658	.708
11	.476	.553	.634	.684
12	.458	.532	.612	.661
13	.441	.514	.592	.641
14	.426	.497	.574	.623
15	.412	.482	.558	.606
16	.400	.468	.543	.590
17	.389	.456	.529	.575
18	.378	.444	.516	.561
19	.369	.433	.503	.549
20	.360	.423	.492	.537
21	.352	.413	.482	.526
22	.344	.404	.472	.515
23	.337	.396	.462	.505
24	.330	.388	.453	.496
25	.323	.381	.445	.487
26	.317	.374	.437	.479
27	.311	.367	.430	.471
28	.306	.361	.423	.463
29	.301	.355	.416	.456
30	.296	.349	.409	.449
31	.291	.344	.403	.442
32	.287	.339	.397	.436
33	.283	.334	.392	.430
34	.279	.329	.386	.424
35	.275	.325	.381	.418
36	.271	.320	.376	.413
37	.267	.316	.371	.408
38	.264	.312	.367	.403
39	.260	.308	.362	.398
40	.257	.304	.358	.393
41	.254	.301	.354	.389
42	.251	.297	.350	.384
43	.248	.294	.346	.380
44	.246	.291	.342	.376
45	.243	.288	.338	.372
46	.240	.285	.335	.368
47	.238	.282	.331	.365
48	.235	.279	.328	.361
49	.233	.276	.325	.358
50	.231	.273	.322	.354
60	.211	.250	.295	.325
70	.195	.232	.274	.302
80	.183	.217	.257	.283
90	.173	.205	.242	.267
100	.164	.195	.230	.254

Table entries computed by author.

Table A3 Transformation of Pearson's Product Moment Correlation Coefficient (r) to Fisher's z

r	z_r	r	z_r	r	z_r	r	z_r
.00	0.000	.25	0.255	.50	0.549	.75	0.973
.01	0.010	.26	0.266	.51	0.563	.76	0.996
.02	0.020	.27	0.277	.52	0.576	.77	1.020
.03	0.030	.28	0.288	.53	0.590	.78	1.045
.04	0.040	.29	0.299	.54	0.604	.79	1.071
.05	0.050	.30	0.310	.55	0.618	.80	1.099
.06	0.060	.31	0.321	.56	0.633	.81	1.127
.07	0.070	.32	0.332	.57	0.648	.82	1.157
.08	0.080	.33	0.343	.58	0.662	.83	1.188
.09	0.090	.34	0.354	.59	0.678	.84	1.221
.10	0.100	.35	0.365	.60	0.693	.85	1.256
.11	0.110	.36	0.377	.61	0.709	.86	1.293
.12	0.121	.37	0.388	.62	0.725	.87	1.333
.13	0.131	.38	0.400	.63	0.741	.88	1.376
.14	0.141	.39	0.412	.64	0.758	.89	1.422
.15	0.151	.40	0.424	.65	0.775	.90	1.472
.16	0.161	.41	0.436	.66	0.793	.91	1.528
.17	0.172	.42	0.448	.67	0.811	.92	1.589
.18	0.182	.43	0.460	.68	0.829	.93	1.658
.19	0.192	.44	0.472	.69	0.848	.94	1.738
.20	0.203	.45	0.485	.70	0.867	.95	1.832
.21	0.213	.46	0.497	.71	0.887	.96	1.946
.22	0.224	.47	0.510	.72	0.908	.97	2.092
.23	0.234	.48	0.523	.73	0.929	.98	2.298
.24	0.245	.49	0.536	.74	0.950	.99	2.647

Table entries computed by author.

Table A.4 Critical values of the *t* Test Statistic

	Alpha Level for One-Tailed Tests					
	.100	.050	.025	.010	.005	.001
	Alpha Level for Two-Tailed Tests					
df	.200	.100	.050	.020	.010	.001
1	3.078	6.314	12.706	31.821	63.657	636.619
2	1.886	2.920	4.303	6.965	9.925	31.599
3	1.638	2.353	3.182	4.541	5.841	12.924
4	1.533	2.132	2.776	3.747	4.604	8.610
5	1.476	2.015	2.571	3.365	4.032	6.869
6	1.440	1.943	2.447	3.143	3.707	5.959
7	1.415	1.895	2.365	2.998	3.499	5.408
8	1.397	1.860	2.306	2.896	3.355	5.041
9	1.383	1.833	2.262	2.821	3.250	4.781
10	1.372	1.812	2.228	2.764	3.169	4.587
11	1.363	1.796	2.201	2.718	3.106	4.437
12	1.356	1.782	2.179	2.681	3.055	4.318
13	1.350	1.771	2.160	2.650	3.012	4.221
14	1.345	1.761	2.145	2.624	2.977	4.140
15	1.341	1.753	2.131	2.602	2.947	4.073
16	1.337	1.746	2.120	2.583	2.921	4.015
17	1.333	1.740	2.110	2.567	2.898	3.965
18	1.330	1.734	2.101	2.552	2.878	3.922
19	1.328	1.729	2.093	2.539	2.861	3.883
20	1.325	1.725	2.086	2.528	2.845	3.850
21	1.323	1.721	2.080	2.518	2.831	3.819
22	1.321	1.717	2.074	2.508	2.819	3.792
23	1.319	1.714	2.069	2.500	2.807	3.768
24	1.318	1.711	2.064	2.492	2.797	3.745
25	1.316	1.708	2.060	2.485	2.787	3.725
26	1.315	1.706	2.056	2.479	2.779	3.707
27	1.314	1.703	2.052	2.473	2.771	3.690
28	1.313	1.701	2.048	2.467	2.763	3.674
29	1.311	1.699	2.045	2.462	2.756	3.659
30	1.310	1.697	2.042	2.457	2.750	3.646
40	1.303	1.684	2.021	2.423	2.704	3.551
50	1.299	1.676	2.009	2.403	2.678	3.496
60	1.296	1.671	2.000	2.390	2.660	3.460
100	1.290	1.660	1.984	2.364	2.626	3.390

Table entries computed by author.

Table A.5 Critical Values of the F Test Statistic: Alpha = .05

	$df_{numerator}$															
$df_{denominator}$	1	2	3	4	5	6	7	8	9	10	15	20	25	30	40	50
1	161.45	199.50	215.71	224.58	230.16	233.99	236.77	238.88	240.54	241.88	245.95	248.01	249.26	250.10	251.14	251.77
2	18.51	19.00	19.16	19.25	19.30	19.33	19.35	19.37	19.38	19.40	19.43	19.45	19.46	19.46	19.47	19.48
3	10.13	9.55	9.28	9.12	9.01	8.94	8.89	8.85	8.81	8.79	8.70	8.66	8.63	8.62	8.59	8.58
4	7.71	6.94	6.59	6.39	6.26	6.16	6.09	6.04	6.00	5.96	5.86	5.80	5.77	5.75	5.72	5.70
5	6.61	5.79	5.41	5.19	5.05	4.95	4.88	4.82	4.77	4.74	4.62	4.56	4.52	4.50	4.46	4.44
6	5.99	5.14	4.76	4.53	4.39	4.28	4.21	4.15	4.10	4.06	3.94	3.87	3.83	3.81	3.77	3.75
7	5.59	4.74	4.35	4.12	3.97	3.87	3.79	3.73	3.68	3.64	3.51	3.44	3.40	3.38	3.34	3.32
8	5.32	4.46	4.07	3.84	3.69	3.58	3.50	3.44	3.39	3.35	3.22	3.15	3.11	3.08	3.04	3.02
9	5.12	4.26	3.86	3.63	3.48	3.37	3.29	3.23	3.18	3.14	3.01	2.94	2.89	2.86	2.83	2.80
10	4.96	4.10	3.71	3.48	3.33	3.22	3.14	3.07	3.02	2.98	2.85	2.77	2.73	2.70	2.66	2.64
11	4.84	3.98	3.59	3.36	3.20	3.09	3.01	2.95	2.90	2.85	2.72	2.65	2.60	2.57	2.53	2.51
12	4.75	3.89	3.49	3.26	3.11	3.00	2.91	2.85	2.80	2.75	2.62	2.54	2.50	2.47	2.43	2.40
13	4.67	3.81	3.41	3.18	3.03	2.92	2.83	2.77	2.71	2.67	2.53	2.46	2.41	2.38	2.34	2.31
14	4.60	3.74	3.34	3.11	2.96	2.85	2.76	2.70	2.65	2.60	2.46	2.39	2.34	2.31	2.27	2.24
15	4.54	3.68	3.29	3.06	2.90	2.79	2.71	2.64	2.59	2.54	2.40	2.33	2.28	2.25	2.20	2.18
16	4.49	3.63	3.24	3.01	2.85	2.74	2.66	2.59	2.54	2.49	2.35	2.28	2.23	2.19	2.15	2.12
17	4.45	3.59	3.20	2.96	2.81	2.70	2.61	2.55	2.49	2.45	2.31	2.23	2.18	2.15	2.10	2.08
18	4.41	3.55	3.16	2.93	2.77	2.66	2.58	2.51	2.46	2.41	2.27	2.19	2.14	2.11	2.06	2.04
19	4.38	3.52	3.13	2.90	2.74	2.63	2.54	2.48	2.42	2.38	2.23	2.16	2.11	2.07	2.03	2.00
20	4.35	3.49	3.10	2.87	2.71	2.60	2.51	2.45	2.39	2.35	2.20	2.12	2.07	2.04	1.99	1.97
21	4.32	3.47	3.07	2.84	2.68	2.57	2.49	2.42	2.37	2.32	2.18	2.10	2.05	2.01	1.96	1.94
22	4.30	3.44	3.05	2.82	2.66	2.55	2.46	2.40	2.34	2.30	2.15	2.07	2.02	1.98	1.94	1.91
23	4.28	3.42	3.03	2.80	2.64	2.53	2.44	2.37	2.32	2.27	2.13	2.05	2.00	1.96	1.91	1.88
24	4.26	3.40	3.01	2.78	2.62	2.51	2.42	2.36	2.30	2.25	2.11	2.03	1.97	1.94	1.89	1.86
25	4.24	3.39	2.99	2.76	2.60	2.49	2.40	2.34	2.28	2.24	2.09	2.01	1.96	1.92	1.87	1.84
26	4.23	3.37	2.98	2.74	2.59	2.47	2.39	2.32	2.27	2.22	2.07	1.99	1.94	1.90	1.85	1.82
27	4.21	3.35	2.96	2.73	2.57	2.46	2.37	2.31	2.25	2.20	2.06	1.97	1.92	1.88	1.84	1.81
28	4.20	3.34	2.95	2.71	2.56	2.45	2.36	2.29	2.24	2.19	2.04	1.96	1.91	1.87	1.82	1.79
29	4.18	3.33	2.93	2.70	2.55	2.43	2.35	2.28	2.22	2.18	2.03	1.94	1.89	1.85	1.81	1.77
30	4.17	3.32	2.92	2.69	2.53	2.42	2.33	2.27	2.21	2.16	2.01	1.93	1.88	1.84	1.79	1.76
40	4.08	3.23	2.84	2.61	2.45	2.34	2.25	2.18	2.12	2.08	1.92	1.84	1.78	1.74	1.69	1.66
50	4.03	3.18	2.79	2.56	2.40	2.29	2.20	2.13	2.07	2.03	1.87	1.78	1.73	1.69	1.63	1.60
60	4.00	3.15	2.76	2.53	2.37	2.25	2.17	2.10	2.04	1.99	1.84	1.75	1.69	1.65	1.59	1.56
120	3.92	3.07	2.68	2.45	2.29	2.18	2.09	2.02	1.96	1.91	1.75	1.66	1.60	1.55	1.50	1.46
500	3.86	3.01	2.62	2.39	2.23	2.12	2.03	1.96	1.90	1.85	1.69	1.59	1.53	1.48	1.42	1.38
1000	3.85	3.00	2.61	2.38	2.22	2.11	2.02	1.95	1.89	1.84	1.68	1.58	1.52	1.47	1.41	1.36

Table entries computed by author.

Table A.6 Critical Values of the *F* Test Statistic: Alpha = .01

	$df_{numerator}$														
$df_{denominator}$	1	2	3	4	5	6	7	8	9	10	15	20	25	30	40	50
1	4052	5000	5403	5625	5764	5859	5928	5981	6022	6056	6157	6209	6240	6261	6287	6303
2	98.50	99.00	99.17	99.25	99.30	99.33	99.36	99.37	99.39	99.40	99.43	99.45	99.46	99.47	99.47	99.48
3	34.12	30.82	29.46	28.71	28.24	27.91	27.67	27.49	27.35	27.23	26.87	26.69	26.58	26.50	26.41	26.35
4	21.20	18.00	16.69	15.98	15.52	15.21	14.98	14.80	14.66	14.55	14.20	14.02	13.91	13.84	13.75	13.69
5	16.26	13.27	12.06	11.39	10.97	10.67	10.46	10.29	10.16	10.05	9.72	9.55	9.45	9.38	9.29	9.24
6	13.75	10.92	9.78	9.15	8.75	8.47	8.26	8.10	7.98	7.87	7.56	7.40	7.30	7.23	7.14	7.09
7	12.25	9.55	8.45	7.85	7.46	7.19	6.99	6.84	6.72	6.62	6.31	6.16	6.06	5.99	5.91	5.86
8	11.26	8.65	7.59	7.01	6.63	6.37	6.18	6.03	5.91	5.81	5.52	5.36	5.26	5.20	5.12	5.07
9	10.56	8.02	6.99	6.42	6.06	5.80	5.61	5.47	5.35	5.26	4.96	4.81	4.71	4.65	4.57	4.52
10	10.04	7.56	6.55	5.99	5.64	5.39	5.20	5.06	4.94	4.85	4.56	4.41	4.31	4.25	4.17	4.12
11	9.65	7.21	6.22	5.67	5.32	5.07	4.89	4.74	4.63	4.54	4.25	4.10	4.01	3.94	3.86	3.81
12	9.33	6.93	5.95	5.41	5.06	4.82	4.64	4.50	4.39	4.30	4.01	3.86	3.76	3.70	3.62	3.57
13	9.07	6.70	5.74	5.21	4.86	4.62	4.44	4.30	4.19	4.10	3.82	3.66	3.57	3.51	3.43	3.38
14	8.86	6.51	5.56	5.04	4.69	4.46	4.28	4.14	4.03	3.94	3.66	3.51	3.41	3.35	3.27	3.22
15	8.68	6.36	5.42	4.89	4.56	4.32	4.14	4.00	3.89	3.80	3.52	3.37	3.28	3.21	3.13	3.08
16	8.53	6.23	5.29	4.77	4.44	4.20	4.03	3.89	3.78	3.69	3.41	3.26	3.16	3.10	3.02	2.97
17	8.40	6.11	5.18	4.67	4.34	4.10	3.93	3.79	3.68	3.59	3.31	3.16	3.07	3.00	2.92	2.87
18	8.29	6.01	5.09	4.58	4.25	4.01	3.84	3.71	3.60	3.51	3.23	3.08	2.98	2.92	2.84	2.78
19	8.18	5.93	5.01	4.50	4.17	3.94	3.77	3.63	3.52	3.43	3.15	3.00	2.91	2.84	2.76	2.71
20	8.10	5.85	4.94	4.43	4.10	3.87	3.70	3.56	3.46	3.37	3.09	2.94	2.84	2.78	2.69	2.64
21	8.02	5.78	4.87	4.37	4.04	3.81	3.64	3.51	3.40	3.31	3.03	2.88	2.79	2.72	2.64	2.58
22	7.95	5.72	4.82	4.31	3.99	3.76	3.59	3.45	3.35	3.26	2.98	2.83	2.73	2.67	2.58	2.53
23	7.88	5.66	4.76	4.26	3.94	3.71	3.54	3.41	3.30	3.21	2.93	2.78	2.69	2.62	2.54	2.48
24	7.82	5.61	4.72	4.22	3.90	3.67	3.50	3.36	3.26	3.17	2.89	2.74	2.64	2.58	2.49	2.44
25	7.77	5.57	4.68	4.18	3.85	3.63	3.46	3.32	3.22	3.13	2.85	2.70	2.60	2.54	2.45	2.40
26	7.72	5.53	4.64	4.14	3.82	3.59	3.42	3.29	3.18	3.09	2.81	2.66	2.57	2.50	2.42	2.36
27	7.68	5.49	4.60	4.11	3.78	3.56	3.39	3.26	3.15	3.06	2.78	2.63	2.54	2.47	2.38	2.33
28	7.64	5.45	4.57	4.07	3.75	3.53	3.36	3.23	3.12	3.03	2.75	2.60	2.51	2.44	2.35	2.30
29	7.60	5.42	4.54	4.04	3.73	3.50	3.33	3.20	3.09	3.00	2.73	2.57	2.48	2.41	2.33	2.27
30	7.56	5.39	4.51	4.02	3.70	3.47	3.30	3.17	3.07	2.98	2.70	2.55	2.45	2.39	2.30	2.25
40	7.31	5.18	4.31	3.83	3.51	3.29	3.12	2.99	2.89	2.80	2.52	2.37	2.27	2.20	2.11	2.06
50	7.17	5.06	4.20	3.72	3.41	3.19	3.02	2.89	2.78	2.70	2.42	2.27	2.17	2.10	2.01	1.95
60	7.08	4.98	4.13	3.65	3.34	3.12	2.95	2.82	2.72	2.63	2.35	2.20	2.10	2.03	1.94	1.88
120	6.85	4.79	3.95	3.48	3.17	2.96	2.79	2.66	2.56	2.47	2.19	2.03	1.93	1.86	1.76	1.70
500	6.69	4.65	3.82	3.36	3.05	2.84	2.68	2.55	2.44	2.36	2.07	1.92	1.81	1.74	1.63	1.57
1000	6.66	4.63	3.80	3.34	3.04	2.82	2.66	2.53	2.43	2.34	2.06	1.90	1.79	1.72	1.61	1.54

Table entries computed by author.

Table A.7 Critical Values of the F Test Statistic: Alpha = .001

	dfnumerator															
dfdenominator	1	2	3	4	5	6	7	8	9	10	15	20	25	30	40	50
1	405284	500000	540379	562500	576405	585937	592873	598144	602284	605621	615764	620908	624017	626099	628712	630285
2	998.50	999.00	999.17	999.25	999.30	999.33	999.36	999.37	999.39	999.40	999.43	999.45	999.46	999.47	999.47	999.48
3	167.03	148.50	141.11	137.10	134.58	132.85	131.58	130.62	129.86	129.25	127.37	126.42	125.84	125.45	124.96	124.66
4	74.14	61.25	56.18	53.44	51.71	50.53	49.66	49.00	48.47	48.05	46.76	46.10	45.70	45.43	45.09	44.88
5	47.18	37.12	33.20	31.09	29.75	28.83	28.16	27.65	27.24	26.92	25.91	25.39	25.08	24.87	24.60	24.44
6	35.51	27.00	23.70	21.92	20.80	20.03	19.46	19.03	18.69	18.41	17.56	17.12	16.85	16.67	16.44	16.31
7	29.25	21.69	18.77	17.20	16.21	15.52	15.02	14.63	14.33	14.08	13.32	12.93	12.69	12.53	12.33	12.20
8	25.41	18.49	15.83	14.39	13.48	12.86	12.40	12.05	11.77	11.54	10.84	10.48	10.26	10.11	9.92	9.80
9	22.86	16.39	13.90	12.56	11.71	11.13	10.70	10.37	10.11	9.89	9.24	8.90	8.69	8.55	8.37	8.26
10	21.04	14.91	12.55	11.28	10.48	9.93	9.52	9.20	8.96	8.75	8.13	7.80	7.60	7.47	7.30	7.19
11	19.69	13.81	11.56	10.35	9.58	9.05	8.66	8.35	8.12	7.92	7.32	7.01	6.81	6.68	6.52	6.42
12	18.64	12.97	10.80	9.63	8.89	8.38	8.00	7.71	7.48	7.29	6.71	6.40	6.22	6.09	5.93	5.83
13	17.82	12.31	10.21	9.07	8.35	7.86	7.49	7.21	6.98	6.80	6.23	5.93	5.75	5.63	5.47	5.37
14	17.14	11.78	9.73	8.62	7.92	7.44	7.08	6.80	6.58	6.40	5.85	5.56	5.38	5.25	5.10	5.00
15	16.59	11.34	9.34	8.25	7.57	7.09	6.74	6.47	6.26	6.08	5.54	5.25	5.07	4.95	4.80	4.70
16	16.12	10.97	9.01	7.94	7.27	6.80	6.46	6.19	5.98	5.81	5.27	4.99	4.82	4.70	4.54	4.45
17	15.72	10.66	8.73	7.68	7.02	6.56	6.22	5.96	5.75	5.58	5.05	4.78	4.60	4.48	4.33	4.24
18	15.38	10.39	8.49	7.46	6.81	6.35	6.02	5.76	5.56	5.39	4.87	4.59	4.42	4.30	4.15	4.06
19	15.08	10.16	8.28	7.27	6.62	6.18	5.85	5.59	5.39	5.22	4.70	4.43	4.26	4.14	3.99	3.90
20	14.82	9.95	8.10	7.10	6.46	6.02	5.69	5.44	5.24	5.08	4.56	4.29	4.12	4.00	3.86	3.77
21	14.59	9.77	7.94	6.95	6.32	5.88	5.56	5.31	5.11	4.95	4.44	4.17	4.00	3.88	3.74	3.64
22	14.38	9.61	7.80	6.81	6.19	5.76	5.44	5.19	4.99	4.83	4.33	4.06	3.89	3.78	3.63	3.54
23	14.20	9.47	7.67	6.70	6.08	5.65	5.33	5.09	4.89	4.73	4.23	3.96	3.79	3.68	3.53	3.44
24	14.03	9.34	7.55	6.59	5.98	5.55	5.23	4.99	4.80	4.64	4.14	3.87	3.71	3.59	3.45	3.36
25	13.88	9.22	7.45	6.49	5.89	5.46	5.15	4.91	4.71	4.56	4.06	3.79	3.63	3.52	3.37	3.28
26	13.74	9.12	7.36	6.41	5.80	5.38	5.07	4.83	4.64	4.48	3.99	3.72	3.56	3.44	3.30	3.21
27	13.61	9.02	7.27	6.33	5.73	5.31	5.00	4.76	4.57	4.41	3.92	3.66	3.49	3.38	3.23	3.14
28	13.50	8.93	7.19	6.25	5.66	5.24	4.93	4.69	4.50	4.35	3.86	3.60	3.43	3.32	3.18	3.09
29	13.39	8.85	7.12	6.19	5.59	5.18	4.87	4.64	4.45	4.29	3.80	3.54	3.38	3.27	3.12	3.03
30	13.29	8.77	7.05	6.12	5.53	5.12	4.82	4.58	4.39	4.24	3.75	3.49	3.33	3.22	3.07	2.98
40	12.61	8.25	6.59	5.70	5.13	4.73	4.44	4.21	4.02	3.87	3.40	3.14	2.98	2.87	2.73	2.64
50	12.22	7.96	6.34	5.46	4.90	4.51	4.22	4.00	3.82	3.67	3.20	2.95	2.79	2.68	2.53	2.44
60	11.97	7.77	6.17	5.31	4.76	4.37	4.09	3.86	3.69	3.54	3.08	2.83	2.67	2.55	2.41	2.32
120	11.38	7.32	5.78	4.95	4.42	4.04	3.77	3.55	3.38	3.24	2.78	2.53	2.37	2.26	2.11	2.02
500	10.96	7.00	5.51	4.69	4.18	3.81	3.54	3.33	3.16	3.02	2.58	2.33	2.17	2.05	1.90	1.80
1000	10.89	6.96	5.46	4.65	4.14	3.78	3.51	3.30	3.13	2.99	2.54	2.30	2.14	2.02	1.87	1.77

Table entries computed by author.

Table A.8 Critical Values of Spearman's Rho Coefficients

	Alpha Level for One-Tailed Test				
	.050	.025	.010	.005	.0005
	Alpha Level for Two-Tailed Test				
N	.100	.050	.020	.010	.001
4	1.000	--	--	--	
5	.900	1.000	1.000	--	
6	.829	.886	.943	1.000	--
7	.714	.786	.893	.929	1.000
8	.643	.738	.833	.881	.976
9	.600	.700	.783	.833	.933
10	.564	.648	.745	.794	.903
11	.536	.618	.709	.755	.873
12	.503	.587	.678	.727	.846
13	.484	.560	.648	.703	.824
14	.464	.538	.626	.679	.802
15	.446	.521	.604	.654	.779
16	.429	.503	.582	.635	.762
17	.414	.488	.566	.618	.748
18	.401	.472	.550	.600	.728
19	.391	.460	.535	.584	.712
20	.380	.447	.522	.570	.696
21	.370	.436	.509	.556	.681
22	.361	.425	.497	.544	.667
23	.353	.416	.486	.532	.654
24	.344	.407	.476	.521	.642
25	.337	.398	.466	.511	.630
26	.331	.390	.457	.501	.619
27	.324	.383	.449	.492	.608
28	.318	.375	.441	.483	.598
29	.312	.368	.433	.475	.589
30	.306	.362	.425	.467	.580
35	.283	.335	.394	.433	.539
40	.313	.368	.405	.439	.507
45	.294	.347	.382	.414	.479
50	.279	.329	.363	.393	.456

Adapted from Zar, J. H. (1972). Significance testing of the Spearman rank correlation coefficient. Journal of The American Statistical Association, 67, 578-580. Reprinted with permission from the American Statistical Association. All rights reserved.

Table A.9 Critical Values of Kendall's Tau Coefficients

	Alpha Level for Two-Tailed Test					
	.200	.100	.050	.020	.010	.002
	Alpha Level for One-Tailed Test					
N	.100	.050	.025	.010	.005	.001
4	1.000	1.000				
5	.800	.800	1.000	1.000		
6	.600	.733	.867	.867	1.000	
7	.524	.619	.714	.810	.905	1.000
8	.429	.571	.643	.714	.786	.857
9	.389	.500	.556	.667	.722	.833
10	.378	.467	.551	.600	.644	.778
11	.345	.418	.491	.564	.600	.709
12	.303	.394	.455	.545	.576	.667
13	.308	.359	.436	.513	.564	.641
14	.275	.363	.407	.473	.516	.604
15	.276	.333	.390	.467	.505	.581
16	.250	.317	.383	.433	.483	.567
17	.250	.309	.368	.426	.471	.544
18	.242	.294	.346	.412	.451	.529
19	.228	.287	.333	.392	.439	.509
20	.221	.274	.326	.379	.421	.495
21	.210	.267	.314	.371	.410	.486
22	.203	.264	.307	.359	.394	.472
23	.202	.257	.296	.352	.391	.455
24	.196	.246	.290	.341	.377	.449
25	.193	.240	.287	.333	.367	.440
26	.188	.237	.280	.329	.360	.428
27	.179	.231	.271	.322	.356	.419
28	.180	.228	.265	.312	.344	.413
29	.172	.222	.261	.310	.340	.404
30	.172	.218	.255	.301	.333	.393

Taken from: http://www.york.ac.uk/depts/maths/tables/kendall.pdf

Table A.10 Critical Values of the Friedman Test Statistic

		Alpha Level					
k	N	.100	.050	.025	.010	.005	.001
3	3	6.000	6.000				
	4	6.000	6.500	8.000	8.000	8.000	
	5	5.200	6.400	7.600	8.400	10.000	10.000
	6	5.330	7.000	8.333	9.000	10.333	12.000
	7	5.429	7.143	7.714	8.857	10.286	12.286
	8	5.250	6.250	7.750	9.000	9.750	12.250
	9	5.556	6.222	8.000	8.667	10.667	12.667
	10	5.000	6.200	7.800	9.600	10.400	12.600
	11	4.909	6.545	7.818	9.455	10.364	13.273
	12	5.167	6.500	8.000	9.500	10.167	12.500
	13	4.769	6.000	7.538	9.385	10.308	12.924
	14	5.143	6.143	7.429	9.000	10.429	13.286
	15	4.933	6.400	7.600	8.933	10.000	12.933
4	2	6.000	6.000				
	3	6.600	7.400	8.200	9.000	9.000	
	4	6.300	7.800	8.400	9.600	10.200	11.100
	5	6.360	7.800	8.760	9.960	10.920	12.600
	6	6.400	7.600	8.800	10.200	11.400	12.800
	7	6.429	7.800	9.000	10.371	11.400	13.800
	8	6.300	7.650	9.000	10.500	11.850	13.950
	9	6.467	7.800	9.133	10.867	12.067	14.467
	10	6.360	7.800	9.120	10.800	12.000	14.640
	11	6.382	7.909	9.327	11.073	12.273	14.891
	12	6.400	7.900	9.200	11.100	12.300	15.000
	13	6.415	7.985	7.369	11.123	12.323	15.277
	14	6.343	7.886	9.343	11.143	12.514	15.257
	15	6.440	8.040	9.400	11.240	12.520	15.400
5	2	7.200	7.600	8.000	8.000		
	3	7.467	8.533	9.600	10.133	10.667	11.467
	4	7.600	8.800	9.800	11.200	12.000	13.200
	5	7.680	8.960	10.240	11.680	12.480	14.400
	6	7.733	9.067	10.400	11.867	13.067	15.200
	7	7.771	9.143	10.514	12.114	13.257	15.657
	8	7.800	9.300	10.600	12.300	13.500	16.000
	9	7.733	9.244	10.667	12.444	13.689	16.356
	10	7.760	9.280	10.720	12.480	13.840	16.480
6	2	8.286	9.143	9.429	9.714	10.000	
	3	8.714	9.857	10.810	11.762	12.524	13.286
	4	9.000	10.286	11.429	12.714	13.571	15.286
	5	9.000	10.486	11.743	13.229	14.257	16.429
	6	9.048	10.571	12.000	13.619	14.762	17.048
	7	9.122	10.674	12.061	13.857	15.000	17.612
	8	9.143	10.714	12.214	14.000	15.286	18.000
	9	9.127	10.778	12.302	14.143	15.476	18.270
	10	9.143	10.800	12.343	14.229	15.600	18.514

Adapted from Martin, L., Leblanc, R., & Toan, N. K. (1993). Tables for the Friedman rank test. *The Canadian Journal of Statistics/La Revue Canadienne de Statistique*, 21(1), 39-43. Reprinted with permission from *The Canadian Journal of Statistics*. Copyright 1993 by the Statistical Society of Canada. All rights reserved.

Appendix B

Data Files

Data files for all demonstration exercises are available for downloading from https://www.keys2statistics.com.

Chapter	Demonstration Datasets: Filenames
Chapter 1	
Chapter 2	
Chapter 3	Carib_Height_Weight_Dups.sav Carib_Height_Weight_Nodups.sav Carib_Height_Weight_2022.sav WVS_TT_Screening_Exer.sav
Chapter 4	Math_Comm_Scores.sav
Chapter 5	Carib_Height_Weight.sav
Chapter 6	NHANES.sav
Chapter 7	Employee_Fin.sav Equal_N_One_Way.sav
Chapter 8	Factorial_ANOVA_Demo.sav
Chapter 9	Two_Way_Comtask.sav Fullness.sav
Chapter 10	Soccer_sprint.sav
Chapter 11	Olympics_16.sav
Chapter 12	Hlth_Vid_NT.sav Hypertension.sav Math_Instr.sav WVS_Trust.sav

References

Abu-Ghaida, D. N., Bundy, D. A. P., Hay, P., Kim, G., Koda, Y., Nagashima, Y., Psifidou, I., Sosale, S., Wang, Y., & Welsh, T. (2005). *Expanding opportunities and building competencies for young people: A new agenda for secondary education.* E. Cuadra, J. M. Moreno, L. Crouch, (Eds.) World Bank Group.

Adedokun, O. A., & Burgess, W. D. (2012). Analysis of paired dichotomous data: A gentle introduction to the McNemar test in SPSS. *Journal of Multi-Disciplinary Evaluation, 8*(17), 125-131. https://www.learntechlib.org/p/76538/.

Adelson, J. L., Dickinson, E. R., & Cunningham, B. C. (2015). Differences in the reading–mathematics relationship: A multi-grade, multi-year statewide examination. *Learning and Individual Differences*, 43, 118-123. https://doi.org/10.1016/j.lindif.2015.08.006.

Allen JS. (2012). *"Theory of food" as a neurocognitive adaptation. American Journal of Human Biology, 24, 123–129.* https://doi:org 10.1002/ajhb.22209.

Altendorf, S. (2019). *Bananas and major tropical fruits in Latin America and the Caribbean: The significance of the region to world supply.* In FAO Food Outlook. www.fao.org/3/ca4526en/ca4526en_sf.pdf.

American Psychological Association. (2020). *Publication manual of the American Psychological Association 2020: the official guide to APA style* (7th ed.). American Psychological Association.

Arbuthnott, J. (1710). An Argument for Divine Providence, Taken from the Constant Regularity Observ'd in the Births of Both Sexes. *Philosophical Transactions (1683-1775), 27*, 186–190. http://www.jstor.org/stable/103111.

Asimov, I. (2009). *I, Asimov: A Memoir.* Bantam Books, p.54.

Baraldi, A. N. & Enders, C. K. (2010). An introduction to modern missing data analyses. *Journal of School Psychology, 48*(1), 5–37. https://doi.org/10.1016/j.jsp.2009.10.001.

Bar-On, R. (1997). The emotional quotient inventory (EQ-i): A test of emotional intelligence. Toronto: Multi-Health Systems.

Bartko, J. J. (1966). The intraclass correlation coefficient as a measure of reliability. *Psychological Reports, 19*(1), 3–11. https://doi.org/10.2466/pr0.1966.19.1.3

Becker, G. S. (1962): "Investment in human capital: A theoretical analysis," *The Journal of Political Economy*, 70(5), 9-49.

Bennett, D. A. (2001). How can I deal with missing data in my study? *Australian and New Zealand Journal of Public Health*, 25, 464-469.

Berman, H. B., *"Transformations in Regression"* (online). https://stattrek.com/regression/linear-transformation

Boffa, J., Moules, N., Mayan, M., & Cowie, R. L. (2013, Summer). More than just great quotes: An introduction to the Canadian Tri-Council's qualitative requirements. *The Canadian Journal of Infectious Diseases & Medical Microbiology = Journal canadien des maladies infectieuses et de la microbiologie medicale, 24*(2), 103–108. https://doi.org/10.1155/2013/253931.

Bonferroni C. E. (1936). Teoria statistica delle classi e calcolo delle probabilita. *Pubblicazioni del R Instituto Superiore de Scienze Economiche e Commerciali de Firenze*, 8, 3–62.

Braun, H. (1994). *The Collected Works of John W. Tukey VIII: Multiple Comparisons Vol. VIII* (1st ed.). Chapman and Hall.

Brown, M. B. & Forsythe, A. B. (1974). The small sample behavior of some statistics which test the equality of several means. *Technometrics*, 16, 129-132.

Brown, M. B. & Forsythe, A. B. (1974). Robust tests for the equality of variances. *Journal of the American Statistical Association*, 69, 364-367

Browne, M. W. (1975). Predictive validity of a linear regression equation. *British Journal of Mathematical and Statistical Psychology*, 28, 79-87.

Burton, R. (2009, August 13). Cogito ergo sum, baby. https://www.salon.com/2009/08/13/philosophical_baby/

Campbell, D.T. and Stanley, J.C. (1963). Experimental and quasi-experimental designs for research on teaching. In N. L. Gage (Ed.), *Handbook of research on teaching.* Rand McNally (pp.171-246).

Campbell, D. T., & Stanley, J. (1963). *Experimental and quasi-experimental designs for research.* Rand McNally.

Carifio, J., & Perla, R. J. (2007). Ten common misunderstandings, misconceptions, persistent myths and urban legends about Likert scales and Likert response formats and their antidotes. *Journal of Social Sciences*, 3, 106-116.

Cohen, B. (2008). *Explaining psychological statistics.* John Wiley & Sons.

Cohen J. (1960). A coefficient of agreement for nominal scales. *Educational and Psychological Measurement.* 20, 37–46.

Cohen, J. (1973). Eta-squared and partial eta-squared in fixed factor ANOVA designs. *Educational and Psychological Measurement, 33*, 107–112.

Cohen J. (1983). The cost of dichotomization. *Applied Psychological Measurement, 7*(3), 249-53.

Cohen, J. (1988). *Statistical power analysis for the behavioral sciences* (2nd ed.). Lawrence Erlbaum.

Cohen, L., Manion, L., & Morrison, K. (2018). *Research methods in education* (8th ed.). Routledge. https://doi.org/10.4324/9781315456539.

Conover, W. J. (1999). Chapter 3.4: The sign test, *Practical nonparametric statistics* (3rd ed.), Wiley, (pp 157-176). ISBN 0-471-16068-7.

Cook, T. D., & Campbell, D. T. (1979*) Quasi-experimentation design and analysis issues for field settings.* Houghton Mifflin.

Cook, R. D. & Weisberg, S. (1982) *Residuals and influence in regression.* Chapman and Hall.

Corder G. W., & Foreman, D. I. (2010). *Nonparametric statistics for non-statisticians: A step-by-step approach.* Wiley.

Cortez, P., & Silva, A. (2008, April). *Using data mining to predict secondary school student performance.* In A. Brito & J. Teixeira (Eds.), Proceedings of 5th Future Business Technology Conference, Porto 5-12. ISBN 978-9077381-39-7. http://archive.ics.uci.edu/ml/datasets/Student+Performance.

Curriculum Planning and Development Division (2015). *Continuous assessment component: English language arts writing exemplars.* Ministry of Education. Trinidad and Tobago.

Delhey, J., Newton, K., & Welzel, C. (2011, October). How general is trust in "most people"? Solving the radius of the trust problem. *American Sociological Review.* 76(5), 786–807. DOI: 10.1177/0003122411420817. http://asr.sagepub.com.

DeVellis, R. F. & Thorpe, C. T. (2022). *Scale development : theory and applications* (Fifth Edition). London: Sage.

Dillman, D. A., Smyth, J. D., & Christian, L. M. (2014). *Internet, phone, mail, and mixed mode surveys: The tailored design method (4th ed.).* John Wiley & Sons Inc.

Djaoui, L., Chamari, K., Owen, A. L., & Dellal, A. (2017). Maximal sprinting speed of elite soccer players during training and matches. *Journal of Strength and Conditioning Research,* 31(6), 1509–1517.

Durbin, J. & Watson, G. S. (1951). Testing for serial correlation in least squares regression. II. *Biometrika*, 38, 159-178.

Durbin, J., & Watson, G. S. (*1971*). Testing for serial correlation in least squares regression. III. *Biometrika*, 58, 1-19.

Einstein, A. (2021, March 17). https://www.easybib.com/guides/quotes-facts-stats/albert-einstein.

Enders, C. K. (2022). *Applied missing data analysis* (2nd ed.). The Guilford Press.

Enders, C. K. (2023, March 16). Missing data: An update on the state of the art. *Psychological Methods.* Advance online publication. https://dx.doi.org/10.1037/met0000563.

Eynizadeha, Z., Amelia, Z., Sahranavarda, M., Daneshparvara, M., Abdollahi Dolagha, M., Roozkhosha, M., Reza Ejtehadib, M., & Azimzadeh Irania, M. (2020, November 23). Biostatistical investigation of correlation between COVID-19 and diabetes mellitus. https://www.medrxiv.org/content/10.1101/2020.11.21.20235853v1.full/.

Fernandez, M. E. (2021). Is mango the luscious superhero of fruit? *American Heart Association News.*

Fieller, E. C., Hartley, H. O., & Pearson, E. S. (1957). *Tests for rank correlation coefficients. Biometrika*, 44(3/4), 470–481.

Firmin, M., Hwang, C., Burger, A., Sammons, J., & Lowrie, R. (2008). Evaluating the concurrent validity of three web-based IQ tests and the Reynolds Intellectual Assessment Scales (RIAS). *Eastern Education Journal*, 37, 20-28.

Fisher, R. A. (1925). *Statistical methods for research workers.* Oliver and Boyd.

Fraenkel, J. R., & Wallen, N. E. (2018). *How to Design and Evaluate Research in Education.* (10th ed.). McGraw-Hill.

Frasca, M., Burucoa, B., Domecq, S., Robinson, N., Dousset, V., Cadenne, M., Sztark, F., & Floccia, M. (2017). Validation of the Behavioural Observation Scale 3 for the evaluation of pain in adults. *European journal of pain (London, England), 21*(9), 1475–1484. https://doi.org/10.1002/ejp.1049

Friedman, M. (1937). The use of ranks to avoid the assumption of normality implicit in the analysis of variance. *Journal of the American Statistical Association.* 32(200): 675–701.

Gauch Jr., H. G. (2003), *Scientific Method in practice.* Cambridge University Press, ISBN 978-0-521-01708-4

Gall, M., Gall, J., & Borg, R. (2007). *Educational research: An introduction* (8th ed.). Pearson Education.

Gay, L. R. (1996). *Educational research: Competences for analysis and application.* Prentice-Hall, Inc.

George, D., & *Mallery* P. (2019). *IBM SPSS Statistics 25 step* by *step*: A *simple guide* and *reference* (15th ed.). Routledge. https://doi.org/10.4324/9781351033909.

Goldberg, L. R. (1992). The development of markers for the big-five factor structure. *Psychological Assessment,* 4(1), 26-42. https://doi.org/10.1037/1040-3590.4.1.26.

Gottfried, M. A. (2010). Evaluating the relationship between student attendance and achievement in urban elementary and middle schools: An instrumental variables approach. *American Educational Research Journal, 47*(2), 434–465.

Gottlieb, S. (2020, April 26). America needs to win the coronavirus vaccine race. *The Wall Street Journal.*

Graham, J. W. (2012). *Missing data: Analysis and design.* Springer.

Graham, S., Kiuhara, S.A. & MacKay, M. (2020). The effects of writing on learning in science, social studies, and mathematics: A meta-analysis. *Review of Educational Research*, 90(2), 179-226. https://doi.org/10.3102/0034654320914744.

Green, S. B. (1991). How many subjects does it take to do a regression analysis? *Multivariate Behavioral Research, 26,* 499-510.

Grimm, K. J. (2008). Longitudinal associations between reading and mathematics achievement. *Developmental Neuropsychology,* 33(3), 410-426.

Gutin, I. (2018, August). In BMI we trust: Reframing the body mass index as a measure of health. *Social Theory and Health.*16(3), 256-271.

Hair, J. F., Black, W. C., Babin, B. J. & Anderson, R. E. (2010) *Multivariate Data Analysis. 7th Edition.* Pearson.

Hall, M. T. (2015). An examination into the validity of secondary school entrance scores in predicting the academic success of secondary aged students. *Current Issues in Education, 18*(1). http://cie.asu.edu/ojs/index.php/cieatasu/article/view/1343.

Harris, R. J. (1985). *A primer of multivariate statistics* (2nd ed.). Academic Press.

Heavenrich, R. M., Murrell, J. D. & Hellman, K. H. (1991). *Light duty automotive technology and fuel economy trends through 1991.* U.S. Environmental Protection Agency. (EPA/AA/CTAB/91-02).

Henry, D. L., Nistor, N., & Baltes, B. (2014). Examining the relationship between math scores and English language proficiency. *Journal of Educational Research and Practice.* 4(1), 11–29.

Hoaglin, D. C., & Welsch, R. E. (1978). The Hat Matrix in Regression and ANOVA. *The American Statistician,* 32(1), 17–22. https://doi.org/10.1080/00031305.1978.10479237

Holland, B. S., & Copenhaver, M. D. (1988). Improved Bonferroni-type multiple testing procedures. *Psychological Bulletin* 104, 145–149.

Holt, S. H., Miller, J. C., Petocz, P., & Farmakalidis, E. (1995). A satiety index of common foods. *European journal of clinical nutrition, 49*(9), 675–690.

The Holy Bible: New King James Version. (1982). Thomas Nelson.

Howell, D. C. (2014). *Fundamental statistics for the behavioral sciences* (8th ed.). Wadsworth Press, Cengage Learning.

Inglehart, R., Haerpfer, C., Moreno, A., Welzel, C., Kizilova, K., Diez-Medrano, J., Lagos, M., Norris, P., Ponarin, E., & Puranen, B. (eds.). 2014. *World Values Survey: Round Six - Country-Pooled Datafile Version.* http://www.worldvaluessurvey.org/WVSDocumentationWV6.jsp. JD Systems Institute.

International Diabetes Federation. (2019, November 25) *International Diabetes Federation Diabetes Atlas.* https://diabetesatlas.org/data/en/.

Cayman iNews. (2015, September 13). *Ivan's 11th Anniversary; remembering the lessons learned.* https://www.ieyenews.com/ivans-11th-anniversary-remembering-the-lessons-learned/.

Jamieson, S. (2004). Likert scales: How to (ab)use them. *Medical Education, 38,* 1212-1218. https://doi.org/10.1111/j.1365-2929.2004.02012.x.

Jornitz, S., & Parreira do Amaral, M. (Eds.). (2021). *The education systems of the Americas.* Springer International Publishing.

Kankaraš, M. (2017). *Personality matters: Relevance and assessment of personality characteristics.* OECD Education Working Paper No. 157.

Keeley, J., Zayac, R., & Correia, C. (2008). Curvilinear relationship between statistics anxiety and performance among undergraduate students: Evidence for optimal anxiety. *Statistics Education Research Journal, 7(1),* 4-15. http://www.stat.auckland.ac.nz/serj.

Kendall, M. (1938). A new measure of rank correlation. *Biometrika*, 30(1–2), 81–89. doi:10.1093/biomet/30.1-2.81. JSTOR 2332226

Khajavi, F., & Hekmat, H. (1971). A comparative study of empathy: The effects of psychiatric training. *Arch Gen Psychiatry.* 25(6), 490-493.

Khamis H. J., Roche, A. F. (1994). Predicting adult stature without using skeletal age: The Khamis-Roche method. *Pediatrics.* 94(4), 504-507.

Kim, Y., & Glassman, M. (2013). Beyond search and communication: Development and validation of the Internet Self-efficacy Scale (ISS). *Computers in Human Behavior*, 29, 1421–1429.

Krzysztof, M., & Mero, A. (2013). A kinematics analysis of the three best 100 m performances ever. *Journal of human kinetics, 36,* 149–160. https://doi.org/10.2478/hukin-2013-0015

Kroes, A. D., & Finley, J. R. (2023). Demystifying omega squared: Practical guidance for effect size in common analysis of variance designs. *Psychological methods.* Advance online publication. https://doi.org/10.1037/met0000581.

Kruskal, W. H., & Wallis, W. A. (1952). Use of ranks in one-criterion variance analysis. *Journal of the American Statistical Association, 47,* 583–621.

Kuo, Y., Tseng, H., & Kuo, Y. T (2020, April). Internet self-efficacy, Self-regulation, and student performance: African-American adult students in online learning. *International Journal on E-Learning*, 19(2), 161-180.

Kuczmarski, R. J., Ogden, C. L., Grummer-Strawn, L. M., Flegal, K. M., Guo, S. S., Wei, R., Mei, Z., Curtin, L. R., Roche, A. F., & Johnson, C. L. (2000, December 4). *CDC growth charts: United States.* Advance data from vital and health statistics, 314. National Center for Health Statistics.

Laerd Statistics. (2020). Pearson's product moment correlation. *Statistical tutorials and software guides.* https://statistics.laerd.com/statistical-guides/pearson-correlation-coefficient-statistical-guide.php.

Lammers, W. J., & Badia, P. (2005). Fundamentals of Behavioral Research. https://uca.edu/psychology/fundamentals-of-behavioral-research-textbook/.

Lane, S., Raymond, M. R., & Haladyna, T. M. (Eds.). (2016). *Handbook of test development (2nd ed.).* Routledge/Taylor & Francis Group.

Lastrucci, C. L. (1963). *The Scientific Approach.* Schenkman.

Leech, N. L., Barrett, K. C., & Morgan, G. A. (2015). *IBM SPSS for intermediate statistics: Use and interpretation.* (5th ed.). Routledge/Taylor & Francis Group.

Likert, R. (1932). A technique for the measurement of attitudes. *Archives of Psychology.* 22(140), 1–55.

Little, R. J. A. & *Rubin,* D. B. (*1987*). *Statistical analysis with missing data.* John Wiley & Sons.

Lohr, S. L. (2022). *Sampling: Design and analysis* (3rd ed.). CRC Press/Taylor & Francis Group.

Lomax, R. G., & Hahs-Vaughn, D. L. (2012). *Statistical concepts: A second course* (4th ed.). Routledge/Taylor & Francis Group.

Losen, D., Whitaker, A., Kizzire, J., Savitsky, Z. & Dunn, K. (2019, June 11) *The striking outlier: The persistent, painful and problematic practice of corporal punishment in schools.* Southern Poverty Law Center & The Center for Civil Rights Remedies. https://www.civilrightsproject.ucla.edu/news/press-releases/press-releases-2019/the-striking-outlier-the-persistent-painful-and-problematic-practice-of-corporal-punishment-in-schools.

Mann, H. B., & Whitney D. R. (1947). On a test of whether one of two random variables is stochastically larger than the other. *The Annals of Mathematical Statistics,*18, 50–60.

Mauchly, J. W. (1940). Significance test for sphericity of a normal *n*-variate distribution. *The Annals of Mathematical Statistics*, 11, 204–209.

McDonald, J. H. (2014). *Handbook of biological statistics* (3rd ed.). Sparky House Publishing. http://www.biostathandbook.com/kruskalwallis.html.

McKnight, P. E., McKnight, K. M., Sidani, S., & Figueredo, A. J. (2007). *Missing data: A gentle introduction.* Guilford Press.

Mertler, C. A. (2016). *Introduction to educational research* (3rd ed.). SAGE Publications, Inc. Meyers, L. S., Gamst, G., & Guarino, A. J. (2016). *Applied multivariate research: Design and interpretation (3rd ed.). Sage Publications, Inc.*

Ministry of Finance and Public Service (Jamaica). (2022, July 26). *Update on the 2022 population and housing census.* https://www.mof.gov.jm/update-on-the-2022-population-and-housing-census/.

Mirshams Shahshahani, P. (2018). *Downloaded IAAF Sprint Results in all Heats for 2004 - 2016 Olympics for both Men and Women.* University of Michigan - Deep Blue. [Dataset] https://doi.org/10.7302/Z20V8B11.

Musu, L., Dohr, S., & Netten, A. (2020.) Quality control during data collection: Refining for rigor. In H. Wagemaker (Ed.). *Reliability and validity of international large-scale assessment, Understanding IEA's Comparative Studies of Student Achievement*, pp.131-150. Springer. https://link.springer.com/book/10.1007/978-3-030-53081-5/.

Myers, J. L., Well, A. D. & Lorch, R. F. (2010). *Research design and statistical analysis.* Routledge.

National Collegiate Athletic Association (NCAA). (2020). *Results from the 2019 GOALS study of the student-athlete experience.* NCAA Convention.

Nishiwaki, M., Kuriyama, A., Ikegami, Y., Nakashima, N., & Matsumoto, N. (2014). A pilot crossover study: Effects of an intervention using an activity monitor with computerized game functions on physical activity and body composition. *Journal of physiological anthropology, 33*(1), 35. https://doi.org/10.1186/1880-6805-33-35

Nola, R., & Sankey, H. (2014). *Theories of Scientific Method* (1st ed.). Taylor and Francis. https://www.perlego.com/book/1559072/theories-of-scientific-method-an-introduction-pdf (Original work published 2014).

Nolen-Hoeksema, S., & Morrow, J. (1991). A prospective study of depression and posttraumatic stress symptoms after a natural disaster: The 1989 Loma Prieta earthquake. *Journal of Personality and Social Psychology*, 61, 115-121.

Norman, G. (2010). Likert scales, levels of measurement and the "laws" of statistics. *Advances in Health Sciences Education*, 15(5), 625–632. https://doi.org/10.1007/s10459-010-9222-y

Nortvedt, G. A., Gustafsson, J. E., Lehre, A. C. W. (2016) The importance of instructional quality for the relation between achievement in reading and mathematics. In T. Nilsen & J. E. Gustafsson (Eds.), *Teacher quality, instructional quality and student outcomes.* IEA Research for Education. Vol. 2. Springer, Cham. https://doi.org/10.1007/978-3-319-41252-8_5.

Nussbaum, E. M. (2015). *Categorical and nonparametric data analysis: Choosing the best statistical technique.* Routledge/Taylor & Francis Group.

Olkin, E. & Pratt, J. W. (1958). Unbiased estimation of certain correlation coefficients. *Annals of Mathematical Statistics*, 29, 201-211.

Osborne, J. W. (2013). *Best practices in data cleaning: A complete guide to everything you need to do before and after collecting your data.* Thousand Oaks, CA: Sage.

Ostchega, Y., Fryar, C. D., Nwankwo, T., & Nguyen, D. T. (2020). *Hypertension prevalence among adults aged 18 and over: United States, 2017–2018.* NCHS Data Brief, no 364. National Center for Health Statistics.

Otis, L. P. (1984). Factors affecting the willingness to taste unusual foods. *Psychological Reports.* 54, 739–745. https://doi.org/10.2466/pr0.1984.54.3.739.

Otsuka, M., Kawahara, T., & Isaka, T. (2016). Acute response of well-trained sprinters to a 100-m race: Higher sprinting velocity achieved with increased step rate compared with speed training. *Journal of Strength Conditioning Research*, 30(3), 635-42.

Pasta, D. J. (2009). Learning when to be discrete: Continuous vs. categorical predictors [Conference session]. Paper 248-2009, SAS Global Forum. http://support.sas.com/resources/papers/proceedings09/248-2009.pdf.

Pearson, K. (1911). On a correction needful in the case of the correlation ratio. *Biometrika, 8*, 254–256.

Pearson, K. (1920). Notes on the history of correlation. *Biometrika,13*, 25-45.

Pedhazur, E. J. (1982). *Multiple regression in behavioral research: Explanation and prediction* (2nd ed.). New York, NY: Holt, Rinehart & Winston.

Pedhazur, E. J. (1997) Multiple regression in behavioral research: Explanation and prediction (3rd ed.). Orlando, FL: Harcourt Brace.

Piaget, J. (1936). *Origins of intelligence in the child.* Routledge & Kegan Paul.

Piaget, J., & Cook, M. T. (1952). *The origins of intelligence in children.* International University Press.

Pigott, T. D. (2001). A review of methods for missing data. *Educational Research and Evaluation, 7*(4), 353–383. https://doi.org/10.1076/edre.7.4.353.8937.

Pinneo, S., O'Mealy, C., Rosas, M., Jr, Tsang, M., Liu, C., Kern, M., Hooshmand, S., & Hong, M. Y. (2022). Fresh mango consumption promotes greater satiety and improves postprandial glucose and insulin responses in healthy overweight and obese adults. *Journal of medicinal food*, *25*(4), 381–388. https://doi.org/10.1089/jmf.2021.0063.

Pituch, K. A. & Stevens, J. P. (2016). Applied multivariate statistics for the social sciences (6th ed.). Routledge.

Pliner, P., & Hobden, K. (1992). Development of a scale to measure the trait of food neophobia in humans. *Appetite*, 19, 105–120.

Quinlan, R. (1993). Auto MPG [Dataset]. UCI Machine Learning Repository. https://doi.org/10.24432/C5859H.

Rammstedt, B., & John, O. P. (2007). Measuring personality in one minute or less: A 10-item short version of the Big Five Inventory in English and German. *Journal of Research in Personality*, 41, 203–212.

Razali, N. & Wah, Y. (2011). Power comparisons of Shapiro-Wilk, Kolmogorov-Smirnov, Lilliefors and Anderson-Darling tests. *Journal of Statistical Modeling and Analytics*, 2, 21-33.

Romero, M. & Lee, Y. (2007). *A national portrait of chronic absenteeism in the early grades: Pathways to early school success project*. National Center for Children in Poverty. Mailman School of Public Health. Columbia University. https://doi.org/10.7916/D89C7650.

Robinson, W. (1950). Ecological correlations and the behavior of individuals. *American Sociological Review, 15*(3), 351-357. https://doi.org/10.2307/2087176.

Rosen, S. (1976). A theory of life learning. *Journal of Political Economy*, 84, S45– S67.

Rubin, D. B. (*1976*). Inference and missing data. *Biometrika*, 63, 581-590.

Ruel, E., Wagner, W., & Gillespie, B. (2016). *Data cleaning*. In E. Ruel, W. E. Wagner III, & B. J. Gillespie. *The practice of survey research: Theory and applications*. Sage.

Ryan, T., Joiner, B., & Ryan, B. (1985). *Minitab student handbook*. Duxbury Press.

Sarid, O., Anson, O., Yaari, A., & Margalith, M. (2004). Academic stress, immunological reaction, and academic performance among students of nursing and physiotherapy. *Research in Nursing and Health*, *27*(5), 370-377.

Sawilowsky, S. S. (1990). Nonparametric tests on interaction in experimental design. *Review of Educational Research, 60*, 91–126.

Schafer, J. L., & Graham, J.W. (2002). Missing data: Our view of the state of the art. *Psychological Methods*, 7(2), 147–177. https://doi.org/10.1037/1082-989X.7.2.147.

Schwartz, S. H. (2006). Les valeurs de base de la personne: Théorie, mesures et applications [Basic human values: Theory, measurement, and applications]. *Revue Française de Sociologie, 47,* 249-288.

Sekiwu, D., Ssempala, F., & Francis, N. (2020, April). The relationship between school attendance and academic performance in universal primary education: The case of Uganda. *African Educational Research Journal*, 8(2), 152-160.

Shadish, W. R., Cook, T. D., & Campbell, D. T. (2002). *Experimental and quasi-experimental designs for generalized causal inference*. Houghton, Mifflin and Company.

Shomaker, L. B., Tanofsky-Kraff, M., Savastano, D. M., Kozlosky, M., Columbo, K. M., Wolkoff, L. E., Zocca, J. M., Brady, S. M., Yanovski, S. Z., Crocker, M. K., Ali, A., & Yanovski, J. A. (2010). Puberty and observed energy intake: Boy, can they eat! *The American Journal of Clinical Nutrition*, *92*(1), 123–129. https://doi.org/10.3945/ajcn.2010.29383.

Sidak, Z. (1967). Rectangular confidence regions for the means of multivariate normal distributions. *Journal of the American Statistical Association*, 62, 626–633.

Spearman, C. (1904, January). The proof and measurement of association between two things. *The American Journal of Psychology, 15(1)* 72–101. doi:10.2307/1412159. JSTOR 1412159.

Spence, C. (2022). What is the link between personality and food behavior? Current Research in Food Science, 5, 19-27. https://doi.org/10.1016/j.crfs.2021.12.001.

Spitzer, R. L., Kroenke, K., Williams, J. B. W., & Löwe, B. (2006). *Generalized Anxiety Disorder 7 (GAD-7)* [Database record]. APA PsycTests. https://doi.org/10.1037/t02591-000.

Stevens, S. S. (1946). On the theory of scales of measurement. *Science*, 103, 677–680. https://doi.org/10.1126/science.103.2684.677.

Stevens, J. P. (*2009*). *Applied multivariate statistics for the social sciences* (5th ed.). Routledge/Taylor & Francis Group.

Stigler, S. M. (1989, May). Francis Galton's account of the invention of correlation. *Statistical Science*, 4(2), 73–79. https://doi.org/10.1214/ss/1177012580.

Tabachnick, B. G., & Fidell, L. S. (2018). *Using multivariate statistics* (6th ed.). Pearson.

The World Bank. (2018). *World Development Report 2018: Learning to Realize Education's Promise.* World Bank. https://openknowledge.worldbank.org/entities/publication/3f1e4f05-f20b-51e0-ad64-a5d6ea2bb994. License: CC BY 3.0 IGO.

The World Bank (2019, February 22). *Infographic. Languages at risk in Latin America and the Caribbean.* https://www.worldbank.org/en/news/infographic/2019/02/22/lenguas-indigenas-legado-en-extincion.

The World Health Organization. (2020, October 23). *WHO Coronavirus Disease (COVID-19) Dashboard.* https://covid19.who.int/.

The World Health Organization. (2018). Estimates of National Immunization Coverage (WUENIC). https://www.who.int/teams/immunization-vaccines-and-biologicals/immunization-analysis-and-insights/global-monitoring/immunization-coverage/who-unicef-estimates-of-national-immunization-coverage.

Thomas, R. M. (2015, April). Education in the Commonwealth Caribbean and Netherlands Antilles. *International Review of Education*, 61(2). DOI:10.1007/s11159-015-9478-9.

Thoren K., Heinig E., & Brunner M. (2016, May 17). Relative age effects in mathematics and reading: Investigating the generalizability across students, time and classes. *Frontiers in Psychology*, 7, 679. https://doi.org/10.3389/fpsyg.2016.00679.

Transformations: An Introduction. http://fmwww.bc.edu/repec/bocode/t/transint.html.

Tukey, J. W. (1977). *Exploratory data analysis.* Addison-Wesley.

Ustün, T. B., Chatterji, S., Mechbal, A., & Murray, C. J. L. (2005). Quality assurance in surveys: Standards, guidelines and procedures. In *Household Surveys in Developing and Transition Countries: Design, Implementation and Analysis.* World Health Organization.

VandenBos, G. R. (Ed.). (2015). *APA dictionary of psychology* (2nd ed.). American Psychological Association. https://doi.org/10.1037/14646-000..

Vogt, W. P., & Johnson, B. (2011). *Dictionary of statistics & methodology: A nontechnical guide for the social sciences.* Sage Publications, Inc.

Warner, R. M. (2013). *Applied statistics: From bivariate through multivariate techniques* (2nd ed.). Sage Publications, Inc.

Welch, B. L. (1951). On the comparison of several mean values: An alternative approach. *Biometrika*, 38, 330-336.

Wilcoxon F. (1945). Individual comparisons by ranking methods. *Biometrics Bulletin,* 1, 80–83.

Williams, E. V. (2020). Investigating the impact of the integrated STEM program on student test scores in Jamaica. [Doctoral dissertation]. William Howard Taft University. ERIC Number: ED606494.

Williams, R. A. (2020). Ordinal independent variables. In P. Atkinson, S. Delamont, A. Cernat, J. W. Sakshaug, & R. A. Williams (Eds.), SAGE Research Methods Foundations. https://www.doi.org/10.4135/9781526421036938055.

Winkens, B., Schouten, H. J., Van Breukelen, G. J., & Berger, M. P. (2005, December 30). Optimal time-points in clinical trials with linearly divergent treatment effects. *Statistics in Medicine*, 24, 3743-56. doi:10.1002/sim.2385. PMID: 16320272.

Woodman, T., Davis, P. A., Hardy, L., Callow, N., Glasscock, I., & Yuill-Proctor, Y. (2009). Emotions and sport performance: An exploration of happiness, hope, and anger. *Journal of Sport & Exercise Psychology,* 31, 169-188.

Wragg, C. B., Maxwell, N. S., & Doust, J. H. (2000). Evaluation of the reliability and validity of a soccer-specific field test of repeated sprint ability. *European Journal of Applied Physiology*, 83(1), 77–83. https://doi.org/10.1007/s004210000246.

Zimmerman, D. (1998). Invalidation of parametric and nonparametric statistical tests by concurrent violation of two assumptions. *The Journal of Experimental Education,* 67(1), 55-68. http://www.jstor.org/stable/20152581.

Zuckerman, I. (1979) *Sensation seeking: Beyond the optimal level of arousal.* Lawrence Erlbaum Associates.

Zumbo, B. D., & Zimmerman, D. W. (1993). Is the selection of statistical methods governed by level of measurement? *Canadian Psychology*, 34, 390-400.

Index

b denotes box; *f* denotes figure; *t* denotes table

A

a priori planned comparisons, 267–269, 268*f*
adjusted mean, 332, 336
adjusted *r*-square (r^2), 193, 196, 202, 244, 306
alpha level, 32, 48, 51
alphanumeric variables, 63, 64
Altendorf, S., 314
alternative hypothesis, 30–31, 31*f*, 32, 33, 48
American Psychological Association (APA) Publication Manual, 50, 53
analysis of covariance (ANCOVA)
    assumptions of, 322–325
    benefits of, 318–319
    Case of Appetite after Accounting for Mangoes, 315–316, 315*f*
    characteristics of covariates, 317–318, 318*f*
    data collection, 322
    defined, 282
    fundamentals of, 316–318
    hypothesis testing and data analysis procedures, 322
    introduction, 314–315
    research design, 321–322
    research plan, Case of Appetite after Accounting for Mangoes, 319–322, 319*b*, 321*f*
    research problem and question, 320–321
    sample report, Case of Appetite after Accounting for Mangoes, 337
    SPSS procedures for one-way ANCOVA, 325–337, 326*f*, 328*f*, 329*f*, 330*f*, 331*f*, 332*f*, 333*f*, 334*f*, 335*f*, 336*f*
    as statistical method based on GLM, 280
    types of, 316–317, 317*f*
    use of, 26
    variables, 321, 321*f*

analysis of variance (ANOVA)
    factorial analysis of variance (ANOVA). *See* factorial analysis of variance (ANOVA)
    one-way analysis of variance (ANOVA). *See* one-way analysis of variance (ANOVA)
    other types of, 282
    within-subjects ANOVA, 366, 367
    use of, 21, 105
ANCOVA (analysis of covariance). *See* analysis of covariance (ANCOVA)
ANOVA (analysis of variance). *See* analysis of variance (ANOVA)
APA (American Psychological Association). *See* American Psychological Association (APA) Publication Manual
applied statistics, consumers of, 2–3
Arbuthnott, J., 391
asking questions, as step in scientific method, 12–13
associative relationships, 4, 6
assumptions
    of analysis of covariance (ANCOVA), 322–325
    of bivariate correlation, 140–142
    factorial analysis of variance (ANOVA), 293
    handling violations of statistical assumptions, 105–106, 106*f*
    homogeneity of variance, 262–263
    Kendall's tau, 429
    Kruskal-Wallis, 405–406
    of linear regression, 179–184
    linearity assumption. *See* linearity assumption
    of mixed design, 373–375
    of multiple regression, 219–220
    near-normal distribution, 392
    normality, 96–101, 262

assumptions *(cont.)*
　of one-way analysis of variance (ANOVA), 261–263
　of repeated measures ANOVA (RMANOVA), 353–354
　Spearman's rho, 422
　sphericity, 354
　SPSS specifications for investigation of, 107
autocorrelation, 105, 180, 190

B

backward elimination, 215
balanced Latin square counterbalancing, 357, 358, 358f
Bangsbo Sprint Test, 339, 347, 365
Bates, B., 1
Becker, G. S., 3
between-group sum of squares (SSB). *See* sum of squares between-group variance (SSB)
between-group variance, 255, 257, 259, 260, 261, 283, 297, 298, 338
between-subjects ANCOVA, 316, 337
between-subjects ANOVA, 249, 250, 263, 284, 316, 350, 353, 375, 399
between-subjects designs, 29, 287
between-subjects effects, 305f, 310, 311f, 328f, 333f, 335, 336f, 362, 363f, 386, 387f
between-subjects factorial ANOVA, 367
between-subjects factorial designs, 287
between-subjects factors, 383f
between-subjects sum of squares (SSB). *See* sum of squares between subjects (SSB)
between-subjects variable, 317, 367, 368, 369, 372, 374, 375, 376, 378, 380–381, 383, 388
between-subjects variance, 338, 349f
bias
　sources of in linear regression, 182–184
　SPSS specifications to identify sources of, 197–201, 199f, 200f, 201f
bivariate correlation
　assumptions of, 140–142
　calculations for Pearson product moment correlation, 134–139
　correlation and causation, 139–140
　eligible data and measurement levels, 124–125
　factors that influence correlation, 152–157
　fundamentals of, 121–122
　hypotheses for, 126
　intercorrelation matrices, 157–158, 158f
　introduction, 119–121, 120f
　magnitude of relationship, 128
　other correlation coefficients, 158–160, 159f, 160f

bivariate correlation *(cont.)*
　Pearson correlation coefficient (Pearson's r), 127–134
　planning study of, 124–125
　research question, 126
　scatterplots and correlation, 122–123, 123f
　scatterplots and prediction, 123–124
　SPSS correlation specifications, 142–152, 143f, 144f, 145f, 146f, 147f, 148f, 149f, 150f, 151f
bivariate monotonic relationship, 416–417, 417f
bivariate regression. *See* simple linear regression
bivariate statistics, 4–5
Boffa, J., 23
Bonferroni, C. E., 267
Bonferroni-corrected pairwise t-tests, 365
boxplots, 84–85, 85f, 89–91, 90f, 101
Box's Test of the Equality of Covariance Matrices, 384, 385f, 386f, 387f
Bravais, Auguste, 127
Brown-Forsythe test, 105, 263, 394
Burton, R., 16–17

C

Campbell, D. T., 23
Carib_Height_Weight_Dups.sav, 68, 68f, 69
Carib_Height_Weight.sav, 162
Caribbean Examinations Council (CXC), 400
Caribbean Secondary Examination Certificate (CSEC) scores, 4, 36, 37, 45–46, 46f, 207, 400
Carifio, J., 21
Case of a Public Health Video Campaign, 432, 434f, 435b, 444
Case of Appetite after Accounting for Mangoes, 315–316, 315f, 319–322, 319b, 321f
Case of Differences in Achievement, 249, 249f, 251b, 253, 284–285, 285t, 290–292, 291b
Case of Employee Salaries, 242–247, 242f, 243f, 244f, 245f, 246f, 247f
Case of Hypertension in Portland, 414, 414t, 418–420, 419b, 422, 426
Case of Olympic Track Athletes' Speed, 367–369, 368t, 387–388, 388f
Case of Performance in Three Instructional Programs, 401–414, 402b, 403t
Case of the Height and Weight of College Students, 162–163, 163f, 176b, 184–194, 201, 203, 264
Case of the Sprinting Speeds of Seven Soccer Players, 339–340, 340t, 346–348, 347b, 358–365, 359f, 360f, 361f, 362f, 363f, 364f, 365
Case of the Weight of U.S. Adolescents, 203–204, 206b, 214–215, 221, 230–231, 231–233, 231f, 232f, 233f

Case of Trust and the Importance of Religion, 414–415, 415*t*, 429, 430*b*
case studies, as quantitative descriptive research designs, 23, 24
casewise diagnostics, 197, 198, 199*f*, 200
categorical variables, 19–20, 54, 71–72, 72*f*, 180
causal-comparative design, 5, 24*f*, 25*f*, 26–27, 322, 401
causation, correlation and, 139–140
census counts, as origin of statistics, 1
Centers for Disease Control and Prevention (CDC), 203
central tendency, 4, 46, 73, 74, 108, 391, 395–396, 396*f*, 398, 423
character variables, 63
Chart Builder (SPSS), 87–88, 87*f*, 88*f*, 90*f*, 91*f*
cherry-picking, 95
cluster random sampling, 43*f*
codebook, 60
coding, 56–61, 57*f*, 180
coefficient of determination (r^2), 51, 52, 131, 137, 165, 193, 207, 214
Cohen, B., 375
Cohen, J., 52, 398, 444
Cohen's *d*, 51, 52, 312
Cohen's kappa, 40
collapsed categories, avoidance of, 59
collinearity, 219–220, 235*f*
complete counterbalancing, 357
compute variable, 115*f*
computer-assisted interview protocols, 35
computer-assisted observations, 35
confidence interval, 50, 108, 109, 110*f*, 117*f*, 131, 138, 139, 142, 153, 154*f*, 276*f*, 278*f*, 279*f*, 304, 307*f*, 308*f*, 309*f*, 312*f*, 336–337, 336*f*, 361, 364*f*, 382–383, 387*f*, 388*f*, 424, 425
constructs
    defining of (underlying variables), 33–34
    measures of, 34–35
continuous variables, 19, 20–21, 54, 179
convenience sampling, 42, 43*f*
Cook's distance (*D*), 93, 183, 184, 197, 198, 199*f*, 200, 201*f*
correlation
    autocorrelation, 105, 180, 190
    bivariate correlation. *See* bivariate correlation
    and causation, 139–140
    direction of relationship, 157
    ecological correlation, 155–157
    Kendall's rank correlation. *See* Kendall's tau (Kendall's rank correlation)
    multiple correlation, 47, 195, 225
    nonparametric correlation. *See* nonparametric correlation
    partial correlations. *See* partial correlations

correlation *(cont.)*
    Pearson correlation coefficient (Pearson's *r*). *See* Pearson correlation coefficient (Pearson's *r*)
    point-biserial correlation, 160*f*
    population correlation. *See* population correlation
    semi-partial correlations, 229, 229*f*
correlational designs, 24*f*, 25–26, 25*f*
correlational relationships, 120, 120*f*
counterbalanced design, 27, 28*f*
counterbalancing, 356–358, 357*f*, 358*f*, 366
covariance, 128. *See also* analysis of covariance (ANCOVA)
covariates, 317–318, 318*f*, 372–373
crossover design, 355–356, 356*f*, 366
cross-tabulation analysis, 77–79, 78*f*, 79*f*
CSEC (Caribbean Secondary Examination Certificate) scores. *See* Caribbean Secondary Examination Certificate (CSEC) scores
cubic trend, 272
curvilinear relationship, 101, 102, 123, 124*f*, 142, 154, 164, 416
curvilinearity, 181
cut-points, 398
CXC (Caribbean Examinations Council), 400

D

data analysis
    for bivariate correlation, 127
    Case of Hypertension in Portland, 418–419
    for multiple regression, 219
    null hypothesis significance testing, 47–52, 49*f*
    objectives of, 45, 46
    software packages for, 57
    types of statistics for, 45–47
data collection
    for Case of Appetite after Accounting for Mangoes, 322
    consequences from improper collection, 40
    defined, 40
    instruments for, 35, 55
    methods of, 33–50, 55
    populations and samples, 40–41
    reliability and validity of instruments for, 35–40, 36*f*
    sample size, 43–45
    sampling techniques, 41–43, 43*f*
    study sample, 40–45
data dictionary, 68, 68*f*
data entry, 56–61
data files, 456
data integrity, 55–56

data screening and cleaning
   advanced, 83–118
   basic review, 61–62
   for basic value cleansing (in SPSS), 67–77
   defined, 61
   for duplicate cases and responses, 62–63
   to identify missing values, 79–82
   for implausible or unlikely responses, 62
   for inconsistent responses, 77–79
   for inconsistent responses across variables, 64
   for incorrect/unexpected value formats, 63–64
   for lack of uniformity, 63
   for missing data, 64–67
   sample report for, 118
   for violation of statistical assumptions, 95–118
data spreadsheet, in SPSS Data Editor, 58f
data transformation, 105–106, 106f
dataset
   checking for errors in, 67–77
   checks before cleaning and screening, 61
   rows and columns of, 58–59
degrees of freedom (df), 97, 112, 133, 134, 138, 171, 178, 196, 225, 248, 258, 259, 260, 261, 262, 268, 276, 277, 296, 298–299, 352, 389
Delhey, J., 415
dependent variables, 18, 54, 161, 179
descriptive research, 23–25
descriptive statistics, 45–46, 54, 75f, 110f
DeVellis, R. F., 55
dichotomous variables, 19f, 20
Dillman, D. A., 55
directional (one-tailed or one-sided) hypothesis, 31, 31f
disordinal interaction, 289
Djaoui, L., 345
'dummy' coding, 180
Dunn-Bonferroni test, 407, 414, 438, 441, 444
Dunn's adjustment, 407
Dunn's test, 403, 406, 443
duplicate cases, 61, 62–63, 69–71, 70f, 71f
Durbin-Watson statistic/test, 104–105, 180, 189, 189f, 190

E

ecological correlation, 155–157
effect size, 50, 51–52, 153, 154, 180, 210, 219, 248, 264–265, 277, 278f, 312, 312f, 338, 389, 404–405, 443–444, 444f
error sum of squares (SSE), 296, 298, 299, 348, 351–352. *See also* sum of squared errors (SSE)
error variance, 166, 260, 261, 296, 299, 301, 305, 309–310, 314, 316, 355, 374, 384. *See also* Levene's test of equality of error variances

errors, independence of. *See* independence of errors
estimated means, 336, 387
eta squared (η^2), 52, 264, 265, 277, 278f, 280, 283, 312, 335, 404, 405
evaluating results, as step in scientific method, 15
Excel, 57
experimental research, 23, 24f, 27–29
experiment-wise error rate, 266, 267
explanations, generating of as step in scientific method, 12–13
explanatory correlational studies, 26

F

F distribution, 260
F ratio, 47f, 260, 298–300, 300f, 352–353
F statistic, 52, 202, 260
F test, 97, 169, 173b, 176, 201, 202, 205f, 214, 262, 265, 450–452
factor, 285
factorial analysis of variance (ANOVA)
   assumptions of, 293
   calculations for, 294–300
   Case of Differences in Achievement, 284–285, 285f
   defined, 282, 338, 366
   differences in achievement using actual test scores, 309–312, 310f, 311f, 312f
   factorial designs beyond two-way ANOVA, 289
   fundamentals of, 285–289
   hypotheses for, 294–295, 295f
   introduction, 284
   post ANOVA tests, 292–293
   research plan, Case of Differences in Achievement, 290–292, 291b
   sample report, Case of Differences in Achievement, 313
   SPSS results, 305–309, 305f, 307f, 308f, 309f
   SPSS specifications for, 300–304, 301f, 302f, 303f, 304f
   two-way factorial ANOVA designs, 286–289, 287f, 288f, 289f
   types of, 286–289, 287f, 288f, 289f
   unequal sample sizes, 293–294
factorial repeated measures design, 344–345, 344f, 345f
family-wise error rate, 266, 300
Fidell, L. S., 4, 67, 83, 97, 98, 102, 106, 221, 267, 370
filters, in data screening and cleaning, 64
first quartile (Q1), in boxplot, 84
fit
   goodness of, 165–166, 166f
   line of best fit, 163–165
Food Neophobia Scale (FNS), 36, 37

formulating measurable hypothesis, as step in scientific method, 13–14
forward selection, 215
four-way mixed design, 371–373, 371*f*, 373*f*
frequency histogram
   for assessing normality, 111, 111*f*
   use of, 99
   of weight with a normal curve overlay, 99*f*
   of weights of selected Caribbean college students, 89*f*
frequency table
   for age, 75*f*
   of created variables, 71*f*
   for gender and educational level, 73*f*
   of missing values, 82*f*
   use of, 72–73, 93
Friedman, M., 105, 431
Friedman's test, 105, 282, 394, 431–444, 434*f*, 435*b*, 436*f*, 437*t*

G

Galton, Francis, 127
Games-Howell procedure, 267, 279–280, 279*f*
general linear model (GLM), 280, 300, 314, 325, 326
General Neophobia Scale (GNS), 37
generalizations, generating of as step in scientific method, 12–13
Generalized Anxiety Disorder 7-item scale, 34
generating explanations, as step in scientific method, 12–13
Glassman, M., 38
goodness of fit, 165–166, 166*f*
Google Forms, 57*f*
Gopnik, A., 16–17
Gossett, W., 48
Gottlieb, S., 432
grand mean (GM), 255, 256, 257, 257*t*, 264, 295, 349, 350
grand variance, 256
Greenhouse-Geisser, 354, 362*f*, 363*f*, 374, 384, 384*f*, 386, 386*f*

H

Helmert contrast, 269
heteroscedasticity, 103–104, 106, 181, 263
hierarchical regression, 213, 216, 217, 218, 240–241, 241*f*, 243, 246, 247*t*, 316
histograms, 84, 85, 86, 87–89, 99, 111*f*
Hoaglin, D. C., 184, 198, 199*f*, 200
Hobden, K., 36, 37
homogeneity of regression, 323–324, 327, 331–333, 333*f*, 337
homogeneity of variance, 83, 96, 102–104, 105, 109, 262–263, 274, 277*f*, 293, 300, 301, 309–310, 323, 326, 327, 334, 335, 354, 374, 375, 378, 379, 381, 391, 392, 394, 405. *See also* Levene's test of homogeneity of variance
homoscedasticity, 96, 102–104, 103*f*, 105, 181, 188, 392
Howell, D. C., 391
Huynh-Feldt correction, 354, 362*f*, 363*f*, 374, 384, 384*f*, 386, 386*f*
hypothesis
   alternative hypothesis. *See* alternative hypothesis
   for bivariate correlation, 126
   development of measurable hypothesis, 30
   formulating of measurable hypothesis as step in scientific method, 13–14
   formulating research hypothesis as step in research administration process, 17–18, 17*f*
   null hypothesis. *See* null hypothesis
   research hypothesis, 30
   testing of, 15, 32–33, 46, 47–52, 49*f*, 112*f*, 169–171, 389

I

IAAF (International Association of Athletics Federation), 387
independence of errors, 96, 104–105, 188, 262
independence of observations, 83, 96, 141, 180, 188–190, 261–262, 274, 300, 322, 406
independent variables, 18, 54, 161, 179
inferential statistics, 46–47, 54
in-person observations, 35
interaction
   as benefit of factorial ANOVA over one-way ANOVA, 285
   disordinal interaction, 289
   ordinal interaction, 289
   variable interaction, 6
interaction effect, 288, 289, 289*f*, 290, 292, 297–298, 298, 303, 306, 314, 315, 319, 332*f*, 385–386
intercorrelation matrices, 157–158, 158*f*
International Association for the Evaluation of Educational Achievement (IEA), 55
International Association of Athletics Federation (IAAF), 387
Internet Self-Efficacy Scale (ISS), 38
interquartile range (IQR), in boxplot, 85
interval, confidence. *See* confidence interval
interval-level variables, 19*f*, 21, 342, 353, 418
interview protocols, 35
IQ tests, 37

J
Jamieson, S., 21

K
Kendall, M., 426
Kendall's coefficient of concordance (Kendall's W), 443, 444, 444f
Kendall's tau (Kendall's rank correlation), 159, 160, 160f, 415–416, 417, 418
Khamis, H. J., 207
Kim, Y., 38
Kolmogorov-Smirnov (K-S) test, 97, 98–99, 102, 112, 113, 181, 408
Kruskal, W., 399
Kruskal-Wallis test, 20, 47f, 105, 282, 394, 395, 399–413, 400t, 402b, 403t, 408f, 417
Krzysztof, M., 372
Kuo, Y., 23, 32, 34
kurtosis, 73, 97, 98, 98f, 101, 106f, 107, 108, 110f, 111, 113, 117f, 118, 146, 262, 323, 354, 393, 396

L
Lane, S, 55
Levene's test, 104, 105, 109, 263, 276, 305, 309. *See also* Levene's test of equality of error variances; Levene's test of homogeneity of variance
Levene's test of equality of error variances, 301, 305f, 311f, 328f, 335f, 383, 385
Levene's test of homogeneity of variance, 300, 323, 327, 335, 374, 378, 384, 385f
leverage, 183–184, 198
Life Satisfaction and Age, 121f
Likert, R., 21
Likert scales, 21–22, 22f, 60, 60f, 66, 390, 390f, 392–393
line of best fit, 163–165, 165f
linear regression
  assumptions of, 179–184
  defined, 314
  goal of, 203
  one-way ANOVA, analysis of variance and linear regression, 280–282, 281r
  simple linear regression. *See* simple linear regression
linear regression equation, 168, 197, 211
linear regression model, 166–168, 203f, 219, 237f, 250, 280, 316
linear relationship, 6, 54, 104, 105, 121, 122, 126, 127, 128, 129f, 131, 132, 133, 134, 137, 141, 142, 143, 152, 162, 163, 177, 181, 184, 186, 187, 219, 225, 229, 318, 323, 327, 328, 329, 330, 337, 415, 417

linearity, 54, 83, 96, 101–102, 105, 141–142, 143–145, 160, 180–181, 184–188, 324
linearity assumption, 105, 106, 152, 416
linearity of regression, 323, 327–330, 330f, 337
lower whisker, in boxplot, 85
Lower-bound correction, 354, 363f, 384, 386f

M
Mahalanobis distance, 93
main effect, 288, 290, 292–293, 307f, 309f, 312f, 314, 332f, 369, 386–387
making prediction, as step in scientific method, 14
Mann-Whitney U Test, 394, 399
matrices, intercorrelation matrices, 157–158, 158f
Mauchly's test of sphericity, 361–365, 362f, 374, 378, 383, 384f
Mauchly's W, 354, 362f, 383
maximum, in boxplot, 85
mean square, 225f, 235f, 244f, 258, 277f, 283, 300f, 305f, 311f, 328f, 333f, 336f, 352, 363f, 386f, 387f
mean square between-groups variance (MSB), 258, 260
mean squared error (MSE), 256, 259, 260, 299, 319, 353
mean squared residual (MSR), 256, 260
mean-square within-group variance (MSW), 256
median (Q2), in boxplot, 84
median test, 410
Mental Measurement Yearbook, 35
Meyers, L. S., 61, 220, 229
minimum, in boxplot, 85
missing data, 60, 61, 79–80, 80f, 122, 194, 323, 325, 355, 369–370, 376
mixed design ANOVA
  benefits of, 375–376
  Case of Olympic Track Athletes' Speed, 367, 368t
  characteristics of simple mixed design ANOVA, 369–370
  disadvantages of, 376
  fundamentals of simple mixed design ANOVA, 367–369, 369f
  introduction, 367
  research plan, Case of Olympic Track Athletes' Speed, 376–378, 377b
  sample report, Case of Olympic Track Athletes' Speed, 387–388, 388f
  SPSS procedures for, 378–387, 379f, 380f, 381f, 382f, 383f, 384f
  types of, 370–373
  underlying assumptions of, 373–375
MSE (mean squared error), 256, 259, 260, 299, 319, 353
MSR (mean squared residual), 256, 260
MSW (mean-square within-group variance), 256
multicollinearity, 208, 219–220, 242, 323, 324

multiple comparisons, 267, 269, 283, 303f, 306, 308, 309f, 312f, 338, 360, 389
multiple comparisons test, 275, 278–280, 279f, 353
multiple correlation, 47, 195, 225. *See also* squared multiple correlation
multiple regression
   Case of Employee Salaries, 204, 206b
   Case of the Weight of U.S. Adolescents, 203–204
   cases for multiple regression SPSS demonstrations, 203–204
   data analysis, 219
   data assumptions of, 219–220
   fundamentals of, 204, 207–217
   hierarchical or sequential multiple regression, 216–217, 217f
   introduction, 203, 203f
   methods of, 213–217
   planning multiple regression inquiry, 217–218, 218f
   properties of, 207–213, 210f, 212f, 213f
   research questions and hypotheses, 218, 218f
   results, Case of Employee Salaries, 242–246, 242f, 243f, 244f, 245f
   sample report, Case of the Weight of U.S. Adolescents, 230–231, 231f
   sample report (hierarchical), Case of Employee Salaries, 246–247, 246f, 247f
   sample report (stepwise), Case of the Weight of U.S. Adolescents, 238–239, 239f
   simple and multiple linear regression similarities, 207
   solving regression equation, 230
   SPSS procedures for standard multiple regression, 221–229, 221f, 222f, 223f, 224f, 225f, 226f, 228f
   SPSS specifications for hierarchical regression, 240–241, 241f
   SPSS specifications for statistical (stepwise) multiple regression, 231–233, 232f, 233f
   standard regression, 214
   statistical (stepwise) results (Case of the Weight of U.S. Adolescents), 233–238, 234f, 235f, 237f, 238f
   statistical multiple regression, 214–216
Musu, L., 56
mutually-exclusive groups, 59, 367

N

National Center for Health Statistics (NCHS), 203
National Health and Nutrition Examination Survey (NHANES), 203, 204
Newton, I., 10

NHST (Null Hypothesis Significance Testing), 46, 47–52, 49f
nominal variables, 19f, 20
nondirectional (two-tailed or two-sided) null hypothesis, 30, 31f
nonexperimental research, 23–27, 24f
nonlinear relationship, 102, 141, 157, 184
nonlinearity, 102, 142, 188
non-normal distribution, 96, 102, 103, 142, 148, 192, 305, 393
nonorthogonal comparisons, 267, 269
nonparametric ANOVA, 282, 401
nonparametric correlation
   Case of Hypertension in Portland, 414, 414t
   Case of Trust and the Importance of Religion, 414–415, 415t
   fundamentals of, 415–417
nonparametric tests/statistical analysis
   central tendency and, 395–396, 396f
   cost of converting continuous scores to ordinal level values, 398
   distribution and measurement level and, 392–393
   Friedman's rank test, 431–444, 433f, 434f, 435b, 436t, 437t
   fundamentals of, 390–399, 391f
   introduction, 389–390
   Kendall's tau (Kendall's rank correlation), 426–431, 427f, 428f, 432f
   Kruskal-Wallis one-day ANOVA (H) test fundamentals, 399–413, 400t, 402b, 403t, 408f
   overall advantages of, 398–399
   overall disadvantages of, 399
   ranks as key secret weapon of, 396–398
   research plan, Case of Performance in Three Instructional Programs, 401–403, 402b
   sample report, Case of Performance in Three Instructional Programs, 413–414
   Spearman's rho, 415–426
   Spearman's rho and Kendall's tau, 414t
   unequal and small sample sizes, 393–394
   unequal group variance and, 394–395, 395f
nonprobability sampling, 42, 43f
nonuniform data, cure for, 63
normal distribution, 54, 95, 96–98, 97f, 99, 101, 102, 112, 113, 146, 148, 152, 181, 190, 192, 262, 276, 300, 323, 326, 379, 390, 391, 392, 393, 394, 397, 399, 405
normal probability plots, 101, 102f
normality
   assessing assumption of, 96–101
   assumption of (one-way ANOVA), 262
   and bivariate correlation, 142

normality *(cont.)*
  graphical measures of, 99
  hypothesis test of, 112*f*
  normal distribution of scores (normal curve), 97*f*
  SPSS specifications to detect normality violations, 107–113
  transforming data for enhancement of, 113–118, 114*f*, 115*f*, 116*f*, 117*f*
normally distributed residuals, 181, 190–192
Norman, G., 21
null hypothesis, 30, 31*f*, 46
Null Hypothesis Significance Testing (NHST), 46, 47–52, 49*f*
numeric variables, 59
Nussbaum, E. M., 391, 393, 420

O

observational studies, as quantitative descriptive research designs, 23, 24
observation(s)
  computer-assisted observations, 35
  independence of. *See* independence of observations
  in-person observations, 35
  as step in scientific method, 11–12
omega squared (ϖ^2), 265, 277, 278*f*, 283, 312
omnibus test, 149, 235, 250, 265, 266, 275, 276, 277*f*, 280, 286, 292, 300, 301–302, 303–305, 310–311, 311*f*, 327, 333, 353, 358, 360, 363, 363*f*, 365, 378, 385–386, 389, 399, 403, 406, 409, 410, 412, 413, 438, 441, 443, 444
one case, one row, 58
one variable, one column, 59
one-factor analysis of variance (ANOVA), 338
one-way analysis of covariance (ANCOVA), SPSS procedures for, 325–337, 326*f*, 328*f*, 329*f*, 330*f*, 331*f*, 332*f*, 333*f*, 334*f*, 335*f*, 336*f*
one-way analysis of variance (ANOVA)
  analysis of variance and linear regression, 280–282, 281*t*
  assumptions of, 261–263
  calculations for, 256–261, 261*f*
  Case of Differences in Achievement, 249, 251*b*
  defined, 338
  eligible data variables and measurement levels, 252
  fundamentals of, 249–250
  hypotheses for, 253–254
  introduction, 248–249
  omnibus one-way ANOVA analysis results, 276–278, 276*f*, 277, 278*f*
  research design, 253
  research question and hypotheses, 252–253, 272–273

results of multiple comparisons tests, 278–280, 279*f*
  sample report, Case of Differences in Achievement, 280
  SPSS procedures for, 272–280
  SPSS specifications for omnibus one-way ANOVA, 273–276, 273*f*, 274*f*, 275*f*
  t test and, 250–251
  variance and group differences, 254–256, 255*f*, 256*t*
  when group means are significantly different, 265–271
  when group sizes are unequal, 263–264
one-way repeated measures ANOVA, 366
one-way repeated measures design, 343–344, 343*f*, 344*f*, 348
order effects, 355, 357, 358, 375–376
ordinal interaction, 289
ordinal variables, 19*f*, 20
orthogonal contrasts/comparisons, 267, 268, 268*f*, 269
Otsuka, K., 372
outliers
  absence of, 142
  defined, 83
  detection of, 84–95, 86*f*, 99, 101
  handling of, 95
  in linear regression, 182–183, 182*f*, 183*f*, 192
  reasons for, 83
out-of-range categorical variables, 71–72, 72*f*, 74*f*
out-of-range quantitative variables, 72–73, 73*f*
out-of-range scores, 73
out-of-range values, 61, 67, 72, 73, 76–77, 80*f*, 96

P

paired comparisons, 283, 307*f*, 338, 413*f*, 438, 441, 442*f*
pairwise comparisons, 307*f*, 308, 308*f*, 361, 363, 364*f*, 382, 387, 388, 388*f*, 406, 407, 413*f*, 414, 442*f*
paper-based interview protocols, 35
partial correlations, 228, 228*f*, 229*f*
Pearson, K., 127
Pearson correlation coefficient (Pearson's *r*), 4, 47*f*, 102, 127–138, 129*f*, 130*t*, 135*t*, 136*f*, 136*t*, 138*f*. *See also* Pearson's *r* (Pearson's rho)
Pearson product moment correlation, 134–139, 135*t*, 136*f*, 136*t*, 415–417
Pearson's *r* (Pearson's rho), 47, 52, 134, 136*f*, 141, 142, 151–152, 160*f*, 165, 175, 197, 239*f*, 247*t*. *See also* Pearson correlation coefficient (Pearson's *r*)
PEP (Primary Exit Profile), 249
performance rating scales, 35
Perla, R. J., 21
phi coefficient, 160*f*

Piaget, J., 10
Pituch, K. A., 221
planned contrasts, 267, 268, 269, 275
Pliner, P., 36, 37
point-biserial correlation, 160f
polynomial contrast, 266, 269, 275
population correlation, 126, 130, 131, 132, 134, 139, 142, 153
populations, in data collection, 40–41
Portrait Values Questionnaire (PVQ), 99, 100
post-hoc tests, 266–267, 275, 275f, 278, 279f, 283, 293, 300, 303, 303f, 306, 308, 310, 311, 319, 353, 378, 382, 387, 403, 407, 409, 433, 439, 441, 442, 442f. *See also* Dunn-Bonferroni test; Dunn's test; Tukey HSD (Honestly Significant Difference) post-hoc test
posttest-only control group design, 27, 28f, 29
predicted value, 104, 166, 167, 169, 170, 181, 182, 184, 187, 188f, 192, 198, 199f, 208, 209, 211, 336
prediction
    making of as step in scientific method, 14
    scatterplots and, 123–124
predictive correlational studies, 26
predictive relationships
    quantification of, 4
    strength of (goodness of fit), 165–166, 166f
predictor variables, 18, 170–171
pre-experimental design, 28, 28f, 401
pretest-posttest control group design, 27, 28f
Primary Exit Profile (PEP), 249
probability plots, 101
probability sampling, 42, 43f
p-value (probability value/significance value), 3, 48, 133, 263, 406, 435
PVQ (Portrait Values Questionnaire), 99, 100

Q

Q-Q plots, 113f
quadratic trend, 270
qualitative research, 23, 45
quality assurance (QA), 55–56
quality control (QC), 56
Qualtrics, 57f
quantitative data, 124–125
quantitative research, 23, 24f, 45, 58
quantitative variables, 20, 72–73, 110f, 111f
quartic trend, 272
quasi-experimental designs, 24f, 25f, 27–29, 28f
questioning
    as step in scientific method, 12–13
    types of research questions, 16
Quinlan, R., 120

quintic trend, 271, 272
quota sampling, 42, 43f

R

random assignment, 6, 25f, 27, 28–29, 290
range restriction, 154–155, 180
ranked data, 392, 418, 429
ranking, 417, 431, 436t, 437t
ratio-level variables, 19f, 21, 22–23, 58, 174, 208, 242, 342, 353, 418
regression
    bivariate regression. *See* bivariate regression
    hierarchical regression. *See* hierarchical regression
    homogeneity of. *See* homogeneity of regression
    linear regression. *See* linear regression
    linearity of. *See* linearity of regression
    multiple regression. *See* multiple regression
    simple linear regression. *See* simple linear regression
    statistical regression. *See* statistical regression
    stepwise regression. *See* stepwise regression
regression constant, 168f, 202
regression equation, 164, 167, 168, 171–172, 176, 188, 197, 202, 207, 208, 209, 211–213, 214, 230
regression line, 129, 161, 163, 165, 167, 168, 169, 170, 172, 175, 176, 177–178, 181, 182, 186f, 202, 323, 331
regression models
    standardized, 168
    testing significance of, 169
    unstandardized regression models, 167–168, 168f
regressor variables, 18
R-E-G-W-Q (Ryan, Einot, Gabriel, and Welsch Q), 267
related theory, as step in research administration process, 16–17
reliability
    Cronbach alpha, 36f, 39
    of data collection instruments, 35–40, 36f
    internal consistency reliability, 36f
    inter-rater reliability, 36f, 39–40
    parallel-forms reliability, 36f, 39
    split-half reliability, 36f
    test-retest reliability, 36f
repeated measures ANOVA (RMANOVA)
    as based on GLM, 280
    benefits of, 355
    calculations for, 348–353, 349f, 349t
    Case of the Sprinting Speeds of Seven Soccer Players, 339–340, 340t
    characteristics of repeated measures designs, 342–343

repeated measures ANOVA (RMANOVA) *(cont.)*
   counterbalancing, 356–358, 357*f*, 358*f*
   data collection and hypothesis testing, 348
   defined, 282
   disadvantages of, 355
   factorial repeated measures design, 344–345, 344*f*, 345*f*
   fundamentals of, 340–342, 341*f*, 342*f*
   introduction, 339
   one-way repeated measures design, 343–344, 343*f*, 344*f*
   research plan, Case of the Sprinting Speeds of Seven Soccer Players, 346–348, 347*b*
   sample report, Case of the Sprinting Speeds of Soccer Players, 365
   simple mixed design, 345–346, 346*f*
   SPSS procedures for one-way repeated ANOVA, 358–365, 359*f*, 360*f*, 361*f*, 362*f*, 363*f*, 364*f*
   strategies to counter repeated measures disadvantages, 355–358
   types of repeated measures designs, 343–346
   underlying assumptions of, 353–354
reporting results, as step in scientific method, 15–16
research administration process, as based on scientific method, 16
research approach, 23, 29
research design
   analysis of covariance (ANCOVA), 321–322
   basic concepts in, 10–53
   for bivariate correlation, 126–127
   case studies, as quantitative descriptive research designs, 23, 24
   causal-comparative. *See* causal-comparative design
   characteristics and statistics in common research designs, 25*f*
   correlational. *See* correlational designs
   descriptive, 23–25, 24*f*, 25*f*
   experimental, 24*f*
   factorial analysis of variance (ANOVA), 291*b*
   Friedman's test, 435*b*
   Kendall's, 430*b*
   Kruskal-Wallis, 401, 402*b*
   multiple regression, 205*f*, 206*b*
   nonexperimental, 24*f*
   one-way analysis of variance (ANOVA), 251*b*, 252
   quasi-experimental. *See* quasi-experimental designs
   repeated measures ANOVA (RMANOVA), 347, 347*b*
   simple linear regression, 175
   simple mixed design ANOVA, 377*b*, 378
   Spearman's rho, 418, 419*b*
   true experimental. *See* true experimental design
   use of term, 23
research hypothesis, 17–18, 17*f*, 30, 266, 292
research method, also called research strategy, 29–31
research plan
   also called research strategy, 29–31
   Case of a Public Health Video Campaign, 433–435, 434*f*, 435*b*
   Case of Appetite after Accounting for Mangoes, 319–322, 320*b*
   Case of Differences in Achievement, 251*b*, 291*b*
   Case of Employee Salaries, 206*b*
   Case of Hypertension in Portland, 418–420, 419*b*
   Case of Olympic Track Athletes' Speed, 376–378, 377*b*
   Case of Performance in Three Instructional Programs, 401–403, 402*b*
   Case of the Height and Weight of College Students, 173*b*
   Case of the Sprinting Speeds of Seven Soccer Players, 346–348
   Case of the Weight of U.S. Adolescents, 205*b*
   Case of Trust and the Importance of Religion, 429, 430*b*
   Case of Writing and Mathematics Skills, 125*b*
research question, 16, 17*f*, 125*b*, 126, 173*b*, 205*b*, 206*b*, 218, 251*b*, 252, 273, 291*b*, 320*b*, 347*b*, 401, 402*b*, 430*b*, 435*b*
research reporting, 52–53
research strategy
   described, 29–31
   use of term, 23
residual, 101, 104–105, 165, 165*f*, 166, 169, 181, 182–183, 184, 187–188, 188*f*, 190–192, 190*f*, 192*f*, 196*f*, 198, 199*f*, 219, 225*f*, 232, 235*f*, 244*f*, 261, 261*f*. *See also* squared residual; standardized residuals; unstandardized residual
results
   evaluation of as step in scientific method, 15
   reporting of as step in scientific method, 15–16
RMANOVA (repeated measures ANOVA). *See* repeated measures ANOVA (RMANOVA)
Roche, A. F., 207
Rosen, S., 3
r-square (r^2), 189*f*, 193, 196, 196*f*, 202, 226*f*, 236, 237*f*, 241, 244*f*. *See also* adjusted r-square (r^2)

S

sample report
   Case of a Public Health Video Campaign, 444
   Case of Appetite after Accounting for Mangoes, 337

sample report *(cont.)*
    Case of Differences in Achievement, 280
    Case of Height and Weight of College Students, 201
    Case of Hypertension in Portland, 426
    Case of Olympic Track Athletes' Speed, 387–388, 388*f*
    Case of Performance in Three Instructional Programs, 413–414
    Case of the Sprinting Speeds of Seven Soccer Players, 365
    Case of the Weight of U.S. Adolescents, 230–231, 231*f*, 238–239, 239*f*
    Case of Writing and Mathematics Skills, 152
sample size
    in correlational studies, 153–155, 153*f*, 154*f*, 155*f*
    in factorial analysis of variance (ANOVA), 293–294
    in linear regression, 180
    and statistical power, 52
samples/sampling
    cluster random sampling, 43*f*
    convenience sampling, 42, 43*f*
    in data collection, 40–45, 43*f*
    nonprobability sampling, 42, 43*f*
    probability sampling, 42, 43*f*
    quota sampling, 42, 43*f*
    simple random sampling, 42, 43*f*
    stratified random sampling, 43*f*
    systematic sampling, 42, 43*f*
sampling error, 47, 130
Sawilowsky, S. S., 401
scales of measurement, characteristics and applications of, 19*f*
scatterplots, 84, 85–86, 91–93, 91*f*, 92*f*, 96, 101, 102, 103*f*, 120*f*, 121*f*, 122–124, 123*f*, 124*f*, 127, 128, 141, 142, 143, 144, 144*f*, 145*f*, 146, 146*f*, 159*f*, 161, 162, 163, 165*f*, 182*f*, 183*f*, 184–187, 186*f*, 188*f*, 198, 201*f*, 213*f*, 323, 324, 327, 330*f*, 374, 417*f*
Scheffé test, 267
Schwartz, S. H., 100
scientific method
    described, 10–11, 11*f*
    research administration process as based on, 16
Secondary Entrance Assessment (SEA), 20, 39
semi-partial correlations, 229, 229*f*
Sensation-Seeking Scale, 36
sequence effects, 356
Shadish, W. R., 23
Shapiro-Wilk (S-W) test, 97, 98–99, 102, 112, 113, 181, 408

shared variability (r^2), 130, 137
significance tests, 54, 98, 104, 113, 126, 128, 130, 131, 132, 140, 148*f*, 169, 176, 181, 188, 262, 292, 322. *See also* Friedman's test; Kolmogorov-Smirnov (K-S) test; Kruskal-Wallis test; Null Hypothesis Significance Testing (NHST); Shapiro-Wilk (S-W) test; *t* test/statistic
simple effects, 293, 300, 303, 304, 306, 307, 308, 369, 383, 386
simple linear regression
    assumptions of linear regression, 179–184
    calculations for, 176–179, 176*b*, 177*t*, 178*f*, 179*f*
    case of height and weight of college students, 162–163, 163*f*, 176*b*
    fundamentals of, 163–172
    hypothesis testing in bivariate regression, 169–171
    introduction, 161–162
    model for, 166–168
    planning simple linear regression study, 172–176, 173*b*
    results for, 194–197, 195*f*, 196*f*, 197*f*
    solving regression equation, 171–172
    SPSS specifications for bivariate regression, 184–194, 185*f*, 186*f*, 187*f*, 188*f*, 189*f*, 190*f*, 191*f*, 192*f*, 193*f*, 194*f*
    SPSS specifications to identify sources of bias, 197–201, 199*f*, 200*f*, 201*f*
simple mixed design, 345–346, 346*f*, 365–370, 378
simple random sampling, 42, 43*f*
single-group design, 28, 28*f*
skewness, 73, 84, 97, 97*f*, 98, 99, 102, 103, 106*f*, 107, 108, 110*f*, 111, 112, 113, 114, 117*f*, 118, 146, 223, 230, 262, 323, 354, 379, 393, 396, 418
social sciences, use of applied statistic in, 3
Spearman, C., 417
Spearman's rank correlation coefficient, 20, 25*f*, 47*f*, 159. *See also* Spearman's rho (ρ or r_s)
Spearman's rho (ρ or r_s), 6, 7, 47, 159–160, 160*f*, 415–416, 417–426, 421*t*, 425*f*, 432*f*, 453*t*. *See also* Spearman's rank correlation coefficient
Spence, C., 13
sphericity, 354, 361–365, 366, 374, 378, 379, 383, 384*f*, 386, 386*f*
SPSS (Statistical Package for the Social Sciences). *See* Statistical Package for the Social Sciences (SPSS)
squared deviations, 257*t*, 259*t*, 283, 350. *See also* sum of squared deviations (SST)
squared multiple correlation, 226, 264, 265
squared residual. *See* mean squared residual (MSR); sum of squared residuals (SSR)
SSB (sum of squared between-group variance), 257–258, 264

SSBA (sums of squares main effects for age), 296–297
SSE (error sum of squares). *See* error sum of squares (SSE)
SSE (sum of squared errors), 256, 258–259, 259*t*. *See also* error sum of squares (SSE)
SSM (sum of squared model variances), 256
SSR (sum of squared residuals), 256
SST (sum of squared deviations), 348, 350, 363, 366. *See also* total sum of squares (SST)
SST (sum of squared variance), 256–257, 257*t*
SST (total sum of squared variance), 256
SST (total sum of squares), 264, 294, 295, 306, 327, 349–350, 366. *See also* sum of squared deviations (SST)
standardized regression coefficient, 202
standardized residuals, 183, 187–188, 188*f*, 190, 190*f*, 191–192, 192*f*, 198
Stanley, J. C., 23
statistical assumptions
   handling violations of, 105–106, 106*f*
   screening for violations of, 95–118
Statistical Package for the Social Sciences (SPSS)
   as data analysis software package, 57, 58
   Data Editor, 58*f*, 70*f*, 76*f*
   descriptives dialog window, 94*f*
   Explore Dialog Windows, 116*f*
   Explore Main Dialog Window, 108*f*
   Explore Statistics and Plots Dialog Windows, 109*f*
   Identify Duplicate Cases wizard, 63
   identifying missing values in, 66
   main menu and chart builder reminder, 87*f*
   procedures for analysis of covariance (ANOVA). *See* analysis of covariance (ANCOVA)
   procedures for Kruskal-Wallis H Test, 407–413, 408*f*, 409*f*, 410*f*, 411*f*, 412*f*, 413*f*
   specifications and results for factorial analysis of variance (ANOVA). *See* factorial analysis of variance (ANOVA)
   specifications and results for Spearman's rho, 422–425, 423*f*, 424*f*, 425*f*
   specifications for bivariate correlation. *See* bivariate correlation
   specifications for bivariate regression. *See* bivariate regression
   specifications for data screening and cleaning, 67–77
   specifications for detecting normality violations, 107–113
   specifications for Friedman's Rank, 438–444, 439*f*, 440*f*, 441*f*, 442*f*, 443*f*, 444*f*
   specifications for identifying sources of bias. *See* bias
   specifications for investigating assumptions, 107
   specifications for Kendall's Tau-b, 430–431, 431*f*
   specifications for mixed design ANOVA. *See* mixed design ANOVA
   specifications for multiple regression. *See* multiple regression
   specifications for one-way analysis of covariance (ANCOVA). *See* one-way analysis of covariance (ANCOVA)
   specifications for one-way analysis of variance (ANOVA). *See* one-way analysis of variance (ANOVA)
   specifications for repeated measures ANOVA (RMANOVA). *See* repeated measures ANOVA (RMANOVA)
   specifications for simple linear regression. *See* simple linear regression
   statistics chart builder dialog menu, 88*f*
   statistics chart builder dialog window, 90*f*
   statistics chart builder for scatterplots, 91*f*
   statistics compute variable dialog window, 114*f*
   statistics main menu, 89*f*
   system-missing value in, 60
statistical power, 45, 50–52, 65, 105, 180, 263, 264, 267, 275, 318–319, 355, 392, 393, 398
statistical regression, 215–216, 217, 218, 219, 222, 232
statistical significance, 31–33, 48–49, 98, 131–134, 132, 134, 137, 142, 150, 154, 157, 196, 196*f*, 243, 244*f*, 250, 420, 443. *See also* significance tests
statistical tables, 446–455
statistics
   bivariate statistics, 4–5
   consumers of applied statistics, 2–3
   defined, 1
   descriptive statistics, 45–46, 54, 110*f*
   inferential statistics, 46–47, 47*f*, 54
   rationale for study of, 3
   univariate statistics, 3–4
   use of term, 2
statistics decision tree, 7, 8*f*
stem-and-leaf plots, 99–101, 100*f*
stepwise method, 215–216
stepwise regression
   estimation of overall model in, 233–236, 234*f*, 235*f*
   model summary for, 237*f*
Stevens, J. P., 105, 221
Stevens, S. S., 19
stratified random sampling, 43*f*
straw man, null hypothesis as often called, 30
string variables, 59, 63, 64
subject-by-subject counterbalancing, 357

sum of squared between-group variance (SSB), 257–258, 264
sum of squared deviations (SST), 348, 350, 363, 366. *See also* total sum of squares (SST)
sum of squared errors (SSE), 256, 258–259, 259*t*. *See also* error sum of squares (SSE)
sum of squared group variance, 258
sum of squared model variances (SSM), 256
sum of squared residuals (SSR), 256
sum of squared variance (SST), 256–257, 257*t*
sum of squares, 196*f*, 225*f*, 235*f*, 244*f*, 264, 277*f*, 283, 297, 300*f*, 348, 349*f*. *See also* error sum of squares (SSE); Type III sum of squares
sum of squares between subjects (SSB), 348, 362, 366
sum of squares between-group variance (SSB), 264, 294, 295, 296, 298, 327
sum of squares within groups (SSW), 294
sum of squares within subjects (SSW), 350–351, 366
sums of squares main effects for age (SSBA), 296–297
SurveyMonkey, 57*f*
SurveyPlanet, 57*f*
surveys
 common software packages for internet surveys, 57*f*
 as quantitative descriptive research designs, 23, 24
S-W (Shapiro-Wilk) test. *See* Shapiro-Wilk (S-W) test
systematic sampling, 42, 43*f*
system-missing values, 60, 80
system-numbered rows, 58

T

t distribution, 178, 248
t test/statistic, 6, 21, 25*f*, 48, 95, 96, 132–133, 137–138, 175, 176, 202, 207, 214, 227, 250–251, 266, 272, 280, 282, 314, 327, 340–341, 342, 361, 365, 391, 399, 449*f*
Tabachnick, B. G., 4, 67, 83, 97, 98, 102, 106, 221, 267, 370
testing hypothesis, as step in scientific method, 15
theoretical constructs, 33
third quartile (Q3), in boxplot, 85
Thorpe, C. T., 55
three-way mixed design, 370–371
time-series design, 28, 28*f*
total sum of squared variance (SST), 256
total sum of squares (SST), 264, 294, 295, 306, 327, 349–350, 366. *See also* sum of squared deviations (SST)
training, in coding, 61
treatment sum of squares (SSM or SS$_{treatment}$), 351, 352, 366
trend analysis, 269–272, 270*f*, 271*f*

true experimental design, 24*f*, 25*f*, 27, 28*f*, 253
trust variables, 113*f*
Tseng, H., 23, 32, 34
Tukey, J., 84, 267
Tukey HSD (Honestly Significant Difference) post-hoc test, 278, 279, 279*f*, 280, 309*f*, 311, 312*f*, 313
Tukey method of multiple comparisons, 267
two-way factorial ANOVA designs, 286–289, 287*f*, 288*f*, 289*f*
Type III sum of squares, 305*f*, 311*f*, 328*f*, 333*f*, 336*f*, 363*f*, 386*f*, 387*f*
Type II/Type 2 errors, 49, 49*f*, 50, 51, 83, 104, 319, 324, 391, 398
Type I/Type 1 errors, 32, 49, 49*f*, 50, 83, 96, 104, 105, 250, 262, 263, 266, 267, 275, 278, 283, 286, 293, 324, 338, 354, 361, 391, 393, 398, 399, 401, 406, 438

U

UNICEF (United Nations International Children's Emergency Fund), 390
univariate statistics, 3–4
Unknown Flavor Sampling Test (UFST), 37
unstandardized regression coefficient, 202
unstandardized residual, 183
user-missing codes, 60, 66
user-missing values, 60, 66, 67, 72, 79–80, 80, 80*f*

V

validity
 concurrent validity, 36–37, 36*f*
 construct validity, 36*f*, 37
 content validity, 36, 36*f*
 convergent validity, 36*f*, 37
 criterion-related validity, 36, 36*f*
 of data collection instruments, 35–40, 36*f*
 discriminant validity, 37–38
 divergent validity, 36*f*
 face validity, 36, 36*f*
 predictive validity, 36*f*, 37
variable interaction, 6
variables
 alphanumeric variables, 63, 64
 in Carib_Height_Weight_Dups.sav, 68, 68*f*
 categorical variables, 19–20, 54, 71–72, 72*f*, 180
 character variables, 63
 characteristics and applications of scales of measurement, 19*f*
 compute variable, 115*f*
 continuous variables, 19, 20–21, 54, 179
 defined, 18

variables *(cont.)*
   dependent variables, 18, 54, 161, 179
   dichotomous variables, 19*f*, 20
   frequency table of created variables, 71*f*
   independent variables, 54, 161, 179–180
   interval-level variables, 19*f*, 21, 342, 353, 418
   level of measurement of, 19–23
   nominal variables, 19*f*, 20
   numeric variables, 59
   ordinal variables, 19*f*, 20
   quantitative variables, 20, 72–73, 110*f*, 111*f*
   ratio-level variables, 19*f*, 21, 22–23, 58, 174, 208, 242, 342, 353, 418
   regressor variables, 18
   for screening exercise, 107*f*
   string variables, 59, 63, 64
   between-subjects variable. *See* between-subjects variable
   trust variables, 113*f*
   types of, 18
   underlying variables, 33–34
   variable values table, 69*f*
variance. *See also* analysis of covariance (ANCOVA); analysis of variance (ANOVA); factorial analysis of variance (ANOVA); one-way analysis of variance (ANOVA); repeated measures ANOVA (RMANOVA)
   error variance. *See* error variance
   grand variance, 256
   within-group variance, 255, 256, 283, 314, 338
   between-group variance. *See* between-group variance
   homogeneity of. *See* homogeneity of variance
   mean square between-groups variance (MSB), 258, 260
   mean-square within-group variance (MSW), 256
   within-subjects variance, 338
   between-subjects variance, 338
   within-subjects variance, 349*f*
   between-subjects variance, 349*f*

W

Wallis, W. A., 399
Welch, B. L., 105
Welch test, 394
Welch's ANOVA, 105, 391
Welsch, R. E., 184, 199*f*, 200, 267
WHO (World Health Organization), 55
Wilcoxon test, 25, 433, 438–439, 440, 442*f*
Williams, E. V., 3, 4
within-group sum of squares. *See* sum of squares within groups (SSW)
within-group variance, 255, 256, 283, 314, 338
within-subjects ANOVA, 366, 367
within-subjects designs, 29
within-subjects variance, 338, 349*f*
Woodman, T., 370
World Bank, 19, 248, 270
World Health Organization (WHO), 55
World Values Survey
   2013, 45
   2016, 99, 100, 100*f*
   no date, 107
World Values Survey Association (WVSA), 390, 414, 415*t*
Wragg, C. B., 339
writing and mathematics skills, case of, 121–127, 122*f*, 125*b*, 152

X

x axis (abscissa), 119

Y

y axis (ordinate), 119

Z

zero-order correlations, 228, 228*f*
Zimmerman, D., 394

www.ingramcontent.com/pod-product-compliance
Lightning Source LLC
Chambersburg PA
CBHW080721230426
43665CB00020B/2576